Functional Proteomics

METHODS IN MOLECULAR BIOLOGY™

John M. Walker, *Series Editor*

484. Functional Proteomics: *Methods and Protocols*, edited by *Julie D. Thompson, Christine Schaeffer-Reiss, and Marius Ueffing, 2008*
483. Recombinant Proteins From Plants: *Methods and Protocols*, edited by *Loïc Faye and Veronique Gomord, 2008*
482. Stem Cells in Regenerative Medicine: *Methods and Protocols*, edited by *Julie Audet and William L. Stanford, 2008*
481. Hepatocyte Transplantation: *Methods and Protocols*, edited by *Anil Dhawan and Robin D. Hughes, 2008*
480. Macromolecular Drug Delivery: *Methods and Protocols*, edited by *Mattias Belting, 2008*
479. Plant Signal Transduction: *Methods and Protocols*, edited by *Thomas Pfannschmidt, 2008*
478. Transgenic Wheat, Barley and Oats: Production and Characterization Protocols, edited by *Huw D. Jones and Peter R. Shewry, 2008*
477. Advanced Protocols in Oxidative Stress I, edited by *Donald Armstrong, 2008*
476. Redox-Mediated Signal Transduction: *Methods and Protocols*, edited by *John T. Hancock, 2008*
475. Cell Fusion: *Overviews and Methods*, edited by *Elizabeth H. Chen, 2008*
474. Nanostructure Design: *Methods and Protocols*, edited by *Ehud Gazit and Ruth Nussinov, 2008*
473. Clinical Epidemiology: *Practice and Methods*, edited by *Patrick Parfrey and Brendon Barrett, 2008*
472. Cancer Epidemiology, Volume 2: *Modifiable Factors*, edited by *Mukesh Verma, 2008*
471. Cancer Epidemiology, Volume 1: *Host Susceptibility Factors*, edited by *Mukesh Verma, 2008*
470. Host-Pathogen Interactions: *Methods and Protocols*, edited by *Steffen Rupp and Kai Sohn, 2008*
469. Wnt Signaling, Volume 2: *Pathway Models*, edited by *Elizabeth Vincan, 2008*
468. Wnt Signaling, Volume 1: *Pathway Methods and Mammalian Models*, edited by *Elizabeth Vincan, 2008*
467. Angiogenesis Protocols: *Second Edition*, edited by *Stewart Martin and Cliff Murray, 2008*
466. Kidney Research: *Experimental Protocols*, edited by *Tim D. Hewitson and Gavin J. Becker, 2008.*
465. Mycobacteria, *Second Edition*, edited by *Tanya Parish and Amanda Claire Brown, 2008*
464. The Nucleus, Volume 2: *Physical Properties and Imaging Methods*, edited by *Ronald Hancock, 2008*
463. The Nucleus, Volume 1: *Nuclei and Subnuclear Components*, edited by *Ronald Hancock, 2008*
462. Lipid Signaling Protocols, edited by *Banafshe Larijani, Rudiger Woscholski, and Colin A. Rosser, 2008*
461. Molecular Embryology: *Methods and Protocols, Second Edition*, edited by *Paul Sharpe and Ivor Mason, 2008*
460. Essential Concepts in Toxicogenomics, edited by *Donna L. Mendrick and William B. Mattes, 2008*
459. Prion Protein Protocols, edited by *Andrew F. Hill, 2008*
458. Artificial Neural Networks: *Methods and Applications*, edited by *David S. Livingstone, 2008*
457. Membrane Trafficking, edited by *Ales Vancura, 2008*
456. Adipose Tissue Protocols, *Second Edition*, edited by *Kaiping Yang, 2008*
455. Osteoporosis, edited by *Jennifer J. Westendorf, 2008*
454. SARS- and Other Coronaviruses: *Laboratory Protocols*, edited by *Dave Cavanagh, 2008*
453. Bioinformatics, Volume 2: *Structure, Function, and Applications*, edited by *Jonathan M. Keith, 2008*
452. Bioinformatics, Volume 1: *Data, Sequence Analysis, and Evolution*, edited by *Jonathan M. Keith, 2008*
451. Plant Virology Protocols: *From Viral Sequence to Protein Function*, edited by *Gary Foster, Elisabeth Johansen, Yiguo Hong, and Peter Nagy, 2008*
450. Germline Stem Cells, edited by *Steven X. Hou and Shree Ram Singh, 2008*
449. Mesenchymal Stem Cells: *Methods and Protocols*, edited by *Darwin J. Prockop, Douglas G. Phinney, and Bruce A. Brunnell, 2008*
448. Pharmacogenomics in Drug Discovery and Development, edited by *Qing Yan, 2008.*
447. Alcohol: *Methods and Protocols*, edited by *Laura E. Nagy, 2008*
446. Post-translational Modifications of Proteins: *Tools for Functional Proteomics, Second Edition*, edited by *Christoph Kannicht, 2008.*
445. Autophagosome and Phagosome, edited by *Vojo Deretic, 2008*
444. Prenatal Diagnosis, edited by *Sinhue Hahn and Laird G. Jackson, 2008.*
443. Molecular Modeling of Proteins, edited by *Andreas Kukol, 2008.*
442. RNAi: *Design and Application*, edited by *Sailen Barik, 2008.*
441. Tissue Proteomics: *Pathways, Biomarkers, and Drug Discovery*, edited by *Brian Liu, 2008*
440. Exocytosis and Endocytosis, edited by *Andrei I. Ivanov, 2008*
439. Genomics Protocols, *Second Edition*, edited by *Mike Starkey and Ramnanth Elaswarapu, 2008*
438. Neural Stem Cells: *Methods and Protocols, Second Edition*, edited by *Leslie P. Weiner, 2008*
437. Drug Delivery Systems, edited by *Kewal K. Jain, 2008*

METHODS IN MOLECULAR BIOLOGY™

Functional Proteomics

Methods and Protocols

Edited by

Julie D. Thompson
Christine Schaeffer-Reiss
Marius Ueffing

Humana Press

Editors
Julie D. Thompson
Laboratoire de Bioinformatique et
Génomique Intégratives
Institut de Génétique et
de Biologie Moléculaire et Cellulaire
Illkirch, France

Christine Schaeffer-Reiss
LSMBO, ECPM
Institut Pluridisciplinaire Hubert Curien
Strasbourg, France

Marius Ueffing
Department of Protein Science
Helmholtz Zentrum München
German Research Center for
Environmental Health
Munich-Neuherberg, Germany

Series Editor
John M. Walker
School of Life Sciences
University of Hertfordshire
Hatfield, Hertfordshire Al10 9 AB
UK

ISBN: 978-1-58829-971-0 e-ISBN: 978-1-59745-398-1
DOI: 10.1007/978-1-59745-398-1

Library of Congress Control Number: 2008921788

Printed on acid-free paper

9 8 7 6 5 4 3 2 1

springer.com

Preface

Recent progress in experimental techniques has led to a revolutionary change in life science research. High-throughput genome sequencing and assembly techniques, together with new information resources, such as structural and functional proteomics, transcriptome data from microarray analyses, or light microscopy images of living cells, have led to a rapid increase in the amount of data available, ranging from complete genome sequences to cellular, structure, phenotype, and other types of biologically relevant information. As a consequence, novel system-level studies are now being performed with the goal of understanding and predicting the behavior of complex systems, such as cells, tissues, organs, and even whole organisms. The field of proteomics plays an essential role in this new systems approach to molecular and cellular studies by identifying the genes involved and determining their functional significance; this makes it possible to understand the complex functional networks and control mechanisms that govern the system's response to perturbations, such as environmental changes or genetic mutations.

Research in the emerging field of proteomics is growing at an extremely rapid rate. The real challenge is the relative quantification of proteins, targeted by their function. Mass spectrometry-based strategies were developed to identify modifications in the proteome profile in correlation with functional changes. In practice, the task involves the identification of peptides in a peptide mixture of extremely high complexity. This identification and relative quantification will allow researchers to study changes in the level of expression, in the processing, or in the post translational modifications of a set of proteins. Recent technical innovations in mass spectrometry-based techniques have resulted in a range of highly sensitive and versatile instruments for high-throughput, high-sensitivity, proteome-scale profiling and the door is now open for a wide range of applications exploiting these approaches. But mass spectrometry is only one among many other techniques that are part of an analytical strategy. These alternative or complementary technologies include two-dimensional gel electrophoresis, protein microarrays, yeast two-hybrid systems, phage display, and immunoprecipitation. However, there is no one technology of choice and the most appropriate method will depend on the size and the nature of the system being studied and the type of results desired. The principal aim of this volume is to describe the latest protocols being developed to address the problems encountered in high-throughput proteomics projects, with emphasis on the factors governing the technical choices for a given application. The volume is aimed at researchers working in the field of proteomics including chemical engineers, analytical chemists, biochemists, cell and molecular biologists, clinical scientists, and bioinformaticians, as well as those who are contemplating using proteomics for functional studies.

In functional proteomics, successful characterization of proteins from mass spectrometry experimental data will depend on the technological choices made during the three main phases of the study:

1. The strategy used for the selection, purification, and preparation of the sample to be analyzed by mass spectrometry.
2. The type of mass spectrometer used and the type of data to be obtained from it.
3. The method used for the interpretation of the mass spectrometry data and the search engine used for the identification of the proteins in the different types of sequence data banks available.

The mass spectrometry part itself is often seen as the most important one because it corresponds to the largest budget. It is also time consuming, being very complex and highly technical. Nevertheless, the sample preparation and the data analysis steps are equally important, if not more important, for the success of a proteomic experiment. Therefore, in this volume, the case studies presented will always insist on the three aspects of the experimental design. In the initial chapters, different mass spectrometry instrumentation will be introduced in the context of various applications, from the study of large multiple protein complexes to complete organism proteomics. The advantages and the best use of the following types of instruments will be discussed: MALDI-TOF for simple mass finger printing protein identifications as well as MALDI-TOF-TOF, LC-MALDI-TOF-TOF, and LC-ESI-MS-MS (at low, average, and high resolution), detailing the characteristics and capabilities of the different types of mass spectrometers in term of sensitivity, resolution, accuracy, and MS-MS. Metabolomic studies, which are also experimentally based on mass spectrometry, will also be presented, since metabolomic changes obviously reveal functional changes. The following chapters describe the use of mass spectrometry for the detection of protein–protein specific interactions and posttranslational modifications.

High-throughput proteomics studies generate huge volumes of data, including gel images, mass spectrometry spectra, and protein identifications. These data have to be collected, stored, organized, and interpreted if they are to be used effectively. Bioinformatics plays an important role by providing common data representation standards to enable the comparison and transfer of information between different systems and laboratories. The last chapters of this volume are therefore dedicated to the most widely used database resources, as well as the new computational techniques being developed to search and analyze proteomic data. Finally, emerging computational systems biology methods are described for the integration of data from multiple sources, in order to model complex structures such as protein networks or regulatory pathways and their response to external perturbations.

Julie D. Thompson
Christine Schaeffer-Reiss
Marius Ueffing

Contents

Contributors

Cédric Atmanene • *Laboratoire de Spectrométrie de Masse Bio-Organique, Institut Pluridisciplinaire Hubert Curien, UMR 7178 CNRS / Université Louis Pasteur, Strasbourg, France*

Harald Barsnes • *Department of informatics, University of Bergen, Bergen, Norway*

Corinne Benay • *Service Régional Vaudois de Transfusion Sanguine, Lausanne, Switzerland*

Guillaume Beslon • *Laboratoire d'InfoRmatique en Images et Systèmes d'information (LIRIS, UMR CNRS 5205), INSA-Lyon, Villeurbanne, France*

Rainer Bischoff • *University of Groningen, Centre of Pharmacy, Analytical Biochemistry, Antonius, Groningen, The Netherlands*

Marie-Pierre Bousquet-Dubouch • *Institut de Pharmacologie et de Biologie Structurale, UMR 5089, CNRS/Université Paul Sabatier, Toulouse, France*

Odile Burlet-Schiltz • *Institut de Pharmacologie et de Biologie Structurale, UMR 5089, CNRS/Université Paul Sabatier, Toulouse, France*

Cosima Damiana Calvano • *Protein Research Group, Department of Biochemistry and Molecular Biology, University of Southern Denmark, Odense M, Denmark*

Andrew Chatr-aryamontri • *Department of Biology, University of Rome "Tor Vergata," Rome, Italy*

Arnaud Ceol • *Department of Biology, University of Rome "Tor Vergata," Rome, Italy*

Gianni Cesareni • *Department of Biology, University of Rome "Tor Vergata," Rome, Italy*

Guillaume Chevreux • *Laboratoire de Spectrométrie de Masse Bio-Organique, Institut Pluridisciplinaire Hubert Curien, UMR 7178 CNRS / Université Louis Pasteur, Strasbourg, France*

Stéphane Claverol • *Pole protéomique, Plateforme Génomique Fonctionelle, Université V. Ségalen Bordeaux, Bordeaux, France*

Holger Conzelmann • *Max Planck Institute for Dynamics of Complex Technical Systems, Magdeburg, Germany*

Richard Côté • *EMBL-European Bioinformatics Institute, Wellcome Trust Genome Campus, Hinxton, Cambridge, UK*

Antoine Coulon • *Université de Lyon, Lyon, France; Université Lyon, Lyon, France; Centre de Génétique Moléculaire et Cellulaire – UMR CNRS 5534, Villeurbanne, France*

David Crettaz • *Service Régional Vaudois de Transfusion Sanguine, Lausanne, Switzerland*

Chuong B. Do • *Computer Science Department, Stanford University, Stanford, CA, USA*

Ingvar Eidhammer • *Department of informatics, University of Bergen, Bergen, Norway*

Henriette Engelken • *EML Research gGmbH, Heidelberg, Germany*

Dmitrij Frishman • *Institute for Bioinformatics, GSF-National Research Center for Environment and Health, Neuherberg, Germany; Department of Genome Oriented Bioinformatics, Technische Universität Munchen, Freising, Germany*

Olivier Gandrillon • *Université de Lyon, Lyon, France; Université Lyon, Lyon, France; Centre de Génétique Moléculaire et Cellulaire – UMR CNRS 5534, Villeurbanne, France*

Monique P. Gelderman • *Laboratory of Cellular Hematology, CBER, FDA, Rockville, MD, USA*

Peter J. Gerber • *Department of Pulmonary Medicine, Erasmus Medical Centre, Rotterdam, The Netherlands*

Kris Gevaert • *Ghent University, Ghent, Belgium*

Ernst-Dieter Gilles • *Max Planck Institute for Dynamics of Complex Technical Systems, Magdeburg, Germany*

Natalia Govorukhina • *University of Groningen, Centre of Pharmacy, Analytical Biochemistry, Antonius, Groningen, The Netherlands*

Attila Gursoy • *Koc University, Center for Computational Biology and Bioinformatics and College of Engineering, Istanbul, Turkey*

Mingyue He • *Technology Research Group, The Babraham Institute, Cambridge, UK*

Joost P.J.J. Hegmans • *Department of Pulmonary Medicine, Erasmus Medical Centre, Rotterdam, The Netherlands*

Henning Hermjakob • *European Bioinformatics Institute, Wellcome Trust Genome Campus, Hinxton, Cambridge, UK*

Peter Horvatovich • *University of Groningen, Centre of Pharmacy, Analytical Biochemistry, Antonius, Groningen, The Netherlands*

Simon J Hubbard • *Michael Smith Building, Faculty of Life Sciences, The University of Manchester, Manchester, UK*

Francis Impens • *Ghent University, Ghent, Belgium*

Jose M. G. Izarzugaza • *Structural Computational Biology Programme, Spanish National Cancer Research Centre (CNIO), Madrid, Spain*

Ole Nørregaard Jensen • *Protein Research Group, Department of Biochemistry and Molecular Biology, University of Southern Denmark, Odense M, Denmark*
Philip Jones • *EMBL-European Bioinformatics Institute, Wellcome Trust Genome Campus, Hinxton, Cambridge, UK*
David Juan • *Structural Computational Biology Programme, Spanish National Cancer Research Centre (CNIO), Madrid, Spain*
Kazutaka Katoh • *Digital Medicine Initiative, Kyushu University, Fukuoka, Japan*
Ozlem Keskin • *Koc University, Center for Computational Biology and Bioinformatics and College of Engineering, Istanbul, Turkey*
Walter Kolch • *Cancer Research Beatson Laboratories, Glasgow, UK*
Bart N. Lambrecht • *Department of Pulmonary Medicine, Erasmus Medical Centre, Rotterdam, The Netherlands*
Martin R. Larsen • *Protein Research Group, Department of Biochemistry and Molecular Biology, University of Southern Denmark, Odense M, Denmark*
Eduardo Leon • *Structural Computational Biology Programme, Spanish National Cancer Research Centre (CNIO), Madrid, Spain*
Stef J. F. Letteboer • *Department of Human Genetics, Nijmegen Centre for Molecular Life Sciences, Radboud University Nijmegen Medical Centre, Nijmegen, The Netherlands*
Jian Liu • *Center for Cellular and Biomolecular Research, University of Toronto, Toronto, Ontario, Canada*
Lennart Martens • *European Bioinformatics Institute, Wellcome Trust Genome Campus, Hinxton, Cambridge, UK*
Raja Mazumder • *Protein Information Resource, Georgetown University Medical Center, Washington, DC, USA*
Petter Melin • *Department of Microbiology, Swedish University of Agricultural Sciences, Uppsala, Sweden*
Svein-Ole Mikalsen • *Institute for Cancer Research, Rikshospitalet-Radiumhospitalet University Hospital, Montebello, Oslo, Norway*
Bernard Monsarrat • *Institut de Pharmacologie et de Biologie Structurale, UMR 5089, CNRS/Université Paul Sabatier, Toulouse, France*
Luisa Montecchi Palazzi • *European Bioinformatics Institute, Wellcome Trust Genome Campus, Hinxton, Cambridge, UK*
Bernd Müller • *Department I Biologie, Ludwig Maximilian University Munich, Munich, Germany*

Rajesh Nair • *CUBIC, Department of Biochemistry and Molecular Biophysics and Center for Computational Biology and Bioinformatics, Columbia University, New York, NY, USA*

Anastasia N. Nikolskaya • *Protein Information Resource, Georgetown University Medical Center, Washington, DC, USA*

Mikkel Nissum • *BD Diagnostics, Martinsried, Germany*

Ruth Nussinov • *Basic Research Program, SAIC-Frederick, Inc. Center for Cancer Research Nanobiology Program NCI-Frederick, Frederick, MD, USA; Sackler Institute of Molecular Medicine, Department of Human Genetics and Molecular Medicine, Sackler School of Medicine, Tel Aviv University, Tel Aviv, Israel*

Florencio Pazos • *Computational Systems Biology Group, National Centre for Biotechnology (CNB-CSIC), Madrid, Spain*

Frédéric Plewniak • *Plate-forme Bio-informatique de Strasbourg, Institut de Génétique et de Biologie Moléculaire et Cellulaire, UMR 7104 – CNRS – Inserm – ULP, Illkirch, France*

Christian Preisinger • *Cancer Research Beatson Laboratories, Glasgow, UK*

Ralf Reski • *Plant Biotechnology, Faculty of Biology, University of Freiburg, Freiburg, Germany*

Uwe Reyle • *Institute for Computational Linguistics, University of Stuttgart, Stuttgart, Germany*

Ronald Roepman • *Department of Human Genetics, Nijmegen Centre for Molecular Life Sciences, Radboud University Nijmegen Medical Centre, Nijmegen, The Netherlands*

Burkhard Rost • *CUBIC, Department of Biochemistry and Molecular Biophysics, Columbia University and Center for Computational Biology and Bioinformatics, Columbia University, New York, NY, USA*

Nathalie Rufer • *NCCR Molecular Oncology; Swiss Institute for Experimental Cancer Research (ISREC), Epalinges, Switzerland*

Sarah Sanglier • *Laboratoire de Spectrométrie de Masse Bio-Organique, Institut Pluridisciplinaire Hubert Curien, UMR 7178 CNRS / Université Louis Pasteur, Strasbourg, France*

Eric Sarnighausen • *Plant Biotechnology, Faculty of Biology, University of Freiburg, Freiburg, Germany*

Jasmin Šarić • *Boehringer Ingelheim Pharma GmbH & Co., Biberach, Germany*

Christine Schaeffer-Reiss • *Laboratoire de Spectrométrie de Masse Bio-Organique, Institut Pluridisciplinaire Hubert Curien, UMR 7178 CNRS / Université Louis Pasteur, Strasbourg, France*

David S. Selby • *Protein Research Group, Department of Biochemistry and Molecular Biology, University of Southern Denmark, Odense M, Denmark*
Julian N Selley • *Michael Smith Building, Faculty of Life Sciences, The University of Manchester, Manchester, UK*
Jennifer A Siepen • *Michael Smith Building, Faculty of Life Sciences, The University of Manchester, Manchester, UK*
Jan Simak • *Laboratory of Cellular Hematology, CBER, FDA, Rockville, MD, USA*
Magdalena Swiatek-de Lange • *Boehringer Ingelheim Pharma GmbH & Co., Biberach an der Riss, Germany*
Loïk Sylvius • *Plate-forme protéomique IFR-100, Etablissement Français du Sang, Dijon, France*
Michael J Taussig • *Technology Research Group, The Babraham Institute, Cambridge, UK*
Lynne Thadikkaran • *Service Régional Vaudois de Transfusion Sanguine, Lausanne, Switzerland*
Jean-Daniel Tissot • *Service Régional Vaudois de Transfusion Sanguine, Lausanne, Switzerland*
Julie D. Thompson • *Institut de Génétique et de Biologie, Moléculaire et Cellulaire, Illkirch, France*
Marius Ueffing • *Institute of Human Genetics, GSF National-Research Center for Environment and Health, Neuherberg, Germany*
Sandrine Uttenweiler-Joseph • *Institut de Pharmacologie et de Biologie Structurale, UMR 5089, Centre National de la Recherche Scientifique/Université Paul Sabatier, Toulouse, France*
Sona Vasudevan • *Protein Information Resource, Georgetown University Medical Center, Washington, DC, USA*
Alfonso Valencia • *Structural Computational Biology Programme, Spanish National Cancer Research Centre (CNIO), C/ Melchor Fernandez Almagro, Madrid, Spain*
Petra Van Damme • *Ghent University, Ghent, Belgium*
Joël Vandekerckhove • *Ghent University, Ghent, Belgium*
Alain Van Dorsselaer • *Laboratoire de Spectrométrie de Masse Bio-Organique, Institut Pluridisciplinaire Hubert Curien, UMR 7178 CNRS / Université Louis Pasteur, Strasbourg, France*
Alex von Kriegsheim • *Cancer Research Beatson Laboratories, Glasgow, UK*
Robert Wildgruber • *BD Diagnostics, Martinsried, Germany*
Philip Wong • *Institute for Bioinformatics, GSF-National Research Center for Environment and Health, Neuherberg, Germany*
Andreas Zanzoni • *Department of Biology, University of Rome "Tor Vergata," Rome, Italy*

I

Introduction

1

A Brief Summary of the Different Types of Mass Spectrometers Used in Proteomics

Christine Schaeffer-Reiss

Summary

Recent technical innovations in mass spectrometry-based techniques have resulted in a range of highly sensitive and versatile instruments for high-throughput, high-sensitive, proteome-scale profiling. This wide diversity of instrumentation commercially available for mass spectrometry-based proteomics makes the choice of instrumentation sometimes difficult. The choice of instruments depends on the biological problem and the proteomic strategy chosen for protein identification. This chapter will give a short overview of the instruments routinely used in proteomic laboratories and the technical criteria that should be considered before instrument selection.

Key Words: Mass spectrometry instrumentation.

1. Introduction: The Special Role of Mass Spectrometry in Proteomics

The goal of proteomics is to identify, characterize, and quantify the whole content of proteins that are present in complex biological materials (tissues, cells in culture, organelles, or fluids). For the past decade, the interest for proteomic studies kept growing exponentially and today, proteomic has reached high-throughput analysis capabilities. This is the result of two major advances: (1) the progress in mass spectrometry (MS) makes possible routine analysis of peptides and proteins with improved sensitivity, reliability, speed, and automation, and (2) the large scale genome sequence programs of the past 10 years provided large protein sequence databases for many organisms which are essential to identify quickly proteins from MS data. As a result, MS has become

From: *Methods in Molecular Biology, vol. 484: Functional Proteomics: Methods and Protocols*
Edited by: J. D. Thompson et al., DOI: 10.1007/978-1-59745-398-1, © Humana Press, Totowa, NJ

a pillar analytical method in proteomic studies for the identification and characterization of the proteins present in complex biological systems. A wide panel of instrumental solutions is now available from several manufacturers and the choice of the appropriate instrumentation can really be puzzling. This chapter will give an overview of the instruments routinely used in proteomic laboratories and the technical criteria that should be considered before instrument selection.

2. General Features and Key Characteristics of Mass Spectrometers

2.1. A Wide Variety of Mass Spectrometers with Very Different Technical Solutions

A broad range of mass spectrometers is used in MS-based proteomic research. Each type of instrument has unique design, data system, and performance specifications, resulting in strengths and weaknesses depending on the types of experiments.

Mass spectrometry is a two-step method: first, the analyte is volatilized and ionized, while keeping intact its integrity, and second, the measurement of the mass-to-charge ratio (*m/z*) of the ionized analyte is obtained. The mass spectrometer is usually made of two distinct parts: the source, where the volatilization/ionization step is performed, and the analyzer/detector, where the ions are separated and the *m/z* ratio is measured by a physical device (**Fig. 1**).

The "heart" of the mass spectrometer is the analyzer. Several analyzers can be combined to perform "two-dimensional" MS. The analyzer separates the

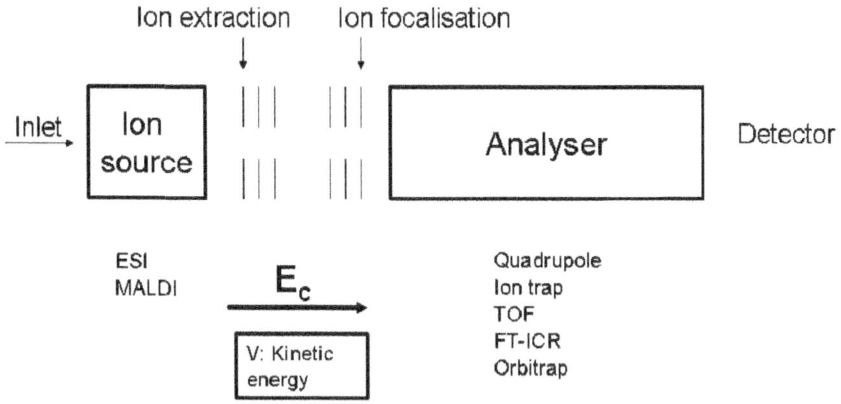

Fig. 1. Simplified configuration of a mass spectrometer. The kinetic energy driving the ions from the source to the analyzer is very different depending on the type of source and analyzer.

gas phase ions. The analyzer uses electrical or magnetic fields, or a combination of both, to move and select the ions from the source to the detector. Because the motion and separation of ions is based on electrical and/or magnetic fields, the *m/z* ratio, and not only the mass, is of importance. The analyzer must be operated under high vacuum, such that ions can travel without colliding with neutral gas atoms and reach the detector with a sufficient yield.

In proteomic analysis, it is important to choose the right source-analyzer association, and also the most adapted combination of analyzers in the case of "two-dimensional" MS. The best mass spectrometer configuration depends on the analytical strategy that will be used for protein identification. The most popular strategies are summarized in the following chapters.

2.2. Key Characteristics of Instruments

For proteomic studies, the key mass spectrometer characteristics that must be considered are (1) mass resolution (or resolving power), (2) mass accuracy, (3) sensitivity, and (4) ability to perform MS/MS. The resolving power (R) measures the ability of the instrument to distinguish between two ions of close masses: if M is the mass of one ion and ΔM the difference between the two ion masses, then R is defined by the ratio $M/\Delta M$. Mass accuracy describes how closely experimental (or measured) mass (M_{exp}) matches theoretical (or expected) mass (M_{th}). The mass accuracy is usually given in parts-per-million (ppm): $10^6 \times (M_{th} - M_{exp})/M_{exp}$. Mass accuracy is directly linked to the resolving power. A low-resolution mass spectrometer cannot provide high accuracy. In addition, several other specifications are important such as the possibility for automation allowing high-throughput analysis and the scan speed of the analyzer. Obviously, it is necessary to keep in mind that resolution, accuracy, scan speed and sensitivity are linked in some ways.

3. Three Main Protein Identification Strategies in Proteomics

The classical strategies for protein identification consist in digesting proteins into peptides that are subsequently analyzed by MS. These strategies are described in detail in a variety of papers ***(1–7)***. Three main methodologies are routinely used for protein identification: peptide mass fingerprinting (PMF), peptide fragment fingerprinting (PFF), and *de novo* sequencing. All these methods use proteolytic enzymes (typically trypsin) to specifically cleave proteins into peptides with a mass suitable for MS and/or MS/MS analysis.

3.1. The Peptide Mass Fingerprinting (PMF) Strategy

In the case of PMF *(8)*, the *m/z* ratio of each peptide obtained after enzymatic digestion of a protein is measured with the highest possible accuracy. The measured masses are then compared with the theoretical masses of all the peptides, which has been obtained after *in silico* proteolytic digestion of a selected protein database (calculated fingerprints). The degree of confidence in protein identification with this approach will strongly depend on the tight correlation between measured and theoretical masses. Therefore, the most important specification of the instrument best suited for that approach is the accuracy of mass measurement.

3.2. The Peptide Fragment Fingerprinting (PFF) Strategy

In the PFF approach, peptides are fragmented using a "two-dimensional" mass spectrometer (MS/MS). Intact peptide ions are selected by a first analyzer (MS1) and then dissociated by collisions, usually by passing through a neutral gas (collision-induced dissociation, CID). This results in the fragmentation of the parent peptide, which occurs at specific bonds of the polypeptide backbone. **Figure 2** presents the six most usual fragmentations obtained in those conditions and the specific nomenclature of each fragment *(9)*. Charged fragments are then separated in a second analyzer (MS2) yielding to a fragmentation fingerprint (**Fig. 3**). Fragment masses obtained experimentally are compared with the theoretical masses of all the fragments, which has been obtained after *in silico* proteolytic digestion and fragmentation of a selected protein database (calculated fingerprints) *(10–12)*. The complexity of the digestion peptide mixture will be important for the choice of the instrument and its tuning. Samples of reduced complexity are obtained when slices cut from one- or two-dimensional polyacrylamide gels are digested. When the total protein extract from the biological sample is digested and directly analyzed by MS (for example, in shotgun proteomics) *(13,14)*, the peptide mixture is extremely complex and scanning parameters will have to be optimized. In this approach, the specifications of the

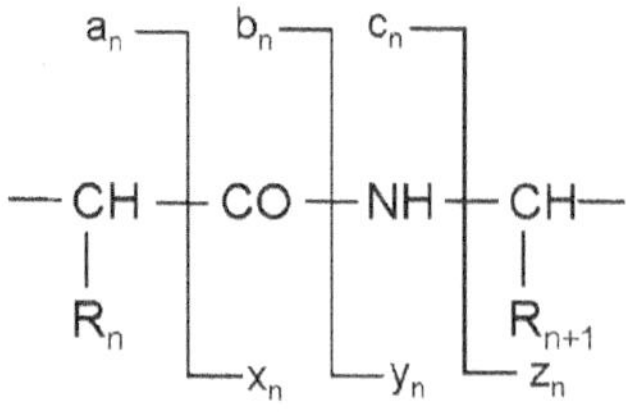

Fig. 2. Nomenclature of the various fragments expected from peptide dissociation *(9)*.

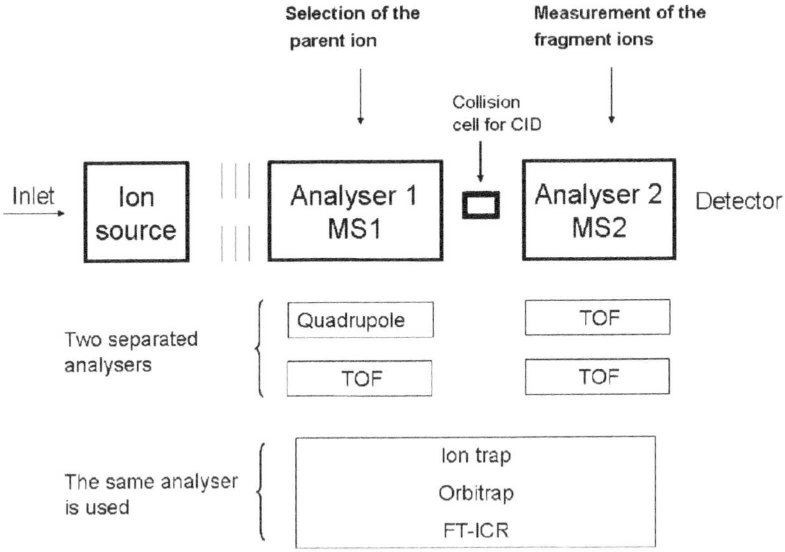

Fig. 3. Most popular analyzer configurations for "two-dimensional" mass spectrometry. Q-TOF and TOF-TOF are real tandem instruments. Ion trap and FT-ICR are using the same analyzer for MS1 and MS2. The Orbitrap is more complex since it is always hyphenated with an ion trap as first analyzer (see text). For simplicity, however, Orbitrap has been compared to IT and FT-ICR.

best suited mass spectrometer must include (1) a collision cell generating a large number of ionized fragments and (2) high accuracy of mass measurements.

These two first strategies require that the exact sequences of the studied proteins are present in the protein databases and require specialized search engines (Mascot, Sequest).

3.3. De Novo Strategy

If the protein database for the studied organism does not contain enough information for the comparison of fragmentation fingerprints, an alternative consists in using the so-called *de novo* sequencing approach. In this case, sequence information is deduced directly from the experimental MS/MS spectra by manual or automatic interpretation of the data. When a sequence of a few amino acids is obtained from an MS/MS spectrum, it can be used in a classical BLAST search to identify the protein(s) *(15)*. For this strategy, the same instrument specifications as the ones for PFF are required, but the highest possible accuracy in MS2 mass measurements is needed.

3.4. Guidelines for Protein Identification by Mass Spectrometry

The three approaches described above allow the identification of proteins, but do not lead to their full characterization, for example in terms of posttranslational modifications. It was previously pointed out that a high number of false protein identifications was observed when experiments used instruments with inadequate performances or when the search criteria in the protein databases were not stringent enough. Unfortunately, this tendency will keep increasing with the number of protein sequences present in databases, making protein identification based on experimental versus calculated "fingerprints" less and less reliable.

A series of guidelines for the identification of proteins in proteomic studies have been proposed ***(16,17)***. Accordingly the most reliable identification of a protein is now obtained using MS/MS strategies. These guidelines helps to select accuracy of mass measurement needed, which depends on the appropriate choice of the MS instrument. Very high resolution instruments still make PMF useful provided the high-resolution mass spectrometer is properly used ***(18)***.

4. Ionization Methods

Matrix-assisted laser desorption ionization (MALDI) and electrospray ionization (ESI) are the two techniques most commonly used to volatize and ionize peptides and proteins in MS analysis ***(19,20)***. Both display femtomolar sensitivity when used in optimal conditions. MALDI is performed on a condensed phase. ESI works on a liquid phase thus allowing an easy coupling with high-performance liquid chromatography (HPLC), which is not the case for MALDI. For peptides and proteins, the charge is generally due to the addition of a variable number of protons. However, the ions observed with MALDI are typically only single charged while ESI adds multiple protons to the basic residues generating multiply charged molecules. In theory all types of analyzers can be adapted to both ionization sources.

4.1. MALDI

The sample is mixed with a saturated solution of matrix (an organic compound with a strong absorption at the laser wavelength) and a microliter drop is laid on the MALDI target ***(19)***. After solvent evaporation and matrix crystallization, the target is positioned in the mass spectrometer source under vacuum and irradiated with pulses of laser light. Once in the vapor phase, proton transfer between matrix and analytes occurs, resulting in ion formation. Ions are subsequently accelerated by applying a high potential (~20 kV) to a series of extraction electrodes and lenses (**Fig. 1**).

4.2. ESI

The sample in solution is infused through a silica capillary (spray capillary) with a typical flow rate between 1 and 100 μL per minute. An electrical field, applied at the extremity of the pneumatically assisted spray capillary, imparts charges to the spray droplets *(20)*. ESI is made at atmospheric pressure. Ions are subsequently transferred in the vacuum of the analyzer after transitioning through the interface, where they are accelerated and desolvated. An ESI source can be readily coupled to liquid-based separation tools (chromatographic or electrophoretic devices). Miniaturization of liquid chromatography (nano-LC) with columns of 50–100 μm internal diameter allows routine subpicomole sensitivities because a high concentration of analytes in the eluted chromatographic peaks is obtained. On line separation prior to MS analysis is an obvious advantage for ESI which is used mainly in the LC-ESI-MS/MS mode *(21)*. In the case of very complex mixtures, initial separation of individual peptides is a strong advantage since "ion suppression" will be mostly avoided. Ion suppression corresponds to the effect of highly ionizable peptides that suppress the signal from less ionizable peptides.

5. Five Types of Analyzers Classically Used

The combination of ESI or MALDI with several types of mass analyzers provides a wide variety of specialized mass spectrometers. Five types of analyzers are currently used in proteomics: quadrupole (Q), ion trap (IT), time-of-flight (TOF), Fourier transform ion-cyclotron resonance (FT-ICR or FT-MS), and Orbitrap (OT). Analyzers are selected as a function of the analytical problems and, obviously, their prices. The choice of a mass spectrometer will strongly depend on the strategy preferred for protein identification and on the biological question. Once these are clearly defined, the key characteristics and performances of the instrument should be considered.

Quadrupoles and TOF are only able to perform "one-dimension" MS analysis. Ion trap and FT-ICR can be used in MS and MS/MS analysis, since the same analyzer is used sequentially as MS1 and MS2. Q-TOF and TOF-TOF are hybrid instruments which are composed of two individual instruments in tandem. The case of the OT is distinct since the available instrument commercialized by Thermo Fisher Scientific is always hyphenated with an ion trap as a first analyzer. **Figure 4** summarizes the most popular source-analyzer configurations routinely used in proteomic laboratories.

The following chapters will briefly present these five types of analyzers. The principle of these techniques is comprehensively described in various reviews and books *(22,23)*.

INSTRUMENTS	ESI	MALDI
TOF (*Time-of-flight*)	X	X
Q-TOF (MS/MS)	X	
TOF-TOF (MS/MS)		X
IT (*Ion trap*) (MS^n)	X	
FT-ICR (*Fourier transform-ion cyclotron resonance*) (MS^n)	X	
Orbitrap (MS-MS)	X	

Fig. 4. Most popular source-analyzer configurations routinely used for proteomics. In proteomic studies, ESI-TOF is not used very often. "Off line" experiments coupling HPLC with MALDI are not mentioned, but they are feasible and can be as powerful as LC-ESI-MS/MS experiments when performed properly. Early on, triple quadrupoles (Q-Q-Q) were widely used despite poor resolution. Currently other instruments are better suited for proteomics.

5.1. Principle of the TOF Analyzer

Ions are maintained in a space as small as possible before being pushed with the same kinetic energy (20–30 kV) through the analyzer (a tube of about 1 m) toward the detector. Since the ions enter the TOF at the same time and with the same kinetic energy, they will reach the detector with speeds directly correlated to their *m/z* ratio. An accurate measurement of the time ions need to travel from the source to the detector allows the ion *m/z* ratio to be determined. The resolution is usually increased when using a reflectron, which has an effect of energy focalization ***(24,25)***. TOF analyzers typically reach a resolution of about 20,000 and allow routine accuracy of ± 10–50 ppm.

5.2. Principle of the Quadrupole and Ion Trap Analyzers

These instruments use electrostatic fields to force ions to oscillate in a very complex way. For quadrupole and ion trap analyzers, the equation of Matthieu describes the movements of the ions and the basis for selecting *m/z* values to allow specific ions to reach the detector and to generate a spectrum ***(26–28)***. Quadrupoles are typically used as a first analyzer (MS1) in MS/MS instruments because their resolution is good enough for molecular ion selection, but too weak to provide an accuracy compatible with PMF identifications. The ion trap-based instruments provide MS/MS capabilities. They are used in PFF identification strategies and sometimes in MS^n analysis of modified peptides (PTM).

5.3. Principle of the FT-ICR

The basic principle of the FT-ICR is to measure ion cyclotronic frequency in a magnetic field, which allows ion mass to be calculated. For this, a pulsed radiofrequency signal is used to excite the ions while they are orbiting. Excited ions generate signals that are processed by a Fourier transform (FT) to obtain the component frequency of the different ions, which correspond to their *m/z* ratio. Because ion frequency can be measured with high accuracy, their corresponding *m/z* ratio is also calculated with high accuracy ***(29)***. One major drawback of these instruments is their high cost, which is partly due to the supramagnetic field required to induce ion circular motion. However, FT-ICR instruments have the highest resolution capabilities.

5.4. Principle of the Orbitrap

This analyzer has some similarities to the FT-ICR, except that it uses complex electrostatic fields instead of a magnetic field ***(30)***. An OT analyzer provides routine resolution of about 60,000 and an accuracy of less than 2 ppm (using internal standard) ***(31)***. OT-based instruments are less expensive than FT-ICR instruments, their running cost is lower, and they are operated more easily. So far, an OT analyzer is used exclusively to measure with high resolution and accuracy the parent ions and the fragment ions selected by an ion trap (MS1). The commercially available OT is therefore always an MS/MS instrument; it is characterized by an excellent versatility, high sensitivity, and high routine resolving power ***(32)***.

5.5. Analyzers Used in PMF Identification

MALDI-TOF is the most widely used instrument for PMF identification in proteomic laboratories because it is easy to operate and very robust. The mass accuracy of the MALDI-TOF is usually between 10 and 50 ppm (with a resolution of about 15,000), which is enough to allow routine identification of most proteins.

PMF analysis using MALDI-TOF is still widespread in many laboratories, although the guidelines published by several journals ***(16,17)*** pointed out the lack of specificity of this technology for protein identification. Its use should be restricted to relatively simple peptide mixtures.

FT-ICR is also used for PMF identification in a nano-LC-MS mode ***(33)***. The resolution of the FT-ICR allows an accuracy of about 1 ppm in routine proteomic analysis. The dynamic range of the FT-ICR is also much higher and low abundant peptides can be detected. FT-ICR analyzers display overall the best performances for proteomic analysis. However, the complexity in operating this system, the price of the machine, and its running cost must be seriously considered before opting for that instrument.

The OT with its high routine resolution also seems well adapted for PMF identification. The OT-based instrument is always hyphenated with an ion trap as MS1. This type of instrument can perform PFF identification at any time.

5.6. MS/MS Analyzers Used in PFF Identification

Classical peptide sequencing (PFF approach) by "two-dimensional" mass spectrometry mainly uses automated instruments including Q-TOF, IT and OT, TOF-TOF, and seldom FT-ICR (**Fig. 3**). MS/MS instruments offer additional possibilities and give access to sophisticated experiments for the characterization of peptide families (phosphopeptides, peptide glycosylation, etc.).

To improve peptide sequencing, fragmentation techniques alternative to classical CID have been developed: electron capture dissociation (ECD) and electron transfer dissociation (ETD).

The advantage of ECD and ETD is to generate fragments that are evenly distributed along the peptide backbone. In contrast, CID-induced fragments are usually restricted to a more limited number of cleavage points in the peptide and, therefore, yield less sequence information. This is a major advantage for the study of PTMs. Indeed, the combination of CID and ECD fragmentation methods *(34)* can be used, for example, to localize PTM on the peptide backbone. However, ECD is not compatible with ion traps or Q-TOF and is limited to FT-ICR instruments.

Electron transfer dissociation (ETD) is compatible with instruments that utilize RF fields to trap ions *(35–37)*. Peptide fragmentation is achieved through gas-phase electron transfer from singly charged anions to multiply protonated peptides and yields fragments that are complementary to the classical CID method.

ETD and ECD are complementary to CID in the determination of sequence information by peptide fragmentation *(38)*. There is no doubt that many MS/MS instruments will soon complement CID with ETD or ECD.

6. The Importance of Chromatography for Sensitivity

In the past few years, the miniaturization of chromatography has been a major innovation to improve the sensitivity of LC-ESI-MS/MS analysis. Nano-LC chromatographic separations are performed on a nanoscale column (75 μm inner diameter) using flow rates in the nanoliter per minute range. This results in high analytical sensitivity due to substantial concentration efficiency of the eluted sample.

The need for increased sensitivity, robustness, and high throughput has led to the recent introduction of nano-HPLC-Chip systems from Agilent

Technologies. The nano-HPLC-Chip system ***(39,40)*** consists of a device that integrates on a single chip: an enrichment column, an analytical column, and the electrospray nozzle. By minimizing the number of connections and dead volumes, the chip offers better chromatographic performances in terms of reproducibility, peak resolution, sensitivity, and spray stability, compared to classical nanocolumns of 75 μm inner diameter. Enhanced sensitivity provided by this system will be particularly interesting for the identification of rare proteins and biomarkers.

It should be mentioned also that "off line" LC-MALDI-TOF-TOF can be readily performed using micro- or nanocollectors, which in some cases may be an interesting alternative to nano-ESI-LC-MS/MS ***(41)***.

7. Conclusions

A wide diversity of instrumentation is commercially available for MS-based proteomics. Instrumentation will probably become more sophisticated in the next years; however, the criteria for selecting the appropriate instrumentation will still depend on the experimental strategy that has been decided to answer the question(s) of the biologist.

Before electing an instrument, the following parameters must be considered: the resolving power, the mass accuracy, the sensitivity, the possibility for "two-dimensional" MS, the dynamic range, the time required for one analysis, the automation possibility, the reliability, the complexity in operating the system,

Analysers	Resolving power	Mass accuracy	Sensitivity	Dynamic range
TOF-TOF	Good	Good	High	Medium
Q-TOF	Good	Good	High	Medium
IT	Low	Low	Very high	Low
Orbitrap	High	High	Very high	Medium
FT-ICR	Very high	Very high	High	Good

Fig. 5. Relative comparison of the resolution, accuracy, sensitivity, and dynamic range of the most popular instrument used in proteomic studies.

and, obviously, the price (**Fig. 5**). The biological problem (material availability, complexity, etc.) and the protein identification approach will decide which of these characteristics are the most important, allowing the appropriate system to be selected accordingly.

It would be misleading to think that only one type of instrument is always the best choice for a specific question. Indeed, the price of the instrument, its running cost, the ease of use, and the robustness have to be evaluated individually in each laboratory that wants to perform proteomic studies. Specialized proteomic platforms may offer interesting options for specific biological questions, which include (1) a combination of MALDI-TOF and nano-LC-ESI-IT, or (2) a combination of nano-LC with Q-TOF or OT.

Finally, looking at the equipment in laboratories specialized in proteomic studies, it is evident that several technical solutions are often needed. Additionally, the training of the scientists performing the experiments is crucial for the success of proteomic research programs. This training must include the correct operation of the instrument(s) and interpretation of MS data as well as and most importantly, the thorough preparation of the biological samples.

References

1. Aebersold, R. and Mann, M. (2003) Mass spectrometry-based proteomics. *Nature* **422**, 198–207.
2. Domon, B. and Aebersold, R. (2006) Mass spectrometry and protein analysis. *Science* **312**, 212–217.
3. Roepstorff, P. (2005) Mass spectrometry instrumentation in proteomics. Encyclopedia of life sciences, John Wiley & Sons, Inc., New York, pp. 1–5.
4. Yates, J. R., Gilchrist, A., Howell, K. E., and Bergeron, J. J. (2005) Proteomics of organelles and large cellular structures. *Nat. Rev. Mol. Cell. Biol.* **6**, 702–714.
5. Sadygov, R. G., Cociorva, D., and Yates, J. R. (2004) Large-scale database searching using tandem mass spectra: looking up the answer in the back of the book. *Nat. Methods* **1**, 195–202.
6. Kicman, A. T, Parkin, M. C., and Iles, R. K. (2007) An introduction to mass spectrometry based proteomics–detection and characterization of gonadotropins and related molecules. *Mol. Cell. Endocrinol.* **260–262**, 212–227.
7. Lubec, G. and Afjedhi-Sadat, L. (2007) Limitations and pitfalls in protein identification by mass spectrometry. *Chem. Rev.* **107**, 3568–3584.
8. Pappin, D. J. C., Hojrup, P., and Bleasby, A. J. (1993) Identification of proteins by peptide-mass fingerprinting. *Curr. Biol.* **3**, 327–332.
9. Biemann, K. (1990) Sequencing of peptides by tandem mass spectrometry and high-energy collision-induced dissociation. *Methods Enzymol.* **193**, 455–479.
10. Mann, M. and Wilm, M. (1994) Error-tolerant identification of peptides in sequence databases by peptide sequence tags. *Anal. Chem.* **66**, 4390–4399.

11. Blueggel, M., Chamrad, D., and Meyer, H. E. (2004) Bioinformatics in proteomics. *Curr. Pharm. Biotechnol.* **5**, 79–88.
12. Steen, H. and Mann, M. (2004) The ABC's (and XYZ's) of peptide sequencing. *Nat. Rev. Mol. Cell. Biol.* **5**, 699–711.
13. Wolters, D. A., Washburn, M. P., and Yates, J. R. III. (2001) An automated multidimensional protein identification technology for shotgun proteomics. *Anal. Chem.* **73**, 5683–5690.
14. Malmström, J., Lee, H., and Aebersold, R. (2007) Advances in proteomic workflows for systems biology. *Curr. Opin. Biotechnol.* **18**, 1–7.
15. Shevchenko, A., Chernushevic, I., Wilm, M., and Mann, M. (2002) "De novo" sequencing of peptides recovered from in-gel digested proteins by nanoelectrospray tandem mass spectrometry. *Mol. Biotechnol.* **20**, 107–118.
16. Bradshaw, R. A., Burlingame A. L., Carr, S., and Aebersold, R. (2006) Reporting protein identification data: the next generation of guidelines. *Mol. Cell. Proteomics* **5**, 787–788.
17. Wilkins, M. R., Appel, R. D., Van Eyk, J. E., Chung, M. C., Görg, A., Hecker, M., Huber, L. A., Langen, H., Link, A. J., Paik, Y. K., Patterson, S. D., Pennington, S. R., Rabilloud, T., Simpson, R. J., Weiss, W., and Dunn, M. J. (2006) Guidelines for the next 10 years of proteomics. *Proteomics* **6**, 4–8.
18. Liu, T., Belov, M. E., Jaitly, N., Qian, W. J., and Smith, R. D. (2007) Accurate mass measurements in proteomics. *Chem. Rev.* **107**, 3621–3653.
19. Karas, M. and Hillenkamp, F. (1988) Laser desorption ionization of proteins with molecular masses exceeding 10,000 daltons. *Anal. Chem.* **60**, 2299–2301.
20. Fenn, J. B., Mann, M., Meng, C. K., Wong, S. F., and Whitehouse, C. M. (1989) Electrospray ionization for mass spectrometry of large biomolecules. *Science* **246**, 64–71.
21. Lane, C. S. (2005) Mass spectrometry-based proteomics in the life sciences. *Cell. Mol. Life Sci.* **62**, 848–869.
22. Baldwin, M. A. (2005) Mass spectrometers for biomolecular analysis. *Methods Enzymol.* **402**, 3–48.
23. Burlingame, A. L., Boyd, R. K., and Gaskell, S. J. (1998) Mass spectrometry. *Anal. Chem.* **70**, 647–716.
24. Karas, M., Bachmann, D., Bahr, U., and Hillenkamp, F. (1987) Matrix-assisted ultraviolet laser desorption of non-volatile compounds. *Int. J. Mass Spectrom. Ion Processes* **78**, 53–68.
25. Standing, K. G. (2000) Timing the flight of biomolecules: a personal perspective. *Int. J. Mass Spectrom.* **200**, 597–610.
26. March, R. E. (1997) An introduction to quadrupole ion trap mass spectrometry. *J. Mass Spectrom.* **32**, 351–369.
27. March, R. E. (1998) Quadrupole ion trap mass spectrometry: theory, simulation, recent developments and applications. *Rapid Commun. Mass Spectrom.* **12**, 1543–1554.
28. Cooks, R. G., Glish, G. L., McLuckey, S. A., and Kaiser, R. E. (1991) Ion trap mass spectrometry. *Chem. Eng News* **25**, 26–41.

29. Marshall, A. G., Hendrickson, C. L., Emmett, M. R., Rodgers, R. P., Blakney, G. T., and Nilsson, C. L. (2007) Fourier transform ion cyclotron resonance: state of the art. *Eur. J. Mass Spectrom.* **13**, 57–59.
30. Hardman, M. and Makarov, A. (2003) Interfacing the orbitrap mass analyzer to an electrospray ion source. *Anal. Chem.* **75**, 1699–1705.
31. Yates, J. R., Cociorva, D., Liao, L., and Zabrouskov, V. (2006) Performance of a linear ion trap-Orbitrap hybrid for peptide analysis. *Anal. Chem.* **78**, 493–500.
32. Scigelova, M. and Makarov, A. (2006) Orbitrap mass analyzer—overview and applications in proteomics. *Proteomics* **6**, S2, 16–21.
33. Martin, S. E., Shabanowitz, J., Hunt, D. F., and Marto, J. A. (2000) Subfemtomole MS and MS/MS peptide sequence analysis using nano-HPLC micro-ESI Fourier transform ion cyclotron resonance mass spectrometry. *Anal. Chem.* **72**, 4266–4274.
34. Zubarev, R. A., Kelleher, N. L., and McLafferty, F. W (1998) Electron capture dissociation of multiply charged protein cations. A nonergodic process. *J. Am. Chem. Soc.* **120**, 3265–3266.
35. Syka, J. E. P., Coon, J. J., Schroeder, M. J., Shabanowitz, J., and Hunt, D. F. (2004) Peptide and protein sequence analysis by electron transfer dissociation mass spectrometry. *Proc. Natl. Acad. Sci. USA* **101**, 9528–9533.
36. Good, D. M., Wirtala, M., McAlister, G. C., and Coon, J. J. (2007) Performance characteristics of electron transfer dissociation mass spectrometry. *Mol. Cell. Proteomics* **6**, 1942–1951.
37. Mikesh, L., Man Chi, B. U., Coon, J. J., Syka, J., Shabanowitz, J., and Hunt, D. F. (2006) The utility of ETD mass spectrometry in proteomic analysis. *Biochim. Biophys. Acta* **1764**, 1811–1822.
38. Creese, A. J. and Cooper, H. J. (2007) Liquid chromatography electron capture dissociation tandem mass spectrometry (LC-ECD-MS/MS) versus liquid chromatography collision-induced dissociation tandem mass spectrometry (LC-CID-MS/MS) for the identification of proteins. *J. Am. Soc Mass Spectrom.* **18**, 891–897.
39. Gauthier, G. and Grimm, G. (2006) Miniaturization: Chip-based liquid chromatography and proteomics. *Drug Discov. Today Technol.* **3**, 59–66.
40. Ghitun, M., Bonneil, E., Fortier, M. H., Yin, H., Killeen, K., and Thibault, P. (2006) Integrated microfluidic devices with enhanced separation performance: application to phosphoproteome analyses of differentiated cell model systems. *J. Sep. Sci.* **29**, 1539–1549.
41. Chen, H. S., Rejtar, T., Andreev, V., Moskovets, E., and Karger, B. L. (2005) Enhanced characterization of complex proteomic samples using LC-MALDI MS/MS: exclusion of redundant peptides from MS/MS analysis in replicate runs. *Anal. Chem.* **77**, 7816–7825.

2

Experimental Setups and Considerations to Study Microbial Interactions

Petter Melin

Summary

Within ecosystems microorganisms coexist and interact. Knowledge of these interactions is of great importance in the fields of ecology, food production, and medicine. Such interactions often involve the synthesis of antibiotic secondary metabolites. Different kinds of s molecules or direct contacts are other forms of microbial interactions. Recently, modern molecular methods such as microarrays and proteomics have been employed to investigate such interactions. In this chapter, the use of proteomics for studies of microbial interactions is discussed. The choice of experimental setup is dependent on the aims of the specific study. One aspect of competition between microbes can be simulated by treatment of one microbe with antibiotics produced by a competing microbe. A more complicated approach involves cocultivation of the competitors, but in order to reveal species-specific protein patterns it is advisable to keep the organisms separated. Alternative techniques are to monitor alterations in the proteomes between the wild-type and mutant strains. The mutant can be either natural or created using random or targeted mutagenesis. Generally, a proteomic study will reveal proteins with both expected and surprising changes in abundance upon competition, but also previously unknown proteins are likely to be identified. A proteomic approach is usually insufficient to obtain a complete data set describing microbial interactions. Therefore, it is essential to follow up identification of proteins with changed abundance by, e.g., the creation of knockout strains for phenotypic analyses. Despite the limitations, proteomics is a useful method, and an important complement to other approaches for studies of microbial interactions.

Key Words: Proteomics; proteome analysis; interactions; microorganisms; fungi; yeasts; bacteria; antibiotics; secondary metabolites.

From: *Methods in Molecular Biology, vol. 484: Functional Proteomics: Methods and Protocols*
Edited by: J. D. Thompson et al., DOI: 10.1007/978-1-59745-398-1, © Humana Press, Totowa, NJ

1. Introduction

In most ecosystems various microorganisms occupy the same habitat and coexist. Microbial interactions differ and can, for example, be mutual, parasitic, and competitive. These events can be studied at different levels, ranging from the whole ecosystem to the gene expression in a single organism. At the ecosystem level, the main concern is to describe variations in the surrounding environment and the content of species present. During the past decade, a very large number of ecological studies have, besides classical methods, been performed using various aspects of the polymerase chain reaction *(1)*. These studies have been aimed at describing discrete microbial communities and monitoring changes in gene expression at the population level. In contrast, only a limited number of studies have been aimed at the responses on the level of protein synthesis. Moreover, most of the protein studies in the area have had a medical rather than an ecological point of view. However, interesting general data concerning microbial interactions can be obtained from these medical studies. Likewise, more general studies of microbial stress responses may be of great interest in medicine, e.g., to elucidate responses to antibiotics. In this chapter, I intend to describe the potential and problems of using proteomics to study responses when different microorganisms interact. It is likely that the protein synthesis in a single microbe will adapt to a competitive environment. These changes in the complement of proteins present in an organism can be assessed by two-dimensional polyacrylamide gel electrophoresis (2D-PAGE). The term proteomics is very wide and can be used in all sorts of protein biology *(2)*, but for simplicity I decided to restrict the term proteomics to the comparison of different protein patterns from a specific organism exposed to different environments. Identified proteins can have an altered abundance due to the interaction. Alternatively, the protein is modified resulting in a different migration on the gel.

2. Why Study Microbial Interactions?

2.1. Antibiotic Secondary Metabolites

Almost all antibiotics used today are of microbial origin. In medicine we experience an increasing problem with pathogenic microbes that becomes resistant to the most commonly used antibiotics *(3,4)*. Thus there is an urgent need to develop new antimicrobial drugs. To use them in a safe way, we have to understand both their mode of action and the pathways and probabilities for development of resistance. Most studies concerning the competition between different microbes have aimed at elucidating the synthesis to antibiotic secondary metabolites, or to reveal the effect on target organisms when encountering these metabolites. The predominant hypothesis is that these secondary metabolites are synthesized to give the producing organism a competitive advantage by killing or inhibiting growth of other microbes ***(5)***. According to that proposal, the biosynthetic genes for a specific antibiotic are usually located in the same gene cluster as the corresponding resistance genes, thus relating synthesis of the antibiotic to competitive advantage ***(6)***. Alternative hypotheses regarding the origin of secondary metabolites have been proposed, e.g., the reduction of abnormally high concentrations of intermediate metabolites during growth arrest. One argument states that the concentrations of secondary metabolites in the field are not high enough to stop growth of other microbes ***(7)***. However, it has been shown that an organism can change the expression of several genes after encountering only subinhibitory concentrations of several different antibiotics ***(8)***.

2.2. Human Health

Bacteria can be both good and bad, and within our bodies we have a large bacterial flora that protects us from infection from pathogenic fungi and bacteria. Bacterial populations play a role in a large number of fungal diseases, e.g., by *Candida albicans* or *Cryptococcus neoformans*. The bacterium can be coinfecting our bodies or play an important role in the defense ***(9)***. Also, the consumption of probiotics, in general strains from the genus *Lactobacillus,* can be a way to protect us from hostile bacteria ***(10)***.

2.3. Microorganisms in Food and Feed

Fungal infection of crops intended for food and feed is a serious agricultural problem. Much effort is going on to replace or decrease the use of fungicides by fungal antagonistic microbes, e.g., *Pseudomonas* species ***(11)***, or by several strains within the filamentous fungi genus *Trichoderma* ***(12)***. When food and feed are stored, some microbes such as lactic acid bacteria ***(13)***, and the yeasts *Candida sake* ***(14)*** and *Pichia anomala* ***(15)*** can be used to protect the food from toxic fungi such as *Aspergillus*, *Botrytis*, and *Penicillium*. Here it is essential not only to decrease fungal growth, but also to know if the production/accumulation of toxic compounds produced is decreased. Some food products actually consist of several microbes, e.g., tempeh, which is a cake of soy beans (or other legumes or cereals), and the fungus *Rhizopus oligosporus* as well as nonpathogenic bacteria ***(16)***.

2.4. Microbial Interactions in Fundamental Ecology

In times with rising threats and an increased concern about the environment it is important to understand how organisms interact within the ecosystems.

Although microbes are small in size, they are present in abundance, are ubiquitous, and play decisive roles in all aspects of ecology. Fungi together with algae or cyanobacteria can live in mutual dependence and form a unique group of symbiotic organisms, the lichens. Fungi and plants can form mycorrhiza; the fungus increases the effective root surface of the plants and facilitates uptake of nutrients. In return, the plant provides the fungus with carbohydrates. It is known that bacteria also have a role in this symbiosis ***(17)***. Since formation of mycorrhiza is crucial for normal growth of many plants, knowledge of the nature of this symbiosis, including all the organisms involved, is not only interesting but also of great economic importance.

3. Materials

3.1. Simple Systems

In my opinion, the most important concern when studying microbial interactions at the laboratory scale is the choice of a system that faithfully mimics the situation of interest. This is independent of the techniques and is relevant regardless of whether the studies are aimed at the proteome, the transcriptome, or the metabolome. The simplest microbial interaction is when only one species is involved. This phenomenon has been observed among bacteria and it is called quorum sensing ***(18)***, and to my knowledge one such proteomics study has been published ***(19)***. To simplify a microbial interaction consisting of two different species, one of the organisms can be replaced by one or more important metabolites produced by that strain. For example, if a researcher wants to elucidate effects on the protein complement when a microbe is subjected to one specific hostile antibiotic, the target organism can be cultivated in the presence and absence of the antibiotic. This kind of proteomic setup has been used to study antibiotic resistance in the pathogenic gram-positive bacterium *Staphylococcus aureus* ***(20)***. Moreover, in medical mycology this experimental approach has been widely used to investigate several antifungals with the potential to replace amphotericin B, which is nephrotoxic for humans ***(21)***. For example, the responses to the antibiotic mulundocandin have been monitored in the human pathogenic yeast *C. albicans* ***(22)***. Grinyer and co-workers performed an interesting alternative approach in the area of biocontrol. They studied changes in the proteome of the biocontrol filamentous fungus *Trichoderma atroviride*. Prior to protein extraction they grew the *Trichoderma* strain with cell wall material from the plant pathogenic fungus *Rhizoctonia solani* as carbon sources compared to glucose in the control. In the study, several cell wall degrading enzymes likely to play a role in the biocontrol were identified ***(23)***.

3.2. Coculturing the Microorganisms

Replacing one interacting microbe with one or several of its metabolites is not always doable. If growth of all the involved microbes is essential, it is practical to keep the organisms separated, e.g., have a membrane that physically separates the organisms but allows metabolites to pass. We successfully used that technique when we cocultured the fungus *Aspergillus nidulans* with an antifungal strain of *Lactobacillus plantarum* **(24)**. Growing the organisms together, coextracting the materials from both organisms, and running the proteins from two or more proteomes on a single gel may be achievable, but it will complicate subsequent experiments, e.g., when identifying the proteins of interest. A potential problem when evaluating the results from a proteomic study from cocultured microorganisms is that not only changes in protein abundances due to metabolites but also responses to the nutritional competition will be monitored.

3.3. Comparing Different Strains

Besides coculturing or replacing a microbe with metabolites, there are several other approaches that can be suitable for proteomic studies of microbial interactions. If the specific target for an antibiotic is known, it is possible to disrupt the gene encoding the target for the antibiotic and then monitor changes in the proteome compared to the wild-type strain. Also, proteomics can be used to characterize mutants with a specific phenotype. For example, this approach was performed to investigate the proteome in a hygromycin-resistant strain of *C. albicans* **(25)**. Moreover, the proteomes of different strains of the same bacteria can be studied, e.g., to find proteins that are unique or absent in strains that are resistant to a specific antibiotic. This approach has been widely used in studies of bacterial proteomes, e.g., in *Lactobacillus sanfranciscensis* **(26)**, *S. aureus* **(27)**, and *Streptococcus pneumonia* **(28)**.

3.4. Experimental Design

All the analytical approaches listed above can and have been used in combination in order to understand the proteomic changes in a microorganism. For example, Yun *et al.* investigated the proteome of tetracycline treated *Pseudomonas putida*, and to understand the antibiobic-induced stress they used a strain that could tolerate high levels of tetracycline but did not carry resistance genes **(29)**. With multiple experiments and combining several different approaches on the same system it should be possible to discriminate responses to a specific antibiotic from the more complicated scenario in cocultures, or more so in complex small ecosystems. This approach was successful in our study when we cocultured *A. nidulans* with *L. plantarum*, we also grown the fungus with each of the known the bacterial metabolites *(24)*.

4. Methods

4.1. Preparation and Separation of the Protein Extract

The main limitation of proteomics is that, on each gel, only a fraction of the proteins will be displayed, i.e., the prominent and successfully extracted proteins, within the experimental parameters. However, more proteins could be made detectable if the parameters are slightly altered. Thus, it is always possible to change the pI intervals in the first dimension and the polyacrylamide concentration in the second. In addition, the method for protein extraction can be adjusted. Another way to improve resolution is to start by separating a specific organelle and then separating its protein components by 2D-PAGE. Accordingly, both cell wall *(30)*, plasma membrane *(31)*, and mitochondrial *(32)* proteins from *S. cerevisiae* have been successfully analyzed on 2D-PAGE. If the number of different proteins is reduced in a preparation, even proteins present in minor quantities can be displayed on the gel by increasing the amount of loaded proteins. Moreover, the field of proteomics is expanding rapidly, and technical improvements will further facilitate extraction, separation, and visualization of proteins *(33)*. It is possible that in the future all proteins in the proteome could be analyzed using 2D-PAGE, although a large number of gels need to be analyzed. The sensitivity of protein detection can also be improved by testing different staining methods. In my experience, working with parallel silver-stained gels and radiolabeled proteins, the latter provided the best resolution and the highest reproducibility. Another advantage of using radiolabeled amino acids is the ability to distinguish between short-term and long-term effects on the proteome. With this approach, only proteins that were synthesized after a specific time point will be visualized using autoradiography. In our experiments we studied proteomic responses in *A. nidulans* when it encountered concanamycin, an inhibitor of V-ATPases produced by *Streptomyces* sp. *(34)*. To achieve a sufficient amount of tissue for protein extraction, we have to preinoculate the fungus before adding the antibiotic. By simultaneously adding labeled amino acids only proteins synthesized after addition of the antibiotic were monitored on 2D-PAGE *(35)*.

4.2. Choices of Microorganisms

Naturally, the use of proteomics alone does not provide comprehensive information about how microbes interact in ecosystems. It is convenient to work with an organism with an available fully sequenced genome. In addition, it is an advantage if the genome is annotated and all hypothetical proteins are deduced. The identification of full-length protein sequences, by blasting the sequences to known protein databases, using only mass spectrometric data is problematic and time consuming. Without a sequenced genome, or a great number of known expressed sequence tags (EST) from a specific microbe, I would not recommend performing proteomics on that organism. Anyhow, if a close relative organism is sequenced, a correct identification of the proteins may be successful. In contrast, different strains of the same bacterial species may be very different and proteins identified by 2D-PAGE may not be fully deduced by blasting identified peptides toward the genome. The same problem can occur if the coverage of the sequence genome is low because parts of the genome are not sequenced. When we performed our first proteomic study using the model fungus *A. nidulans* ***(34)***, the genome was sequenced only with a 3× coverage; thus the full sequence of one identified protein could only be partially deduced and the sequence of one other protein could not be deduced at all. Another obstacle was that several peptides (identified with mass spectrometry) were located on different exons making the full detection of the complete protein and DNA sequences very time consuming.

4.3. How to Interpret the Results?

Most proteomic reports describe up- or downregulation of proteins due to a specific environmental change, e.g., a microbial interaction. Usually, several of these proteins are already identified in previous studies. However, there is often no logical explanation as to why these proteins should be involved in the actual response. It is obvious that the mechanisms behind protein synthesis are complicated events, and it is often impossible to predict secondary effects that alter the synthesis of a specific protein. Additional experiments are often required to provide answers. To learn more about an unknown protein, the most straightforward approach is to disrupt the encoding gene and investigate phenotypical consequences. Repeating the proteomic approach using the mutant strain is one method to study the new phenotype. Since additional studies are required to understand observed changes in the proteomic pattern, I would recommend, in addition to a complete genomic sequence, using a model organism with developed molecular techniques, including a functional transformation system.

4.4. Comparison with Transcriptomics

In principal, the system designed for studying responses in the proteome, using proteomics, can also be used to study gene expression, i.e., transcriptomics. The observed changes in the proteome are the result of the interaction, but since only the most abundant proteins will be displayed it is likely that minor proteins, being very important in the response to other microorganisms, may not be monitored. In this respect monitoring the transcriptome, e.g., with microarrays, is a more suitable approach. The important difference in favor of proteomics is due to stability. Proteins tend to be stable whereas mRNAs are relatively short-lived molecules. Therefore, short-term changes in the expression/synthesis are probably most conveniently studied at the mRNA level. On the other hand, since regulation often also occurs at posttranscriptional levels, mRNA levels may be misleading, and a determination of the final gene product, the protein, may be more instructive for general metabolic potential.

5. Conclusions

In this chapter I have summarized the use of proteomics to study microbial interactions. Although proteomics is a comparatively new approach in functional biology, it has been proven useful when elucidating molecular responses in microorganisms upon microbial interactions. There are, however, several inherent limitations with the technique. One fundamental problem with proteomics is the choice of a system that faithfully mimics the interaction of choice. However, this problem is encountered in any microbial study at the laboratory scale. Another aspect more specifically connected to proteomics is that the microbe may not change its protein production during competition to detectable levels. For example, the molecular response to an antibiotic may be extreme during laboratory conditions, but, in the field, the concentrations of antibiotic secondary metabolites may not be high enough to cause the same changes in protein synthesis. Despite these limitations I think the proteomic approach in ecological studies is a useful complement to other techniques, although the potential of proteomics is probably greater in medicine. The knowledge of responses at the protein level to antibiotics is important in understanding the full mode of action as well as secondary responses in both the target microbe and in the host.

References

1. Kirk, J. L., Beaudette, L. A., Hart, M., Moutoglis, P., Khironomos, J. N., Lee, H., *et al.* (2004) Methods of studying soil microbial diversity. *J Microbiol. Met.* **58**, 169–188.
2. Pandey, A. and Mann, M. (2000) Proteomics to study genes and genomes. *Nature* **405**, 837–846.
3. Cowen, L. E. (2001) Predicting the emergence of resistance to antifungal drugs. FEMS *Microbiol Let.* **204**, 1–7.
4. Lipsitch, M. (2001) The rise and fall of antimicrobial resistance. *Trends Microbiol.* **9**, 438–444.

5. Maplestone, R. A., Stone, M. J., and Williams, D. H. (1992) The evolutionary role of secondary metabolites—-a review. *Gene* **115**, 151–157.
6. Stone, M. J. and Williams, D. H. (1992) On the evolution of functional secondary metabolites (natural-products). *Mol. Microbiol.* **6**, 29–34.
7. Gottlieb, D. (1976) The production and role of antibiotics in soli. *J. Antibiot.* **29**, 987–1000.
8. Goh, E. B., Yim, G., Tsui, W., McClure, J., Surette, M. G., and Davies, J. (2002) Transcriptional modulation of bacterial gene expression by subinhibitory concentrations of antibiotics. *Proc. Natl. Acad. Sci. USA* **99**, 17025–17030.
9. Wargo, M. J. and Hogan, D. A. (2006) Fungal-bacterial interactions: a mixed bag of mingling microbes. *Curr. Opin. Microbiol.* **9**, 359–364.
10. Reid, G. and Burton, J. (2002) Use of *Lactobacillus* to prevent infection by pathogenic bacteria. *Microb. Infect.* **4**, 319–324.
11. Gerhardson, B. (2002) Biological substitutes for pesticides. *Trends Biotech.* **20**, 338–343.
12. Harman, G. E., Howell, C. R., Viterbo, A., Chet, I. and Lorito, M. (2004) *Trichoderma* species—-opportunistic, avirulent plant symbionts. *Nature Rev. Microbiol* **2**, 43–56.
13. Lindgren, S. E. and Dobrogosz, W. J. (1990) Antagonistic activities of lactic-acid bacteria in food and feed fermentations. *FEMS Microbiol. Rev.* **87**, 149–163.
14. Vinas, I., Usall, J., Teixido, N., and Sanchis, V. (1998) Biological control of major postharvest pathogens on apple with *Candida sake*. *Int. Food Microbiol.* **40**, 9–16.
15. Passoth, V., Fredlund, E., Druvefors, U. Ä., and Schnurer, J. (2006) Biotechnology, physiology and genetics of the yeast *Pichia anomala*. *FEMS Yeast Res.* **6**, 3–13.
16. Feng, X. M., Eriksson, A. R. B., and Schnurer, J. (2005) Growth of lactic acid bacteria and *Rhizopus oligosporus* during barley tempeh fermentation. *Int. J. Food Microbiol.* **104**, 249–256.
17. Garbaye, J. (1994) Helper bacteria—-a new dimension to the mycorrhizal symbiosis. *New Phyt.* **128**, 197–210.
18. Miller, M. B. and Bassler, B. L. (2001) Quorum sensing in bacteria. *Annu. Rev. Microbiol.* **55**, 165–199.
19. Riedel, K., Arevalo-Ferro, C., Reil, G., Gorg, A., Lottspeich, F., and Eberl, L. (2003) Analysis of the quorum-sensing regulon of the opportunistic pathogen *Burkholderia cepacia* H111 by proteomics. *Electrophoresis* **24**, 740–750.
20. Hecker, M., Engelmann, S., and Cordwell, S. J. (2003) Proteomics of *Staphylococcus aureus*—-current state and future challenges. *J. Chromatogr. B Analyt. Technol. Biomed. Life Sci.* **787**, 179—-195.
21. Finquelievich, J. L. and Odds, F. C., Queiroz-Telles, F., and Wheat L. J. (2000) New advances in antifungal treatment. *Med. Mycol.* **8**, 317–322.
22. Bruneau, J. M., Maillet, I., Tagat, E., Legrand, R., Supatto, F., Fudali, C., *et al.* (2003) Drug induced proteome changes in *Candida albicans*: comparison of the effect of beta(1,3) glucan synthase inhibitors and two triazoles, fluconazole and itraconazole. *Proteomics* **3**, 325–336.

23. Grinyer, J., Hunt, S., McKay, M., Herbert, B. R., and Nevalainen, H. (2005) Proteomic response of the biological control fungus *Trichoderma atroviride* to growth on the cell walls of *Rhizoctonia solani*. *Curr. Genet.* **47**, 381–388.
24. Ström, K., Schnürer, J., and Melin, P. (2005) Co-cultivation of antifungal *Lactobacillus plantarum* MiLAB 393 and *Aspergillus nidulans*, evaluation of effects on fungal growth and protein expression. *FEMS Microbiol. Lett.* **246**, 119–124.
25. De Backer, M. D., de Hoogt, R. A., Froyen, G., Odds, F. C., Simons, F., Contreras, R., *et al.* (2000) Single allele knock-out of *Candida albicans* CGT1 leads to unexpected resistance to hygromycin B and elevated temperature. *Microbiology* **146**, 353–365.
26. De Angelis, M., Bini, L., Pallini, V., Cocconcelli, P. S., and Gobbetti, M. (2001) The acid-stress response in *Lactobacillus sanfranciscensis* CB1. *Microbiology* **147**, 1863–1873.
27. Cordwell, S. J., Larsen, M. R., Cole, R. T., and Walsh, B. J. (2002) Comparative proteomics of *Staphylococcus aureus* and the response of methicillin-resistant and methicillin-sensitive strains to Triton X-100. *Microbiology* **148**, 2765–2781.
28. Cash, P., Argo, E., Ford, L., Lawrie, L., and McKenzie, H. (1999) A proteomic analysis of erythromycin resistance in *Streptococcus pneumoniae*. *Electrophoresis* **20**, 2259–2268.
29. Yun, S. H., Kim, Y. H., Joo, E. J., Choi, J. S., Sohn, J. H., and Kim, S. (2006) Proteome analysis of cellular response of *Pseudomonas putida* KT2440 to tetracycline stress. *Curr. Microbiol.* **53**, 95–101.
30. Pardo, M., Ward, M., Bains, S., Molina, M., Blackstock, W., Gil, C., *et al.* (2000) A proteomic approach for the study of *Saccharomyces cerevisiae* cell wall biogenesis. *Electrophoresis* **21**, 3396–3410.
31. Navarre, C., Degand, H., Bennett, K. L., Crawford, J. S., Mortz, E., and Boutry, M. (2002) Subproteomics: identification of plasma membrane proteins from the yeast *Saccharomyces cerevisiae*. *Proteomics* **2**, 1706–1714.
32. Zischka, H., Weber, G., Weber, P. J. A., Posch, A., Braun, R. J., Buhringer, D., Schneider, U., Nissum, M., Meitinger, T., Ueffing, M., and Eckerskorn, C. (2003) Improved proteome analysis of *Saccharomyces cerevisiae* mitochondria by free-flow electrophoresis. *Proteomics* **3**, 906–916.
33. Harry, J. L., Wilkins, M. R., Herbert, B. R., Packer, N. H., Gooley, A. A., and Williams, K. L. (2000) Proteomics: Capacity versus utility. *Electrophoresis* **21**, 1071–1081.
34. Bowman, E. J., Siebers, A., and Altendorf, K. (1988) Bafilomycins: a class of inhibitors of membrane ATPases from microorganisms, animal cells, and plant cells. *Proc. Natl. Acad. Sci. USA* **85**, 7972–7976.
35. Melin, P., Schnürer, J., and Wagner, E. G. H. (2002) Proteome analysis of *Aspergillus nidulans* reveals proteins associated with the response to the antibiotic concanamycin A, produced by *Streptomyces* species. *Mol. Genet. Genom.* **267**, 695–702.

II

Proteomics

3

Plant Proteomics

Eric Sarnighausen and Ralf Reski

Summary

An understanding of gene function requires a complementation of gene and gene expression analysis by the systematic analysis of proteins. Progress in plant proteomics has been lagging behind animal and microbial proteomics due to the lack of plant genome data and the problems involved in successful protein extraction from plant material. With the sequencing of more and more plant genomes, this slow progress will soon be overcome. The moss *Physcomitrella patens* is a model organism in the field of plant functional genomics. *P. patens* is the first seedless plant for which the complete genome was sequenced. Genome annotation is currently in progress. While identification of proteins requires knowledge of all coding genes of the organism under study, gene annotation and functional characterization benefit greatly from the findings of proteome analysis. The proteome of *P. patens* is accessible and approaches are under way to increase the spectrum of proteomic methods applied to this plant. Here we provide a protocol for the extraction of proteins from *P. patens* and describe the basic and still most important method of proteome analysis, two-dimensional polyacrylamide electrophoresis of proteins. As this technique (not entirely unjustifiably) has the reputation of being unpredictably complicated, we provide a detailed protocol intended to reduce the reluctance that many scientists may have in using this technique.

Key Words: Plant proteomics; *Physcomitrella patens*; protein extraction; two-dimensional electrophoresis; isoelectric focusing; SDS–PAGE.

1. Introduction

Progress in the field of plant proteomics has always lagged behind research in the animal or microbial field *(1)*. There are numerous reasons for this. Compared with multicellular organisms, proteomes of unicellular prokaryotes

From: *Methods in Molecular Biology, vol. 484: Functional Proteomics: Methods and Protocols*
Edited by: J. D. Thompson et al., DOI: 10.1007/978-1-59745-398-1, © Humana Press, Totowa, NJ

and eukaryotes are of reduced complexity and therefore more easily accessible; at the same time these were the first organisms for which the genome sequences were available. Furthermore, there is hardly any material that is more reluctant to proteome analysis than plant tissue. The presence of a rigid cell wall, which is often enforced through deposition of strengthening substances, like lignin (wood), suberin (cork), or inorganic salts (calcification), can render tissue disruption problematic. Compared to animal tissue, protein content in most parts of the plant is rather low. On the other hand, plants contain a multitude of substances that interfere strongly with a successful protein extraction process; foremost among these are phenolic compounds, organic acids, and proteases—compounds that tend to modify, inactivate, precipitate, aggregate, or degrade proteins in crude extracts. Consequently, special techniques are required to disrupt the cell walls and to protect proteins from damaging components released on breakage. A direct single-step extraction of proteins, which is a general procedure when working with bacteria *(2)*, yeast, or animal tissue *(3)*, is therefore hardly ever the best choice for workers in the plant field *(4)*. The ultimate goal is to separate the total proteome from substances that interfere with proteome analysis while at the same time avoiding quantitative or qualitative modification of the proteome during this process. As protein extraction procedures can hardly be automated, plant proteomics requires extensive processing at a step that is considered most critical for the generation of reproducible results. Protein purification procedures, required for the analysis of the plant proteome, will inevitably be selective for certain proteins and will at the same time discriminate others *(5)*. Among the most commonly used plant protein extraction procedures are acetone/trichloroacetic acid (TCA) precipitation *(6)*, phenolic extraction *(7)*, and extraction of soluble proteins in combination with acetone or TCA precipitation *(8)*. While all these procedures can render high quality separations of proteins on two-dimensional gels, protein spot patterns obtained from the same tissues display considerable variations if extraction methods are varied *(9,10)*. Another problem researchers in plant proteomics have to face is the unequal distribution of the concentration of distinct protein species among the plant proteome. Proteins related to the photosynthetic apparatus can represent far more than 50% of the total protein mass in plants and will always dominate in the separation patterns while low abundant proteins are likely to escape detection *(5)*.

The moss *Physcomitrella patens* (**Fig. 1A**) has emerged as a model organism in the field of functional genome analysis. *P. patens* is unique among land plants as its nuclear genes can be directly targeted due to highly efficient homologous recombination *(11)*. In reverse genetics approaches, a gene of interest is disrupted and the resulting phenotypical aberrations subsequently allow conclusions to be drawn on the function of the gene *(12)*. Due to its

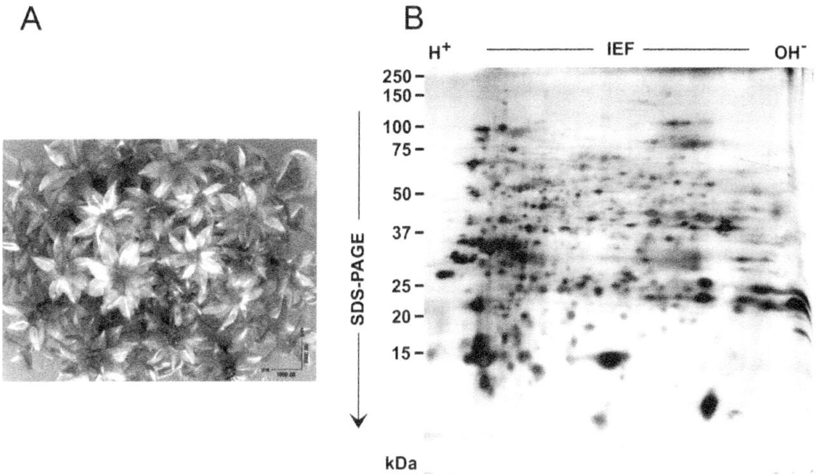

Fig. 1. Proteome analysis of *Physcomitrella patens*. (**A**) The moss *P. patens* is a model organism in plant functional genomics. (Courtesy of Dr. Julia Schulte.) (**B**) Proteins of *P. patens* were extracted with acetone/TCA and were subsequently separated via isoelectric focusing in the first dimension and via SDS–PAGE in the second dimension. (Courtesy of Anika Erxleben.)

outstanding features as a model organism ***(13)***, *P. patens* has been chosen as the first seedless plant to have its full genome sequenced (http://www.jgi.doe.gov/sequencing/why/CSP2005/physcomitrella.html). Knowledge of all coding genes now adds additional weight to proteome analysis as a tool of functional genomics in *P. patens*. Complementation of phenotypical analysis by differential or functional proteomics studies allows for the elucidation of regulatory networks and a precise classification of gene functions in the context of complex living systems.

From the repertoire of proteomic techniques used in our laboratory, this chapter will focus on those methods of classical proteome analysis that will most likely describe the most accessible approach for researchers interested in the field. Plant protein extraction by acetone/TCA precipitation is straightforward, fast, and simple and yields samples of high purity. However, it should be mentioned that sometimes (depending on the source tissue) the price that needs to be paid for this degree of purity is reduced extractability, not only of impurities but also of proteins ***(14)***. We describe a two-dimensional (IEF/SDS–PAGE) electrophoresis system routinely used in our laboratory. The high separation power of this system lies in the combination of two independent protein separation techniques. Isoelectric focusing (IEF) as the first dimension separates the proteins according to their intrinsic charge (their isoelectric points).

In the second dimension proteins are subsequently separated on the basis of their molecular masses using sodium dodecyl sulfate polyacrylamide gel electrophoresis (SDS–PAGE) (**Fig. 1B**). While at first glance it might appear unprogressive to not use ready-cast immobilized pH gradient gels for isoelectric focusing ***(15)***, our experience shows that a larger number of protein spots can be resolved two-dimensionally if self-cast gels containing carrier ampholytes are used in the first dimension. Visualization of proteins is accomplished via colloidal Coomassie staining. This is a very reliable method of protein staining that combines a good sensitivity and an acceptable dynamic range of staining intensities with the advantage that it is compatible with further identification of proteins via mass spectrometry.

2. Materials

2.1. Growth of Plant Material

1. *P. patens* protonema is grown in Knop medium: 250 mg/L KH_2PO_4, 250 mg/L KCl, 250 mg/L $MgSO_4 \times 7\ H_2O$, 1000 mg/L $Ca(NO_3)_2 \times 4\ H_2O$, 12.5 mg/L $FeSO_4 \times 7\ H_2O$, adjust pH to 5.8 with KOH. Knop medium is autoclaved twice at an interval of 2 days.

2.2. Protein Extraction

1. Acetone/TCA solution: 10% (w/v) TCA, 0.2% (w/v) dithiothreitol (DTT), in acetone. Store at –20°C.
2. IEF lysis buffer: 8 *M* urea, 4% (w/v) 3-[(3-cholamidopropyl)dimethylammonio]-l-propane-sulfonate (CHAPS), 100 m*M* DTT, 40 m*M* Tris-base, 0.16% (w/v) Biolyte Ampholytes, pH 5–8 (Bio-Rad, Richmond, CA), 0.04% (w/v) Biolyte Ampholytes pH 3–10 (Bio-Rad). Urea should be of highest purity (e.g., Roche, EP-MB Grade). Water should be of high-performance liquid chromatography (HPLC) quality. IEF lysis buffer is stored in 1-mL aliquots at –20°C.

2.3. Protein Assay

1. 0.4% (w/v) bovine serum albumin (BSA) stock solution in IEF lysis buffer, stored in small aliquots at –20°C.
2. 0.1 *N* hydrochloric acid, stored at room temperature.
3. Bradford reagent (stock solution): 0.05% (w/v) Coomassie brilliant blue G 250, 25% (v/v) methanol, 72.25% orthophosphoric acid. Stored at 4°C. 100 mg Coomassie brilliant blue is dissolved in 50 mL methanol. 100 mL of 85% orthophosphoric acid is added and finally the volume is adjusted to 200 mL with water. Bradford stock solution is stable at 4°C.

2.4. Isoelectric Focusing

1. All solutions are made with HPLC grade water (bidistilled).
2. Urea should be of highest purity (Roche, EP-MB Grade).
3. Biolytes 3/10 and 5/8 Ampholytes (Bio-Rad) are stored as aliquots of 500 μL at 4°C protected from light.
4. 10% (w/v) CHAPS in water is stored in 1-mL aliquots at –20°C.
5. Acrylamide stock solution for IEF (30% T, 5.3% C): 28.4% (w/v) acrylamide (Bio-Rad), 1.6% (w/v) piperazine diacrylamide (Bio-Rad) (*see* **Note 1**). The solution is deionized via Serdolit MB-1 mixed bed ion exchanger resin (Serva, Heidelberg) (*see* **Note 2**). Acrylamide stock solution is stirred with 1% (w/v) Serdolit at room temperature protected from the light for at least 10 min. Serdolit is removed by paper filtration and finally the acrylamide stock solution is passed through a 0.22-μm membrane filter. Acrylamide stock solution is stored in 0.7-mL aliquots at –20°C. Acrylamide monomers are potent neurotoxins and should be handled with appropriate safety measures. The easiest way to detoxify acrylamide is polymerization to polyacrylamide (see below).
6. 10% (w/v) ammonium persulfate, prepared freshly.
7. Gel overlay solution: 6.5 *M* urea, stored in 500-μL aliquots at –20°C.
8. Lysis buffer: *see* 2.2.2.
9. Sample overlay solution: 7 *M* urea, 0.8% (w/v) Biolytes 5/8 Ampholytes, 0.2% (w/v) Biolytes 3/10 Ampholytes (Biolytes come as a 40% [w/v] stock solution), stored in 200-μL aliquots at –20°C.
10. Cathode electrolyte solution: 0.02 *M* NaOH (degassed, prepared freshly).
11. Anode electrolyte solution: 0.01 *M* H_3PO_4 (prepared freshly).
12. Bromophenol blue solution in water (at the point of saturation), 1 mL stored at 4°C.

2.5. Sodium Dodecyl Sulfate Polyacrylamide Gel Electrophoresis (SDS–PAGE)

1. IEF gel equilibration buffer: 6 *M* urea, 30% (w/v) glycerol, 50 m*M* Tris–HCl, pH 8.3 (*see* **Note 3**), 4% (w/v) SDS (from a 20% stock solution, Bio-Rad). In addition

   (a) for the first step (reduction) 2% (v/v) tributylphosphine (*see* **Note 4**) and
   (b) for the second step (alkylation) 2.5% iodoacetamide.

2. Acrylamide stock solution for SDS–PAGE (30% T, 2.7% C) 29.2% (w/v) acrylamide (Bio-Rad), 0.8% (w/v) piperazine diacrylamide (Bio-Rad). The solution is filtered through a 0.45-μm membrane filter and stored at 4°C protected from light.
3. 1.5 *M* Tris–HCl, pH 8.8; the solution is filtered through a 0.45-μm membrane filter and stored in 50-mL aliquots at –20°C (long time storage) or 4°C (short time storage).

4. 0.5 *M* Tris–HCl, pH 6.8; the solution is filtered through a 0.45-μm membrane filter and stored in 50-mL aliquots at –20°C (long time storage) or 4°C (short time storage).
5. 10% (w/v) ammonium persulfate, prepared freshly.
6. SDS–PAGE running buffer (cathode buffer): 25 m*M* Tris base, 192 m*M* glycine, 0.02% (w/v) sodium thiosulfate (anhydrous), 0.4% (w/v) SDS (*see* **Note 5**) (from a 20% stock solution, Bio-Rad); do NOT adjust the pH (*see* **Note 6**).
7. SDS–PAGE anode buffer: 25 m*M* Tris–HCl, pH 8.3 (*see* **Note 7**).

2.6. Colloidal Coomassie Staining

1. Solution A: 1.7% (w/v) orthophosphoric acid, 10% (w/v) ammonium sulfate.
2. Solution B: 5% (w/v) Coomassie brilliant blue G250 in water (colloid); stir or shake vigorously prior to use (*see* **Note 8**).
3. Solution C: 49 vol solution A, 1 vol solution B.
4. Solution D: 4 vol solution C, 1 vol methanol; methanol must be added slowly (*see* **Note 9**); prepare freshly prior to use.

3. Methods

3.1. Growth of Plant Material

1. *P. patens* protonema is cultivated in 500- mL Erlenmeyer flasks in 180 mL of Knop medium at 25°C and a light intensity of 55 μmol/m^2 under long day conditions (16 h light, 8 h darkness) with shaking at 121 rpm. The filamentous protonema is transferred to fresh medium and disintegrated weekly with an Ultra Turrax T 25 (IKA Labortechnik, Staufen, Germany). Inoculation density is 150 mg dry weight per liter. The material is harvested by paper filtration using a Buchner funnel with suction and immediately frozen in liquid nitrogen. Moss material is stored at –80°C until use.

3.2. Protein Extraction

1. Frozen moss protonema is disrupted in a ball mill (*see* **Note 10**) equipped with Stainless-steel grinding jars and grinding balls for 90 s at 1800 rpm. To prevent the material from thawing during this process, balls and jars are precooled in liquid nitrogen.
2. Using a spatula precooled in liquid nitrogen, 300 mg of ground moss material is transferred to a precooled 2-mL reaction tube.
3. 1.5 mL of ice-cold acetone/TCA is added immediately to the plant material. The mixture is vortexed briefly and allowed to stand at –20°C for 1 h (*see* **Note 11**).
4. Samples are centrifuged at 19,000 × *g* for 15 min at –5°C and the supernatant is discarded.
5. The pellet is washed three times with 1.5 mL of ice-cold acetone containing 0.2% (w/v) DTT. The samples are allowed to stand for 1 h at –20°C between the washes

and the tubes are centrifuged at 19,000 × *g* for 15 min at –5°C prior to the removal of the acetone. The final pellet should be deprived of chlorophyll.

6. The pellet is dried in a speed vac. To this end the lids of the reaction tubes are perforated with a needle in order to allow the evaporation of the acetone. The pellet should not be dried with the reaction tubes opened, as there is a high risk of loosing the sample during venting of the rotor chamber.
7. The proteins are extracted from the dried material in 600 μL of lysis buffer; the slurry is transferred to a 1.5-mL reaction tube and protein extraction is performed by vortexing the sample at room temperature for 30 min (*see* **Note 12**).
8. Cell debris is removed by centrifuging the sample twice at 19,000 × *g* for 15 min at room temperature (*see* **Notes 13 and 14**).
9. Protein samples are stored at –80°C. Repeated thawing and freezing is not recommended!

3.3. Protein Assay

We use a modification of the Bradford protein assay optimized for protein samples in urea buffer ***(16)***.

1. 4 μL 0.1 *N* HCl is added to 4 μL of protein extract. The acidified extract is diluted with 80 μL of water.
2. 6 μL of 0.1 *N* HCl is added to 6 μl of BSA stock solution (4 mg/mL). The acidified solution is diluted with 120 μL of water.
3. A 1:1:20 mixture of lysis buffer, 0.1 *N* HCl, and water is used to further dilute the sample and the BSA solution. Dilutions of the BSA solution are required to build a calibration curve. Dilutions of the moss protein sample are prepared to ensure absorption values that are within the range of the calibration curve.
4. Bradford reagent stock solution is diluted 5-fold with water and filtered through paper.
5. 300 μL of Bradford reagent is added to 20 μL protein solution in the wells of a 96-well microtiter plate.
6. Absorbance at 595 nm is determined within 30 min in a microtiter plate reader.
7. Protein concentrations of the moss protein samples are calculated from the calibration curve.

3.4. Isoelectric Focusing

Initially, we used commercially available IPG (immobilized pH gradient) strips for isoelectric focusing ***(17)***. While the use of these precast gels considerably simplifies the procedure of isoelectric focusing and is known to yield gels of high reproducibility even in the hands of rather inexperienced labworkers ***(3)***, the problems associated with this method are well known. Separation of large, basic, acidic, or hydrophobic proteins, in particular, is problematic when IPG strips are used. We are able to resolve a larger number of proteins using carrier ampholyte tube gels as described by O'Farrell ***(18)***.

1. The IEF gel solution is prepared in a 100-mL side arm flask: 2.25 g urea, 665 μL IEF acrylamide stock solution, 1 mL 10% (w/v) CHAPS, 500 μl Biolytes 5/8 Ampholytes, 125 μL Biolytes 3/10 Ampholytes, 1.17 mL of water.
2. As oxygen interferes with the acrylamide polymerization, the IEF gel solution is degassed for 15 min (*see* **Note 15**). The side arm flask is connected to a membrane vacuum pump. A Wolffs bottle should be inserted between the side arm flask and the pump in order to avoid contamination of the latter with acrylamide solution. The pump should be used in a fume hood. The urea should not be dissolved in the gel solution immediately, as the crystals will act as nucleation centers for gas bubbles. Eventually, the solution should be mixed while still under vacuum by gentle movements of the side arm flasks. The urea should be dissolved without warming of the solution as increased temperature will promote acrylamide polymerization. The walls of the side arm flasks should not be wetted by the solution as this might induce the precipitation of urea.
3. Clean glass tubes 20 cm in length with an inner diameter of 2.3 mm are labeled to a height of 16.5 cm. The bottom of the tube is sealed tightly with Parafilm. Avoid covering large parts of the tubes' surface with Parafilm as the gel solution must be visible through the glass during the casting process.
4. Glass tubes are mounted in an upright position in a casting stand.
5. Polymerization of the IEF acrylamide solution is initiated by the addition of 4 μL TEMED and 8 μL 10% (w/v) ammonium persulfate solution. Note that the polymerization process will start immediately. The gel solution is mixed gently and is then aspirated into a (self made) 10-L syringe equipped with a thin Teflon tubing of 22 cm. Aspiration of air must be avoided.
6. The glass tubes are filled to the label. The Teflon tubing must be inserted to the bottom of the glass tube prior to the injection of the gel solution or air bubbles will form. Keep the tip of the tubing approximately 0.5 cm below the meniscus while filling the tubes (*see* **Note 16**).
7. Each gel is immediately carefully overlaid with 130 μL of overlay solution. The tubes are filled to the rim with water.
8. After 2 h, overlay solution is replaced by 100 μL of IEF lysis buffer. The lysis buffer is overlaid with water to completely fill the tubes. Polymerization is complete after an additional 2 h.
9. Cathode electrolyte (500 mL if a Bio-Rad Protean xi II cell is used) is degassed under vacuum with stirring for 1 h (*see* **Note 17**).
10. The parafilm is removed from the glass tubes. The gels need to be secured from sliding out of the tubes during electrofocusing by sealing the tubes at the bottom with dialyses membranes wetted in anode electrolyte. The membrane pieces are fixed with O-rings that are cut from rubber tubing. No air bubbles should be trapped between the gel and the membrane. This is achieved by wetting the bottom of the IEF gels with anode electrolyte prior to the application of the membrane (*see* **Note 18**).
11. The lower buffer chamber is filled with anode electrolyte (1.5 L if a Bio-Rad Protean xi II cell is used). The glass tubes are installed in the electrophoresis chamber.

12. The protein concentration of the sample is adjusted to 500 μg/100 μL with IEF lysis buffer.
13. The overlaying IEF buffer is removed and replaced by the protein sample in 100 μL of IEF buffer (*see* **Note 19**). The samples are overlaid with 20 μL of sample overlay solution. Finally the glass tubes are filled to the rim with cathode electrolyte. The upper electrophoresis chamber is filled with cathode electrolyte. The tubes must be completely covered by the cathode electrolyte.
14. Isoelectric focusing is run at 10°C for 30 min at 200 V, for 18 h at 500 V, for 1 h at 800 V, and finally for 1 h at 1000 V.
15. After the disassembly of the electrophoresis chamber, liquid is removed from the IEF gels and the surface of the gels is rinsed with water once. Subsequently, the glass tubes are filled with water and the gels are released into a disposable Petri dish by air pressure applied via a (self made) 10-mL syringe equipped with a silicone tubing that fits tightly over the glass tubes. Extreme care must be taken not to damage the gels during this process (this requires some practice). The force required to press the gels from the glass tubes decreases rapidly as the gel is released. The pressure applied to the glass tube must be adjusted accordingly or the gel will be destroyed when being ejected from the tube.
16. The basic (former upper) end of the gel is labeled with one droplet of saturated bromophenol blue solution.
17. The gels can be stored indefinitely at –80°C.

3.5. Sodium Dodecyl Sulfate Polyacrylamide Gel Electrophoresis (SDS–PAGE)

1. Glass plates and spacers must be cleaned thoroughly with ethanol and water prior to the assembly of the gel sandwich. Glass plates must be dried with lint-free cloth only. We use the Bio-Rad Protean xi II chamber to cast two gels of 1 mm × 185 mm × 185 mm simultaneously.
2. To cast gels with a linear acrylamide gradient, two gel solutions containing different amounts of acrylamide need to be prepared (*see* **Note 20**). For the lower acrylamide solution (high amount of acrylamide) mix in a 500-mL side arm flask 25 mL of SDS–PAGE acrylamide stock solution, 11.25 mL 1.5 *M* Tris-HCl, pH 8.8, 225 μl 20% (w/v) SDS, 4.5 g glycerol and adjust to 45 mL with water. For the upper acrylamide solution (low amount of acrylamide) prepare in a 500-mL side arm flask 10.5 mL of SDS–PAGE acrylamide stock solution, 11.25 mL Tris–HCl, pH 8.8, 225 μl 20% (w/v) SDS, and adjust to 45 mL with water.
3. Both solutions are degassed via a membrane pump for 15 min. A Wolffs bottle should again be used to prevent the acrylamide solution from contaminating the pump.
4. The gradient mixer is placed on a magnetic stirrer and a stir bar is placed in the mixing chamber (front beaker).
5. The lower (high density) acrylamide solution is poured into the mixing chamber of the gradient mixer.

6. 110 µL 10% ammonium persulfate and 30 µL TEMED are added to the upper (low density) acrylamide solution, which is then mixed by gentle shaking and is poured into the reservoir chamber of the gradient mixer.
7. The magnetic stirrer is switched on and 110 µL 10% ammonium persulfate and 30 µL TEMED are added to the lower acrylamide solution.
8. The stopcock is opened and acrylamide solution is released form the gradient mixer (*see* **Note 21**). The flow is either driven by a peristaltic pump (which is preferable) or via hydrostatic pressure by placing the gradient mixer on a shelf above the gel sandwich. Two gels are cast simultaneously by inserting a T-piece into the tubing (*see* **Note 22**).
9. The valve stem between the two chambers is opened in order to start gradient formation.
10. Both gels should be cast at the same speed. It might be necessary to adjust the simultaneous flow of the acrylamide solution by squeezing one or the other of the tubing to decrease the flow.
11. The running gel is cast to a height 18 mm below the top of the lower glass plate.
12. Each gel is overlaid carefully with water-saturated isobutanol. The gels are allowed to polymerize for 2 h.
13. To prepare the stacking gel solution for two gels, mix 1.3 mL SDS–PAGE acrylamide stock solution, 2.5 mL 0.5 *M* Tris–HCl, pH 6.8, 50 µL 20% (w/v) SDS, and add water to a volume of 10 mL.
14. The solution is degassed via a membrane pump for 15 min.
15. The isobutanol and a layer of unpolymerized acrylamide solution are removed from the running gels and disposed of. The surface of the gels is rinsed with water and dried carefully with filter paper.
16. 50 µL 10% (w/v) ammonium persulfate and 10 µL TEMED are added to the stacking gel solution. The running gels are overlaid with the stacking gel solution to a height 6 mm below the top of the lower plate.
17. The stacking gel solution is carefully overlaid with water-saturated isobutanol. The gels are allowed to polymerize for 2 h.
18. The IEF gels are equilibrated with gentle shaking in equilibration buffer supplemented with 2% (v/v) tributylphosphine for 20 min and subsequently in equilibration buffer supplemented with 2.5% (w/v) iodoacetamide for 20 min (*see* **Note 23**).
19. Pieces of filter paper approximately 4 mm × 6 mm are wetted with 5 µL protein standards (PageRuler™ Unstained Protein Ladder, Fermentas or Precision Plus Protein™ Standards, Bio-Rad). The filter papers are allowed to dry.
20. Melt 1% low melting point agarose in stacking gel buffer (5 mL 0.5 *M* Tris–HCl, pH 6.8, 100 µL 20% [w/v] SDS, add water to a volume of 20 mL).
21. Melt 1% standard agarose in running gel buffer. Add bromophenol blue to give the agarose a deep blue color.
22. The IEF gels are placed on pieces of Parafilm 17 cm × 5 cm that have been folded lengthwise in order to increase stability. IEF gels are straightened and the excess of equilibration buffer is drained off.

23. Isobutanol and unpolymerized acrylamide solution are removed from the stacking gels. The surface of the stacking gel is rinsed with stacking gel buffer (5 mL 0.5 *M* Tris–HCl, pH 6.8, 100 μL 20% [w/v] SDS, add water to a volume of 20 mL) and carefully dried with filter paper.
24. The gel sandwich is filled with 1% low melting point agarose in stacking gel buffer. Immediately allow the IEF gel to glide from the Parafilm onto the melted agarose. It is a good idea to always use the same orientation of the gel (i.e., basic [=blue] end to the right). Avoid trapping air bubbles under the IEF gel. Allow enough space at one side of the gel to insert the marker paper.
25. Insert the marker filter paper into the melted agarose at one side of the IEF gel.
26. Once the low melting point agarose has solidified, cover the IEF gel with melted agarose in SDS–PAGE running buffer.
27. The lower buffer chamber of the electrophoresis unit is filled with 1.5 L of SDS–PAGE anode buffer.
28. Once the electrophoresis unit is assembled, the upper buffer chamber is filled with 400 mL SDS–PAGE running buffer.
29. SDS–PAGE is performed overnight (16 h) at a constant current of 12 mA/gel with cooling to 15°C. Electrophoresis is stopped when the bromophenol blue front is about to migrate out of the gel.

Colloidal Coomassie Staining (see *Note 24*)

1. The electrophoresis unit is disassembled and the gel sandwiches are opened carefully. The stacking gel and the IEF gel are cut from the running gel. The best way to do this is to use a pizza cutter!
2. The gels are directly transferred from one glass plate to a staining dish filled with 250 mL of colloidal Coomassie staining solution D. The gel is incubated in the solution for 24 h.
3. The staining solution is discarded and the gel is washed in water with frequent changes until the background is clear.

Stained gels are ready for manual or automated image analysis and subsequent isolation of protein spots (*see* **Note 25**).

4. Notes

1. Piperazine diacrylamide is rather expensive but is nevertheless preferred to *N,N'*- methylene bisacrylamide as a crosslinker for two-dimensional protein gel electrophoresis, because it confers increased strength to the polyacrylamide gels, leads to increased resolution of proteins, and reduces silver stain background *(19)*.
2. This procedure will remove any traces of acrylic acid from the solution, which would otherwise interfere with the generation of a pH gradient during isoelectric

focusing of proteins. While this step is probably dispensable whenever high-purity grade chemicals are used (e.g., Bio-Rad) it is an absolute requirement if acrylamide purity grade is questionable.

3. It should be noted that most protocols for two-dimensional electrophoresis include equilibration of IEF gels at a pH of 6.8 (which corresponds to the pH of the SDS stacking gel) rather than pH 8.3 (which corresponds to the pH of the SDS running buffer). However, reduction and alkylation of –SH groups are rather inefficient at pH 6.8 (the optimum pH is between 8.5 and 8.9) so the use of a more alkaline equilibration buffer is highly recommended ***(20)***.
4. Tributylphosphine is used instead of 2-mercaptoethanol or DTT because it has been reported to be more active, increases protein resolution, and results in an increased transfer of proteins to the second dimension. Tributylphosphine is inactivated by oxygen and should be handled and stored accordingly ***(21)***.
5. Classical SDS–PAGE running buffer contains only 0.1% SDS. Increasing the concentration to 0.4% efficiently reduces vertical streaking in the second dimension.
6. Separation of proteins in SDS–PAGE gels depends on the presence of different anions in the gel buffer (chloride) and the running buffer (glycinate). Stacking of proteins occurs because the chloride anions (leading ions) will move more easily through the stacking gel than glycinate ions (trailing ions), which results in the formation of a high-voltage gradient where all proteins pile up to form a tight disc between the glycinate and chloride ions ***(22)***. The presence of chloride ions in the running buffer would interfere with this process.
7. Glycine and SDS are required to separate the proteins in the acrylamide gel but their presence is not required at the end point of separation (the lower end of the gels). Therefore, SDS can be omitted and glycine is replaced by the much cheaper hydrochloric acid as a counterion to the Tris base.
8. Coomassie brilliant blue will hardly (and is not supposed to!) dissolve in water. The dye will therefore form a colloid and will sediment to the bottom upon storage.
9. It is essential to the staining procedure that the Coomassie brilliant blue remains in a colloidal state but is not dissolved in the staining solution. If methanol is added too quickly, temporarily high concentrations of the solute will dissolve the dye. This will result in high background staining of the gels.
10. Moss protonema cells are extremely resistant to mechanical disruption. Tissue disruption using a mortar and pestle is tedious and rather inefficient as was observed when the cells were analyzed microscopically. Proteonema disruption in a ball mill is fast and results in breakage of all cells in a protonema thread.
11. Samples should not be kept in acetone/TCA solutions for prolonged periods of time as modifications or cleavage of proteins might occur.
12. In contrast to SDS-Laemmli buffer, protein samples in urea buffer must never be heated to temperatures higher than 37°C. High temperatures will promote the formation of ammonium cyanate from the urea, which will induce carbamylation of protein amine groups. This covalent modification will affect the charge of the proteins and hence their migration during isoelectric focusing.

13. Samples in urea buffer should not be stored or centrifuged at low temperatures or precipitation of urea will occur.
14. The insoluble pellet is discarded. Jacobs *et al.* *(23)* describe a procedure for sequential solubilization of plant proteins precipitated with acetone/TCA. They perform a reextraction of the pellet with another IEF lysis buffer containing thiourea. This treatment results in the resolubilization of additional proteins that were not released from the pellet under mild extraction conditions. While this method works for cultured *Catharantus roseus* cells, it could not be successfully applied to *P. patens* as the thiourea extracts did not yield 2D gels of satisfactory quality.
15. If degassing of acrylamide solution is omitted, the amount of polymerization initiators ammonium persulfate and TEMED needs to be increased. High amounts of initiators will affect formation of the pH gradient during isoelectric focusing and excess amounts of ammonium persulfate and TEMED may interact with (and modify) proteins.
16. Air bubbles can be removed by reinserting the tubing down to the position of the bubble. This will cause the air bubble to rise.
17. Degassing must be performed in order to remove carbon dioxide from the cathode electrolyte thereby preventing the formation of sodium carbonate, which would decrease the pH of the electrolyte.
18. The easiest way to apply the dialysis membrane to the bottom of the tubes without trapping air bubbles is to turn the tubes upside down. In this case, the water overlaying the IEF lysis buffer must be removed first or the fluids will mix.
19. The sample volume should be kept as small as possible to allow solubilization of the proteins but should always be the same between gels to ensure reproducibility. Separation of proteins will occur over the length of the gel *including* the IEF buffer. If large volumes of IEF buffer are used to apply the sample, proteins in the basic range will not enter the gel at all and will be lost. It has to be mentioned, though, that isoelectric focusing of basic proteins in the presence of urea is problematic, which is why the sample is applied at the basic end of the gradient where separation is not expected to be excellent.
20. Migration of proteins is approximately inversely proportional to the logarithms of their masses. In nongradient gels, this will lead to a high separation in the low-molecular-weight range, whereas separation of proteins is rather poor in the high-molecular-weigh range. A gradient gel with concentration of polyacrylamide increasing from top to bottom will counter this effect and result in a satisfactory separation of proteins over a wide range of masses.
21. The stopcock should be opened prior to the valve stem or the high-density solution will flow "backward" into the reservoir chamber.
22. Plastic pipette tips (200 μL) should be attached to the end of the tubings. The tips should slowly be moved back and forth over the whole length of the gel sandwich or the gradient will be distorted.
23. Equilibration is necessary to transfer the proteins from one electrophoretic separation technique that requires the proteins to maintain their native charges

to another technique that requires them to be covered with the anionic detergent SDS. To ensure complete unfolding of the proteins, disulfide bonds must be split. This is accomplished via the addition of tributylphosphine in the first step of equilibration. Iodoacetamide, which is added to the equilibration buffer in a next step, performs alkylation of free –SH groups, thereby preventing reformation of disulfide bonds.

24. Colloidal Coomassie staining detects protein amounts down to 10 ng in a spot. While the sensitivity of silver staining is higher by a factor of 10, silver staining protocols are usually laborious. The dynamic range of silver staining methods is rather narrow, which limits protein quantitation and most silver staining methods are not compatible with mass spectrometric identification of proteins. The exact mechanism of silver staining is still unknown. It is, however, very obvious that efficiency of staining differs between protein spots with quite a large number of proteins not being stained by silver at all. Lower protein loads, however, usually result in better resolution during isoelectric focusing, so silver stained gels usually appear to be of a higher quality than gels stained with Coomassie brilliant blue. Recently two protocols describing highly sensitive silver staining methods that are compatible with mass spectrometry analysis have been published ***(24,25)***.
25. As the name implies, "differential proteomics" aims at finding qualitative and quantitative differences between proteomes. In the case of two-dimensional protein electrophoresis, patterns of protein spots need to be compared. It is evident that similarities in protein patterns must outweigh the differences in order to make comparisons possible. Visual analysis and comparison of gel patterns (each consisting of around 1000 protein spots) is rather cumbersome and the development of 2D gel analysis software has made this job easier. However, spot detection is still a critical point in software-aided gel image analysis and requires manual intervention, which is time consuming and inevitably introduces subjectivity. Protein spots of interest are excised from the gel (either manually or by a robot, which is much more convenient). Proteins are destained and specifically cleaved (usually by in gel trypsin digestion) prior to identification by mass spectrometry (*see* **Chapter 1**). Via peptide mass fingerprinting and *de novo* peptide sequencing by tandem mass spectrometry we were able to identify 306 proteins from *P. patens* after two-dimensional electrophoresis and colloidal Coomassie staining ***(17)***. Cho and colleagues predicted the identities of 90 protein spots on 2D gels from protonema and gametophores and observed differences in the proteome patterns in these two tissues of *P. patens* ***(26)***.

References

1. Rossignol, M., Peltier, J. B., Mock, H. P., Matros, A., Maldonado, A. M., and Jorrin, J. V. (2006) Plant proteome analysis: A 2004–2006 update. *Proteomics* **6**, 5529–5548.
2. Pasquali, C., Frutiger, S., Wilkins, M. R., Hughes, G. J., Appel, R. D., Bairoch, A., Schaller, D., Sanchez, J. C., and Hochstrasser, D. F. (1996) Two-dimensional gel

electrophoresis of *Escherichia coli* homogenates: the *Escherichia coli* SWISS-2DPAGE database. *Electrophoresis* **17**, 547–555.

3. Gorg, A., Obermaier, C., Boguth, G., Harder, A., Scheibe, B., Wildgruber, R., and Weiss, W. (2000) The current state of two-dimensional electrophoresis with immobilized pH gradients. *Electrophoresis* **21**, 1037–1053.
4. Cho, K., Torres, N. L., Subramanyam, S., Deepak, S. A., Sardesai, N., Han, O., Williams, C. E., Ishii, H., Iwahashi, H., and Rakwal, R. (2006) Protein extraction/solubilization protocol for monocot and dicot plant gel-based proteomics. *J. Plant Biol.* **49**, 413–420.
5. Rose, J. K. C., Bashir, S., Giovannoni, J. J., Jahn, M. M., and Saravanan, R. S. (2004) Tackling the plant proteome: practical approaches, hurdles and experimental tools. *Plant J.* **39**, 715–733.
6. Damerval, C., Devienne, D., Zivy, M., and Thiellement, H. (1986) Technical improvements in two-dimensional electrophoresis increase the level of genetic variation detected in wheat seedling proteins. *Electrophoresis* **7**, 52–54.
7. Hurkman, W. J. and Tanaka, C. K. (1986) Solubilization of plant membrane proteins for analysis by two-dimensional gel electrophoresis. *Plant Physiol.* **81**, 802–806.
8. Sarhan, F. and Perras, M. (1987) Accumulation of a high molecular weight protein during cold hardening of wheat (*Triticum aestivum* L). *Plant Cell Physiol.* **28**, 1173–1179.
9. Granier, F. (1988) Extraction of plant proteins for two-dimensional electrophoresis. *Electrophoresis* **9**, 712–718.
10. Saravanan, R. S. and Rose, J. K. C. (2004) A critical evaluation of sample extraction techniques for enhanced proteomic analysis of recalcitrant plant tissues. *Proteomics* **4**, 2522–2532.
11. Schaefer, D. G. and Zryd, J. P. (1997) Efficient gene targeting in the moss *Physcomitrella patens*. *Plant J.* **11**, 1195–1206.
12. Frank, W., Holtorf, H., and Reski, R. (2005) Functional genomics in *Physcomitrella*. In *Plant Functional Genomics* (Leister, D., ed.). The Harworth Press, Binghamton, NY, pp. 203–234.
13. Reski, R. and Cove, D. J. (2004) Quick guide: *Physcomitrella patens*. *Curr. Biol.* **14**, R261–R262.
14. Chen, S. X. and Harmon, A. C. (2006) Advances in plant proteomics. *Proteomics* **6**, 5504–5516.
15. Bjellqvist, B., Ek, K., Righetti, P. G., Gianazza, E., Gorg, A., Westermeier, R., and Postel, W. (1982) Isoelectric focusing in immobilized pH gradients—principle, methodology and some applications. *J. Biochem. Biophys. Methods* **6**, 317–339.
16. Ramagli, L. S. and Rodriguez, L. V. (1985) Quantitation of microgram amounts of protein in two-dimensional polyacrylamide gel electrophoresis sample buffer. *Electrophoresis* **6**, 559–563.
17. Sarnighausen, E., Wurtz, V., Heintz, D., Van Dorsselaer, A., and Reski, R. (2004) Mapping of the *Physcomitrella patens* proteome. *Phytochemistry* **65**, 1589–1607.
18. O'Farrell, P. H. (1975) High resolution two-dimensional electrophoresis of proteins. *J. Biol. Chem.* **250**, 4007–4021.

19. Hochstrasser, D. F., Patchornik, A., and Merril, C. R. (1988) Development of polyacrylamide gels that improve the separation of proteins and their detection by silver staining. *Anal. Biochem.* **173**, 412–423.
20. Herbert, B., Galvani, M., Hamdan, M., Olivieri, E., MacCarthy, J., Pedersen, S., and Righetti, P. G. (2001) Reduction and alkylation of proteins in preparation of two-dimensional map analysis: why, when, and how? *Electrophoresis* **22**, 2046–2057.
21. Herbert, B. R., Molloy, M. P., Gooley, A. A., Walsh, B. J., Bryson, W. G., and Williams, K. L. (1998) Improved protein solubility in two-dimensional electrophoresis using tributyl phosphine as reducing agent. *Electrophoresis* **19**, 845–851.
22. Gallagher, S. R. (1995) One-dimensional SDS gel electrophoresis of proteins. In *Current Protocols in Protein Science* (Coligan, J. E., *et al.*, eds.). John Wiley & Sons, Inc., New York, pp. 10.1.1–10.1.34.
23. Jacobs, D. I., van Rijssen, M. S., van der Heijden, R., and Verpoorte, R. (2001) Sequential solubilization of proteins precipitated with trichloroacetic acid in acetone from cultured *Catharanthus roseus* cells yields 52% more spots after two-dimensional electrophoresis. *Proteomics* **1**, 1345–1350.
24. Jin, L. T., Hwang, S. Y., Yoo, G. S., and Choi, J. K. (2006) A mass spectrometry compatible silver staining method for protein incorporating a new silver sensitizer in sodium dodecyl sulfate-polyacrylamide electrophoresis gels. *Proteomics* **6**, 2334–2337.
25. Chevallet, M., Diemer, H., Luche, S., Van Dorsselaer, A., Rabilloud, T., and Leize-Wagner, E. (2006) Improved mass spectrometry compatibility is afforded by ammoniacal silver staining. *Proteomics* **6**, 2350–2354.
26. Cho, S. H., Hoang, Q. T., Kim, Y. T., Shin, H. Y., Ok, S. H., Bae, J. M., and Shin, J. S. (2006) Proteome analysis of gametophores identified a metallothionein involved in various abiotic stress responses in *Physcomitrella patens*. *Plant Cell Rep.* **25**, 475–488.

4

Methods for Human CD8+ T Lymphocyte Proteome Analysis

Lynne Thadikkaran, Nathalie Rufer, Corinne Benay, David Crettaz, and Jean-Daniel Tissot

Summary

T lymphocytes, including cytotoxic CD8+ T cells, are important cells involved in immunology, as they can destroy infected or tumor cells. We describe here a detailed protocol starting from CD8+ T lymphocytes isolation for T cell culture followed by total protein extraction or subcellular fractionation, like nuclei isolation. We also describe well-defined biochemistry and cell biology methods adapted to T lymphocytes, showing the importance of using the method best suited to answering the question addressed. All these techniques would be very helpful to immunologists willing to study underlying biological processes related to T lymphocytes.

Key Words: T lymphocyte; proteomics; nuclear extraction; confocal immunofluorescence; Western blot.

1. Introduction

Cytotoxic T cells, also called CD8+ T cells, can recognize and kill virus-infected or tumor cells. They have been identified as potent effectors of the adaptive antitumor immune response and therefore represent an important tool for adoptive immunotherapy *(1)*. Cytotoxic T cells have a finite life span and the challenge for the coming years is to study their mechanisms of growth control as well as the parameters contributing to their expansion.

There are several studies on T lymphocytes proteome analysis *(2,3)*. The advantage of proteomics is that it allows a global protein pattern analysis. Moreover, posttranslational modifications can be pointed out by this technique.

From: *Methods in Molecular Biology, vol. 484: Functional Proteomics: Methods and Protocols*
Edited by: J. D. Thompson et al., DOI: 10.1007/978-1-59745-398-1, © Humana Press, Totowa, NJ

However, one critical point of proteomics is the sample preparation, particularly when subcellular fractionation is required. In a recent study, we compared the proteome pattern of human $CD8^+$ T lymphocytes overexpressing or not overexpressing telomerase, a reverse transcriptase able to add a telomeric repeat at the end of the chromosomes, resulting in elongation of the life span. Overexpression of telomerase into human T lymphocytes results in the extension of their replicative life span *(4)*, but it still remains unclear whether these cells are physiologically indistinguishable from normal ones. To address this question, we compared the proteome of young and aged $CD8^+$ T lymphocytes with that of T cells transduced with hTERT and found that the latter cells displayed an intermediate protein pattern, sharing similar protein expression with young, but also with elderly cells *(5)*. These results are in agreement with our overall gene transcription profiling *(4)*. This study opened several new perspectives, one of these being nuclei isolation in order to point out more accurately changes in the nucleus associated with telomerase overexpression. For this reason, we describe here detailed methods for $CD8^+$ T lymphocyte isolation and cell culture followed by total protein extraction or nuclear isolation.

2. Materials

2.1. $CD8^+$ T Lymphocytes Isolation

1. Plastic bags (Baxter, La Châtre, France).
2. Sepacell RZ-2000 filters (Baxter, Asahi, Japan).
3. Citric acid-dextrose-adenine (ACD-A, Haemonetics, MA).
4. Ficoll-Paque from GE Healthcare (previously Amersham Biosciences, Uppsala, Sweden).
5. Buffer used for isolation: phosphate-buffered saline (PBS), pH 7.2, supplemented with 0.5% bovine serum albumin (BSA, Sigma-Aldrich, St. Louis, MO) and 2 m*M* ethylenediaminetetraacetic acid (EDTA) (Merck, Glattbrugg, Switzerland).
6. MACS $CD8^+$ Microbeads, MS^+/LS^+ columns, and MiniMACS magnet were from Miltenyi Biotec (Gladbach, Germany).

2.2. Fluorescence-Activated Cell Sorter (FACS)

1. RPMI 1640 (Gibco, Invitrogen, Carlsbad, CA).
2. Fetal calf serum (FCS, Gibco, Invitrogen).
3. Anti-CD8-FITC, anti-CD3-PE, anti-IgG_1-FITC, and anti-IgG_1-PE were purchased from Becton Dickinson (BD Biosciences, Allschwil, Switzerland).
4. PBS-azide: prepare a solution with 0.1% (w/v) sodium azide (Merck); it can be stored at 4°C for 1 month.
5. Paraformaldehyde (Merck): prepare a 1% (w/v) solution in PBS-azide fresh for each experiment. The solution may need to be carefully heated (a stirring hot plate

in a fume hood should be used) to dissolve. The solution should be cooled down to room temperature and filtered with a 0.22-µm filter before use. The solution can be kept at 4°C for 1 week.

6. FACS Scan or Calibur flow cytometer from Becton Dickinson.

2.3. Cell Culture of CD8⁺ T Lymphocytes

1. RPMI 1640 with sodium bicarbonate, without HEPES (Gibco, Invitrogen, Carlsbad, CA).
2. Complete RPMI 1640 medium: RPMI 1640 supplemented with 1% L-glutamine (Gibco, Invitrogen), 1% sodium pyruvate (Gibco, Invitrogen), 1% nonessential amino acids (Gibco Invitrogen), 1% penicillin/streptomycin (Gibco, Invitrogen), and 5×10^{-5} *M* 2-mercaptoethanol (Sigma).
3. A stock solution of phytohemagglutinin (PHA; Sodiag, Losone, Switzerland) is prepared at 1 mg/mL in PBS.
4. Recombinant human interleukin-2 (rIL-2, Roche, Mannheim, Germany): a stock solution at 10,000 U/mL is prepared in PBS supplemented with 2% FCS.

2.4. Two-Dimensional Gel Electrophoresis (2-DE)

1. Isobuffer: 8 *M* urea (MP Biomedicals, previously ICN Biomedicals, Illkirch, France), 4% (3-[(3-cholamidopropyl)dimethylammonio]-1-propane sulfonate (CHAPS; MP Biomedicals), 40 m*M* Tris (MP Biomedicals), 65 m*M* DTE (MP Biomedicals), and 5 U endonuclease (Sigma).
2. Immobiline Dry-Strip, pH range 4–7, 18 cm from GE Healthcare.
3. Rehydration solution: 8 *M* urea, 2% CHAPS, 10 m*M* DTE, 2% Pharmalyte, pH 3–10 (GE Healthcare), 1% Servalyte, pH 4–7 (Serva, Heidelberg, Germany), and traces of bromophenol blue.
4. The first equilibration solution contains 6 *M* urea, 50 m*M* Tris, 30% glycerol, 2% sodium dodecyl sulfate (SDS), 2% 1,4-dithioerythritol (DTE) and the second one contains traces of bromophenol blue and 2.5% iodoacetamide instead of DTE. All products are from MP Biomedicals.
5. Piperazine diacrylamide (PDA; Bio-Rad, Hercules, CA).
6. Ethanol, acetic acid, and sodium acetate are from Merck (Dietikon, Switzerland). 2,7-Naphthalenedisulfonic acid (NDS) is from Acros Organics (NJ). Ammoniacal silver nitrate solution contains 8% (w/v) silver nitrate (Fluka), 13% (v/v) ammoniacal solution 25% (Merck), and 20 m*M* sodium hydroxide (Merck).
7. Citric acid and formaldehyde are from Merck.

2.5. Nuclei Isolation

1. Hypotonic buffer A: 10 m*M* Tris–HCl, pH 7.5 (USB, Cleveland, OH), 10 m*M* KCl (Merck), 0.1 m*M* EDTA (Merck). One tablet of cocktail protease inhibitor (Roche) per 50 mL of hypotonic buffer. The tablet should be added just prior to use.

2. Buffer B: 0.34 *M* sucrose (Fluka), 0.05 m*M* $MgCl_2$ (Merck).
3. A 10 % solution of Nonidet P40 (NP-40) is prepared by mixing 1 mL of NP-40 (Merck) with 9 mL H_2O.
4. Protein assay reagent (Bio-Rad).

2.6. SDS–PAGE and Western Blot

1. SDS loading buffer: 150 m*M* Tris–HCl, pH 6.8 (Bio-Rad), 6% SDS (MP Biomedicals), 0.3% bromophenol blue, 30% glycerol (MP Biomedicals).
2. Thirty percent acrylamide/0.8% bisacrylamide (both from MP Biomedicals) solution (this is a neurotoxin when unpolymerized and so care should be taken not to be exposed to it) is prepared in water and stored at 4°C protected from light. A 10% ammonium persulfate (MP Biomedicals) solution is freshly made. *N,N,N,N'*-Tetramethylethylenediamine (TEMED) is from GE Healthcare. Tris–HCl 1.5 *M*, pH 8.8 is from Bio-Rad.
3. Running buffer: 25 m*M* Tris, 192 m*M* glycine, 0.1% (w/v) SDS.
4. Prestained molecular weight marker: BenchMark™ Prestained Protein Ladder (Invitrogen, Carlsbad, CA).
5. Transfer buffer: 25 m*M* Tris, 192 m*M* glycine, 20% (v/v) methanol.
6. Polyvinylidine difluoride (PVDF) membrane: Millipore (Bedford, MA).
7. Blocking buffer: PBS 1×, 0.1% (v/v) Tween-20 (Roche, Mannheim, Germany), 5% (w/v) milk (Fluka), 1% (w/v) BSA (Sigma).
8. PBS-T: PBS 1×, 0.1% Tween-20.
9. Novex tank system (Invitrogen).
10. Primary antibodies: mouse monoclonal antinucleolin (Santa Cruz, Santa Cruz, CA) and mouse monoclonal antiactin (Sigma-Aldrich, St. Louis, MO).
11. Secondary goat antimouse HRP-conjugated antibody: Dako, Baar, Switzerland.
12. Enhanced chemiluminescent (ECL) reagents and Hyperfilm™ are from GE Healthcare.

2.7. Mass Spectrometry

1. Colloidal Coomassie blue: GelCode©, Pierce, Socochim, Lausanne, Switzerland.
2. 96-well plate: Perkin Elmer Life Sciences, Wellesley, MA.
3. Sequencing-grade trypsin: Promega, Madison, WI.
4. Robotic workstation Investigator ProGest: Perkin Elmer Life Sciences, Wellesley, MA.
5. SCIEX QSTAR Pulsar: Concord, Ontario, Canada.
6. LC-Packings Ultimate HPLC system: Amsterdam, Netherlands.

2.8. Confocal Microscopy

1. Confocal microscope: Zeiss LSM 510 Meta (Carl Zeiss AG, Feldbach, Switzerland).
2. Microscope slides: SuperFrost® Plus (Menzel-Glaser, Braunschweig, Germany).

3. 1% (w/v) paraformaldehyde (Merck) solution is prepared with PBS.
4. Antibody dilution buffer: PBS with 1% Triton X-100 (v/v), 2% BSA (w/v), 10% goat serum (Sigma). The solution should be filtered before use and kept at 4°C.
5. Primary antibodies: mouse monoclonal antihuman nucleolin (Santa Cruz), mouse monoclonal antiactin (Sigma-Aldrich), rabbit polyclonal antihuman CD45 (Santa Cruz).
6. Secondary antibodies: Alexa Fluor® 488 goat antirabbit IgG (H+L) and Alexa Fluor® 546 goat antimouse IgG_1 (γ1) are from Molecular Probes (OR).
7. Mounting medium with 4',6-diamidino-2-phenylindole (double-stranded DNA staining [DAPI]): Vectashield® (Burlingame, CA).

3. Methods

3.1. CD8+ T Lymphocyte Isolation

1. Peripheral blood mononuclear cells (PBMCs) should first be isolated. PBMCs are obtained from healthy donors. About 450 mL of blood, obtained from volunteer donors, is collected in plastic bags containing citrate, phosphate, and dextrose. White blood cell reduction is systematically performed by filtration on all blood units between 1 and 15 h after collection, according to Swiss law and the regulations of the Swiss Blood Transfusion Service. The filtration is performed at room temperature, using Sepacell RZ-2000 filters according to the manufacturer's instructions. White blood cells as well as platelets are trapped in the multiple layers of synthetic nonwoven fibers of the filters. Leukocytes are recovered by injecting, in the reverse sense of the filters, 3 × 10 mL of PBS containing 10% citric ACD-A. Platelets, monocytes, and lymphocytes are separated from residual red blood cells and granulocytes using Ficoll-Paque gradient centrifugation. Briefly, 30 mL of the cell suspension is put on 15 mL of Ficoll-Paque (density 1.077) and is centrifuged 30 min at 690 × *g* at 20°C. The ring between the Ficoll-Paque and the PBS is gently recuperated and washed three times in PBS using different centrifugation protocols (10 min at 710 × *g* at 20°C, twice, to eliminate residual Ficoll-Paque as well as plasma, and then 10 min at 220 × *g* at 20°C, once, to eliminate platelets). The cell count is done using trypan blue.
2. PBMCs should be washed and suspended in 80 µL of buffer per 10^7 cells.
3. 20 µL of MACS CD8 MicroBeads per 10^7 cells is added and the mix is incubated 30 min on ice.
4. The cells are washed by adding 5 mL of buffer and suspended with 500 µL of buffer.
5. The MS^+/LS^+ column is put on the MiniMacs magnet and washed twice with 500 µL of buffer.
6. The cell suspension is applied on the column and washed once.
7. After removing the column from the magnet, 1 mL of buffer is added and pressed through the column by using the provided plunger.
8. Isolated cells are counted with trypan blue.

3.2. FACS

1. The purity of the isolation is determined by FACS analysis. 2×10^6 cells are spun down and suspended in 400 μL of RPMI 1640 medium supplemented with 10% FCS. 0.5×10^6 cells (100 μL) are used per condition.
2. Cells are incubated for 30 min at 4°C.
3. 20 μL of antibody is used to label 10^6 cells/100 μL. Four conditions are prepared: ***(1)*** IgG_1-FITC/IgG_1-PE, ***(2)*** CD8-FITC, ***(3)*** CD3-PE, and ***(4)*** CD8-FITC/CD3-PE (*see* **Note 1**).
4. Incubate 20 min at 4°C protected from light.
5. Wash twice with 1 mL of cold PBS-azide.
6. Suspend in 200 μl of cold paraformaldehyde 1% and complete to 1 mL with cold PBS-azide.
7. The samples are then analyzed on FACS Calibur. Results are shown in **Fig. 1**.

3.3. Cell Culture of CD8⁺ T Lymphocytes

1. Culture of T cells is obtained by seeding them onto 24-well culture plates (2×10^6 cells in 2 mL/well) in complete RPMI 1640 medium supplemented with 8% HS and 150 U/mL of recombinant human IL-2.
2. T cells are stimulated with 1 μg/mL PHA plus 1×10^6/mL irradiated allogeneic PBMCs (3000 rad) as feeder cells. Culture medium should be checked daily and changed when required.
3. Population doublings (PDs) are determined by periodic counting of living cells using trypan blue to exclude dying cells, and according to the following formula: PD (day x; day y) = (log [average cell count at day y] – log [average cell seeded at day x])/log 2. **Figure 2** represents an example of growth kinetics (PD versus time).

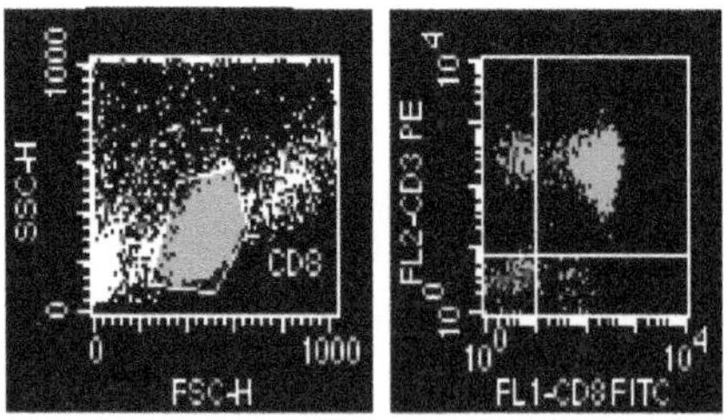

Quad	Events	% Gated	% Total
UL	138	1.37	0.70
UR	9813	97.43	49.71
LL	88	0.87	0.45
LR	33	0.33	0.17

Fig. 1. Fluorescence-activated cell sorter (FACS) dot plot of CD8⁺ T cells labeled with α-CD3-PE and α-CD8-FITC antibodies. The purity of the isolated T cells (upper right panel) reaches 98%.

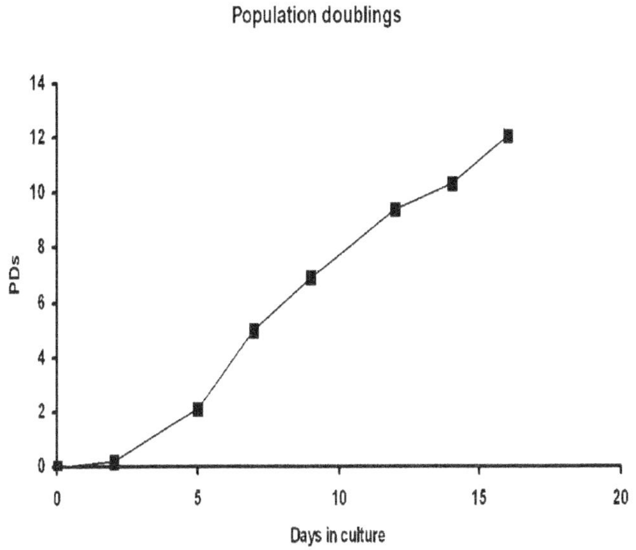

Fig. 2. Growth kinetics (population doubling [PD] versus time) of $CD8^+$ T lymphocytes after stimulation with phytohemagglutinin (PHA). Population doubling was calculated by periodic cell counting.

3.4. Two-Dimensional Gel Electrophoresis (2-DE)

2-DE methods for freshly isolated T lymphocytes from peripheral blood were already described in our laboratory by Vuadens *et al.* ***(6)***.

1. 1×10^6 cells are solubilized in 80 μL of isobuffer.
2. Isoelectric focusing (IEF) is performed under paraffin oil, using linear immobilized pH gradients (Immobiline Dry-Strip, pH range 4–7, 18 cm from GE Healthcare). The strips are rehydrated overnight in 340 μL of rehydration solution.
3. 40 μg of sample is loaded on the cathodic side of the gels. The voltage is progressively increased from 300 V to 3000 V during the first 3 h, followed by 1 h at 3500 V and finally stabilized at 5000 V, for a total of 100 kVh.
4. Before the second dimension, strips were equilibrated in the first equilibration solution for 12 min, and then in the second equilibration solution for 5 min.
5. Strips are placed on the top of 9–16% gradient polyacrylamide second dimensional gels that were copolymerized with piperazine diacrylamide (PDA) as a cross-linker. The migration is performed with a current of 40 mA/gel.
6. Ammoniacal silver staining is done according to standard protocols ***(7)***. At the end of the run, the gels are washed in H_2O, then soaked in ethanol:acetic acid:water (40:10:50) for 1 h and ethanol:acetic acid:water (10:5:85) overnight. After a water wash, the gels are soaked 30 min in glutaraldehyde (1%) buffered with sodium acetate (0.5 *M*) and the glutaraldehyde is removed by deionized water washes. The gels are then soaked in a 2,7-naphthalenedisulfonic acid fresh solution (0.05%,

w/v) for 30 min and rinsed again with deionized water. The gels are stained in a freshly made ammoniacal silver nitrate solution for 30 min and then rinsed with deionized water.

7. The images are finally developed in a solution containing citric acid (0.01%, w/v) and formaldehyde (0.1%, w/v). Development is stopped with an acetic acid:water (5:95) solution. All incubations are performed on an orbital shaker. **Figure 3** shows a 2-DE map of cultured CD8$^+$ T lymphocyte with an extended life span (overexpressing telomerase, *see* **Note 2**). Arrows indicate the proteins identified either by matrix assisted laser desorption/ionization time of flight (MALDI-TOF-TOF) or after comparison with our lymphocyte 2-DE map (http://www.expasy.ch/cgi-bin/map1). The detailed list is shown in **Table 1**.

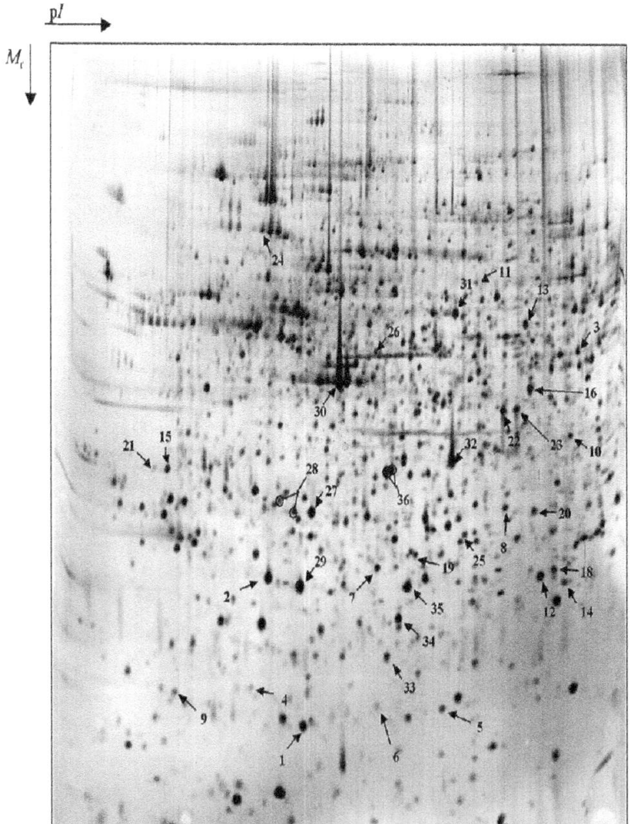

Fig. 3. High-resolution silver-stained two-dimensional polyacrylamide gel of CD8$^+$ T lymphocytes in culture. The numbers indicate the localization of the identified proteins either after spot picking or comparison with our 2-DE map (http://www.expasy.ch/cgi-bin/map1). [Adapted with copyright permission from Thadikkaran *et al.* (*5*).]

Table 1
Spots Identified by MALDI-TOF-TOF[a]

Spot no.	Accession no. SWISS-PROT	Protein name	M*r* (Da)	p*I*	Mascot score	Coverage (%)	Number of peptides matched
1	P63241	Eukaryotic translation initiation factor (eIF5A)	16918	5.08	393	35	12
2	P52565	Rho-GDP-dissociation inhibitor 1 (Rho-GDI)	23250	5.03	369	37	16
3	P06733	α-Enolase	47350	6.99	382	37	16
4	Q8TDP1	RNase H1 small subunit (AYP1)	17943	4.95	110	34	4
5	P16949	Stathmin	17161	5.77	391	60	18
6	P16949	Stathmin (*phosphorylated*)	17161	—	200	37	16
7	P09936	Ubiquitin carboxyl-terminal hydrolase L1 (UCHL1)	25151	5.33	117	41	12
8	Q9H0R4	Hypothetical protein DKFZp564D1378	28476	5.84	185	23	6
9	P19105	Myosin regulatory light chain 2 (MRLC)	19707	4.67	281	51	10
10	O00170	AH-receptor-interacting protein (AIP)	38096	6.1	126	26	6
11	P12004	Proliferating cell nuclear antigen (PCNA)	29092	4.6	382	31	7
12	Q9ULZ3	Apoptosis-associated speck-like protein	21670	6.0	228	61	13
	P30048	Thioredoxin-dependent peroxide reductase	28017	7.7	190	31	12

(Continued)

Table 1
(Continued)

Spot no.	Accession no. SWISS-PROT	Protein name	*Mr* (Da)	p*I*	Mascot score	Coverage (%)	Number of peptides matched
13	P23381	Tryptophanyl-tRNA synthetase	53474	5.8	195	39	23
14	P61758	Prefoldin subunit 3	21435	6.6	99	43	11
15	P13674	Prolyl 4-hydroxylase α_1 subunit precursor	61296	5.7	121	28	12
16	P30740	Leukocyte elastase inhibitor	42829	5.9	212	38	13
17	P31949	Calgizzarin	11847	6.6	444	56	13
	Q9UDP3	Putative S100 calcium-binding protein H_NH0456N16.1	11673	8.8	187	35	6
18	P49720	Proteasome subunit β type 3	23219	6.1	406	60	22
19	Q93125	Green fluorescent protein mutant 3	26937	5.7	470	39	19
20	P78417	Glutathione transferase omega 1	27833	6.2	225	39	10
21	P12004	Proliferating cell nuclear antigen (PCNA)	29092	4.6	46	23	7
22	P40121	Macrophage capping protein	38779	5.9	200	17	11
23	P40121	Macrophage capping protein	38779	5.9	204	22	15
24	P11021	78-kDa glucose-regulated protein precursor	72402	5.1	702	58	45

25	O95336	6-Phosphogluconolactonase	27815	5.7	327	53	18
26	O14745	Ezrin-radixin-moesin binding phosphoprotein 50	38999	5.6	299	42	21
Spots identified by comparison with our lymphocyte 2-DE map (Swiss 2D-PAGE)							
27	O00299	Chloride intracellular channel protein 1	26792	5.1	—	—	—
28	Q96C19	EF-hand domain-containing protein 2 (Swiprosin 1)	26697	5.2	—	—	—
29	P52566	Rho-GDP-dissociation inhibitor 2 (Rho-GDI 2)	22857	5.1	—	—	—
30	P60709	Actin cytoplasmic 1 (β-actin)	41737	5.3	—	—	—
31	P30101	Protein disulfide isomerase A3	56782	6.0	—	—	—
32	P07339	Cathespin D	44552	6.1	—	—	—
33	P07741	Adenine phosphoribosyltransferase	19477	5.8	—	—	—
34	P32119	Peroxiredoxin 2	21761	5.7	—	—	—
35	P09211	Gluthatione *S*-transferase P	23225	5.4	—	—	—
36	P52907	F-actin capping protein α_1 subunit	32792	5.5	—	—	—

[a] Adapted from Thadikkaran *et al.* (**5**) with permission.

3.5. Nuclei Isolation

1. 3×10^7 cells are centrifuged for 10 s at 15,000 × *g* on a benchtop centrifuge.
2. 1 mL of hypotonic buffer A is added to the pellet and mixed by pipeting.
3. The cells are incubated on ice for 15 min to let them swell out and at the end 10 μL of NP-40 10% is added.
4. Vortex 10 s at 75% speed.
5. Cells are centrifuged at 4°C for 30 s and the supernatant is quickly taken out. It represents the cytoplasmic fraction and should be kept at 4°C until protein quantification (see below) and then stored at –80°C. The pellet contains the nuclei.
6. 200 μL of buffer B is added to the pellet and the nuclei suspension is then disrupted by three sonications of 10 bursts each. The suspension should become homogeneous with no viscous elements. Foam should be avoided (the amplitude of the sonicator can be reduced or the volume of the sample increased by adding 50 μL of buffer B).
7. The nuclei suspension is centrifuged at 15,000 × *g* for 5 min at 4°C.
8. The supernatant containing the nuclear extract is taken out and kept at 4°C until protein quantification.
9. Protein concentrations are measured by a standard protein-dye binding coloring method (Bio-Rad) according to the manufacturer's instructions. Usually, a recovery of about 1 mg of nuclear proteins is expected.
10. The samples are finally stored at –80°C until use.

3.6. SDS–PAGE and Western Blot

1. 20 μg of proteins from the nuclear and cytoplasmic fractions is solubilized in SDS loading buffer and heated at 95°C for 5 min. Samples are prepared twice, once for Western blot and once for Coomassie staining and identification by mass spectrometry (*see* **Note 3**).
2. A 9% SDS polyacrylamide minigel is prepared by mixing 6 mL acrylamide/bis solution, 5 mL Tris–HCl 1.5 *M* (pH 8.8), 8.6 mL H_2O, 200 μl SDS 10%, 200 μL APS, and 50 μL TEMED (the amount is enough for pouring two minigels). Pour the gel, leaving space for a stacking gel, and overlay with isobutanol 10%. The gel should polymerize in about 20 min.
3. Pour off the isobutanol and rinse twice with water.
4. The stacking gel is prepared by mixing 550 μL of acrylamide/bis solution with 1.25 mL Tris–HCl (pH 6.8), 3.1 mL H_2O, 50 μl SDS 10%, 60 μL APS 10%, and 30 μL TEMED. Pour the gel and insert the comb. The stacking gel should polymerize in 30 min.
5. The gels are then soaked in running buffer. The samples are loaded onto the minigels. The migration is carried out at constant voltage (200 V).
6. Upon completion of electrophoresis, proteins are transferred to PVDF membranes (prewetted in methanol) using a wet Novex tank system for 1 h and 30 min at fixed voltage (30 V) according to the manufacturer's instructions.

7. After transfer, blots were left to dry for 2 min, wetted in methanol, and blocked overnight with blocking buffer. After two washes of 3 min each with PBS-T, antinucleolin and antiactin antibodies were used both at a dilution 1:1000 for 1 h at room temperature (*see* **Note 4**).
8. The secondary goat antimouse HRP-conjugated antibody was used at a dilution of 1:10,000 for 30 min.

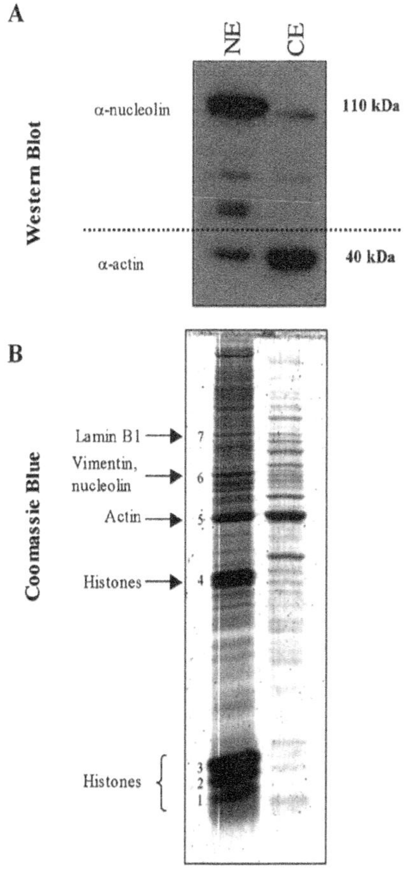

Fig. 4. (**A**) Western blot performed on CD8$^+$ T cell cytoplasmic and nuclear extracts. Antibodies α-nucleolin and α-actin were used as markers of, respectively, the nucleus and cytoplasm. (**B**) Nuclear and cytoplasmic extracts (NE and CE, respectively) were stained with Coomassie blue and relevant bands were cut out for mass spectrometry analysis. Nonexhaustive identified proteins are shown here. Refer to **Table 1** for the complete list of identified proteins.

9. After six washes of 5 min each, subsequent visualization was performed using ECL (GE Healthcare). 1 mL of each reagent was mixed and applied on the membrane, which is then rotated by hand for 1 min.
10. The blot is removed from the ECL reagents and placed between leaves of an acetate sheet protector.
11. A hyperfilm is applied on the membrane for a suitable exposure time, typically a few minutes. An example of the result is shown in **Fig. 4A**.

3.7. Protein Identification by Tandem Mass Spectrometry

1. SDS–PAGE is performed as described in **Subheading 3.6, steps 1–5**.
2. Upon completion of electrophoresis, the gel is rinsed twice with deionized water and stained with colloidal Coomassie blue overnight. The gel is then washed twice with water. An example of the result is shown in **Fig. 4B.**
3. Coomassie blue-stained bands are excised from SDS–PAGE with a scalpel and transferred to special 96-well plates.
4. In-gel proteolytic cleavage with sequencing-grade trypsin is performed automatically in the robotic workstation Investigator ProGest according to the protocol of Wilm *et al.* ***(8)***. Supernatants containing proteolytic peptides are concentrated by evaporation and analyzed by LC-MS/MS on a SCIEX QSTAR Pulsar hybrid quadrupole time-of-flight instrument equipped with a nanoelectrospray source and interfaced to an LC-Packings Ultimate HPLC system (Amsterdam, Netherlands).
5. Peptides are separated on a PepMap reversed-phase capillary C_{18} (75 mm i.d. 615 cm) column at a flow rate of 200 nL min^{-1} along a 52 min gradient of acetonitrile (0–40%).
6. The Analyst software is used for peak detection and automatically select peptides for collision-induced fragmentation.
7. Noninterpreted peptide tandem mass spectra are used for direct interrogation of the Uniprot (Swissprot + TrEMBL) database using Mascot 2.0 (http://www.matrix-science.com). MASCOT search parameters are as follows: trypsin cleavage specificity with maximum one missed cleavage; carbamidomethyl cysteine as fixed modification, and methionine single oxidation as variable modification. Mass tolerances for database searches were 0.5 Da for LC-MS data. MASCOT was set up to report only peptide matches with a score above 14. With the parameters used, the threshold for statistical significance ($p < 0.05$) corresponded to a total (protein) MASCOT score of 33. Proteins scoring above 80 are automatically considered valid, while all protein identifications with a total MASCOT score between 33 and 80 are manually validated. Validation included examination of the peptide rms mass error (<120 ppm) and of individual peptide matches. Peptide matches are validated only if at least an ion series of four consecutive *y* ions are matched, in addition to ions belonging to other series. Generally, only proteins matched by at least two peptides are accepted. The identified proteins are described in **Table 2**.

3.8. Confocal Microscopy

1. T lymphocytes are spun down, washed with PBS, and resuspended in 500 μL of PBS; 1 × 10^6 CD8+ will give enough material for two conditions.
2. Cell suspension is spread onto a SuperFrost slide and incubated for 30 min to let the cells adhere.
3. 1 mL of paraformaldehyde (1%) solution is then added for 20 min at room temperature to fix the cells.
4. The paraformaldehyde is discarded (into a hazardous waste container) and the sample is washed four times for 1 min each with PBS.
5. Cells are labeled and permeabilized by incubation in antibody dilution buffer containing antiactin (1:100) or antinucleolin (1:200) antibody. The incubation is performed for 1 h in a wet chamber at room temperature.
6. The primary antibody is removed and the sample is washed three times for 2 min each with PBS. For subsequent steps, the sample should be protected from light.
7. The secondary antibody is prepared at 1:50 in antibody dilution buffer and added to the sample for 30 min at room temperature.
8. The sample is washed three times for 2 min each and is then ready to be mounted.
9. A drop of mounting medium containing DAPI is spread over the cells and immediately covered with a coverslip. Nail varnish is used for sealing. The sample can be viewed immediately or stored in the dark at 4°C for up to 1 week.
10. Slides are viewed under confocal microscopy. Excitation at 488 (laser argon) nm induces the AlexaFluor 488 fluorescence (green emission) for actin and nucleolin, while excitation at 543 nm (laser HeNe) induces AlexaFluor 546 fluorescence (red emission) for CD45. Excitation at 405 nm (laser diode) induces DAPI fluorescence (blue emission) (*see* **Note 5**). Software can be used to separate the different fluorescence layers as shown in **Fig. 5** (*see* Color Plate 1).

4. Notes

1. The first condition corresponds to the negative control and allows the setting of the voltage in the FACS. Then, conditions 2 and 3 are needed to compensate for the fluorescence. The last condition is the double positive and should be read in the FACS without changing any parameters.
2. The global protein pattern of CD8+ T lymphocytes overexpressing telomerase very much resembles that of the control CD8+ T lymphocytes. Nine proteins were found differentially expressed *(5)*.
3. 1D-SDS PAGE is better adapted than 2-DE for analyzing nuclear proteins that are often basic. In fact, proteins with high p*I* will not migrate properly in 2-DE and therefore could be easily missed.
4. We have found these antibodies to be excellent for both Western blotting and immunofluorescence. Numerous competitive reagents are available from other commercial sources.

Table 2
1D-SDS PAGE Band Identification by LC-MS/MS

Band no	Protein name (SWISS-PROT accession no.)	*Mr*	Mascot score
1	H4_HUMAN (P62805) Histone H4	11229	361
	H2A1H_HUMAN (Q96KK5) Histone H2A type 1-H	13767	173
	H2AZ_HUMAN (P0C0S5) Histone H2A.z (H2A/z)	13414	150
	H2A2B_HUMAN (Q81UE6) Histone H2A type 2-B	13856	129
	H2BA_HUMAN (P62807) Histone H2B.a/g/h/k/l	13767	75
	RETBP_HUMAN (P02753) Plasma retinol-binding protein precursor	2337	74
	H3T_HUMAN (Q16695) Histone H3.4	15482	58
2	H2BA_HUMAN (P62807) Histone H2B.a/g/h/k/l	13767	277
	H2BF_HUMAN (P33778) Histone H2B.f	13811	254
	H2A1C_HUMAN (Q93077) Histone H2A type 1-C	13966	229
	H3T_HUMAN (Q16695) Histone H3.4	15482	204
	H2A2B_HUMAN (Q8IUE6) Histone H2A type 2-B	13856	170
	H4_HUMAN (P62805) Histone H4	11229	156
	H2AZ_HUMAN (P0C0S5) Histone H2A.z	13414	138
	RL35A_HUMAN (P18077) 60S ribosomal protein L35a	12587	114
	RS19_HUMAN (P39019) 40S ribosomal protein S19	15919	82
	SMD2_HUMAN (P62316) Small nuclear ribonucleoprotein Sm D2	13632	77
	RL23_HUMAN (P62829) 60S ribosomal protein L23	14970	68
	RS16_HUMAN (P62249) 40S ribosomal protein S16	16418	66
	RS15A_HUMAN (P62244) 40S ribosomal protein S15a	14813	63
	RS14_HUMAN (P62263) 40S ribosomal protein S14	16303	56

	RL22_HUMAN (P35268) 60S ribosomal protein L22	14704	54
	ATPD_HUMAN (P30049) ATP synthase delta chain	17479	45
	SELH_HUMAN (Q8IZQ5) Selenoprotein H	13512	38
3	H2BA_HUMAN (P62807) Histone H2B.a/g/h/k/l	13767	473
	H2BF_HUMAN (P33778) Histone H2B.f	13811	457
	H2BX_HUMAN (Q8N257) Histone H2B type 12	13768	454
	H31_HUMAN (P68431) Histone H3.1	15377	225
	H2A1A_HUMAN (Q96QV6) Histone H2A type 1-A	14094	135
	SMD3_HUMAN (P62318) Small nuclear ribonucleoprotein Sm D3	14021	91
	H4_HUMAN (P62805) Histone H4	11229	85
	RLA2_HUMAN (P05387) 60S acidic ribosomal protein P2	11658	57
	RL31_HUMAN (P62899) 60S ribosomal protein L31	14454	54
	TCP4_HUMAN (P53999) Activated RNA polymerase II transcriptional coactivator	14255	51
	RS20_HUMAN (P60866) 40S ribosomal protein S20	13478	42
	COX41_HUMAN (P13073) Cytochrome *c* oxidase subunit 4 isoform 1	19621	39
4	H15_HUMAN (P16401) Histone H1.5	22435	479
	PHB2_HUMAN (Q99623) Prohibitin-2	33276	462
	H12_HUMAN (P16403) Histone H1.2	21221	381
	ANXA2_HUMAN (P07355) Annexin A2	38677	360
	H13_HUMAN (P16402) Histone H1.3	22205	349
	ROA1_HUMAN (P09651) Heterogeneous nuclear ribonucleoprotein A1	38805	294
	HNRPC_HUMAN (P07910) Heterogeneous nuclear ribonucleoproteins C1/C2	33725	216
	HCC1_HUMAN (P82979) Nuclear protein Hcc-1	23582	138
	EF1A1_HUMAN (P68104) Elongation factor 1-α 1	50451	130
	VDAC2_HUMAN (P45880) Voltage-dependent anion-selective channel protein 2	38639	92
	VDAC1_HUMAN (P21796) Voltage-dependent anion-selective channel protein 1	30737	65

(Continued)

Table 2
(Continued)

Band no	Protein name (SWISS-PROT accession no.)	M*r*	Mascot score
	H4_HUMAN (P62805) Histone H4	11229	50
	GBBI_HUMAN (P62873) Guanine nucleotide-binding protein	38020	47
	LDHA_HUMAN (P00338) L-Lactate dehydrogenase A chain	36819	43
	PCNA_HUMAN (P12004) Proliferating cell nuclear antigen	29092	41
	SSRA_HUMAN (P43307) Translocon-associated protein α subunit precursor	32215	38
5	ACTB_HUMAN (P60709) Actin, cytoplasmic 1	42052	548
	EF1A1_HUMAN (P68104) Elongation factor 1-α 1	50451	189
	VIME_HUMAN (P08670) Vimentin	53545	167
	HNRPD_HUMAN (Q14103) Heterogeneous nuclear ribonucleoprotein D0	38581	163
	HNRPG_HUMAN (P38159) Heterogeneous nuclear ribonucleoprotein G	42306	161
	Q562Z4_HUMAN (Q562Z4) Actin-like protein	11548	76
	PA2G4_HUMAN (Q9UQ80) Proliferation-associated protein 2G4	43970	48
	H4_HUMAN (P62805) Histone H4	11229	46
	SEPTI_HUMAN (Q8WYJ6) Septin-I	42400	45
	HNRPK_HUMAN (P61978) Heterogeneous nuclear ribonucleoprotein K	51230	33
6	VIME_HUMAN (P08670) Vimentin	53545	1115
	COR1A_HUMAN (P31146) Coronin-1A	51678	212
	RCC2_HUMAN (Q9P258) Protein RCC2	56790	126
	PTBP1_HUMAN (P26599) Polypyrimidine tract-binding protein 1	57357	122
	HNRPK_HUMAN (P61978) Heterogeneous nuclear ribonucleoprotein K	51230	91
	PRP19_HUMAN (Q9UMS4) Pre-mRNA-splicing factor 19	55603	85
	TBAK_HUMAN (P68363) Tubulin α-ubiquitous chain	50804	66
	H4_HUMAN (P62805) Histone H4	11229	52

	PAIRB_HUMAN (Q8NC51) Plasminogen activator inhibitor 1 RNA-binding protein	44995	52
	NUCL_HUMAN (P19338) Nucleolin	76494	41
	LBR_HUMAN (Q14739) Lamin-B receptor: weak identification	71057	36
7	LMNB1_HUMAN (P20700) Lamin-B1	66522	1144
	RIB1_HUMAN (P04843) Dolichyl-diphosphooligosaccharide—protein	68641	390
	HNRPR_HUMAN (O43390) Heterogeneous nuclear ribonucleoprotein R	71184	188
	KU70_HUMAN (P12956) ATP-dependent DNA helicase 2 subunit 1	69953	187
	HNRPM_HUMAN (P52272) Heterogeneous nuclear ribonucleoprotein M	77618	134
	DDX5_HUMAN (P17844) Probable ATP-dependent RNA helicase DDX5	69618	121
	NOP56_HUMAN (O00567) Nucleolar protein Nop56	66394	102
	DHSA_HUMAN (P31040) Succinate dehydrogenase	73672	83
	CMC1_HUMAN (O75746) Calcium-binding mitochondrial carrier protein Aralar 1	75108	48
	RL1D1_HUMAN (O76021) Ribosomal L1 domain-containing protein 1	55167	44
	HNRPL_HUMAN (P14866) Heterogeneous nuclear: weak identification	60719	37

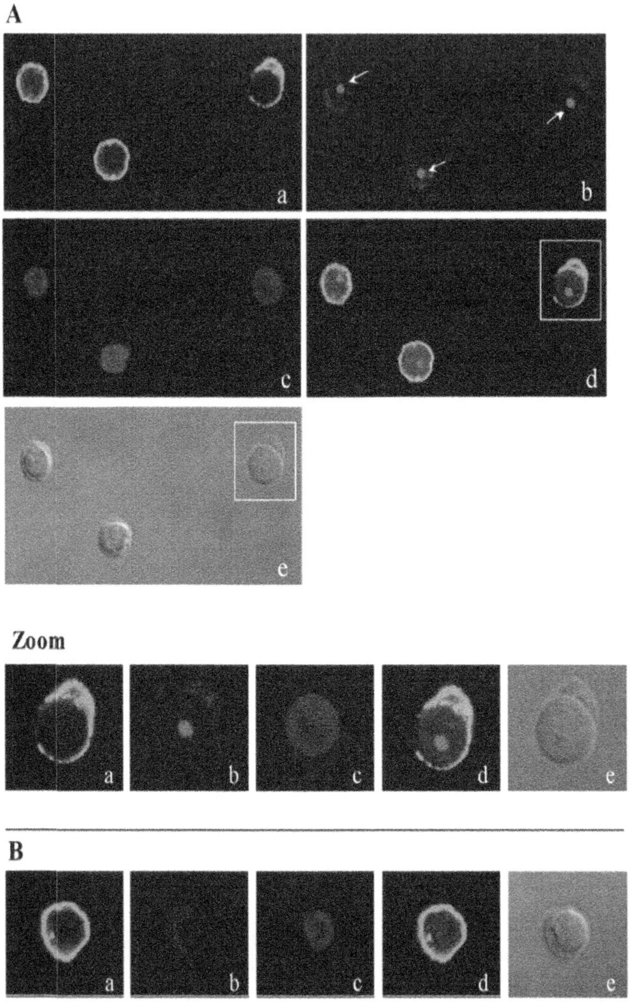

Fig. 5. Confocal microscopy of $CD8^+$ T cells: (**A**) membrane is stained in green with α-CD45 (**a**), nucleolus in red with α-nucleolin (**b**, arrows), and nucleus in blue with DAPI (**c**). (**d**) The merged and (**e**) the differential interference contrast (DIC) images are shown. A zoom of the cell inside the white square is also shown. (**B**) Membrane of another $CD8^+$ T cell is stained in green with α-CD45 (**a**), actin in red with α-actin (**b**), and nucleus in blue with DAPI (**c**). (**d**) The merged and (**e**) the differential interference contrast (DIC) images are shown. (*See* Color Plate 1)

5. DAPI is a fluorescent stain that binds strongly to DNA, therefore labeling cell nuclei. Antiactin and antinucleolin are used to label cytoplasm and nucleolus, respectively. Anti-CD45 (leukocyte common antigen) labels the cell membrane of human leukocytes.

Acknowledgments

The authors would like to thank the people from the Cellular Imaging Facility (CIF) and the Proteomic Analysis Facility (PAF) for excellent technical assistance in confocal microscopy and MS analysis, respectively. The authors also thank the Fondation CETRASA for financial support.

References

1. Rosenberg, S. A. (2001) Progress in human tumour immunology and immunotherapy. *Nature* **411**, 380–384.
2. Rautajoki, K., Nyman, T. A., and Lahesmaa, R. (2004) Proteome characterization of human T helper 1 and 2 cells. *Proteomics* **4**, 84–92.
3. Kronfeld, K., Hochleitner, E., Mendler, S., *et al.* (2005) B7/CD28 costimulation of T cells induces a distinct proteome pattern. *Mol. Cell. Proteomics* **4**, 1876–1887.
4. Menzel, O., Migliaccio, M., Goldstein, D. R., Dahoun, S., Delorenzi, M., and Rufer, N. (2006) Mechanisms regulating the proliferative potential of human CD8⁺ T lymphocytes overexpressing telomerase. *J. Immunol.* **177**, 3657–3668.
5. Thadikkaran, L., Menzel, O., Tissot, J. D., and Rufer, N. (2007) Proteomic and transcriptomic analysis of human CD8⁺ T lymphocytes over-expressing telomerase. *Proteomics Clin. Appl.* **1(3)**, 299–311.
6. Vuadens, F., Gasparini, D., Deon C., *et al.* (2002) Identification of specific proteins in different lymphocyte populations by proteomic tools. *Proteomics* **2**, 105–111.
7. Tissot, J. D. and Spertini, F. (1995) Analysis of immunoglobulins by two-dimensional gel electrophoresis. *J. Chromatogr. A* **698**, 225–250.
8. Shevchenko A., Wilm M., Vorm O., and Mann M. (1996) Mass spectrometric sequencing of proteins silver-stained polyacrylamide gels. *Anal. Chem.* **68**, 850–858.

5

Label-Free Proteomics of Serum

Natalia Govorukhina, Peter Horvatovich, and Rainer Bischoff

Summary

In this chapter we describe a method to analyze human serum with the goal of discovering disease-related changes in the serum proteome. The methodology is based on the removal of the six most abundant serum proteins by immunoaffinity chromatography. This step is followed by trypsin digestion and reversed-phase high-performance liquid chromatography (HPLC) coupled on-line to mass spectrometry (MS) using either a capillary HPLC or a microfluidics chip HPLC system. The obtained, highly complex data sets are processed and statistically analyzed to discover significant differences between groups of samples. The complete analytical procedure will be described with serum samples, to which a given amount of horse heart cytochrome *c* has been added as well as with serum samples from early stage cervical cancer patients prior to and after therapy. The use of reversed-phase HPLC to separate serum proteins at 80°C with subsequent analysis by sodium dodecyl sulfate-polyacrylamide gel electrophoresis (SDS–PAGE) in order to lower the concentration sensitivity will also be briefly described.

Key Words: Serum; label-free profiling; depletion; HPLC; mass spectrometry; bioinformatics; nearest shrunken centroids; principal component analysis; cervical cancer.

1. Introduction

The comparative analysis of serum samples from patients and healthy controls requires highly standardized operating procedures that produce reproducible data *(1)*. The generated data need to be processed so as to bring the significant, disease-related differences in protein or peptide profiles forward and to reduce nonrelated noise *(2)*. Processed data have to be analyzed in a statistically rigorous fashion and subjected to both statistical and biological validation.

From: *Methods in Molecular Biology, vol. 484: Functional Proteomics: Methods and Protocols*
Edited by: J. D. Thompson et al., DOI: 10.1007/978-1-59745-398-1, © Humana Press, Totowa, NJ

In this chapter we present a protocol to perform proteomics of serum samples obtained from cancer patients but the protocol is generic enough to also be applicable to sera from patients with other diseases. This is obviously just one way of proceeding and there are quite a number of other approaches, some of which can be found in this book. To enhance the concentration sensitivity of this method, we remove high-abundance proteins by immunoaffinity chromatography. We have recently shown that this can be done efficiently and with high repeatability *(3,4)*. The subsequent trypsin digestion step and reversed-phase high-performance liquid chromatography-mass spectrometry (HPLC-MS) analysis are performed in a reproducible manner and controlled with standard samples at regular intervals. Concentration sensitivity of this method is approximately 0.5 μ*M* with respect to the added cytochrome *c*. To enhance concentration sensitivity further, it is optional to include an additional protein separation step. We describe the use of a recently developed reversed-phase material that can be run at 80°C *(5)*.

Although we touch upon data processing and statistical analysis, we cannot go into the methodological details due to limited space. We refer the reader to the cited references as well as to a dedicated book in this series focusing on bioinformatics.

2. Materials

2.1. Depletion of the Six Most Abundant Proteins on a Multiple Affinity Removal Column

1. Store serum samples at –80°C in aliquots.
2. Buffer A (# 5185-5987, Agilent, Palo Alto, CA).
3. Buffer B (#5185-5988, Agilent, Palo Alto, CA).
4. 0.22-μm spin filters (Part #5185-5990, Agilent).
5. Multiple affinity removal column (Agilent, 4.6 × 50 mm, Part #5185-5984, Palo Alto, CA).

2.2. SDS–PAGE

1. All chemicals for polyacrylamide gels were from Bio-Rad (Bio-Rad, www.biorad.com).
2. PageRuler™ Prestained Protein Ladder (Fermentas, #SM0671).
3. Coomassie brilliant blue R concentrate (Sigma, www.sigmaaldrich.com).

2.3. HPLC-MS

1. AtlantisTM dC_{18} (1.0 × 150 mm, 3 μm) column for cap-LC-MS (Waters, Milford, MA, www.waters.com).
2. Atlantis™ dC_{18} in-line trap column for cap-LC-MS (Waters, Milford, MA, www.waters.com).

3. Chip for chip-LC-MS with a 40 nL trap column (75 μm × 11 mm) and a 75 μm × 43 mm analytical column both containing C-18SB-ZX 5 μm chromatographic material (Cat. #G4240-62001, Agilent, Palo Alto, CA). The chip is equipped with a nanoelectrospray tip of 2 mm length with conical shape: 100 μm o.d. × 8 μm i.d.
4. Micro BCA™ protein assay reagent kit (www.piercenet.com).
5. Sequencing grade modified trypsin (Promega, Cat. #V5111).
6. Acetonitrile HPLC-S (ACN) gradient grade (Biosolve, Valkenswaard, The Netherlands).
7. Formic acid, FA, 98–100% pro analysis (Cat. #1.00264.1000., Merck, Darmstadt, Germany).
8. Ultrapure water (conductivity: 18.2 MΩ), Maxima System (Elga Labwater, Ede, The Netherlands).

2.4. Prefractionation of Proteins in Depleted Serum by High-Temperature Reversed-Phase HPLC

1. Macroporous reversed-phase mRP-C_{18} column (Agilent, 4.6 × 50 mm, Part #5188-5231, Palo Alto, CA).
2. Trifluoroacetic acid (TFA), sequencing grade (#28902, Pierce).
3. Ultrapure water (conductivity: 18.2 MO), Maxima System (Elga Labwater, Ede, The Netherlands).
4. Urea (#084K0063, Sigma, www.sigmaaldrich.com).
5. Glacial acetic acid (Cat. #1.00063.1000, Merck, Darmstadt, Germany).
6. Solvent A for mRP column (97% water/0.1% TFA).
7. Solvent B for mRP column (97% AcN/0.1% TFA).

3. Methods

3.1. Preparation of Samples

1. Mix 20 μL of crude serum with 80 μL of buffer A (Agilent). Filter through 0.22-μm spin filters at 13,000 × *g* and 4°C for 10 min to remove particulates.
2. Inject 80 μL (80% of the total amount of diluted crude serum) on a multiple affinity removal column for depletion according to the manufacturer's instructions (with detection at 280 nm using the following timetable: 0–9 min, 100% buffer A (0.25 mL/min); 9.0–9.1 min, linear gradient 0–100 B % (1 mL /min), 9.1–12.5 min, 100% buffer B (1 mL /min); 12.5–12.6 min, linear gradient 100–0% buffer B (1 mL /min); 12.6–20 min, 100% buffer A (1 mL /min). Removal of abundant proteins, as described above, was performed on a LaChrom HPLC System (Merck Hitachi, www.merck.com) or on an AKTA FPLC system (GE Healthcare).
3. Collect the flowthrough fraction (depleted serum collected between 2 and 6 min) of a total volume of approximately 1 mL.
4. Determine protein concentrations with the Micro BCA™ protein assay reagent kit (www.piercenet.com) and calculate for an average protein molecular weight of 50 kDa. Use bovine serum albumin (BSA) as the calibration standard.

5. Digest 100 μL (~10% of the total amount, which corresponds to ~8 μg or 160 pmol of total protein) of depleted serum with trypsin (1:20 w/w enzyme to substrate) at 37°C overnight with shaking at 400 rpm.

3.2. Sodium Dodecyl Sulfate Polyacrylamide Gel Electrophoresis (SDS–PAGE)

1. SDS–PAGE was performed in a Mini-Protein III cell (Bio-Rad, www.biorad.com) using 12% gels with 0.1% SDS according to the manufacturer's instructions.
2. Boil samples with sample buffer containing 0.02 *M* dithiothreitol (DTT) for 1 min, cool down, and apply directly to the gel.
3. Perform staining with Coomassie brilliant blue R concentrate (Sigma, www.sigmaaldrich.com) diluted and used as prescribed by the manufacturer.

3.3. HPLC-MS

3.3.1. Capillary- and Chip-LC-MS

1. All LC-MS analyses were performed on an Agilent 1100 capillary HPLC system coupled on-line to an SL ion trap mass spectrometer (www.home.agilent.com). In the case of cap-LC-MS the instrument was equipped with an AtlantisTM dC_{18} (1.0 × 150 mm, 3 μm) column that was protected by an Atlantis™ dC_{18} in-line trap column. Then 40 μL of the pretreated (depleted and digested) serum corresponding to ~8 μg or 160 pmol of total protein digest (calculation based on a 50 kDa protein) was injected. An autosampler (Cat. #G1367A) equipped with a 100-μL injection loop was used for cap-LC-MS. For chip-LC-MS the same mass spectrometer was used but equipped with a microfluidics (chip-cube) interface (Cat. # G4240A) including a chip microfluidic device. The injected sample amount was 0.25 μg (3.4–5.1 μL; 5 pmol) of depleted, trypsin-digested serum, 10 times diluted with 0.1% aq. FA. Injections were performed with an autosampler (Agilent, Cat. #G1389A) equipped with an injection loop of 8 μL (this also includes the dead volume up to the trapping column). In both cases the autosampler was temperature controlled using a cooler (Cat. #G1330A) maintaining the samples at 4°C.

   The HPLC system for cap-LC-MS had the following additional modules: capillary pump (Cat. #G1376A), solvent degasser (Cat. #G1379A), UV detector (Cat. #G1314A), and column holder (Cat. #G1316A). The sample was injected and washed in the back flush mode for 30 min (0.1% aq. FA and 3% acetonitrile at a flow rate of 50 μL/min). Peptides were eluted in a linear gradient from 0 to 70% (0.5%/min) acetonitrile with 0.1% formic acid at a flow rate of 20 μL/min. After each injection the in-line trap and the analytical column were equilibrated with eluent A for 20 min prior to the next injection. The chip-LC-MS system contained the following additional modules: nanopump (Cat. #G2226A), capillary loading pump, and solvent degasser. The sample was injected and washed in the back flush mode for 4 min (0.1% aq. FA, 2 μL/min) and then the on-chip trapping column was switched in-line with the analytical column on the microfluidics device. For these separations the same eluents A and B as for the cap-LC-MS system were used at a flow rate of 0.3 μL/min. After elution for 6 min with eluent A, a linear gradient from 0 to 50% eluent B at 0.5%/min followed by a step gradient from 50 to 70% at 1%/min of eluent B was run; 70% eluent B was maintained for 10 min. After each injection the in-line trap and the analytical column were equilibrated with eluent A for 20 min at 2 and 0.3 μL/min, respectively.

2. In the MS acquisition parameters only the ionization voltage and the use of nebulizer gas were different between the two systems (1800–2000 V of ionization voltage and no use of nebulizer gas for chip-LC-MS; 16.0 psi N_2 nebulizer gas and 3500 V of ionization voltage for cap-LC-MS). The following general settings were used for MS during LC-MS: drying gas, 6.0 L/min N_2; skimmer, 40.0 V; cap. Exit, 158.5 V; Oct. 1, 12.0 V; Oct. 2, 2.48 V; Oct. RF, 150 Vpp (voltage, peak power point); lens 1, –5.0 V; lens 2, –60.0 V; trap drive, 53.3; T°, 325°C; scan resolution, enhanced (5500 *m/z* per second scan speed). Target mass, 600. Scan range, 100–1500 *m/z*. Spectra were saved in centroid mode. LC-MS chromatographic data were analyzed with Bruker Data Analysis software, version 2.1 (Build 37).

3.4. Data Processing

The original Bruker Daltoniks LC-MS data files were converted into ASCII format with the Bruker data analysis software. For further data analysis Matlab (version 7.2.0.232 [R2006a], Mathworks, Natick, MA) and the PLS toolbox (version 3.5.2, Eigenvector Research Inc., Wenatchee, WA) were used. Centroid data were smoothed and reduced using a normalized two-dimensional Gaussian filter with rounding of the nominal *m/z* ratios to 1 *m/z* (the original data had a resolution of 0.1 *m/z*). After meshing the data files of all chromatograms, they were time aligned to a reference data file using correlation optimized warping (COW) based on total ion currents (TICs) constructed from signals in the range of 100–1500 *m/z*.

A modified *M-N* rule was applied for peak detection by first calculating a median local baseline using a sliding window technique separately for each *m/z* trace. A median window size of 1200 data points, corresponding to 20.84 min for chip-LC-MS and 20.17 min for cap-LC-MS, was used with a moving rate of 10 points and a minimum median value of 200 counts. According to the *M-N* rule, a threshold of *M*-times the local baseline was used and a peak was assigned if, within one *m/z* trace, the signal exceeded this threshold for at least *N* consecutive points. For each detected peak the *m/z* value, the mean retention times of the three highest measured intensities (within the same peak reduced by the local baseline) were stored in a peak list created for every chromatogram.

To combine the peak lists from different samples with each other, one-dimensional peak matching was achieved by using the sliding window technique, in which the same *m/z* traces were evaluated for peaks that are proximate in time (step size 0.1 min; search window 1.0 min; maximal accepted standard deviation for all retention times within a group of matched peaks was 0.75 min). Missing peak locations were filled with extracted local signals reduced with the local baseline at a given *m/z* retention time location. The generated peak matrix, created from the peak lists of the individual samples, consisted of a peak (row)–sample (column)–intensity (value) matrix. This peak matrix was used for multivariate statistical analysis.

A nearest shrunken centroid (NSC) supervised classification algorithm in conjunction with leave-one-out cross-validation (LOOCV) was applied to select the most discriminating compounds. The selected compounds were then subjected to autoscaled principal component analysis (PCA) and visualized using biplots of the first two principal components. All data processing and statistical analyses were done on a personal computer equipped with a dual core +3800 MHz AMD 64 X2 processor equipped with 4 GB of RAM. **Figure 1** shows an example of data obtained by chip- and capillary-LC-MS.

3.5. Prefractionation of Depleted Serum by Reversed-Phase HPLC on an mRP Column at 80°C

1. Add to ~300 μg (about 300 μL) of depleted serum 0.48 g urea and 13 μL of glacial acetic acid, according to the manufacturer's instructions (www.agilent.com/chem/bioreagents).
2. Add solvent A to a final volume of 1 mL and inject the total volume with a 1 mL loop onto the column.
3. Fractionate at 80°C (pH < 5.0) at a flow rate of 0.75 mL/min with UV detection at 280 nm.
4. Run the gradient from 3 to 30% B in 6 min, to 55% solvent B in 40 min, and up to 100% in 53 min.
5. Collect fractions of 0.75 mL (*see* **Fig. 2a**).
6. Compare fractions after prefractionation by SDS–PAGE (in our case pairwise before and after medical treatment for each patient) (*see* **Fig. 2b**).

4. Notes

Prefractionating proteins in depleted serum/plasma on a recently described macroporose reversed-phase C_{18} column (mRP) prior to digestion is a reproducible step and makes it possible to reduce sample complexity significantly. The fractionated proteins can be identified by trypsin digestion and LC-MS with higher confidence than in the original sample. Based on our own experience and the work of Martosella *et al.* ***(5)*** it is critical to maintain an elevated column temperature (80°C), since poor temperature control could result in lack of reproducibility and low chromatographic resolution.

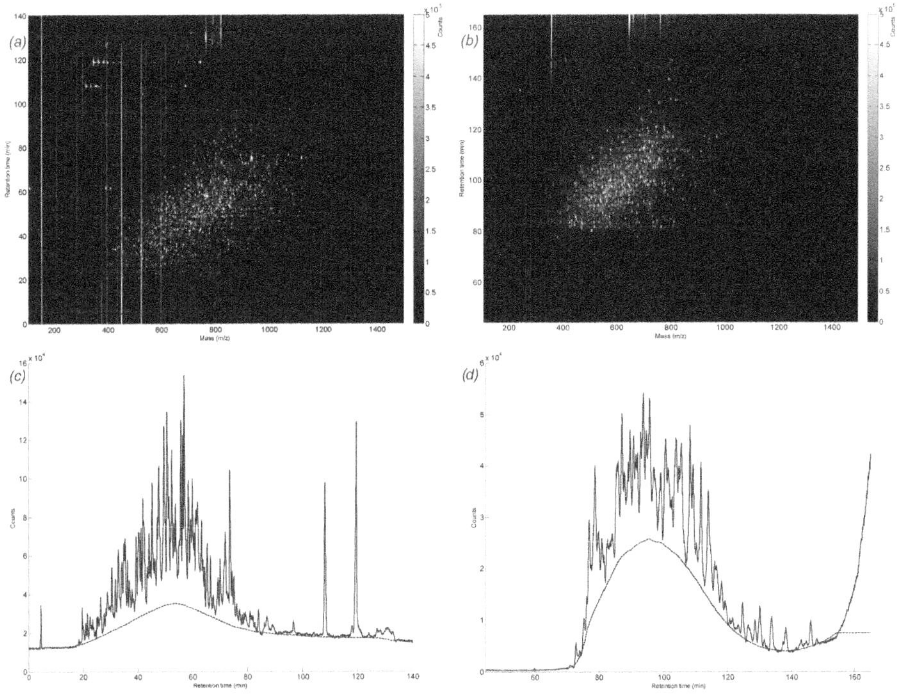

Fig. 1. Raw LC-MS data of depleted and trypsin-digested serum analyzed by chip- (**a**) or capillary LC-MS (**b**) represented as a "heat map." The horizontal axis represents the *m/z* values in amu and the vertical axis shows the retention time in minutes. Peak intensity is coded as indicated (white: high; black: low). (**c**, **d**) The same data in the conventional representation as total ion chromatograms (TICs). The dashed lines depict the calculated baseline. Data were collected in centroid mode and meshed using a data reduction of 1:10.

The performance of the described methodology was evaluated by comparing the ability of cap- and chip-LC-MS to find discriminating features *(6)*. For this purpose five serum samples, spiked with 21 pmol of horse heart cytochrome *c* in 2 μl serum, were analyzed next to five nonspiked serum samples. Due to losses during immunoaffinity depletion of high-abundance proteins, the actual amount of cytochrome *c* that was analyzed was 4.2 pmol *(3)*, corresponding to about 3% (mol/mol) of the total protein content. The obtained raw data were subjected to data processing as described *(6)* followed by supervised classification and selection of discriminating features using the NSC algorithm *(7)*. The shrinkage parameter was optimized using an LOOCV strategy with the aim of reaching the lowest cross-validation error. Although we applied a rather low threshold ($M = 2$, $N = 5$) for peak picking, which introduced more noise in the

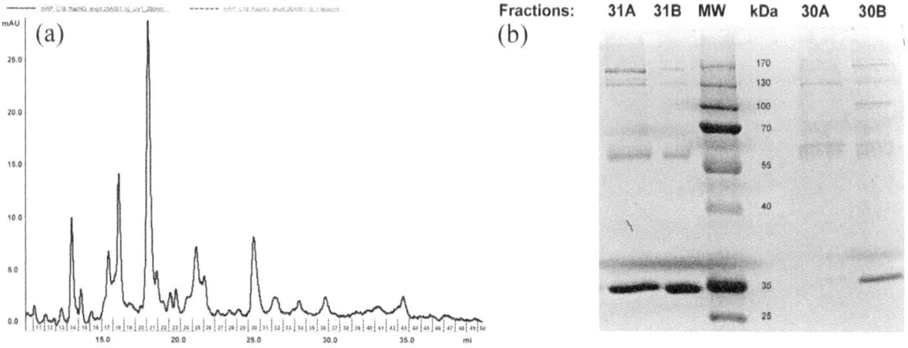

Fig. 2. (**a**) A total of 300 μg of depleted serum was prefractionated on an mRP column at 80°C (example of a sample from a cervical cancer patient before treatment). (**b**) SDS–PAGE (12%) of serum from a cervical cancer patient before (A) and after (B) medical treatment. 30A and 31A: fractions 30 and 31 from the mRP column of patient serum before medical treatment; 30B and 31B: fractions 30 and 31 of the serum from the same patient after treatment. Note the clear difference at about 35 kDa in fraction 30.

peak list, a large domain of shrinkage showed no cross-validation error (51– for chip- and 0.61–16.80 for cap-LC-MS; **Fig. 3a** and **b**, respectively) indicating a robust classification model. Evaluating the 16 most discriminating features selected at shrinkages of 10 and 8.5 for chip- and cap-LC-MS, respectively, resulted in six different peptides. Six peptides selected from the chip-LC-MS and five of the six peptides selected from the cap-LC-MS data corresponded to *in silico* predicted tryptic peptides of horse heart cytochrome *c*. **Figure 3** shows that correct discrimination between spiked versus nonspiked serum samples was easily possible based on the selected peaks (**Fig. 3c** and **d**). PCA analysis of the selected features (**Fig. 3e** and **f**) revealed that almost all variability in the data can be explained by principal component 1 (PC 1) (99% for chip- and 98% for cap-LC-MS). Visualization of the extracted ion chromatograms (EICs) of some of the selected peaks (**Fig. 4**) confirmed that highly discriminating peaks had been correctly found within the complex mixture of digested serum proteins. **Figure 4** also shows the generally good time alignment using COW.

The results show that integration of nano-LC into a microfluidic device makes it possible to perform quantitative, comparative profiling studies of serum and to detect differences in protein profiles down to a level of about 0.5 μ*M*. Microfluidics nano-LC uses ~30 times less sample than capillary LC. Further prefractionation with high-temperature reversed-phase HPLC improves concentration-sensitivity significantly.

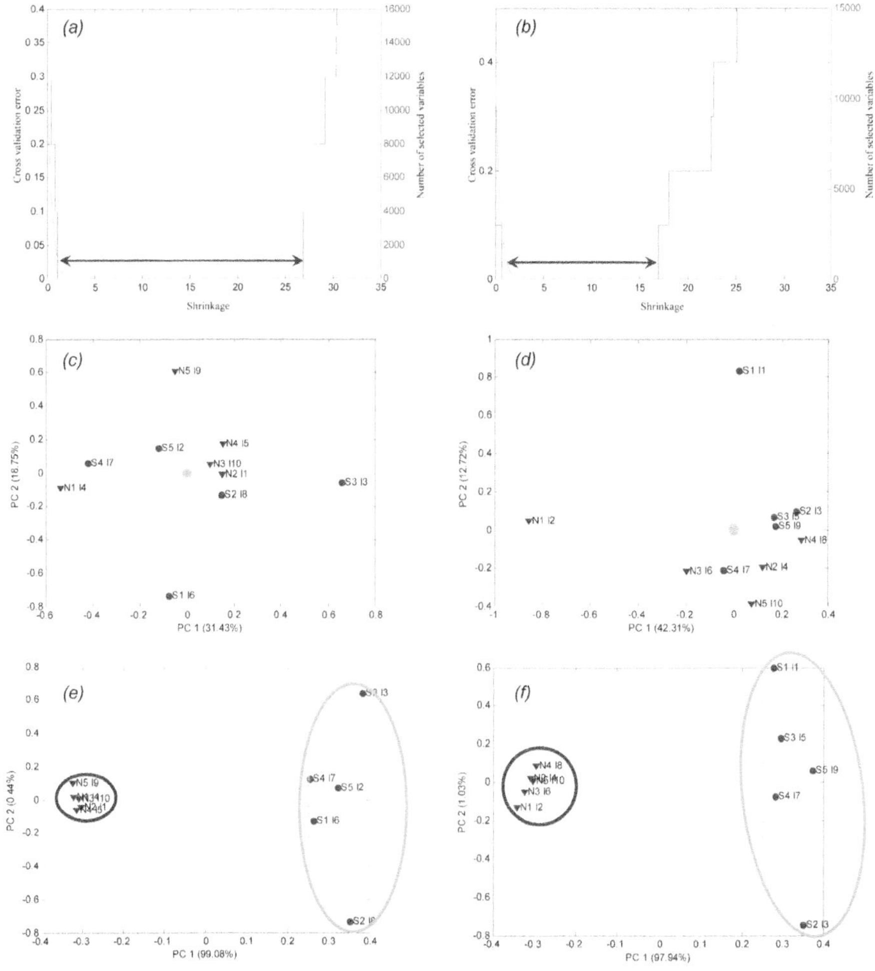

Fig. 3. Representation of the "leave-one out" cross-validation (LOOCV) error and the number of selected variables as a function of the shrinkage for chip-LC-MS (**a**) and cap-LC-MS (**b**). The selected variables, where the shrinkage domain has no cross-validation error, are indicated with arrows. For these domains the selected variables enabled a perfect separation of the two classes (**e** and **f**). PCA plots using all peaks obtained with $M = 2$, $N = 5$ for chip-LC-MS (**c**) and cap-LC-MS (**d**) (14091 for chip-LC-MS and 11256 for cap-LC-MS) did not allow discrimination between the classes. PC 1 and PC 2 refer to the principal components 1 and 2.

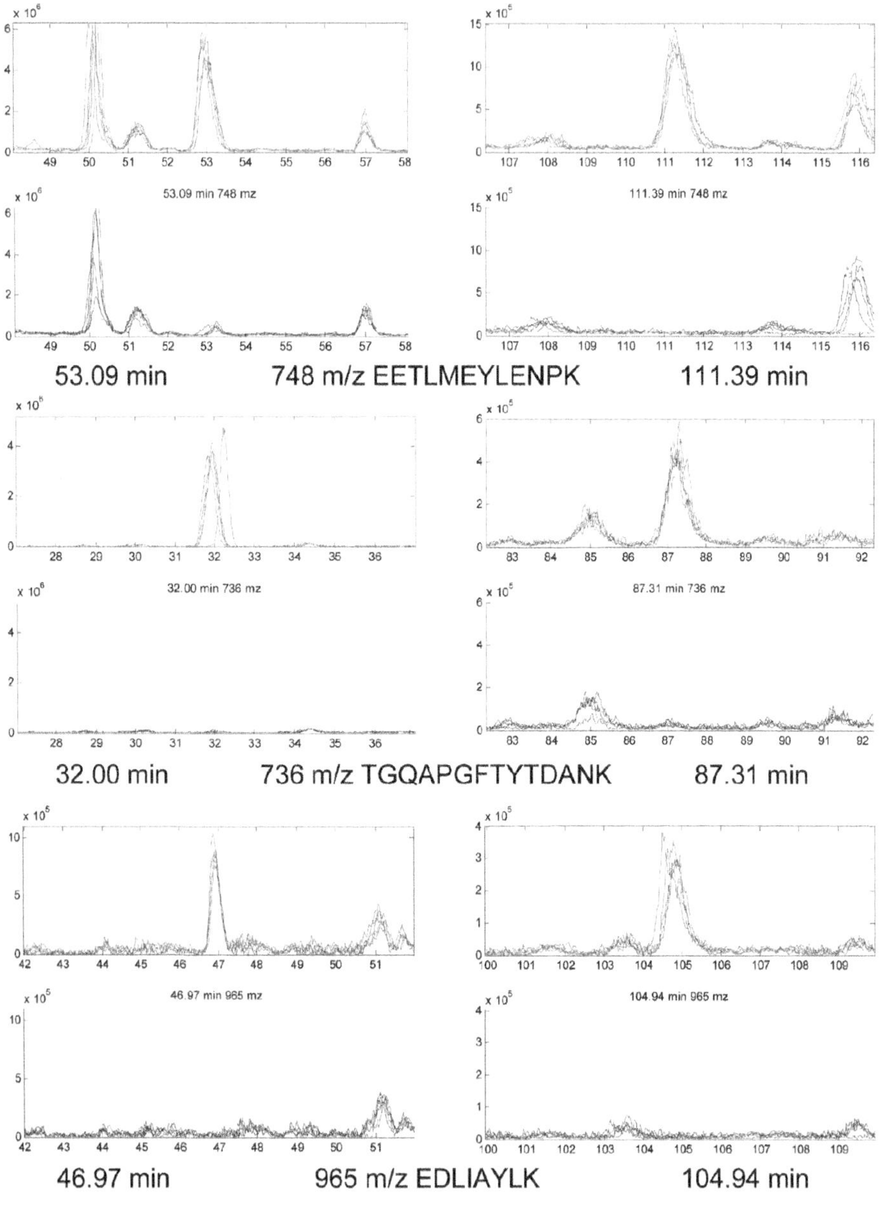

Fig. 4. Examples of extracted ion chromatograms (EICs) of NSC-selected peaks corresponding to tryptic fragments of horse heart cytochrome *c* from datasets obtained with chip-LC-MS (left) and cap-LC-MS (right). The upper traces were obtained from spiked samples and the lower traces were obtained from nonspiked samples.

Acknowledgments

The work in the Department of Analytical Biochemistry described in this chapter is being supported by grants from the Dutch Cancer Society (KWF; RUG2004-3165), the Netherlands Proteomics Center (NPC; Bsik03015), and the Netherlands Bioinformatics Center (NBIC; Biorange 2.2.3). The authors thank all colleagues who contributed to this work.

References

1. Villanueva, J., Philip, J., Chaparro, C. A., Li, Y., Toledo-Crow, R., DeNoyer, L., Fleisher, M., Robbins, R. J., and Tempst, P. (2005) Correcting common errors in identifying cancer-specific serum peptide signatures. *J. Proteome Res.* **4,** 1060–1072.
2. Kemperman, R. F., Horvatovich, P. L., Hoekman, B., Reijmers, T. H., Muskiet, F. A., and Bischoff, R. (2007) Comparative urine analysis by liquid chromatography-mass spectrometry and multivariate statistics: method development, evaluation, and application to proteinuria. *J. Proteome Res.* **6,** 194–206.
3. Govorukhina, N. I., Reijmers, T. H., Nyangoma, S. O., van der Zee, A. G. J., Jansen, R. C., and Bischoff, R. (2006) Analysis of human serum by LC-MS: improved sample preparation and data analysis. *J. Chromatogr. A* **110,** 142–150.
4. Dekker, L. J., Bosman, J., Burgers, P. C., van Rijswijk, A., Freije, R., Luider, T., and Bischoff, R. (2007) Depletion of high-abundance proteins from serum by immunoaffinity chromatography: a MALDI-FT-MS study. *J. Chromatogr. B* **847,** 65–69.
5. Martosella, J., Zolotarjova, N., Liu, H., Nicol, G., and Boyes, B. E. (2005) Reversed-phase high-performance liquid chromatographic prefractionation of immunodepleted human serum proteins to enhance mass spectrometry identification of lower-abundant proteins. *J. Proteome Res.* **4,** 1522–1537.
6. Horvatovich, P., Govorukhina, N. I., Reijmers, T. H., van der Zee, A. G. J., and Bischoff, R. (2007) Evaluation of HPLC-chip/MS platform for label-free profiling for biomarker discovery, *Electrophoresis* **28,** 4493–4505.
7. Tibshirani, R., Hastie, T., Narasimhan, B., and Chu, G. (2002) Diagnosis of multiple cancer types by shrunken centroids of gene expression. *Proc. Natl. Acad. Sci. USA* **99,** 6567–6572.

6

Flow Cytometric Analysis of Cell Membrane Microparticles

Monique P. Gelderman and Jan Simak

Summary

Cell membrane microparticles (MPs) are phospholipid microvesicles shed from the plasma membrane of most eukaryotic cells undergoing activation or apoptosis. The presence of MPs is common in healthy individuals. However, an increase in their release is a controlled event and is considered a hallmark of cellular alteration. Microparticles display cell surface proteins that indicate their cellular origin. In addition, they may also express other markers, e.g., markers of cellular activation. Elevated levels of circulating MPs are associated with various vascular pathologies and their pathogenic potential has been widely documented. MPs have been analyzed in plasma and cell cultures by means of flow cytometry or solid phase assays. Here we present a three-color flow cytometric assay for immunophenotyping of MPs in plasma. This assay has been used to study elevated counts of different phenotypes of circulating endothelial MPs in several hematological and vascular diseases. A modified version of this assay can also be used for MP analysis in blood products and cell cultures.

Key Words: Cell membrane microparticles; membrane microvesicles; ectosomes; flow cytometry; monoclonal antibodies; endothelial cells; platelets; vascular diseases.

1. Introduction

Cell membrane microparticles (MPs), 0.05–1.0 μm in size, are also referred to as microvesicles released from the plasma membrane of eukaryotic cells. Microparticles express surface antigens of their originating cells and are free of nucleic acids. They should not be mistaken for exosomes, which are derived from multivesicular bodies and overlap in size with the smaller MPs (40–80 nm).

From: *Methods in Molecular Biology, vol. 484: Functional Proteomics: Methods and Protocols*
Edited by: J. D. Thompson et al., DOI: 10.1007/978-1-59745-398-1, © Humana Press, Totowa, NJ

The larger MPs (>1.0 μm in size) may be difficult to distinguish from platelets, MP aggregates, or apoptotic bodies. Therefore, 1 μm is considered by most authors as the size limit when defining MPs. As a result of different types of stimulation, MPs are shed from the cellular membrane of a variety of eukaryotic cells. The following are examples of, but not limited to, different stimuli: shear stress, complement attack, or proapoptotic triggers. Long considered "cell dust," MPs derived from various cells are normally present in the circulation of healthy individuals. The elevated counts of MPs in various diseases indicate their potentially diagnostic importance, particularly in vascular pathologies. Several comprehensive reviews discussing MPs are available ***(1–5)***.

MPs have been shown to exhibit a variety of activities. They may facilitate cell-to-cell interactions, induce cell signaling, or even transfer receptors between different cell types. A physiological role of MPs in several tissue defense processes has been suggested. In addition, pathophysiological implications of MPs in thrombosis, inflammation, and cancer metastasis, or their role in responding to pathogens have been proposed ***(1,5–12)***. Thus, assessing the presence and counts of circulating MPs in blood seems important, not only for their possible diagnostic importance, but also for understanding the potential role of MPs in the pathogenesis of various diseases.

We have developed a three-color flow cytometric assay for immunophenotyping MPs that are present in plasma. The assay has been used to study MPs in plasma of healthy donors and in patients with paroxysmal nocturnal hemoglobinuria, sickle cell disease, and also in patients with acute ischemic stroke ***(13–15)***. A modified version of this assay has been used for MP analysis in blood transfusion products, such as apheresis platelets, and also in endothelial cell cultures ***(16,17)***.

2. Materials

2.1. Blood Collection, Blood Sample Processing, and Platelet-Free Plasma Storage

1. BD Vacutainer® blood collection tubes (13 × 100 mm) containing acid citrate dextrose solution A (Becton Dickinson Labware, Franklin Lakes, NJ).
2. BD Vacutainer® blood collection sets, holders, and sharp collectors (Becton Dickinson Labware, Franklin Lakes, NJ).
3. Adams™ Nutator Mixer (Becton Dickinson Labware).
4. 1.5-mL microcentrifuge tubes (Fisher Scientific).
5. 2-mL Sarstedt screw cap microtubes (Fisher Scientific).
6. BLUE MAX Jr. 15-mL polypropylene conical tubes (Becton Dickinson Labware).
7. 3.5-mL Samco fine tip transfer pipettes (MG Scientific, Pleasant Prairie, WI).

2.2. Flow Cytometry

1. 5-mL polystyrene round-bottom tubes (352052) (Becton Dickinson Labware).
2. Calibrite Beads (Becton-Dickinson, Franklin Lakes, NJ).
3. TruCount Tubes (Becton-Dickinson, Franklin Lakes, NJ).
4. Beads 0.2–3 μm: Molecular Probes Flow Cytometry Size Calibration Kit (F-13838) (Molecular Probes, Eugene, OR) and Sigma Latex Beads LB-3, LB-8, and LB-30 (Sigma).
5. Hanks' balanced salt solution (HBSS) (Sigma) supplemented with 0.35% albumin from bovine serum (BSA, Sigma) (referred to in the Methods section as "HBSS/BSA") (*see* **Note 1**).
6. HBSS (Sigma) without calcium chloride, magnesium sulfate, and phenol red (referred to in the Methods section as "HBSS, w/o Ca^{2+}").
7. EDTA (Sigma).
8. $CaCl_2$ (Sigma).
9. Annexin V and antibodies (*see* **Note 2**): phycoerythrin (PE) and fluorescein isothiocyanate (FITC) conjugated IgG_1, IgG_{2a} isotype controls (IgIC), peridinin chlorophyl protein (PerCP), conjugated monoclonal antibody (Mab) to CD45 (clone TÜ116), Mab to human CD41a (FITC or PerCP-Cy5.5-conjugated, clone HIP8), Mab to human CD 144, and annexin V (FITC-conjugated) from BD PharMingen (San Diego, CA). Mab to human CD54 (FITC-conjugated, clone MEM111) and Mab to human CD235a (FITC-conjugated, clone CLB-ery-1) from Caltag Laboratories (Burlingame, CA). Mab to human CD105 (PE-conjugated, clone N1-3A1) and rabbit polyclonal antibody to human CD144 (FITC-conjugated) were from Ancell/Alexis (San Diego, CA). Rabbit IgG (FITC-conjugated) was from U.S. Biological (Swampscott, MA) (*see* **Note 3**).

3. Methods

Several different experimental approaches have been used to analyze MPs ***(18)***. In general, the majority of investigators use either solid phase assays (microplate affinity) or flow cytometric assays for MP analysis. Flow cytometry is the most commonly used and the basic method for MP analysis. It allows for the analysis of large numbers of MPs (to the order of tens of thousands), and in addition, makes it possible to collect information about their corpuscular characteristics. The size of MPs correlates with forward scatter (FS) and their granularity is reflected by the side scatter (SS) parameter. Standard beads of different diameters may be used for size calibration. A known count of larger beads (Tru Count beads) as an internal standard, or assayed in a parallel sample, is commonly used for flow rate calibration. Thus, the count of MPs per analyzed volume can easily be calculated. With the use of antibodies conjugated to different chromophores, a combination of three or even more antigens can be analyzed on a single MP. In a similar fashion, annexin V conjugated to a chromophore can be used to detect accessible phosphatidylserine (PS) on MPs.

Some investigators count and analyze only MPs that are able to bind annexin V in their assay. MPs can bind to annexin V only when they expose accessible PS on their surface. However, it has been shown that only a limited portion of MPs in blood binds to annexin V. With that type of approach, a significant population of MPs, particularly of endothelial origin, is missed from the analysis. Currently, there is no acceptable method available for the detection of all MPs in blood to calculate a total MP count. Various methods using lipophilic fluorescent dyes, chromophore-labeled lectins, or antibodies to ubiquitous antigens were unable to provide satisfactory results to resolve this issue.

There are several requirements for target antigens when detecting MPs: cell specificity, an abundance of the antigen on both parent cells and MPs, stability of the antigen, commercial availability of avid antibodies (preferably monoclonal), and conjugated to a chromophore. The titration of antibodies using MPs prepared from their parental cells *in vitro* as well as using MPs in plasma is recommended. The use of two clones against different epitopes of an antigen is a good confirmation of detection specificity. In addition, relevant isotype immunoglobulin controls raised against an irrelevant antigen should be used.

With regard to the identification of the MP's cellular origin, glycophorin A (CD235a) is used almost exclusively for the identification of red blood cell-derived MPs. The leukocyte common antigen (CD45) is usually used to identify white blood cell-derived MPs. Monoclonal antibodies to CD14, CD66b, CD4, CD8, and CD20 are used to detect MPs originating from monocytes, granulocytes, T_H, T_S, and B lymphocytes, respectively ***(19)***. Platelet-derived MPs are detected using monoclonal antibodies to GPIIb (CD41), glycoprotein complex GPIIb/IIIa (CD41a), GPIX (CD42a), $GPIb_\alpha$ (CD42b), or GPIIIa (CD61). It has been suggested that $CD41^+$MP and $CD42^+$MP populations are not identical and may reflect different pathophysiological phenomena ***(18)***. The analysis of both phenotypes is therefore recommended. Different endothelial antigens have been used for the detection of endothelial cell-derived MPs in blood: integrin α_v (CD51) ***(20)***, S-Endo/Muc 18 antigen (CD146) ***(21)***, E-selectin (CD62E) ***(22)***, VE-cadherin (CD144) ***(23)***, or PECAM-1 (CD31) with simultaneous exclusion of MPs expressing the platelet antigen CD42 ***(24)***. Since VE-cadherin (CD144) is the most specific marker for endothelial cells currently available, it is probably the most suitable marker for endothelial cell-derived MPs (EMPs). Another marker for EMPs used in our laboratory is endoglin (CD105). In addition to being strongly expressed on vascular endothelial cells, endoglin is weakly expressed on hematopoietic stem cells, monocytes, fibroblasts, stromal cells, and vascular smooth muscle cells. While we are able to exclude the contribution of activated monocytes in our endothelial MP assay by counting $CD105^+CD45^-$MPs (or preferably $CD105^-CD14^-$MPs), still other cell types could contribute. A small subset of hematopoietic stem cells and endothelial progenitors probably expresses CD105 in levels high enough to be detectable on MPs. Also $CD105^{+}$MPs derived from smooth muscle cells may be present in blood. In our laboratory, the best combination of antigens that suggest a true endothelial-derived MP population is $CD105^{+}CD144^{+}$. The potential diagnostic importance of plasma $CD105^{+}CD144^{+}$MPs as a marker of endothelial injury is supported by our studies showing a significant elevation of $CD105^{+}CD144^{+}$MPs in plasma of patients with paroxysmal nocturnal hemoglobinuria (PNH), sickle cell disease (SCD) ***(14)***, or acute ischemic stroke ***(15)***.

Antigens and clones of monoclonal antibodies used for the identification of cellular origin of MPs in blood are summarized in **Table 1**. It is important to

Table 1
Blood Cell, Platelet, and Endothelial Antigens Used for the Detection of MPs *(1)*

Cellular origin of MPs	Antigen	Alternative names	Mab clones
Red blood cell	CD235a	Glycophorin A	JC159; CLB-ery-1
Leukocyte	CD45	LCA, T200, B220	TÜ116; HI30
Monocyte	CD14	LPS-R	CRIS-6; MØP9; RMO52
Granulocyte	CD66b	CD67, CGM6, NCA-95	80H3; CLB-gran/10
T_H lymphocyte	CD4	T4, L3T4 (mouse), W3/25 (rat)	CLB-T4/2
T_S lymphocyte	CD8	T8, Leu-2, Lyt 2,3	SK1
B lymphocyte	CD20	B1, Bp35	L27
Platelet	CD41	GPIIb, aIIβ integrin	P2
	CD41a	GPIIbIIIa, aIIββ_3 integrin	HIP8
	CD42a	GPIX	KMP9
	CD42b	GPIbα	HIP1; SZ2
	CD61	GPIIIa, β_3 integrin	Y2/51
Endothelial MP phenotypes			
$CD31^{+}CD42b^{-}$	CD31	PECAM-1	MBC782; WM59
$CD34^{+}$	CD34	gp105-120	8G12
$CD62E^{+}$	CD62E	E-selectin	CI26CIOB7; 1.2B6
$CD51^{+}$	CD51	α_v integrin	AMF7; 23C6
$CD105^{+}CD144^{+}$	CD105	Endoglin	N1-3A1
$CD105^{+}CD45^{-}$			

note that the presence of an antigen on an MP does not exclusively identify its cellular origin. For example, in blood, soluble antigens derived from one cell type may adhere to MPs derived from another cell type. Moreover, MPs derived from one cell type may fuse with the membrane of different cell types. These cells may subsequently release MPs with an "adopted" antigen. Keeping these possibilities in mind, it is necessary to be cautious when interpreting the results of immunophenotyping experiments.

Other antigens have been used to characterize different MP phenotypes that can be present in blood such as von Willebrand factor (vWF) ***(25)***, p-selectin glycoprotein ligand 1 (PSGL-1) ***(26)***, or cellular prion protein (PrPc) ***(13)***. In addition, the analysis of MPs in blood derived from tumor cells or extravascular tissues could have a high diagnostic potential ***(27)***.

The expression of several antigens, which may reflect either the stimulation or the cytokine activation status of the parental cells, has been studied. One example of a frequently studied antigen is P-selectin (CD62P). In stimulated platelets and endothelial cells, P-selectin is rapidly upregulated on plasma membrane from intracellular sources. Another activation marker is the intercellular adhesion molecule 1 (ICAM-1, CD54). ICAM-1 belongs to the immunoglobulin gene superfamily of receptors and is constitutively expressed at low levels on endothelial cells, leukocytes, fibroblasts, and epithelial cells. However, its expression is dramatically upregulated by proinflammatory cytokines. Thus, the presence of $CD54^+$MPs could indicate inflammatory stimulation of leukocytes or endothelial cells ***(14,28)***. Other potential markers are E-selectin (CD62E) or VCAM-1 (CD106), both expressed on endothelial cells after stimulation with proinflammatory cytokines. However, both $CD62E^+$MPs and $CD106^+$MPs are difficult to analyze in plasma, because of the low number of molecules of these antigens present on MPs ***(25,29,30)***.

As far as MPs affecting hemostasis and thrombosis, PS^+MPs detected by annexin V should be considered as MPs with a prothrombotic phenotype, because they may provide PS for the assembly of FX- and prothrombin activation complexes. On the other hand, we can speculate that in healthy individuals, the presence of PS^+MPs in plasma may actually promote low thrombin generation required for the protein C system activation and thus have a possible antithrombotic effect ***(22)***. In general, highly elevated counts of PS^+MPs should definitely be considered as prothrombotic. There are several studies that analyze the expression of tissue factor (TF, CD142) on MPs ***(23,31–33)***. It should be taken into consideration that immunodetection of CD142 on MPs can be associated with a high level of nonspecificity. Therefore, the selection of correct monoclonal antibodies and their careful titration are essential. Finally, complementary functional assays should be used to confirm the prothrombotic or the proinflammatory nature of MPs.

The limitation of flow cytometric analysis of MPs is that current commercially available flow cytometers are not capable of analyzing MPs smaller than approximately 200–300 nm. This results in an inability to analyze the population of smaller sized MPs. In addition, this technique is not able to distinguish between small cell debris and MPs. The upper size limit of 1.0 μm in MP analysis serves to avoid analyzing too much cell debris, platelets, MP aggregates, or apoptotic bodies. Nevertheless, flow cytometry is still the best candidate to be considered as the "gold standard" for MP analysis.

3.1. Whole Blood Sample Preparation and Platelet-Free Plasma Storage after Blood Collection

1. Collect 10 mL whole blood in a BD Vacutainer® blood collection tube (13 × 100 mm) containing acid citrate dextrose solution A following standard phlebotomy procedures. Keep the tubes at room temperature on an Adams™ Nutator Mixer until step 2 (*see* **Note 4**).
2. Transfer the complete blood sample, using a Pasteur pipette, into a 15-mL polypropylene tube.
3. Centrifuge the sample for 15 min at 10°C and 2600 × *g*.
4. Transfer approximately 4.5 mL of platelet-poor plasma (PPP), using a transfer pipette, into three microcentrifuge tubes (approximately 1.5 mL PPP in each microcentrifuge tube).
5. Centrifuge these three tubes in a microcentrifuge for 5 min at 10°C and 9900 × *g*.
6. Transfer 1.4 mL of the supernatant (platelet free plasma, PFP) from each tube into three 2-mL Sarstedt screw cap microtubes.
7. Proceed with the preparation of the MP suspension or immediately snap freeze platelet-free plasma samples in the liquid phase of nitrogen and store the samples in a liquid nitrogen storage tank (*see* **Note 5**) until further analysis.

3.2. Preparation of MP Suspension from Platelet-Free Plasma

1. Thaw the PFP samples quickly in a 37°C waterbath (*see* **Note 6**). Once the samples are thawed, transfer them into microcentrifuge tubes.
2. Centrifuge the samples for 10 min at 10°C and 19,800 × *g* (*see* **Note 7**).
3. Remove the supernatant using a blunt, 4-inch-long 14-gauge suction needle attached to a vacuum apparatus (set the vacuum regulator to 5 in. Hg), leaving 100 μL in the tube (*see* **Note 8**).
4. Resuspend the 100 μL sediment with 1 mL of HBSS, w/o Ca^{2+}.
5. Centrifuge the samples for 10 min at 10°C and 19,800 *x g*.
6. Repeat Step 3.
7. Resuspend the 100 μL sediment with 700 μL HBSS/BSA.
8. Store on wet ice and use within 1 h.

3.3. Labeling of MPs

1. Transfer 50 μL, after gentle mixing of the resuspended MPs, into individual microcentrifuge tubes.
2. Add to each microcentrifuge tube, containing 50 μL of the MP suspension, 5 μL of three different antibodies or annexin V, each at saturating concentrations (*see* **Note 9**), each conjugated to a different fluorescent tag (FITC-, PE-, and PerCP- conjugated antibodies or FITC- or PE-conjugated annexin V). In parallel, prepare nonlabeled samples and samples labeled with relevant isotype controls and controls with annexin V in the presence of 20 m*M* EDTA.
3. Incubate all tubes for 20 min at room temperature in the dark by covering the tubes with aluminum foil.
4. After this incubation, add 1 mL HBSS/BSA to each tube.
5. Centrifuge sample(s) for 10 min at 10°C and 19,800 × *g*.
6. Repeat Step 3.
7. Add 500 μL of HBSS/BSA, resuspend the pellet, and transfer all samples to polystyrene round-bottom tubes. Keep the tubes covered with aluminum foil for the duration of sample acquisition.

3.4. Flow Cytometry of MPs

Three-color flow cytometry was performed on a FACS Calibur flow cytometer equipped with CellQuestPro software (Becton Dickinson, San Jose, CA). However, MP analysis may be performed on any competitive instrument. MPs should be analyzed in a protocol with both forward scatter (FSC) and side scatter (SSC) set to the logarithmic mode. Double fluorescence plots from flow cytometric analysis demonstrating the presence of MPs of different cellular origin in normal human plasma are shown in **Fig. 1**. An example of the size distribution of $CD105^+$MPs in normal plasma is shown in **Fig. 2**.

1. Adjust the instrument setting and fluorescence compensation using Calibrite 3 fluorescence beads (Becton Dickinson), following the manufacturer's instructions.
2. Run beads, 0.2–3 μm in diameter (Sigma, St. Louis, MO; Molecular Probes, Eugene, OR), resuspended in HBSS/BSA for the estimation of MP size in the FSC setting. The generally accepted upper size limit for MPs is 1 μm.
3. Before acquisition of the samples, perform flow calibration. To calibrate, use one TruCount tube (Becton Dickinson) and add 500 μL of HBSS/BSA. Mix the beads by pipetting up and down twice. Transfer the total volume from the TruCount tube to a polystyrene round-bottom tube. Set the acquisition time for 60 s and run TruCount beads three consecutive times at different flow rates (low, medium, and high). For optimal flow rate monitoring, three TruCount tubes should be run before and after each set of samples. The sample flow volume per minute at different flow speeds can be calculated from the total number of beads in the tube (provided for

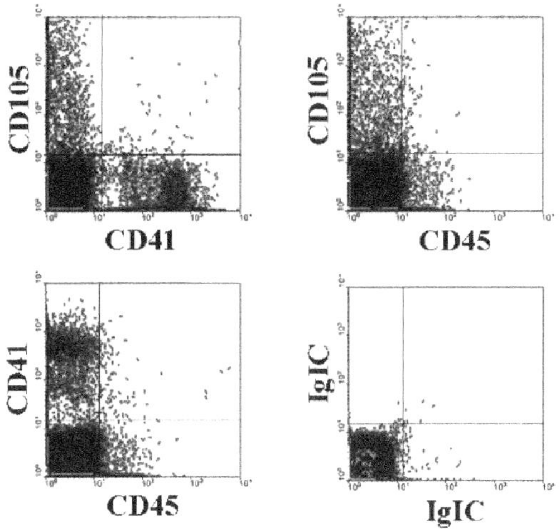

Fig. 1. Flow cytometric analysis of cell-specific MPs in normal human plasma. Double fluorescence plots demonstrate distinct populations of platelet ($CD41^+CD105^-$), white blood cell ($CD45^+CD41^-$), and endothelial ($CD105^+CD45^-$) MPs in plasma of a representative healthy donor. To confirm the endothelial origin of $CD105^+CD45^-$ MPs, the exclusion of monocyte-derived $CD14^+CD105^+$MPs and/or analysis of the coexpression of CD144 on $CD105^+$MPs may be used. IgIC, isotype control. [Reprinted with permission from *Br. J. Haematol.* ***(14)***.]

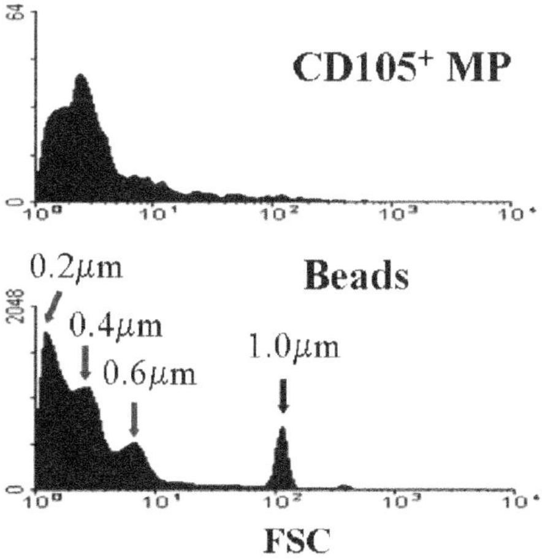

Fig. 2. Size distribution of $CD105^+$MP in normal human plasma. Flow cytometry of $CD105^+$MPs and standard beads. The forward scatter (FSC) histograms show the size distribution of $CD105^+$MPs in plasma of a representative healthy donor (top) relative to standard beads (bottom). [Reprinted with permission from *Br. J. Haematol.* ***(14)***.]

each lot), in combination with the volume used for beads resuspension, and the number of beads counted by the instrument per minute (*see* **Note 10**).

4. Acquire the samples at low or medium rate for 60 or 120 s, depending on the concentration of all events. The optimal count of events per second is 300–900, depending on the type of flow cytometer used. The total count of acquired events is usually 20,000–60,000. We acquire all events including background and fluorescence-negative MP populations.
5. Use double fluorescence plots and SSC versus fluorescence plots for the analysis of samples labeled with isotype controls (or annexin V + EDTA) in order to gate for negative and positive MP populations. For standard evaluation, use quadrant gating when possible.
6. Use double fluorescence plots and SSC versus fluorescence plots to evaluate counts of specific MP phenotypes per run. Since MPs are very heterogeneous in FSC/SSC characteristics, we do not apply elimination of doublets using FSC and SSC geometry (*see* **Note 11**).
7. Keep all dilution factors and a sample flow volume/minute in mind when calculating MP counts/μL of plasma (*see* **Note 12**).

4. Notes

1. For optimal binding of annexin V to PS^+MPs, the Ca^{2+} concentration in HBSS should be increased to 3 m*M* using $CaCl_2$. HBSS/BSA should be filtered using a 0.22-μm filter attached to a sterile bottle (90 mm Filter Unit, Nalgene, Rochester, NY). When aseptically manipulated in a biological safety cabinet under laminar flow, the solution may be stored up to 3 weeks at 4°C. The solution should be checked before use by flow cytometry for the presence of precipitated albumin microparticles, particularly when higher Ca^{2+} concentrations are used.
2. There are numerous competitive antibodies available from several other commercial sources.
3. We have used FITC-conjugated rabbit polyclonal antibody to CD144 in the past. However, the chromophore-conjugated Mabs to CD144 are now commercially available.
4. The collection of venous blood and the subsequent sample processing steps may have a dramatic impact on the results of MP analysis in clinical samples. The following are variables that need to be considered: the sampling site (cubital vein or central venous catheter), needle diameter or catheter, discharge of the first portion of blood, manner of collection (vacutainer, syringe, tube), and the type of anticoagulant (ACD, citrate, or heparin) used. In general, blood samples should not be chilled, overheated, or extensively shaken, because temperature changes or shear stress may induce MP release from blood cells. We believe that freshly filled vacutainer tubes can be stored at room temperature in combination with a very slow and gentle agitation in order to bridge the period between sample collection and processing. This period should be kept as short as possible. Less than 1 h is best. However, this is not always possible. Although no supporting

data are available, the addition of enzyme inhibitors or other preservatives to the blood samples at time of collection might be beneficial. In particular, inhibitors of proteases or phospholipases could be helpful when analysis is focused on an unstable population of MPs, or on an MP antigen sensitive to proteolysis. However, it is necessary to take into consideration that for some antigens or epitopes, the redox status affecting disulfide bonds and the presence of chelators affecting Ca^{2+}- or Mg^{2+}-dependent complexes are critical factors.

5. The practice of freezing plasma samples before MP analysis is definitely associated with a high risk of generating artifacts. In most clinical studies, it is not possible to process the samples and perform MP assays in the desired short time frame. Therefore, some investigators freeze and store plasma samples before MP analysis ***(23,34)***. When freezing plasma samples, they should be true platelet-free plasma (PFP) and not platelet- poor plasma (PPP). Different protocols for freezing and thawing may substantially affect the results of MP analysis. We recommend snap freezing of PFP in the liquid phase of nitrogen, followed by immediate storage in liquid nitrogen. While the freezing temperature is of importance, we believe that storage for a couple of weeks at –70°C may be acceptable. However, we do not have any data to support this claim.
6. The process of thawing is as important as freezing. Some investigators thaw MPs samples on wet ice ***(34)***. In our laboratory, we do a quick thaw at 37°C with gentle shaking, which is immediately followed by cooling the sample to 10°C. Quick thawing at 37°C should prevent intermediate formation of large ice crystals; however, prolonged incubation of a sample at 37°C leads to the deterioration of MPs and the degradation of sensitive antigens. Our data showed that counts of different endothelial cell MP populations ($CD105^+$MPs, $CD105^+PS^+$MPs, and $CD105^+CD54^+$MPs) in plasma after a freeze–thaw cycle were not significantly different from samples stored for 1 h at 4°C. For each study it is important to investigate how MPs of specific phenotypes of interest are affected by a single freeze–thaw cycle. The freezing of MPs should be further investigated, since freezing samples before analysis would be a great advantage for the potential diagnostic use of MP assays.
7. Among the potential deleterious effects of centrifugation is the possibility of MP loss during processing in the discarded sediment with blood cells and platelets or in the supernatant if MPs are sedimented. In addition, there is a risk of MP release from blood cells and platelets during centrifugation and other associated manipulations. However, the preparation of PPP or PFP is usually an essential step. We analyze MPs obtained from PFP after a 10 min spin at 19,800 × *g*, which quantitatively sediments particles of 0.2 μm diameter ***(14)***. Since a particle of this size is at the detection limit of the flow cytometer, a more extensive ultracentrifugation is not needed. Our assay includes repeated washing steps before and after immunolabeling, which may increase the specificity and minimize the formation of artifactual immunocomplexes. There is always the risk of losing some MPs during several washing steps when not done carefully. This protocol requires an experienced operator and is time consuming. Other

investigators use direct immunolabeling of plasma and flow cytometry analysis without isolation and washing of MPs. This method showed very promising results in different clinical studies ***(35–37)*** and would be very useful for clinical diagnostic purposes. However, the size of the analyzed MP, the contribution of plasma soluble antigens, and the formation of immunocomplexes by different antibodies in this assay would be of interest.

8. If the vacuum is set to greater than 5 in. Hg, the pelleted/precipitated microparticles will be disturbed and lost when removing the supernatant. The supernatant can also be removed by using a fine tip transfer pipette or a long tip regular pipette.
9. All platelet-specific and blood cell-specific antibodies used for MP detection were titrated using platelets, red blood cells, and white blood cells isolated from blood from healthy donors. In addition, for each cell type membrane microparticles were generated *in vitro* and tested to ensure specificity of the assay. Specificity and saturating concentrations of antibodies against endothelial antigens were evaluated using resting and tumor necrosis factor (TNF)-α-stimulated cultured human umbilical vein endothelial cells.
10. In preliminary experiments the flow rate variation from tube to tube was evaluated using TruCount beads resuspended in HBSS/BSA and analyzed in 5-mL Falcon (352052) polystyrene tubes. Analysis of 30 consecutive samples at a medium rate showed the flow rate to be 33.3±0.8 μL/min. The resulting coefficient of variation was 2.4%. In our experience, TruCount beads are not an accurate internal standard. We observed that the accuracy of counting of these beads using a separate gate was influenced by the presence of different counts of MPs in the samples.
11. It has been demonstrated that MP analysis using a BD FACSAria digital flow cytometer offers an improved resolution and greater ability to discriminate, characterize, and sort MP populations ***(38)***. In this study, a dot plot with FSC-height (FSC-H) vs. FSC-width (FSC-W) was used to eliminate doublets by FSC geometry by drawing a gate around the dominant population. These gated events were then displayed in an SSC-H vs. SSC-W dot plot that further eliminated doublets through side scatter geometry.
12. Our assay, similar to other flow cytometric methods of MP analysis developed in different laboratories, is associated with various artifacts. Therefore, standardization of all sample processing and analytic steps is essential to allow interlaboratory comparison of absolute counts of different phenotypes of MPs in plasma, other biological fluids, blood products, and cell cultures. It is our expectation that novel technologies and instruments with higher resolution will soon substantially improve the sensitivity and specificity of MP assays.

Acknowledgments

The findings and conclusions in this chapter have not been formally disseminated by the Food and Drug Administration and should not be construed to represent any Agency determination or policy.

References

1. Simak, J. and Gelderman, M. P. (2006) Cell membrane microparticles in blood and blood products: potentially pathogenic agents and diagnostic markers. *Transfus. Med. Rev.* **20,** 1–26.
2. Nomura, S. (2001) Function and clinical significance of platelet-derived microparticles. *Int. J. Hematol.* **74,** 397–404.
3. Horstman, L. L., Jy, W., Jimenez, J. J. and Ahn, Y. S. (2004) Endothelial microparticles as markers of endothelial dysfunction. *Front Biosci.* **9,** 1118–1135.
4. Greenwalt, T. J. (2006) The how and why of exocytic vesicles. *Transfusion* **46,** 143–152.
5. Freyssinet, J. M. (2003) Cellular microparticles: what are they bad or good for? *J. Thromb. Haemost.* **1,** 1655–1662.
6. Morel, O., Toti, F., Hugel, B., Bakouboula, B., Camoin-Jau, L., Dignat-George, F., and Freyssinet, J. M. (2006) Procoagulant microparticles: disrupting the vascular homeostasis equation? *Arterioscler. Thromb. Vasc. Biol.* **26,** 2594–2604.
7. Martinez, M. C., Tesse, A., Zobairi, F., and Andriantsitohaina, R. (2005) Shed membrane microparticles from circulating and vascular cells in regulating vascular function. *Am. J. Physiol. Heart Circ. Physiol.* **288,** H1004–1009.
8. Ahn, Y. S., Jy, W., Jimenez, J. J., and Horstman, L. L. (2004) More on cellular microparticles: what are they bad or good for? *J. Thromb. Haemost.* **2,** 1215–1216.
9. Diamant, M., Tushuizen, M. E., Sturk, A., and Nieuwland, R. (2004) Cellular microparticles: new players in the field of vascular disease? *Eur. J. Clin. Invest.* **34,** 392–401.
10. Distler, J. H., Huber, L. C., Gay, S., Distler, O., and Pisetsky, D. S. (2006) Microparticles as mediators of cellular cross-talk in inflammatory disease. *Autoimmunity* **39,** 683–690.
11. Hugel, B., Martinez, M. C., Kunzelmann, C., and Freyssinet, J. M. (2005) Membrane microparticles: two sides of the coin. *Physiology (Bethesda)* **20,** 22–27.
12. Morel, O., Toti, F., Hugel, B., and Freyssinet, J. M. (2004) Cellular microparticles: a disseminated storage pool of bioactive vascular effectors. *Curr. Opin. Hematol.* **11,** 156–164.
13. Simak, J., Holada, K., D'Agnillo, F., Janota, J., and Vostal, J. G. (2002) Cellular prion protein is expressed on endothelial cells and is released during apoptosis on membrane microparticles found in human plasma. *Transfusion* **42,** 334–342.
14. Simak, J., Holada, K., Risitano, A. M., Zivny, J. H., Young, N. S., and Vostal, J. G. (2004) Elevated circulating endothelial membrane microparticles in paroxysmal nocturnal haemoglobinuria. *Br. J. Haematol.* **125,** 804–813.
15. Simak, J., Gelderman, M. P., Yu, H., Wright, V., and Baird, A. E. (2006) Circulating endothelial microparticles in acute ischemic stroke: a link to severity, lesion volume and outcome. *J. Thromb. Haemost.* **4,** 1296–1302.
16. Simak, J., Holada, K., and Vostal, J. G. (2002) Release of annexin V-binding membrane microparticles from cultured human umbilical vein endothelial cells after treatment with camptothecin. *BMC Cell Biol.* **3,** 11.

17. Gelderman, M. P., Carter, L. B., and Simak, J. (2004) High counts of potentially pathogenic cell membrane microparticles in apheresis platelets. *Blood* **104,** 988a.
18. Horstman, L. L., Jy, W., Jimenez, J. J., Bidot, C., and Ahn, Y. S. (2004) New horizons in the analysis of circulating cell-derived microparticles. *Keio J. Med.* **53,** 210–230.
19. Nieuwland, R., Berckmans, R. J., McGregor, S., Boing, A. N., Romijn, F. P., Westendorp, R. G., Hack, C. E., and Sturk, A. (2000) Cellular origin and procoagulant properties of microparticles in meningococcal sepsis. *Blood* **95,** 930–935.
20. Combes, V., Simon, A. C., Grau, G. E., Arnoux, D., Camoin, L., Sabatier, F., Mutin, M., Sanmarco, M., Sampol, J., and Dignat-George, F. (1999) In vitro generation of endothelial microparticles and possible prothrombotic activity in patients with lupus anticoagulant. *J. Clin. Invest.* **104,** 93–102.
21. Mallat, Z., Benamer, H., Hugel, B., Benessiano, J., Steg, P. G., Freyssinet, J. M., and Tedgui, A. (2000) Elevated levels of shed membrane microparticles with procoagulant potential in the peripheral circulating blood of patients with acute coronary syndromes. *Circulation* **101,** 841–843.
22. Berckmans, R. J., Neiuwland, R., Boing, A. N., Romijn, F. P., Hack, C. E., and Sturk, A. (2001) Cell-derived microparticles circulate in healthy humans and support low grade thrombin generation. *Thromb. Haemost.* **85,** 639–646.
23. Shet, A. S., Aras, O., Gupta, K., Hass, M. J., Rausch, D. J., Saba, N., Koopmeiners, L., Key, N. S., and Hebbel, R. P. (2003) Sickle blood contains tissue factor-positive microparticles derived from endothelial cells and monocytes. *Blood* **102,** 2678–2683.
24. Jimenez, J. J., Jy, W., Mauro, L. M., Horstman, L. L., and Ahn, Y. S. (2001) Elevated endothelial microparticles in thrombotic thrombocytopenic purpura: findings from brain and renal microvascular cell culture and patients with active disease. *Br. J. Haematol.* **112,** 81–90.
25. Jimenez, J. J., Jy, W., Mauro, L. M., Horstman, L. L., Soderland, C., and Ahn, Y. S. (2003) Endothelial microparticles released in thrombotic thrombocytopenic purpura express von Willebrand factor and markers of endothelial activation. *Br. J. Haematol.* **123,** 896–902.
26. Falati, S., Liu, Q., Gross, P., Merrill-Skoloff, G., Chou, J., Vandendries, E., Celi, A., Croce, K., Furie, B. C., and Furie, B. (2003) Accumulation of tissue factor into developing thrombi in vivo is dependent upon microparticle P-selectin glycoprotein ligand 1 and platelet P-selectin. *J. Exp. Med.* **197,** 1585–1598.
27. Taylor, D. D. and Gercel-Taylor, C. (2005) Tumour-derived exosomes and their role in cancer-associated T-cell signalling defects. *Br. J. Cancer* **92,** 305–311.
28. Ogura, H., Tanaka, H., Koh, T., Fujita, K., Fujimi, S., Nakamori, Y., Hosotsubo, H., Kuwagata, Y., Shimazu, T., and Sugimoto, H. (2004) Enhanced production of endothelial microparticles with increased binding to leukocytes in patients with severe systemic inflammatory response syndrome. *J. Trauma* **56,** 823–830; discussion 830–831.
29. Sabatier, F., Roux, V., Anfosso, F., Camoin, L., Sampol, J., and Dignat-George, F. (2002) Interaction of endothelial microparticles with monocytic cells in vitro induces tissue factor-dependent procoagulant activity. *Blood.* **99,** 3962–70.

30. Brogan, P. A. and Dillon, M. J. (2004) Endothelial microparticles and the diagnosis of the vasculitides. *Intern. Med.* **43,** 1115–1119.
31. Diamant, M., Nieuwland, R., Pablo, R. F., Sturk, A., Smit, J. W., and Radder, J. K. (2002) Elevated numbers of tissue-factor exposing microparticles correlate with components of the metabolic syndrome in uncomplicated type 2 diabetes mellitus. *Circulation.* **106,** 2442–2447.
32. Chou, J., Mackman, N., Merrill-Skoloff, G., Pedersen, B., Furie, B. C., and Furie, B. (2004) Hematopoietic cell-derived microparticle tissue factor contributes to fibrin formation during thrombus propagation. *Blood* **104,** 3190–3197.
33. Aras, O., Shet, A., Bach, R. R., Hysjulien, J. L., Slungaard, A., Hebbel, R. P., Escolar, G., Jilma, B., and Key, N. S. (2004) Induction of microparticle- and cell-associated intravascular tissue factor in human endotoxemia. *Blood* **103,** 4545–4553.
34. Abid Hussein, M. N., Meesters, E. W., Osmanovic, N., Romijn, F. P., Nieuwland, R., and Sturk, A. (2003) Antigenic characterization of endothelial cell-derived microparticles and their detection ex vivo. *J. Thromb. Haemost.* **1,** 2434–2443.
35. Bernal-Mizrachi, L., Jy, W., Jimenez, J. J., Pastor, J., Mauro, L. M., Horstman, L. L., de Marchena, E., and Ahn, Y. S. (2003) High levels of circulating endothelial microparticles in patients with acute coronary syndromes. *Am. Heart J.* **145,** 962–970.
36. Minagar, A., Jy, W., Jimenez, J. J., Sheremata, W. A., Mauro, L. M., Mao, W. W., Horstman, L. L., and Ahn, Y. S. (2001) Elevated plasma endothelial microparticles in multiple sclerosis. *Neurology* **56,** 1319–1324.
37. Preston, R. A., Jy, W., Jimenez, J. J., Mauro, L. M., Horstman, L. L., Valle, M., Aime, G., and Ahn, Y. S. (2003) Effects of severe hypertension on endothelial and platelet microparticles. *Hypertension* **41,** 211–217.
38. Perez-Pujol, S., Marker, P. H., and Key, N. S. (2007) Platelet microparticles are heterogeneous and highly dependent on the activation mechanism: studies using a new digital flow cytometer. *Cytometry Part A* **71A**, 38–45.

III

Protein Expression Profiling

7

Exosomes

Joost P. J. J. Hegmans, Peter J. Gerber, and Bart N. Lambrecht

Summary

Exosomes are small natural membrane vesicles released by a wide variety of cell types into the extracellular compartment by exocytosis. The biological functions of exosomes are only slowly unveiled, but it is clear that they serve to remove unnecessary cellular proteins (e.g., during reticulocyte maturation) and act as intercellular messengers because they fuse easily with the membranes of neighboring cells, delivering membrane and cytoplasmic proteins from one cell to another. Recent findings suggests that cell-derived vesicles (exosomes are also named membranous vesicles or microvesicles) could also induce immune tolerance, suppression of natural killer cell function, T cell apoptosis, or metastasis. For example, by secreting exosomes, tumors may be able to accomplish the loss of those antigens that may be immunogenic and capable of signaling to immune cells as well as inducing dysfunction or death of immune effector cells. On the other hand, dendritic cell-derived exosomes have the potential to be an attractive powerful immunotherapeutic tool combining the antitumor activity of dendritic cells with the advantages of a cell-free vehicle. Although the full understanding of the significance of exosomes requires additional studies, these membrane vesicles could become a new important component in orchestrating responses between cells.

Key Words: Dexosomes; electron microscopy; exosomes; MALDI-TOF; mesothelioma; SDS–PAGE; Western blot.

1. Introduction

Cells communicate with other cells not only through direct cell–cell contact or cytokine production, but also through secretion of exosomes ***(1–16)***. Exosomes are small membrane vesicles (60–150 nm in diameter) of endosomal origin, which are secreted upon fusion of multivesicular bodies with the plasma

From: *Methods in Molecular Biology, vol. 484: Functional Proteomics: Methods and Protocols*
Edited by: J. D. Thompson et al., DOI: 10.1007/978-1-59745-398-1, © Humana Press, Totowa, NJ

membrane *(1,17)*. Exosomes display a discrete set of proteins involved in antigen presentation, such as major histocompatibility complex (MHC)-I and MHC-II *(18)*.

Dendritic cell-derived exosomes (dexosomes) can transfer antigen-loaded MHC class I and II molecules, and other associated molecules, to other dendritic cells (DCs) and T cells, potentially leading to the amplification of immune responses *(13)*. They are able to elicit potent antitumor immune responses in tumor-bearing mice *(19)*. Because of this, exosomes may be a novel source of cell-free therapeutic cancer vaccines *(11–13,20)*. The first two phase I trials evaluated in the clinic consisted of autologous dexosomes (patient-specific exosomes released by DCs and loaded with tumor antigen-derived peptides) as immunotherapeutic regimens for melanoma and non-small-cell lung cancer *(21,22)*. These studies revealed that dexosome immunotherapy is well tolerated and led to the induction of immune responses and disease stabilization for several patients.

Tumor cell types have also been shown to secrete exosomes *(23,24)*. These exosomes are morphologically analogous to exosomes produced by DCs. However, the production of exosomes by tumor cells appears to be lower than that of DCs. The tumor-derived exosomes are capable of transferring MHC-I-peptide complexes to DCs, inducing a $CD8^+$ T cell-dependent cross-immunization in tumor-bearing mice *(24)*. Exosomes are capable of doing so since they display, among others, proteins containing native tumor antigens. Even exosomes derived from poorly immunogenic cancers are therapeutically effective, while the tumor lysate is not capable of inducing antitumor responses *(19)*. More surprisingly, tumor-derived exosomes, from mesothelioma, colon, mammary, and other carcinomas, loaded on DCs triggered T cell-mediated antitumor immune responses leading to a strong intertumor cross-protection *(23)*. This suggests that the exosomes probably contain shared tumor-rejection antigens.

In this chapter we describe the isolation of exosomes from cell lines *in vitro* to gain information on their potential biological functions. Exosomes obtained after high-speed centrifugations are immunolabeled and visualized by electron microscopy (*see* **Fig. 1**). Sodium dodecyl sulfate polyacrylamide gel electrophoresis (SDS–PAGE) separation followed by matrix-assisted laser desorption ionization time-of-flight (MALDI-TOF) mass spectrometry is used to characterize the protein composition of these exosomes. Western blot analysis is performed to confirm the proteins detected by MALDI-TOF. Using these technologies, developmental endothelial locus-1 (DEL-1) was detected in mesothelioma-derived exosomes *(25)*, which can act as a strong angiogenic factor *(26,27)* and may increase the vascular development in the neighborhood of the tumor. Therefore, mesothelioma-derived exosomes may favor the tumor

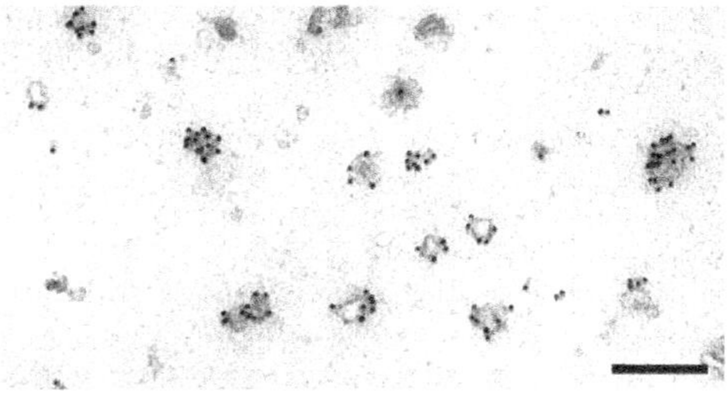

Fig. 1. Electron micrograph of the 100,000 × *g* pellet of tumor cell supernatant, showing cup-shaped membrane vesicles rather homogeneous in size and not exceeding 150 nm in diameter. Exosomes were fixed in 2% paraformaldehyde and immunolabeled for CD63, a tetraspanin on late endosomes characteristic of these vesicles (black dots) (bar, 200 nm).

growth. Earlier we have shown that mouse mesothelioma-derived exosomes can be used as a source of tumor antigens for DCs, which then mediated $CD8^+$ T cell-dependent antitumor effects *(28)*. Our current knowledge of exosomes is still in its infancy and because their protein composition will vary from the origin of the producing cells, further proteomic research may elucidate some of the functions of exosomes *in vivo*.

2. Materials

2.1. Cell Culture

1. Roswell Park Memorial Institute (RPMI) with HEPES and Glutamax (GIBCO) supplemented with 50 µg/mL gentamicin and 10% fetal bovine serum (FBS, Sigma-Aldrich).
2. Solution of trypsin (0.05%) and ethylenediamine tetraacetic acid (EDTA) (0.53 m*M*) in phosphate-buffered saline (PBS) (all from GIBCO).
3. Serum replacer TCH (use 1× working strength [ICN]) (*see* **Note 2**).
4. Protein quantification using the CBQCA kit (Molecular Probes, Leiden, The Netherlands).
5. Fluorescence microplate reader (CytoFluor 4000, PerSeptive Biosystems, Foster City, CA).

2.2. Transmission Electron Microscopy

1. Formvar/carbon-coated nickel grids.
2. Paraformaldehyde: prepare a 2% (w/v) paraformaldehyde solution in PBS fresh for each experiment. The solution may need to be carefully heated (use a stirring hot-plate in the fume hood) to dissolve, and then cool to room temperature for use.
3. 10-nm protein A gold particles (Aurion, Wageningen, The Netherlands).

2.3. One-Dimensional Sodium Dodecyl Sulfate Polyacrylamide Gel Electrophoresis (1D SDS–PAGE)

1. 1.5 *M* Tris–HCl, pH 8.8: 18.15 g Tris base is dissolved in 60 mL water and adjusted to pH 8.8 with 1 *N* HCl. Add to 100 mL with water. Store at room temperature.
2. 0.5 *M* Tris–HCl, pH 6.8: 6 g Tris base is dissolved in 60 mL water and adjusted to pH 6.8 with 1 *N* HCl. Add water to a total volume of 100 mL. Store at room temperature.
3. SDS solution: prepare a 10% (w/v) solution by dissolving 10 g of SDS in 100 mL water. Store at room temperature.
4. For the sample buffer preparation *see* **Subheading 3.3.2**.
5. Water-saturated isobutanol: shake equal volumes of water and isobutanol in a glass bottle and allow to separate. Use the top layer. Store at room temperature
6. Running buffer: 25 m*M* Tris base, 192 m*M* glycine, 0.1% SDS, adjust pH to 8.3.
7. Acrylamide/Bis 30% (37.5:1 mixture, Bio-Rad). Store at 4°C (*see* **Note 3**).
8. Ammonium persulfate (APS): prepare a 10% (w/v) solution in water and immediately freeze in single use aliquots and store at –20°C or prepare fresh.
9. *N,N,N',N'*-Tetramethylethylenediamine (TEMED) (Sigma-Aldrich) (*see* **Note 4**).
10. Coomassie staining solution (Invitrogen).
11. Destaining solution I: 10% (v/v) methanol, 5% (v/v) acetic acid in water.

2.4. Matrix-Assisted Laser Desorption Ionization-Time-of Flight Analysis (MALDI-TOF)

1. Destaining solution II: 0.125 g ammonium hydrogen carbonate is dissolved in 22 mL water and 9.4 mL acetonitrile (CH_3CN). The solution is stored for a maximum of 1 week at room temperature in a Teflon bottle (*see* **Note 5**).
2. Trypsin work solution: 100 μg trypsin (Promega Benelux) is dissolved in 1 mL filtered (0.45 μm filter) water and 60 μL of filtered 50 m*M* Tris–HCl, pH 8.8. Aliquot the solution in 50 μL and store at –20°C.
3. Matrix solution: dissolve 2 mg of α-cyano-4-hydroxycinnamic acid (ACCA, Bruker Daltonics, Billerica, MA) in 1 mL acetonitrile. Sonicate for 30 min. The solution is stored in a brown, light-sealed centrifuge tube (ACCA is light sensitive). Matrix solution can be used for about 1 week. Tip: matrix solution prepared 2–3 days in advance works better than freshly made.

2.5. Western Blot Analysis

1. Blotbuffer: dissolve 3.03 g Tris base and 14.4 g glycine in 500 mL water, add 200 mL methanol (Sigma-Aldrich), and adjust the volume to 1 L with water. Do not add acid or base to adjust the pH. Prechill at 4°C before use.
2. Immobilon P membrane (polyvinylidene fluoride [PVDF]) (Millipore, 45 μm).
3. Ponceau-S red (Sigma-Aldrich).

4. TBS (Tris-buffered saline): dissolve 8.8 g NaCl and 20 mL of 0.5 *M* Tris–HCl, pH 8.0, in 800 mL water. Adjust to pH 8.0 and bring the final volume to 1 L.
5. TBS-T (Tris-buffered saline with Tween-20): 0.05% Tween-20 in TBS.
6. Low fat milk powder (Campina, ELK).
7. Antibodies and secondary horseradish peroxidase (HRP) conjugate.
8. Enhanced chemiluminescent (ECL) reagents (Pierce, SuperSignal, West Pico).
9. Chemiluminescense film: Bio-Max ML film (Kodak, Rochester, NY).

3. Methods

3.1. Isolation of Exosomes

1. Adherent cell lines are cultured in RPMI/10% fetal bovine serum (FBS) and passaged when approaching confluence with trypsin/EDTA to provide new maintenance cultures in T75-cm^2 culture flasks (*see* **Note 6**).
2. When a flask reaches 80% confluency, cells are washed twice with PBS to remove traces of FBS.
3. Cells are incubated in 12 mL of RPMI medium (containing HEPES, Glutamax, and gentamicin) supplemented with serum replacer TCH (1 × working strength) for 48 h at 37°C in a humidified atmosphere of 5% CO_2, 95% air.
4. Cell culture supernatants are subjected to three successive centrifugations to remove cells and debris: 300 × *g* for 10 min, 2000 × *g* for 20 min, and finally at 10,000 × *g* for 30 min, all at 4°C.
5. Exosomes are then pelleted at 64,000 × *g* for 100 min using an SW28 rotor (Beckman Coulter Instruments).
6. Pellets are resuspended in PBS and centrifuged at 100,000 × *g* for 1 h (SW60 rotor).
7. Exosomes are resuspended in PBS. The quantification of recovered exosomal proteins is performed using the ATTO-TAG CBQCA kit according to the manufacturer's recommendations. This kit works well even in the presence of lipids and detergents. The fluorescence emission is measured at ~550 nm (filter 530 ± 30 nm) with excitation at ~465 nm (filter 485 ± 20 nm) in a fluorescence microplate reader (gain 40).
8. Exosomes are aliquoted and stored at –80°C.

3.2. Transmission Electron Microscopy

1. Exosomes obtained after centrifugation of cell culture supernatants are adsorbed onto Formvar/carbon-coated nickel grids for 15 min.
2. Adsorbed exosomes are fixed with 2% paraformaldehyde in PBS.
3. Grids are rinsed three times in PBS for 5 min each and then blocked in 1% BSA in PBS for 15 min.
4. The grids are floated upside down on top of drops of diluted antibody overnight at 4°C (e.g., CLB-gran1/2, 435 [anti-CD63] CLB, Amsterdam, The Netherlands). Incubation times and dilutions should be determined for each particular primary

antibody being used. During the immunolabeling process, be careful not to let the grids dry out.

5. Wash twice by floating on drops of PBS.
6. Visualization is performed by floating on drops of diluted 10 nm colloidal gold coupled to staphylococcal protein A (protein A-gold) particles for 2 h at room temperature or overnight at 4°C (this size of the gold does not require enhancement).
7. After rinses in PBS followed by distilled water, grids are stained for contrast with aqueous uranyl acetate for 10 min on ice. Grids are allowed to dry and are examined with a Philips CM 100 electron microscope at 80 kV (Philips Industries, Eindhoven, The Netherlands).

3.3. Sample Preparation and 1D SDS–PAGE

1. The following procedure presumes the use of the Bio-Rad electrophoresis system PROTEAN II xi Cell and is performed according to the manufacturer's recommendations (Bio-Rad), the Bio-Rad powerpac 3000 as power supply, as well as the carefully cleaning and assembling of its parts (*see* **Note 7**). All steps of protein sample preparation should proceed fast and on ice, unless otherwise specified.
2. Sample preparation is performed as follows: exosome preparations are diluted into 8 *M* urea (Sigma-Aldrich), 2% CHAPS (Amersham Pharmacia Biotech), 20 m*M* dithiothreitol (DTT, Sigma-Aldrich), 0.01% bromophenol blue (Sigma-Aldrich) to obtain 50 μg per lane of a gel (*see* **Note 8**). The protein sample should be diluted at least 1:4 with this sample buffer. Before loading, the sample is heated for 5 min at 95°C to denaturate the proteins, and then immediately placed on ice.
3. The separation gel solution is prepared as follows: 20 mL of distilled water, 12.5 mL of 1.5 *M* Tris–HCl (pH 8.8), 0.5 mL of 10% SDS, and 16.75 mL of acrylamide/bis (30%) are mixed together. The amounts of reagents indicated are sufficient for the preparation of two 16 × 16 cm gels, 1.0 mm thick. Degas under vacuum for approximately 10–20 min until air bubbles are no longer released. Then 250 μL of 10% APS is added together with 25 μL of TEMED to the solution just before use. The solution is then carefully poured between the assembled glass plates, avoiding the inclusion of air bubbles. Leave sufficient space at the top (at least 1 cm) for the stacking gel to be added later.
4. Gently overlay the gel mix with water-saturated isobutanol, and allow the gel to polymerize for at least 30 min.
5. After polymerization, remove the isobutanol and rinse the surface of the separating gel with water.
6. The solution for 4% stacking gel is prepared as follows: 6.1 mL of water, 2.5 mL of 0.5 *M* Tris–HCl (pH 6.8), 0.1 mL of 10% SDS, and 1.3 mL of acrylamide/bis (30%) are mixed together; 250 μL of 10% APS together with 25 μL of TEMED are added to this solution just before use. The solution is then carefully layered on top of the separating gel between the glass plates. Insert the comb immediately after filling the remaining space with the stacking gel solution. Avoiding the

inclusion of air bubbles is crucial. Polymerization should be completed within 30 min. Avoid drying of the stacking gel after removing the comb. Mark the position of the slots with a permanent marker on one glass plate before removing the comb to make the loading easier.

7. The gel sandwich is assembled with the upper buffer chamber with the cooling core and place it into the lower buffer chamber. The cooling core is connected to the cooling system. Running buffer is placed into the inner chamber. The remaining buffer is diluted 1:1 with water and placed in the lower buffer chamber. The sample(s) and a protein weight marker are loaded into the slots of the stacking gel using a thin and extra long pipette tip.
8. The gel is run at constant current conditions of 7 mA per gel at 10°C.
9. After 15–18 h when the blue front marking reaches the end of the separation space, gels are stained with a general staining protocol, e.g., Coomassie blue staining kit according to the manufacturer's instructions (*see* **Note 9**). For this, gels to be stained are placed into the staining solution immediately after electrophoresis. Allow the gels to stain at room temperature with gentle agitation for at least 30 min, but no longer than 3 h.
10. After staining, pour off the staining solution
11. Add destaining solution I, and agitate gently at room temperature for 20 min.
12. Repeat step 10 until the background is clear (normally two or three times).

3.4. Enzymatic Digestion of Protein Spots

1. The colloidal blue-stained protein spots of interest are manually excised with a scalpel or plastic plunger (*see* **Note 10**).
2. Each gel plug is then transferred into a well of a 96-well low protein binding microtiter plate (Nunc).
3. Gel plugs are washed with 100 μL of water for 5 min with shaking at 650 rpm.
4. Gel plugs are destained using destaining solution II for 20 min at room temperature with shaking.
5. Repeat step 4.
6. Gel plugs are washed with water.
7. Plugs are lyophilized in a rotary evaporator (Savant, Farmingdale, NY) for 30 min. Do not use heat (*see* **Note 11**).
8. Protein digestion is performed by the addition of 4 μL of 100 μg/mL sequencing grade-modified trypsin (Promega, Madison, WI) to each well.
9. The plate is sealed with an adhesive aluminum foil.
10. Incubate overnight at room temperature (20–25°C) or for 3 h at 37°C.

3.5. MALDI-TOF

1. After the specific hydrolysis at the carboxylic sides of lysine and arginine residues by trypsin, 7 μL of 1 part acetonitrile:0.1% and 2 parts trifluoroacetic acid is added to the gel plugs.
2. 1 μL of the tryptic digest is taken and mixed with 2.5 μL of matrix solution.

3. 0.5 μL of this tryptic digest-matrix solution is pipetted onto a 400-μm 384-well anchor chip MALDI-TOF plate and air-dried for 5 min. We acquire peptide mass spectra on a Biflex III MALDI-TOF mass spectrometer equipped with a 337-nm nitrogen laser (Bruker Daltonics, Bremen, Germany). The instrument is calibrated with a peptide calibration standard in the mass range of 500–3500 kDa (Bruker Daltonics). Spectra are compared using autolytic fragments from trypsin. A mass list is obtained from the spectra and submitted to Matrix Science Mascot UK software to identify the proteins in the MSDB database of the NCBI.
4. The criteria for identification of proteins are determined as follows: top scores are given by software higher than 61 ($p < 0.05$; however, this is dependent on the size of the database used), a maximum allowed peptide mass error of 200 ppm, and at least five matching peptide masses; the molecular weight of the identified protein should match estimated values by comparing with protein weight markers.

3.6. Western Blot

1. To confirm exosome-derived proteins detected by MALDI-TOF, proteins can be transferred onto a membrane and detected by specific antibodies (*see* **Note 12**).
2. The following procedure presumes the use of the Trans-Blot Electrophoretic Transfer Cell system (Bio-Rad). The stacking gel from the 1D SDS-PAGE (*see* **Subheading 3.3**) is removed and one corner of the separation gel is cut to mark the gel's orientation and to distinguish it from other gels.
3. The Immobilon-P PVDF membranes are cut to a size similar to the 1-D gels of interest and prewetted in 100% methanol for 15–30 s (*see* **Note 13**). Then they are rinsed in water for 2 min.
4. The membranes are equilibrated in blotbuffer for at least 5 min. The "gel side" of the membrane is marked with a pencil.
5. The blot sandwich is prepared as follows: black plastic support (cathode), with blotbuffer soaked sponge, thick filter paper soaked with blotbuffer, gel, preequilibrated Immobilon-P PVDF membrane, thick filter paper soaked with blotbuffer, with a blotbuffer-soaked sponge, and the white plastic support (anode). Remove air bubbles between gel and PVDF membrane before closing and position the sandwich correctly into the transfer system.
6. The transfer tank is filled with prechilled blotbuffer, with an additional cooling container (–20°C). Cool to 4°C with the super cooling coil and a refrigerated recirculator.
7. Transfer is performed for 1–2 h at 100 V (0.36 A) or overnight at 10 V (0.1 A).
8. After termination of the transfer, the sandwich is carefully disassembled and the membrane taken out (*see* **Note 14**). The proteins on the membrane are visualized by 0.5% Ponceau-S red and 1% acetic acid for 1 min followed by a rinse with distilled water.
9. The membrane is cut between the visualized lanes. To saturate nonspecific protein binding sites, incubate the membrane in TBS-T/5% (w/v) milk powder for 1 h at room temperature or at 4°C overnight (*see* **Note 15**).

10. Replace the blocking solution with TBS-T containing the appropriate dilution of primary antibody and incubate for at least 1 h at room temperature with gentle agitation. Small membranes can be put into a 15-mL tube to save antibodies (*see* **Note 16**).
11. To remove unbound antibody, wash the membrane three times in TBS-T for 5–10 min each.
12. Transfer the membrane to TBS-T containing the appropriate dilution of anti-IgG HRP conjugate and incubate the membrane for 60 min at room temperature.
13. Wash the membranes three times for 5–10 min each to remove unbound secondary antibody.
14. Rinse briefly in two changes of TBS or water to remove Tween-20 from the membrane surface (*see* **Note 17**)
15. The ECL reagents (substrate) are prepared according to the manufacturer's instructions. The membrane (proteins upward) is placed onto a piece of parafilm (or plastic wrap) and then just covered with substrate, carefully avoiding air bubbles. Incubate for 5 min at room temperature.
16. Carefully tilt the parafilm to decant the excess substrate.
17. Place the membranes on the overhead foil into an X-ray film cassette. The exposure time has to be checked for each individual case, but is usually from seconds to a few minutes.

4. Notes

1. Unless otherwise stated, all solutions should be prepared in water that has a resistency of less than 18.2 MΩ-cm and a total organic content of less than five parts per billion. This standard is referred to as "water" in this text.
2. Fetal bovine serum may also contain exosomes. We describe the use of a serum replacer, but FBS may also be used after depletion of the exosomes using high-speed centrifugation.
3. Acrylamide powder could also be used, but is not recommended because of health hazards during weighing. Acrylamide and bis are toxic in the monomer form. Avoid skin contact and dispose of the remains ecologically. Polymerize the remains with an excess of ammonium persulfate.
4. TEMED is best stored at room temperature in a desiccator. Buy small bottles as it may decline in quality after opening.
5. In general, solutions for mass spectrometry are stored in Teflon tubes to prevent contamination coming from glass or plastics.
6. Exosomes can also be isolated from body fluids but you may encounter difficulties during the purification process and the source of these exosomes is unknown.
7. Meticulously cleaned instruments and glass plates as well as high-quality reagents are of great importance. Mass spectrometry analysis is very sensitive, and every contamination will show up in the mass spectrum and might lead to

wrong results. Keratins derived from hair and skin are the most frequently found protein contamination.

8. Sample preparation is optimized for whole cell lysates. To dissolve specific proteins, i.e., membrane proteins, the sample preparation needs to be adapted.
9. Many of the proteomics laboratories use Coomassie blue staining because of its compatibility with mass spectrometry analysis. A spot, which is visible with this staining, contains enough protein for identification and characterization with mass spectrometry. Silver staining can detect more proteins (higher sensitivity) but often interferes with mass spectrometry measurements.
10. Stained gels can be stored in a refrigerator when sealed in plastic zip lock bags between transparencies for up to 1 year. Use as little fluids as possible to prevent yeast growth on the gels. Do not use glycerol because this will interfere with mass spectrometry.
11. Using radiant cover when lyophilizing the gel plugs might spoil your experiment. Never use heating.
12. In general, proteins immobilized on membranes are detected with antibodies in a three step process. First, the primary antibody, an immunoglobulin G (IgG) directed against the protein in question, is added to bind potential antigenic sites. In the second step, a secondary antibody-enzyme conjugate that recognizes general features of all IgGs is added to find locations where the primary antibody is bound. The enzyme HRP conjugated to the secondary antibody catalyzes a reaction in the third step, when the appropriate substrate is added, and provides a visual indication of potential primary antibody recognition.
13. Always wear gloves when handling membranes to avoid localized background problems.
14. If PVDF membranes are dried after electrophoretic blotting, they must be rewet in methanol followed by TBS before continuing with the blocking step. However, if PVDF membranes dry, there will be some loss of sensitivity.
15. Do not allow the membranes to dry out during any of the subsequent steps. Perform all the washing and incubation steps at room temperature with gentle shaking.
16. For antibody incubations and the substrate reaction, use enough solution to submerge the membrane, protein side up. Usually, this volume is about 0.1–0.15 mL/cm^2 of membrane surface. Use at least twice this volume for blocking and washing steps.
17. Residual Tween-20 can affect depositing of the precipitated substrate and lead to smearing of bands.

Acknowledgments

We thank our colleague Margaretha Lambers for fruitful discussions and for critically reading the manuscript.

References

1. Denzer, K., Kleijmeer, M. J., Heijnen, H. F., Stoorvogel, W., and Geuze, H. J. (2000) Exosome: from internal vesicle of the multivesicular body to intercellular signaling device. *J. Cell Sci.* **113**, 3365–3374.
2. Johnstone, R. M., Adam, M., Hammond, J. R., Orr, L., and Turbide, C. (1987) Vesicle formation during reticulocyte maturation. Association of plasma membrane activities with released vesicles (exosomes). *J. Biol. Chem.* **262**, 9412–9420.
3. Ratajczak, J., Wysoczynski, M., Hayek, F., Janowska-Wieczorek, A., and Ratajczak, M. Z. (2006) Membrane-derived microvesicles: important and underappreciated mediators of cell-to-cell communication. *Leukemia* **20**, 1487–1495.
4. Peche, H., Heslan, M., Usal, C., Amigorena, S., and Cuturi, M. C. (2003) Presentation of donor major histocompatibility complex antigens by bone marrow dendritic cell-derived exosomes modulates allograft rejection. *Transplantation* **76**, 1503–1510.
5. Taylor, D. D. and Gercel-Taylor, C. (2005) Tumour-derived exosomes and their role in cancer-associated T-cell signalling defects. *Br. J. Cancer* **92**, 305–311.
6. Liu, C., Yu, S., Zinn, K., Wang, J., Zhang, L., Jia, Y., Kappes, J.C., Barnes, S., Kimberly, R. P., Grizzle, W. E., and Zhang, H. G. (2006) Murine mammary carcinoma exosomes promote tumor growth by suppression of NK cell function. *J. Immunol.* **176**, 1375–1385.
7. Frangsmyr, L., Baranov, V., Nagaeva, O., Stendahl, U., Kjellberg, L., and Mincheva-Nilsson, L. (2005) Cytoplasmic microvesicular form of Fas ligand in human early placenta: switching the tissue immune privilege hypothesis from cellular to vesicular level. *Mol. Hum. Reprod.* **11**, 35–41.
8. Janowska-Wieczorek, A., Marquez-Curtis, L. A., Wysoczynski, M., and Ratajczak, M. Z. (2006) Enhancing effect of platelet-derived microvesicles on the invasive potential of breast cancer cells. *Transfusion* **46**, 1199–1209.
9. Janowska-Wieczorek, A., Wysoczynski, M., Kijowski, J., Marquez-Curtis, L., Machalinski, B., Ratajczak, J., and Ratajczak, M. Z. (2005) Microvesicles derived from activated platelets induce metastasis and angiogenesis in lung cancer. *Int. J. Cancer* **113**, 752–760.
10. Whiteside, T. L. (2005) Tumour-derived exosomes or microvesicles: another mechanism of tumour escape from the host immune system? *Br. J. Cancer* **92**, 209–211.
11. Delcayre, A. and Le Pecq, J. B. (2006) Exosomes as novel therapeutic nanodevices. *Curr. Opin. Mol. Ther.* **8**, 31–38.
12. Delcayre, A., Estelles, A., Sperinde, J., Roulon, T., Paz, P., Aguilar, B., Villanueva, J., Khine, S., and Le Pecq, J. B. (2005) Exosome display technology: applications to the development of new diagnostics and therapeutics. *Blood Cells Mol. Dis.* **35**, 158–168.
13. Delcayre, A., Shu, H., and Le Pecq, J. B. (2005) Dendritic cell-derived exosomes in cancer immunotherapy: exploiting nature's antigen delivery pathway. *Expert Rev. Anticancer Ther.* **5**, 537–547.

14. Thery, C., Zitvogel, L., and Amigorena, S. (2002) Exosomes: composition, biogenesis and function. *Nat. Rev. Immunol.* **2**, 569–579.
15. van Niel, G. and Heyman, M. (2002) The epithelial cell cytoskeleton and intracellular trafficking. II. Intestinal epithelial cell exosomes: perspectives on their structure and function. *Am. J. Physiol. Gastrointest. Liver Physiol.* **283**, G251–255.
16. Wubbolts, R. W., Leckie, R. S., Veenhuizen, P. T., Schwartzmann, G., Moebius, W., Hoernschemeyer, J., Slot, J. W., Geuze, H. J., and Stoorvogel, W. (2003) Proteomic and biochemical analyses of human B cell-derived exosomes: potential implications for their function and multivesicular body formation. *J. Biol. Chem.* **7,** 7.
17. Denzer, K., van Eijk, M., Kleijmeer, M. J., Jakobson, E., de Groot, C., and Geuze, H. J. (2000) Follicular dendritic cells carry MHC class II-expressing microvesicles at their surface. *J. Immunol.* **165**, 1259–1265.
18. Thery, C., Regnault, A., Garin, J., Wolfers, J., Zitvogel, L., Ricciardi-Castagnoli, P., Raposo, G., and Amigorena, S. (1999) Molecular characterization of dendritic cell-derived exosomes. Selective accumulation of the heat shock protein hsc73. *J. Cell. Bio.l* **147**, 599–610.
19. Zitvogel, L., Regnault, A., Lozier, A., Wolfers, J., Flament, C., Tenza, D., Ricciardi-Castagnoli, P., Raposo, G., and Amigorena, S. (1998) Eradication of established murine tumors using a novel cell-free vaccine: dendritic cell-derived exosomes. *Nat. Med.* **4**, 594–600.
20. Mignot, G., Roux, S., Thery, C., Segura, E., and Zitvogel, L. (2006) Prospects for exosomes in immunotherapy of cancer. *J. Cell. Mol. Med.* **10**, 376–388.
21. Escudier, B., Dorval, T., Chaput, N., Andre, F., Caby, M. P., Novault, S., Flament, C., Leboulaire, C., Borg, C., Amigorena, S., Boccaccio, C., Bonnerot, C., Dhellin, O., Movassagh, M., Piperno, S., Robert, C., Serra, V., Valente, N., Le Pecq, J.B., Spatz, A., Lantz, O., Tursz, T., Angevin, E., and Zitvogel, L. (2005) Vaccination of metastatic melanoma patients with autologous dendritic cell (DC) derived-exosomes: results of the first phase I clinical trial. *J. Transl. Med.* **3**, 10.
22. Morse, M. A., Garst, J., Osada, T., Khan, S., Hobeika, A., Clay, T. M., Valente, N., Shreeniwas, R., Sutton, M. A., Delcayre, A., Hsu, D. H., Le Pecq, J. B., and Lyerly, H. K. (2005) A phase I study of dexosome immunotherapy in patients with advanced non-small cell lung cancer. *J. Transl. Med.* **3**, 9.
23. Andre, F., Schartz, N. E., Movassagh, M., Flament, C., Pautier, P., Morice, P., Pomel, C., Lhomme, C., Escudier, B., Le Chevalier, T., Tursz, T., Amigorena, S., Raposo, G., Angevin, E., and Zitvogel, L. (2002) Malignant effusions and immunogenic tumour-derived exosomes. *Lancet* **360**, 295–305.
24. Wolfers, J., Lozier, A., Raposo, G., Regnault, A., Thery, C., Masurier, C., Flament, C., Pouzieux, S., Faure, F., Tursz, T., Angevin, E., Amigorena, S., and Zitvogel, L. (2001) Tumor-derived exosomes are a source of shared tumor rejection antigens for CTL cross-priming. *Nat. Med.* **7**, 297–303.
25. Hegmans, J. P., Bard, M. P., Hemmes, A., Luider, T. M., Kleijmeer, M. J., Prins, J. B., Zitvogel, L., Burgers, S. A., Hoogsteden, H. C., and Lambrecht, B. N. (2004) Proteomic analysis of exosomes secreted by human mesothelioma cells. *Am. J. Pathol.* **164**, 1807–1815.

26. Rajagopalan, S., Olin, J.W., Young, S., Erikson, M., Grossman, P. M., Mendelsohn, F. O., Regensteiner, J. G., Hiatt, W. R., and Annex, B. H. (2004) Design of the Del-1 for therapeutic angiogenesis trial (DELTA-1), a phase II multicenter, double-blind, placebo-controlled trial of VLTS-589 in subjects with intermittent claudication secondary to peripheral arterial disease. *Hum. Gene Ther.* **15**, 619–624.
27. Zhong, J., Eliceiri, B., Stupack, D., Penta, K., Sakamoto, G., Quertermous, T., Coleman, M., Boudreau, N., and Varner, J. A. (2003) Neovascularization of ischemic tissues by gene delivery of the extracellular matrix protein Del-1. *J. Clin. Invest.* **112**, 30–41.
28. Hegmans, J. P., Hemmes, A., Aerts, J. G., Hoogsteden, H. C., and Lambrecht, B. N. (2005) Immunotherapy of murine malignant mesothelioma using tumor lysate-pulsed dendritic cells. *Am. J. Respir. Crit. Care Med.* **171**, 1168–1177.

8

Toward a Full Characterization of the Human 20S Proteasome Subunits and Their Isoforms by a Combination of Proteomic Approaches

Sandrine Uttenweiler-Joseph, Stéphane Claverol, Loïk Sylvius, Marie-Pierre Bousquet-Dubouch, Odile Burlet-Schiltz, and Bernard Monsarrat

Summary

The 20S proteasome is a multicatalytic protein complex, present in all eukaryotic cells, that plays a major role in intracellular protein degradation. In mammalian cells, this symmetrical cylindrical complex is composed of two copies each of seven different α and β subunits arranged into four stacked rings ($\alpha_7\beta_7\beta_7\alpha_7$). Separation by two-dimensional (2D) gel electrophoresis of the human erythrocytes 20S proteasome subunits and mass spectrometry (MS) identification of all the observed spots reveal the presence of multiple isoforms for most of the subunits. These isoforms could correspond to protein variants and/or posttranslational modifications that may influence the 20S proteasome proteolytic activity. Their characterization is therefore important to establish the rules governing structure/activity relationships of the human 20S proteasome. This chapter describes the use of a combination of proteomic approaches to characterize the human 20S proteasome subunit isoforms separated by 2D gel electrophoresis. A "top-down" strategy was developed to determine by electrospray MS the molecular mass of the intact protein after its passive elution from the gel. Comparison of the experimental molecular mass to the theoretical one can reveal the presence of possible modifications. "Bottom-up" proteomic approaches are then performed and, after protein digestion, tandem MS analyses of the modified peptides allow the characterization and location of the modification. These methods are discussed for the study of the human erythrocytes 20S proteasome subunit isoforms.

Key Words: 2D gel electrophoresis; protein gel elution; top-down and bottom-up proteomic approaches; mass spectrometry; protein modifications; catalytic protein complex.

From: *Methods in Molecular Biology, vol. 484: Functional Proteomics: Methods and Protocols*
Edited by: J. D. Thompson et al., DOI: 10.1007/978-1-59745-398-1, © Humana Press, Totowa, NJ

1. Introduction

The 20S proteasome constitutes the catalytic core complex of the 26S proteasome, which is the primary machinery responsible for intracellular protein degradation. In eukaryotes, the 20S proteasome is composed of four stacked heptameric rings, each of which is organized from seven homologous but nonidentical subunits *(1)*. The two outer rings contain seven α subunits (α_1–α_7) that direct the assembly of the complex and form the gate through which substrates enter and products are released. The two inner rings are made up of seven β subunits (β_1–β_7) including three, β_1, β_2, and β_5, that are catalytically active *(2)*. In higher eukaryotes, the composition of the 20S proteasome complex is variable. For example, interferon-γ induces the replacement of the three standard catalytic subunits by three other ones, β_{1i}, β_{2i}, and β_{5i}, leading to the immunoproteasome. Standard 20S proteasome and immunoproteasome vary in their protein degradation efficiency, resulting in the production of different sets of peptides. This may, for example, dramatically affect the generation of MHC class I antigenic peptides *(3–6)*. In the literature evidence has accumulated that the 20S proteasome subunit composition is heterogeneous and that the distribution of the different proteasome forms varies in different cells and tissues *(7–9)*.

The proteasome has been extensively studied by proteomic approaches *(10)*. In particular, the separation of 20S proteasome subunits by two-dimensional (2D) gel electrophoresis followed by their identification by mass spectrometry (MS) has been used successfully to identify subunits of 20S proteasomes purified from different species *(11–16)*. These analyses reveal, in addition to various proportions of standard catalytic subunits and immunosubunits, the presence of isoforms for most subunits, which suggests an increased heterogeneity of 20S proteasome complexes. The various 20S subunit isoforms reported in these studies mostly differ in p*I*s but some also in molecular weights. These isoforms could correspond to a protein variant (resulting from truncated proteins, alternative splice forms, or single nucleotide polymorphisms) or could indicate the presence of posttranslational modifications (PTMs). For example, phosphorylation and N-terminal acetylation were already reported on several 20S proteasome subunits *(11,15,17–21)*. From a physiological point of view, the effects of these PTMs are not clearly defined. To clarify the individual roles of the 20S proteasome subunits and their isoforms, it is essential to investigate the primary structures of each individual subunit.

This chapter focuses on the characterization by MS-based approaches of human 20S proteasome subunits and their isoforms separated by 2D gel electrophoresis. The first part describes a proteomic "top-down" strategy used to determine by electrospray MS the molecular mass of the entire protein after its passive elution from the gel. This measurement can suggest the presence of modifications on the protein studied. The second part shows how "bottom-up" proteomic approaches can then be performed to allow the characterization and

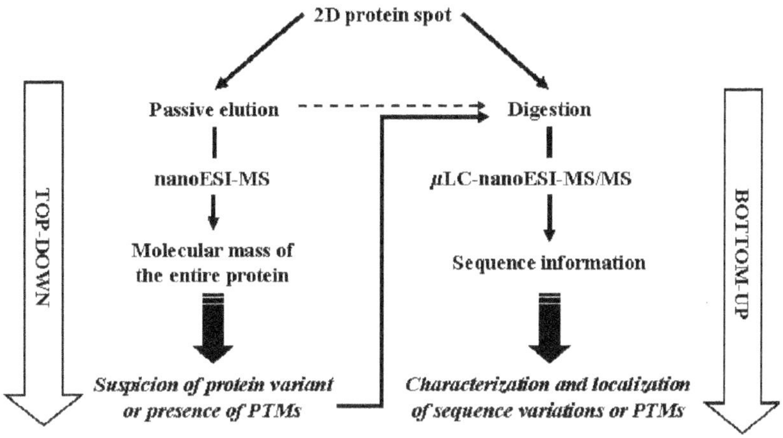

Fig. 1. Combination of "top-down" and "bottom-up" proteomic approaches to characterize the human erythrocytes 20S proteasome subunits and their isoforms.

localization of the putative modifications (**Fig. 1**). The whole procedure is illustrated in a third part by the analysis of various subunits from the standard 20S proteasome purified from human erythrocytes.

2. Materials

2.1. Passive Elution of Proteins from Polyacrylamide Gels

1. Milli-Q water.
2. Elution solution: 0.1 *M* sodium acetate, pH 8.2 (adjusted with 0.1 *M* NaOH), 0.1% sodium dodecyl sulfate (SDS).
3. Thermomixer.

2.2. Nanoscale Hydrophilic Phase Chromatography

1. Zip-Tip$_{HPL}$ (Hydrophilic Interaction Chromatography; Millipore).
2. Rehydrating solution: water/acetonitrile (ACN)/acetic acid, 50/50/0.1 (v/v/v), pH 5 (adjusted with 2 *M* NaOH).
3. Equilibrating solution: water/ACN/acetic acid, 10/90/0.1 (v/v/v), pH 5.5 (adjusted with 0.1 *M* NaOH).
4. Elution solution: water/ACN/formic acid (FA), 49/50/1 (v/v/v).

2.3. Nano-Electrospray Ionization Mass Spectrometry (ESI-MS) of Intact Proteins

1. Nano-ESI spray needle for off-line analysis.
2. GELoader® tip (Eppendorf, reference: 0030 001.222).

3. ESI-MS instrument.
4. Deconvolution software.

2.4. In-Gel Protein Digestion and Peptide Extraction

1. Trypsin, sequencing grade (Sequencing Grade Modified Trypsin, Promega).
2. Trypsin solution: 12.5 ng/μL of trypsin in 12.5 m*M* ammonium bicarbonate prepared from a stock solution at 0.1 μg/μL of trypsin in the Promega resolubilization buffer provided by the supplier.
3. Endoproteinase Glu-C.
4. Endoproteinase Glu-C solution: 20 ng/μL of endoproteinase Glu-C in 25 m*M* ammonium bicarbonate.
5. Acetonitrile.
6. Formic acid.
7. Thermomixer.
8. Speed-vacuum centrifuge.
9. Ultrasonic bath.

2.5. Separation of Proteolytic Peptides by Nano-High-Performance Liquid Chromatography (HPLC)

1. Nano-HPLC system.
2. Micro precolumn cartridge (300 μm i.d. × 5 mm).
3. C_{18} capillary column (75 μm i.d. × 15 cm).
4. HPLC solutions prepared with HPLC grade water, HPLC grade ACN, FA, trifluoroacetic acid (TFA): desalting solution = water/ACN/TFA, 98/2/0.05 (v/v/v); Solution A = water/ACN/FA, 95/5/0.2 (v/v/v); and Solution B = water/ACN/FA, 10/90/0.2 (v/v/v).

2.6. ESI-MS/MS Analysis of Proteolytic Peptides

1. Electrospray needle for on-line coupling.
2. ESI-MS/MS instrument.

2.7. Interpretation of MS/MS Spectra

1. Database search software for MS/MS data.
2. Optional, *de novo* sequencing software.

3. Methods

The methods developed in the following subheadings describe how the combination of "top-down" and "bottom-up" proteomic approaches can be used to characterize 20S proteasome subunits and their isoforms previously separated by 2D gel electrophoresis (**Fig. 1**). The purification and 2D gel separation of human 20S proteasome have already been described elsewhere *(22)*. The whole procedure is then illustrated with the proteomic analysis of human erythrocytes 20S proteasome subunits.

3.1. "Top-Down" Proteomic Approach to Measure Intact Protein Molecular Mass

The "top-down" proteomic strategy chosen to characterize the 20S proteasome subunit isoforms consists of a passive elution of the proteins from 2D gel spots using 0.1% SDS, SDS removal by nanoscale hydrophilic phase chromatography, and ESI-MS analyses of the intact proteins. The protocol used was published *(23)*, but modifications were brought to improve the overall method and to adapt it for direct infusion ESI-MS analysis.

This strategy allows the determination of the molecular mass of 5–10 pmol of a protein up to circa 50 kDa loaded on a gel. Thus, it can be applied to the study of the human 20S proteasome subunits which molecular masses range from 22 to 31 kDa. Separation of the proteasome subunits and their isoforms is performed by 2D gel electrophoresis and the 2D gel obtained is stained with Coomassie blue (*see* **Note 1**).

3.1.1. Passive Elution of Proteins from Polyacrylamide Gels

1. Excise gel spots using a clean scalpel blade (*see* **Note 2**) and wash the gel pieces in 0.5-mL centrifuge tubes containing Milli-Q water (300 μL, 2 h, room temperature, under shaking).
2. Remove water and add 20–50 μL of elution solution (enough to cover the gel pieces). Incubate overnight at 37°C under shaking (*see* **Note 3**).

3.1.2. Nanoscale Hydrophilic Phase Chromatography

Coomassie blue, SDS, and salts have to be removed from the eluted protein sample because these compounds interfere with ESI. This can be achieved using hydrophilic (HPL) chromatography. The amount of HPL stationary phase in a Zip-Tip$_{HPL}$ is well suited for the low quantity of sample available in this case. This step also allows the concentration of the protein sample and its elution in the ESI solvent; it is critical because the sensitivity of ESI ionization is concentration dependent.

1. Add 200 μL of equilibrating solution to the tube containing the supernatant of the passive elution and the gel piece. Vortex.
2. Connect a ZipTip$_{HPL}$ to a 20 μL delivering pipette and wash the stationary phase with rehydrating solution (20 μL, three times). Equilibrate it with the equilibrating

solution (20 μL, three times). Load the passive elution sample onto the $ZipTip_{HPL}$ stationary phase in several steps of 20 μL admission/delivery cycles: take 20 μL of samples (tube 1) and dispense it in another tube (tube 2). At the end, repeat the admission/delivery cycles from tube 2 to tube 1. Then, wash the $ZipTip_{HPL}$ stationary phase 10 times with 20 μL of equilibrating solution to eliminate all the undesired compounds. Finally, elute with 3–4 μL of elution solution by applying several admission/delivery cycles to maximize the elution efficiency (*see* **Note 4**).

3.1.3. Nano-ESI-MS of Intact Proteins

Nano-ESI-MS analysis allows a fast measurement of the molecular mass (MM) of a single protein from a low amount of starting material. To obtain high mass accuracy, high-resolution mass spectrometers are required. For this reason, Fourier transform ion cyclotron resonance (FT-ICR) instruments would be the most adapted, but their operation and maintenance are more challenging compared to other mass spectrometers, which limits their application. Hybrid quadrupole time-of-flight (QqTOF) mass spectrometers offer interesting alternative approaches as they are easier to use and their resolution is high enough to obtain a mass accuracy of 1 or 2 Da for an MM around 20–30 kDa. The raw data spectrum obtained corresponds to a series of multicharged ions and a deconvolution process gives access to the MM of the protein analyzed.

1. Set the ESI mass spectrometer in positive ion mode.
2. Calibrate the instrument just before protein measurement.
3. Introduce the 3–4 μL protein sample eluted from the $ZipTip_{HPL}$ into the nano-ESI spray needle using a GELoader® tip. Ensure that no bubbles are trapped at the spraying extremity of the needle.
4. Connect the needle to the nano-ESI source.
5. Adjust all necessary parameters including the position and the voltage applied on the nano-ESI needle to obtain a spray and the highest intensity MS signal. Acquire mass spectra from *m/z* 400 to *m/z* 2000.
6. Perform the deconvolution step using the dedicated software to calculate the experimental MM. Interpret the deconvoluted spectrum (*see* **Note 5**).
7. Calculate the theoretical MM of the protein with adapted software based on information present in protein databases or you may have acquired during its study (*see* **Note 6**). Do not forget to take into account the chemical modifications introduced during all the purification and analytical steps, for example, if reduction-alkylation was performed on the proteins before 2D gel separation. For some proteins, information present in databases can also include annotations referring to putative or demonstrated features such as protein variants, spliced isoforms, PTMs, sequence conflicts. You have to take into account these features to generate a list of theoretical MMs that combines all possible modifications.
8. Compare the experimental mass with the theoretical one. If the MMs are identical, taking into account the mass accuracy of your instrument, it is probable that the

protein corresponds to the expected one (*see* **Note 7**). If the MMs are different, this could be explained by the fact that some information present in protein databases is false or that the protein is modified. The difference of MM observed between the theoretical and the experimental MMs can permit us to draw hypotheses on the nature of the modification. For example, a difference of 80 Da can be correlated to a phosphorylation, and one of 42 Da to an N-acetylation, which occurs very often at the N-terminus of eukaryotic proteins. If so, an appropriate proteomic strategy can then be performed to verify these assumptions using enrichment of phosphorylated peptides and choosing an appropriate proteolytic enzyme that will give access to an MS-compatible N-terminal peptide, respectively (*see* **Subheading 3.3**). In any case, further experiments will be needed to confirm the presence of and to localize the putative modifications. A "bottom-up" proteomic approach based on LC-MS/MS analyses of the proteolytic peptides of the protein of interest can be the most straightforward approach.

3.2. "Bottom-Up" Proteomic Approach to Characterize Protein Modifications

"Bottom-up" proteomic approaches were introduced since the beginning of the proteomic era. Such approaches are now usually performed to identify proteins from more or less complex mixtures, but it can also be applied to single proteins. In this case, the MS/MS analyses are more likely to cover a large part of the protein sequence. The main steps of a typical "bottom-up" proteomic strategy performed to characterize a protein are described in this subheading and are as follows: (1) protein digestion with one or several appropriate proteolytic enzymes, (2) separation of the resulting peptides by reversed-phase liquid chromatography, (3) MS and MS/MS analyses of the peptides, and (4) interpretation of the MS/MS spectra. These steps can be achieved by several techniques; the instruments and tools we are using in our laboratory to study the human 20S proteasome subunits is described in detail below.

3.2.1. In-Gel Protein Digestion and Peptide Extraction

The protein solution used to measure the protein MM by nano-ESI-MS could be further used for digestion if the sample consumption was low during the MS analysis. In this case, digestion in solution can be performed after solvent evaporation (go directly to step 6 and follow with step 9) with the possible advantage of retrieving more proteolytic peptides and therefore covering a larger part of the protein sequence. But if no protein sample is available, the "bottom-up" strategy has to be accomplished with a new sample. This part describes the protein digestion using two enzymes, trypsin and endoproteinase Glu-C (also called *Staphylococcus aureus* protease V8) because they are both suitable for in-gel digestion (*see* **Note 8**).

1. Excise gel spots using a clean scalpel blade (*see* **Note 2**) and wash the gel pieces in 0.5-mL centrifuge tubes containing deionized water (100 μL).
2. Remove the water, add 30–50 μL of acetonitrile to dehydrate the gel pieces, and discard the supernatant.
3. Incubate in 100 m*M* ammonium bicarbonate (30–50 μL, 15 min, 37°C, under shaking). Add 30–50 μL of acetonitrile (15 min, 37°C, under shaking) to destain the gel pieces. Discard the supernatant.
4. Repeat the incubation step (step 3) until the gel pieces become colorless.
5. Lyophilize to dryness (around 10 min in a speed-vacuum centrifuge; gel pieces become white and hard).
6. Add 5–10 μL of trypsin solution or endoproteinase Glu-C solution (enough to cover the gel pieces) and incubate overnight at 37°C for trypsin and at 25°C for Glu-C under shaking.
7. Sonicate for 5–10 min and transfer the peptide-containing supernatant to another tube.
8. Incubate the tube containing the gel pieces with 15 μL of 25 m*M* ammonium bicarbonate for 15 min at 37°C under gentle shaking and sonicate for 5 min. Add 15 μL of acetonitrile and reincubate for 15 min at 37°C under gentle shaking. Resonicate and pool the supernatant with the previous one from step 7. Add 15 μL of 5% (v/v) formic acid solution to the tube containing the gel pieces and incubate for 15 min at 37°C under shaking. Sonicate 5 min. Add 15 μL of acetonitrile and reincubate for 15 min at 37°C under shaking. Resonicate and combine all supernatants.
9. Lyophilize to dryness using a speed-vacuum.

3.2.2. Separation of Proteolytic Peptides by Nano-HPLC

Separation of the peptides obtained after enzymatic digestion is highly recommended in order to maximize the number of peptides analyzed by MS/MS and therefore have access to the highest sequence coverage. As the quantity of sample is very low, this has to be done with a nano-HPLC system on a capillary column. Reversed-phase chromatography is a very efficient and easy to use technique to separate peptides and the most suited phase is C_{18}. The nano-HPLC separation can be performed separately (for example, if MALDI-MS/MS is further used) or coupled on-line with an ESI-MS/MS instrument as described below.

1. Set the nano-HPLC system with a C_{18} capillary column and, if possible, with a guard column to desalt your sample before chromatographic separation. This latest configuration is considered in the protocol described below where two pumping systems are used: one for the desalting solution and the other for the A and B solutions.
2. Resuspend the proteolytic peptides in 15 μL desalting solution; this will allow you to perform at least two LC-MS/MS analyses with a 5 μL injection.

3. Program the LC run. The method described below is the one we usually apply, but it can be adapted for the sample of interest. The flow rate of the desalting solution through the guard column is set at 20 μL/min and the flow rate of solutions A and B through both the guard column and the analytical column is set at 200 nL/min. These two solutions are used for the gradient elution of peptides from both columns with the following program:

Time (min)	Solution A (%)	Solution B (%)
0	95	5
5	95	5
45	50	50
50	5	95
60	5	95
80	95	5

After sample injection on the guard column, this column is disconnected from the analytical column for 7 min during which it is rinsed with the desalting solution. Valves are then switched to connect the guard column to the analytical column. To optimize analysis time, the guard column is disconnected again after 70 min of the program to equilibrate with the desalting solution before the next sample injection.

3.2.3. ESI-MS/MS Analysis of Proteolytic Peptides

MS/MS analyses of the peptides issued from the enzymatic digestion of the protein of interest provide data on its primary sequence and potential PTMs.

1. Set the mass spectrometer in positive ion mode (*see* **Note 9**).
2. Calibrate the mass spectrometer.
3. Connect the nano-HPLC to the source of the ESI-MS instrument with an electrospray needle and adjust the voltage applied on the needle and its position to obtain the most intense MS signal.
4. Acquire the MS data during the LC run. In our laboratory, this is performed in an information-dependent mode in which each full MS scan (acquired from *m/z* 300–2000) is followed by two MS/MS scans (acquired from *m/z* 80–2000) of the two most abundant peptide molecular ions. They are dynamically selected and then temporarily excluded for collision-induced dissociation (CID) to generate tandem mass spectra.

3.2.4. MS/MS Spectra Interpretation

Peptide identification from the tandem mass spectra can be performed by database search in a first attempt. Identification will then depend on the information present in the database (for example, are all the already known protein variants present in this database and, if so, are they taken into account in the search?) and on the parameters used for the search including the selected variable modifications. However, manual interpretation could be necessary to verify unexpected or unusual peptide sequences. All these analyses have to be processed stepwise by keeping in mind the hypotheses drawn after the "top-down" proteomic analyses. A general guideline is proposed below.

1. Perform a database search (*see* **Note 10**) using dedicated software, such as Protein Prospector and Mascot; this will confirm the identity of the protein in your sample and give a first series of information. Check the sequence coverage obtained after this first search.
2. If the modification suspected from the "top-down" proteomic analysis is not found after the first database search, perform a second search including this modification as variable modification, if possible (*see* **Note 11**). Several other trials can be performed by changing the search parameters. For example, possible chymotrypsin cleavages can be included because chymotryptic activity is often observed in the proteolytic peptides generated by trypsin, even though the latter enzyme is treated to inhibit this activity. If the modified peptide is found after one of these searches, manually check the corresponding MS/MS spectrum to ascertain its identity.
3. If the suspected modification is not found after these first two steps, manual search and interpretation have to be performed. Check the protein sequence already covered by the first database searches. Calculate the theoretical MM of the remaining putative modified peptides; check if corresponding ions are present in the MS spectra and if subsequent MS/MS has been performed. If it is the case, manually interpret the tandem mass spectra to confirm its identity. The search can still be unsuccessful for several reasons (for example, the suspected modification is not the right one) and in this case, the last thing to do is to analyze all uninterpreted MS/MS spectra, with the help of *de novo* sequencing software, if possible.

At the end of all these assays, if the modification is still not found, analyze the sequence coverage of your protein. Parts of the sequence were perhaps missed during MS analyses. This can be due to several factors including the enzyme chosen for the protein digestion generating too big or too short peptides; if so, perform one or several other "bottom-up" proteomic analyses with appropriate enzymes until the characterization of the modification or the complete sequence coverage.

3.3. Characterization of the Human Erythrocytes 20S Proteasome Subunits by Proteomic Analysis

The "top-down" and "bottom-up" proteomic approaches described in **Subheadings 3.1** and **3.2**, respectively, have been carried out to characterize the human erythrocytes 20S proteasome subunits and their isoforms after 2D

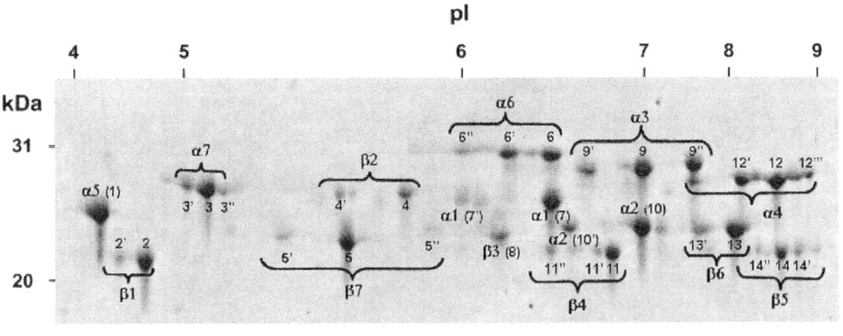

Fig. 2. 2D reference map of human erythrocytes 20S proteasome. The 2D gel image has been obtained with 40 μg of 20S proteasome purified from human erythrocytes after Coomassie blue staining. All spots have been identified by MALDI-TOF mass spectrometry and database search after tryptic digestion. This analysis reveals the presence of numerous isoforms for most α and β 20S proteasome subunits.

gel electrophoresis separation and Coomassie blue staining (**Fig. 2**). MALDI-TOF analysis of the spots already allowed the identification of all the standard proteasome subunits (*see* **Note 12**) and the detection of subunit isoforms as indicated in **Fig. 2** ***(11)***. The whole procedure described in this chapter is first illustrated with the detailed proteomic analyses of two chosen subunits before giving a summary of the results obtained so far on all of them.

3.3.1. Identification of β_7 Variants

During our studies on the human erythrocytes 20S proteasome, several 2DE gels presented an additional spot, named 5bis (**Fig. 3**), just below the usual position of spot 5 (**Fig. 2**), corresponding to one isoform of the β_7 subunit. This doublet is not observed systematically but is reproducible within same samples of erythrocytes (i.e., coming from particular individual volunteers). As indicated in **Fig. 3**, at the time of this study, the annotations present in the Swiss-Prot database for the human β_7 proteasome subunit (accession number P28070) reported that a propeptide (from amino acids 1 to 45) was removed to give the mature form and that two polymorphisms, described in the human Single Nucleotide Polymorphism database (dbSNP), generated two possible variants at positions 95 and 184. The theoretical average MM of the four proteins resulting from the combination of these two polymorphisms was calculated taking into account the cysteine alkylation with iodoacetamide performed before the second dimension of the 2D gel. The MMs of the proteins eluted from spot 5 and 5bis were measured by ESI-MS after gel elution and nanoscale hydrophilic phase chromatography (**Fig. 3**). Both deconvoluted spectra presented several peaks by

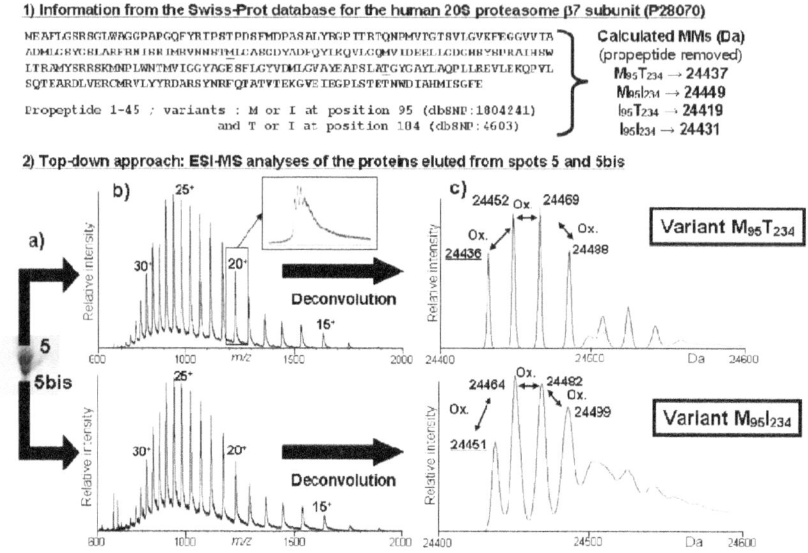

Fig. 3. Identification of β_7 variants in human erythrocytes 20S proteasome. Two isoforms of β_7 showing a shift in molecular weight are sometimes observed on the 2DE gels of the human 20S proteasome (**2a**). Information present in the Swiss-Prot protein database about this proteasome subunit (**1**) and measurement of their MMs after protein gel elution (**2b and c**) suggest that they may correspond to two variants that differ by only one amino acid at position 234. Ox, oxidation.

which MMs differed by approximately 16 Da; they corresponded to the same protein presenting several oxidation states that occurred mostly at methionine residues (*see* **Note 13**). The MMs of the proteins present in spot 5 and 5bis were measured at 24,436 and 24,451 Da, respectively; they could therefore correspond to the $M_{95}T_{234}$ and $M_{95}I_{234}$ variants, respectively (**Fig. 3**). MS/MS analyses of the corresponding tryptic peptides confirmed these hypotheses. These proteomic analyses demonstrate for the first time the existence of these variants at the protein level. It would now be interesting to test the proteolytic activity of both subtypes of 20S proteasome complexes to determine if this compositional variation can have a functional impact.

3.3.2. N-Acetylation and Sequence Validation for the α_5 Subunit

The α_5 subunit is present in human erythrocytes 20S proteasome in only one form (**Fig. 2**). At present, the annotations present in the Swiss-Prot database for this subunit (P28066) reported two conflicts in the sequence: A or D at position 27 and V or L at position 184 (**Fig. 4**). The "top-down" proteomic

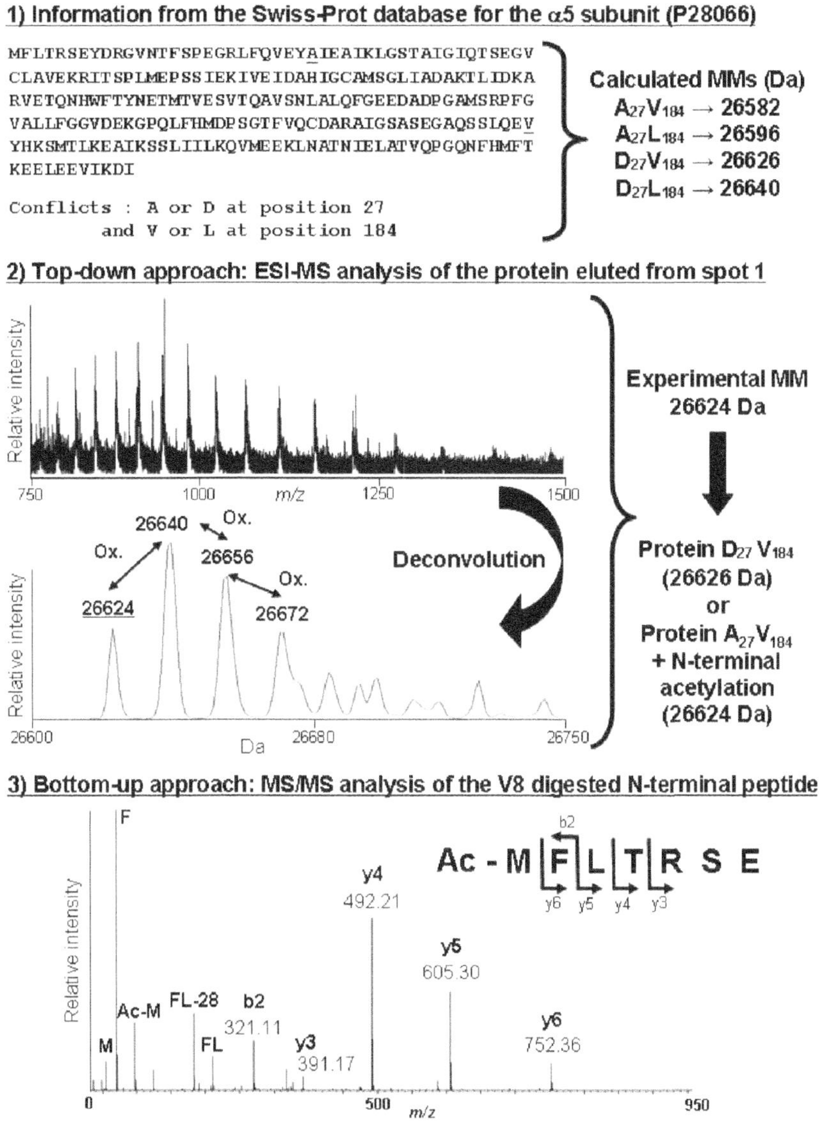

Fig. 4. Characterization of the human erythrocytes 20S proteasome α_5 subunit by a combination of "bottom-up" and "top-down" proteomic approaches. According to the information present in the Swiss-Prot protein database (1), the MM measurement of the single form of the α_5 subunit (2) suggests two hypotheses for its characterization. After digestion of this protein with the V8 endoproteinase, MS/MS analysis of the N-terminal peptide confirmed the presence of an N-terminal acetylation on the first methionine (3).

approach described in this chapter allowed the measurement of the MM of this protein at 26,624 Da. Taking into account the possible error in MM according to the mass accuracy of the instrument (1–2 Da for a protein around 20–30 kDa with the QqTOF used), this MM could correspond to the variant $D_{27}V_{184}$ (26,626 Da), but also to the variant $A_{27}V_{184}$ (26,582 Da) modified by an N-terminal acetylation (26,582 + 42 = 26,624 Da). This second hypothesis was suggested by the fact that Lee and collaborators ***(24)*** have shown that only five subunits of human erythrocytes 20S proteasome were susceptible to Edman degradation (β_1, β_2, β_5, β_6, β_7). The theoretical tryptic N-terminal peptide is composed of five amino acids (**Fig. 4**) and could be too short to be analyzed by LC-MS/MS. We then took advantage of the presence of a glutamic acid residue at the proximity of the N-terminus to perform a V8 digestion instead of a tryptic digestion, thus generating a heptapeptide instead of a pentapeptide (**Fig. 4**). The MS/MS spectrum obtained for this peptide confirmed the presence of the N-acetylation on the initiator methionine residue (**Fig. 4**). The identification of the variant $A_{27}V_{184}$ was also confirmed by MS/MS analyses of the V8 digest peptides containing the amino acids A_{27} and V_{184}. Since this study, the N-terminal acetylation was also observed on this subunit by Wang and collaborators in human embryonic kidney 293 cells ***(21)***. Phosphorylation at S_{56} ***(25)*** and S_{16} ***(26)*** was also described in human HeLa cells, but the ESI-MS measurement performed in this study on the unique form of α_5 visible on the 2D map of human erythrocytes 20S proteasome does not suggest the presence of these PTMs in these blood cells.

3.3.3. Summary of all Subunits Characterization

The combination of "top-down" and "bottom-up" proteomic approaches described in this chapter was applied to all human erythrocytes 20S proteasome subunits and their isoforms shown in **Fig. 2**; a summary of the results obtained is presented in **Table 1**. The MM of a few isoforms was not measurable because the quantity was too low, even if 250 μg of 20S proteasome was separated on a 2D gel. For the β_4 subunits, a series of LC-MS/MS analyses was performed on the most intense isoform (number 11) that covered 94% of the sequence. These analyses demonstrated an N-terminal acetylation of the initiator methionine. Taking this modification into account, an unexplained mass discrepancy of 55 Da is still observed between the measured (22,994 Da) and the theoretical (23,049 Da) MM. This suggests unexpected modifications at one or several amino acids among the 12 not detected during MS analyses.

This study permits us to confirm the primary sequence of most of the subunits, to verify sequence conflicts, to identify protein variants, and to observe PTMs like phosphorylations and N-terminal acetylations. We determined the

Table 1
Characterization of the Human Erythrocytes 20S Standard Proteasome Subunits and Their Isoforms by Proteomic Approaches

Subunit	Swiss-Prot	Isoform[a]	Experimental MM (Da)[b]	Characterization[c]
α_1	P60900	7, 7′	27,765, 27,770	– Met_{ini} + N-acetylation on Ser_1; Lys_{58} (conflict)
α_2	P25787	10, 10′	25,922, 25,924	– Met_{ini} + N-acetylation on Ala_1
α_3	P25789	9, 9′, 9″	29,680, 29,685, 29,678	– Met_{ini} + N-acetylation on Ser_1
α_4	O14818	12, 12′, 12″, 12‴	27,974, 27,975, 27,975, 27,976	– Met_{ini} + N-acetylation on Ser_1
α_5	P28066	1	26,624	N-Acetylation on Met_{ini}; Ala_{27} and Val_{184} (conflict)
α_6	P25786	6, 6′, 6″	29,893, N.D., N. D.	N.D.
α_7	P25788	3, 3′	28,650, 28,657	– Met_{ini} + N-acetylation on Ser_1; phosphorylation on Ser_{249}; Ile_{90} (conflict)
		3″	28,575	– Met_{ini} + N-acetylation on Ser_1; Ile_{90} (conflict)
β_1	P28072	2	22, 131	Variant Pro_{107}; Val_{145} (conflict)
		2′	N. D.	N.D.[d]
β_2	Q99436	4, 4′	25,582, 25,585	No modification
β_3	P49720	8	23,143	– Met_{ini} + N-acetylation on Ser_1; Met_{33} (conflict)
β_4	P49721	11, 11′, 11″	22,994, 22,998, 23,005	N-Acetylation on Met_{ini} but not entirely characterized
β_5	P28074	14	22,633	Ile_{30}; Ala_{54}; Thr_{103} (conflicts)
		14′, 14″	N.D., N.D.	N.D.
β_6	P20618	13, 13′	23,776, 23,777	Pro_{11} (conflict)
β_7	P28070	5	24,436 Da	Variant Met_{95} and Thr_{234}
		5bis	24,451 Da	Variant Met_{95} and Ile_{234}
		5′, 5″	N.D.	N.D.

[a] Indicated on the 2D gel presented in Fig. 2.
[b] Average MM.
[c] Removal of propeptide was confirmed for subunits β_1, β_2, β_5, β_6, and β_7 as indicated in the Swiss-Prot database.
[d] N.D., not determined.

N-terminal sequence of all subunits, except for α_6. We thus describe for the first time the N-acetylation of the α_1, α_2, and α_3 subunits. In summary, our results combined with the results obtained by Wang *et al.* lead to the following conclusions on the N-terminal sequence of the human standard 20S proteasome subunits: the β_1, β_2, β_5, β_6, and β_7 subunits are not blocked by N-acetylation after propeptide removal as already suggested by the experiment of Lee *et al.*, whereas all other subunits are acetylated either at the first methionine residue (α_5, α_6, β_4) or at the second amino acid after the removal of the first methionine residue (α_1, α_2, α_3, α_4, α_7, β_3).

One aim of this study was to characterize the subunit isoforms differing in their isoelectric point (p*I*). This feature could suggest the presence of a phosphorylation that induces a p*I* shift. However, the presence of a phosphorylation was demonstrated only for the two more acidic isoforms of α_7 and was not detected for other subunit isoforms. Indeed, the MMs that could be determined for different isoforms of the same subunit were quite similar with a mass difference of 1–5 Da. One hypothesis that could then explain this p*I* shift between isoforms with such a small mass difference is the presence of deamidation. This modification induces a mass shift of 1 Da and converts basic residues into acidic ones (Asn to Asp; Gln to Glu). To confirm the presence of deamidation on the 20S subunit isoforms, MM measurement has to be performed on a high-resolution instrument like FT-ICR. This measurement has to be done on the entire protein and not on the digested peptides because deamidation occurs very often during proteolytic digestion ***(27)***.

The functional impact of structural variations between different subpopulations of standard 20S proteasome is not well known. Therefore, the MS-based strategy developed here for the identification of proteasome subunits is important to better understand the molecular mechanisms involved in the activity and function of this essential molecular machine. Moreover, the combination of proteomic approaches described in this chapter can be applied for the study of other complexes.

4. Notes

1. This strategy has never been applied on silver-stained gel because the quantity would be too low.
2. We recommend the use of a laminar flow hood to prevent keratin contamination. The use of disposable gloves, a face mask, and a hat is mandatory.
3. After incubation, you can sonicate the centrifuge tube for 10–20 min. This step could help the protein diffusion.
4. Measure the desired volume of eluting solution in a centrifuge tube prior to performing the elution step because the back pressure created by the stationary

phase may result in pipetting a lower volume. Take care never to dehydrate the phase except at the end of the elution step.

5. The deconvoluted spectrum may display several peaks that could correspond to several proteins due to protein mixture or several forms of the same protein due to different modifications like series of oxidation.
6. To calculate the theoretical MM of a protein, specific programs included in the MS instrument software can be used. If not, proteomic tools are available at the ExPASy website (http://www.expasy.org/). The PeptideMass tool calculates the entire MM of a protein when the "no cutting" parameter is used for the enzyme window selection. Protein modification MMs are available in the UNIMOD database (http://www.unimod.org).
7. An exception is if your protein is modified by a combination of changes that results in no difference in MM, for example, if an L/I and a G (113 + 57 = 171 Da) are replaced by a V and an A (99 + 71 = 171 Da). Depending on the mass accuracy of the MS instrument used, these combinations can be discriminated or not.
8. Small-sized proteolytic enzymes (with MM under approximately 30 kDa) are compatible for in-gel digestion because they can reach the protein in the gel matrix.
9. We perform the nano-LC-MS/MS analysis on an ESI-Q-TOF mass spectrometer (QSTAR XL, Applied Biosystems). The mass resolution we obtain on this instrument is around 15,000 in the selected mass range.
10. To analyze the human 20S proteasome, we use the following parameters for the initial database search: Swiss-Prot database restricted to human species, trypsin able to cleave before proline residues, two possible missed cleavages, carbamidomethylation as a systematic cysteine modification (the 20S proteasome sample is alkylated with iodoacetamide before the second dimension of 2D gel electrophoresis), methionine oxidation, and protein N-acetylation as possible modifications (most 20S proteasome subunits possess an acetylated N-terminus). The mass tolerance for MS and MS/MS analyses depends on the instrument used. When searching the Swiss-Prot database for proteasome subunits, note that known subunit variants will not be directly identified, as only one sequence per entry is used for database search. However, all known sequence modifications are listed in the protein description entry. The Swiss-Prot Varsplic database has recently become available; it includes all isomeric proteins recorded in the features (FT) section of the SwissProt database.
11. Depending on the search engine used, your suspected modifications can be already included in the list of modifications. If not, check if you can introduce new modifications in the software.
12. In this chapter, the proteasome subunits are named with the nomenclature previously described ***(28)***.
13. Oxidation is commonly encountered in polyacrylamide gel-separated proteins ***(29)***. This modification involves mostly the methionine residues, but can also affect tryptophan residues ***(30)***. It is important to limit protein oxidation because

it divides the MS signal and therefore diminishes the sensitivity. For this, work at 4°C during the sample preparation if possible and, after 2D gel separation, you can store the gel or excised spots in 1% ascorbic acid to prevent further oxidation. It is, however, better to follow the entire process without sample storage to prevent oxidation as much as possible.

Acknowledgments

We thank all the group members and students who contributed to the characterization of the human erythrocytes 20S proteasome. This work was supported by the Centre National de la Recherche Scientifique (CNRS) and in part by the Réseau National des Génopoles, the Région Midi-Pyrénées, and the ASG program from the French Ministry of Research.

References

1. Unno, M., Mizushima, T., Morimoto, Y., Tomisugi, Y., Tanaka, K., Yasuoka, N., *et al.* (2002) The structure of the mammalian 20S proteasome at 2.75 Å resolution. *Structure* **10**, 609–618.
2. Orlowski, M. and Wilk, S. (2000) Catalytic activities of the 20S proteasome, a multicatalytic proteinase complex. *Arch. Biochem. Biophys.* **383**, 1–16.
3. Rock, K. L. and Goldberg, A. L. (1999) Degradation of cell proteins and the generation of MHC class I-presented peptides. *Annu. Rev. Immunol.* **17**, 739–779.
4. Morel, S., Levy, F., Burlet-Schiltz, O., Brasseur, F., Probst-Kepper, M., Peitrequin, A. L., *et al.* (2000) Processing of some antigens by the standard proteasome but not by the immunoproteasome results in poor presentation by dendritic cells. *Immunity* **12**, 107–117.
5. Van den Eynde, B. J. and Morel, S. (2001) Differential processing of class-I-restricted epitopes by the standard proteasome and the immunoproteasome. *Curr. Opin. Immunol.* **13**, 147–153.
6. Burlet-Schiltz, O., Claverol, S., Gairin, J. E., and Monsarrat, B. (2005) The use of mass spectrometry to identify antigens from proteasome processing. *Methods Enzymol.* **405**, 264–300.
7. Macagno, A., Gilliet, M., Sallusto, F., Lanzavecchia, A., Nestle, F. O., and Groettrup, M. (1999) Dendritic cells up-regulate immunoproteasomes and the proteasome regulator PA28 during maturation. *Eur. J. Immunol.* **29**, 4037–4042.
8. Noda, C., Tanahashi, N., Shimbara, N., Hendil, K. B., and Tanaka, K. (2000) Tissue distribution of constitutive proteasomes, immunoproteasomes, and PA28 in rats. *Biochem. Biophys. Res. Commun.* **277**, 348–354.
9. Husom, A. D., Peters, E. A., Kolling, E. A., Fugere, N. A., Thompson, L. V., and Ferrington, D. A. (2004) Altered proteasome function and subunit composition in aged muscle. *Arch. Biochem. Biophys.* **421**, 67–76.
10. Drews, O., Zong, C., and Ping, P. (2007) Exploring proteasome complexes by proteomic approaches. *Proteomics* **7**, 1047–1058.

11. Claverol, S., Burlet-Schiltz, O., Girbal-Neuhauser, E., Gairin, J. E., and Monsarrat, B. (2002) Mapping and structural dissection of human 20S proteasome using proteomic approaches. *Mol. Cell. Proteomics* **1**, 567–578.
12. Iwafune, Y., Kawasaki, H., and Hirano, H. (2002) Electrophoretic analysis of phosphorylation of the yeast 20S proteasome. *Electrophoresis* **23**, 329–338.
13. Kurucz, E., Ando, I., Sumegi, M., Holzl, H., Kapelari, B., Baumeister, W., *et al.* (2002) Assembly of the Drosophila 26S proteasome is accompanied by extensive subunit rearrangements. *Biochem. J.* **365**, 527–536.
14. Yang, P., Fu, H., Walker, J., Papa, C. M., Smalle, J., Ju, Y. M., *et al.* (2004) Purification of the Arabidopsis 26S proteasome: biochemical and molecular analyses revealed the presence of multiple isoforms. *J. Biol. Chem.* **279**, 6401–6413.
15. Huang, L. and Burlingame, A. L. (2005) Comprehensive mass spectrometric analysis of the 20S proteasome complex. *Methods Enzymol.* **405**, 187–236.
16. Froment, C., Uttenweiler-Joseph, S., Bousquet-Dubouch, M. P., Matondo, M., Borges, J. P., Esmenjaud, C., *et al.* (2005) A quantitative proteomic approach using two-dimensional gel electrophoresis and isotope-coded affinity tag labeling for studying human 20S proteasome heterogeneity. *Proteomics* **5**, 2351–2363.
17. Gomes, A. V., Zong, C., Edmondson, R. D., Li, X., Stefani, E., Zhang, J., *et al.* (2006) Mapping the murine cardiac 26S proteasome complexes. *Circ. Res.* **99**, 362–371.
18. Schmidt, F., Dahlmann, B., Janek, K., Kloss, A., Wacker, M., Ackermann, R., *et al.* (2006) Comprehensive quantitative proteome analysis of 20S proteasome subtypes from rat liver by isotope coded affinity tag and 2-D gel-based approaches. *Proteomics* **6**, 4622–4632.
19. Castano, J. G., Mahillo, E., Aritzi, P., and Arribas, J. (1996) Phosphorylation of C8 and C9 subunits of the multicatalytic proteinase by casein kinase II and identification of the C8 phosphorylation sites by direct mutagenesis. *Biochemistry* **35**, 3782–3789.
20. Kimura, Y., Takaoka, M., Tanaka, S., Sassa, H., Tanaka, K., Polevoda, B., *et al.* (2000) N(alpha)-acetylation and proteolytic activity of the yeast 20S proteasome. *J. Biol. Chem.* **275**, 4635–4639.
21. Wang, X., Chen, C. F., Baker, P. R., Chen, P. I., Kaiser, P., and Huang, L. (2007) Mass spectrometric characterization of the affinity-purified human 26S proteasome complex. *Biochemistry* **46**, 3553–3565.
22. Bousquet-Dubouch, M. P., Uttenweiler-Joseph, S., Ducoux-Petit, M., Matondo, M., Monsarrat, M., and Burlet-Schiltz, O. (2008) Organelle proteomics. In: Pflieger, D., Rossier, J. (eds.). *Methods Mol. Biol.* **432**, 301–320.
23. Claverol, S., Burlet-Schiltz, O., Gairin, J. E., and Monsarrat, B. (2003) Characterization of protein variants and post-translational modifications: ESI-MSn analyses of intact protein eluted from polyacrylamide gels. *Mol. Cell. Proteomics* **2**, 483–493.
24. Lee, L. W., Moomaw, C. R., Orth, K., McGuire, M. J., DeMartino, G. N., and Slaughter, C. A. (1990) Relationship among the subunits of the high molecular weight proteinase, macropain (proteasome). *Biochim. Biophys. Acta.* **1037**, 178–185.
25. Beausoleil, S. A., Jedrychowski, M., Schwartz, D., Elias, J. E., Villen, J., Li, J., *et al.* (2004) Large-scale characterization of HeLa cell nuclear phosphoproteins. *Proc. Natl. Acad. Sci. USA* **101**, 12130–12135.

26. Beausoleil, S. A., Villen, J., Gerber, S. A., Rush, J., and Gygi S. P. (2006) A probability-based approach for high-throughput protein phosphorylation analysis and site localization. *Nat. Biotechnol.* **24**, 1285–1292.
27. Krokhin, O. V., Antonovici, M., Ens, W., Wilkins, J. A., and Standing, K. G. (2006) Deamidation of -Asn-Gly- sequences during sample preparation for proteomics: consequences for MALDI and HPLC-MALDI analysis. *Anal. Chem.* **78**, 6645–6650.
28. Baumeister, W., Walz, J., Zühl, F., and Seemüller, E. (1998) The proteasome: paradigm of a self-compartmentalizing protease. *Cell* **92**, 367–380.
29. Hamdan, M., Galvani, M., and Righetti, P. G. (2001) Monitoring 2-D gel-induced modifications of proteins by MALDI-TOF mass spectrometry. *Mass Spectrom. Rev.* **20**, 121–141.
30. Shevchenko, A., Loboda, A., Ens, W., Schraven, B., and Standing, K. G. (2001) Archived polyacrylamide gels as a resource for proteome characterization by mass spectrometry. *Electrophoresis* **22**, 1194–1203.

9

Free-Flow Electrophoresis of the Human Urinary Proteome

Mikkel Nissum and Robert Wildgruber

Summary

Prefractionation of complex protein samples prior to mass spectrometry provides a method for the isolation of low-abundance proteins into specific fractions thereby enabling their identification. Free-flow electrophoresis in the isoelectric focusing mode (IEF-FFE) presents a complementary approach to established prefractionation methodologies. Proteins are separated in solution according to their isoelectric point (p*I*) with a high throughput of sample volume. The separation may be performed under denaturing or nondenaturing conditions and detergents may be added to promote protein solubilization. A protocol covering the pH range from pH 3 to 9 under denaturing conditions was used to illustrate the method of IEF-FFE including sample preparation prior to reversed-phase liquid chromatography and tandem mass spectrometry. The IEF-FFE separation was applied to a sample of human urine.

Key Words: Free-flow electrophoresis; isoelectric focusing; prefractionation; liquid chromatography, tandem mass spectrometry; human urine; proteomics.

1. Introduction

The comprehensive analysis of complex protein mixtures based on mass spectrometry posts a major challenge in proteomics due to the wide dynamic ranges of protein abundances, which may vary from 10^5 to 10^6 for cellular extracts and 10^9 to 10^{10} for body fluids *(1)*. A variety of prefractionation methods have been applied to simplify the complex protein mixtures prior to mass spectrometry. The initial separation approach in proteomics was two-dimensional gel electrophoresis *(2,3)* where the proteins in the sample

From: *Methods in Molecular Biology, vol. 484: Functional Proteomics: Methods and Protocols*
Edited by: J. D. Thompson et al., DOI: 10.1007/978-1-59745-398-1, © Humana Press, Totowa, NJ

are separated by isoelectric focusing (IEF) in the first dimension and by molecular weight in the second, resulting in a two-dimensional map of proteins. More recently, chromatographic methods were introduced such as strong cation-exchange chromatography in combination with reversed-phase liquid chromatography (RPLC) ***(4)*** and affinity chromatography ***(5,6)***. Here, we describe free-flow electrophoresis (FFE), which is a liquid-based separation method where bioparticles such as proteins are separated according to their net charges ***(7)***. Complex protein mixtures are continuously injected into a carrier ampholyte solution flowing as a thin laminar film (0.4 mm) between two parallel plates. In the IEF mode, proteins are separated according to their p*I* values into a number of well-defined fractions by the introduction of an electric field perpendicular to the flow direction (**Fig. 1**). The number of

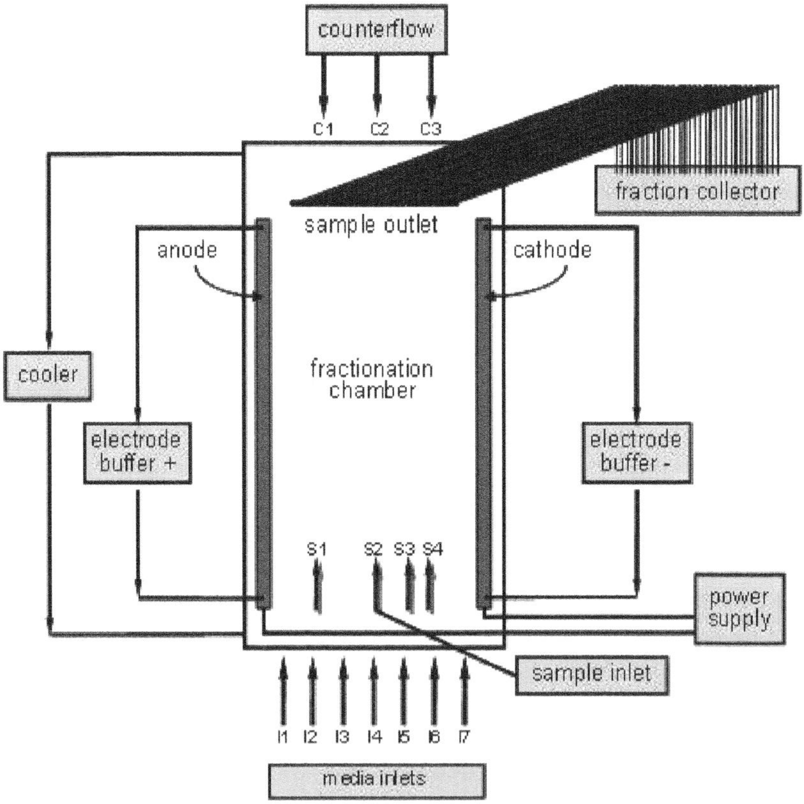

Fig. 1. Schematic representation of the FFE apparatus. Dimensions of the separation chamber are 500 × 100 × 0.4 mm. The inlets I1–I7, S1–S4, and C1–C3 represent ports for stabilization and separation media, for sample delivery, and for counterflow, respectively.

protein-containing fractions depends on the shape and range of the pH gradient that are determined by the applied separation buffers ***(8–10)***. A fraction to fraction resolution better than 0.01 pH unit may be obtained. IEF-FFE may be performed under nondenaturing conditions or denaturing conditions by the addition of a chaotropic agent such as urea or thiourea to the separation buffers. Reducing conditions may be obtained by the addition of dithiothreitol (DTT) or tris(2-carboxyethyl) phosphine (TCEP). Nonionic and zwitterionic detergents promote protein solubilization and most of the detergents within these two classes are compatible with IEF-FFE. A protocol covering the pH range from approximately pH 3 to 9 under denaturing conditions is used to illustrate the method of IEF-FFE including sample preparation, instrument set-up/shut-down, and sample separation. In addition, the sample preparation prior to RPLC and tandem mass spectrometry (MS/MS) is described and we demonstrate that all components of the media are compatible with RPLC-MS/MS. The IEF-FFE separation is applied to a sample of human urine. Urine is a desirable material for the identification of biomarkers to be used in the diagnosis and classification of diseases. Although it can conveniently be collected in large amounts, the urinary proteome has not yet been fully characterized ***(11,12)***. Here, we demonstrate the usefulness of the urea/thiourea-containing separation buffers to maintain membrane proteins from urine in solution throughout the separation process.

2. Materials

2.1. Sample Preparation

1. Protein sample to be separated and analyzed (urine from healthy donor).
2. Protease inhibitors (leupeptin, benzamidine, and phenylmethylsulfonyl fluoride).
3. BCA Protein Assay Kit (Pierce, Rockford, IL).
4. Vivaspin 6 ultrafiltration spin columns, MWCO 5 kDa (Vivascience, Hannover, Germany).

2.2. FFE-IEF Separation

1. FFE device (BD Diagnostics, Munich, Germany).
2. External water chiller to maintain a cooled separation chamber (10°C) during IEF.
3. FFE supplies: spacer (0.4 mm), electrode membranes, and electrode filter paper (BD Diagnostics).
4. IEF-FFE pH 3–9 buffer solutions at 4°C (**Table 1**) (BD Diagnostics).
5. SPADNS (2-[4-sulfophenylazo]-1,8-dihydroxy-3,6-naphthalenedisulfonic acid) (BD Diagnostics).
6. Hydroxypropyl methyl cellulose (HPMC, 86 kDa), 0.2% (w/v) at 4°C (BD Diagnostics).
7. p*I* Marker Stock Solution (BD Diagnostics).

Table 1
IEF-FFE Solutions and Buffers, pH 3–9 Gradient[a]

FFE solution	Preparation[b]
Anodic electrolyte solution	5.0 g 5 *M* H_2SO_4 245.0 g Milli-Q water
Cathodic electrolyte solution	25.0 g 1 *M* NaOH 225.0 g Milli-Q water
Anodic stabilization medium	58.5 g Anodic Stabilization Stock, pH 3–9 37.0 g urea 13.5 g thiourea 4.0 g mannitol
Cathodic stabilization medium	87.8 g Cathodic Stabilization Stock, pH 3–9 55.5 g urea 20.3 g thiourea 6.0 g mannitol
Separation buffer	15.4 g Prolyte 3–9 43.1 g Milli-Q water 37.0 g urea 13.5 g thiourea 4.0 g mannitol
Counterflow solution	300.0 g Milli-Q water 189.0 g urea 69.0 g thiourea 20.4 g mannitol
p*I* marker solution	315.0 mg urea 115.0 mg thiourea 34.5 mg mannitol 500.0 µl pı0t I Marker Stock 500.0 µl Separation buffer

[a]The listed quantities allow 4 h of operation (*see* **Note 12**).
[b]All component are measured by weight.

8. pH/conductivity multimeter.
9. UV/VIS microplate reader.
10. Sodium dodecyl sulfate polyacrylamide gel electrophoresis (SDS–PAGE) equipment.

2.3. Enzymatic Digestion of FFE Fractions and RPLC-MS/MS

1. Trypsin (porcine, modified sequencing grade, Promega, Madison, WI).
2. 200 m*M* TCEP in Milli-Q water.
3. 200 m*M* iodoacetamide in Milli-Q water.

4. 100 m*M* ammonium bicarbonate in Milli-Q water.
5. 0.1% trifluoroacetic acid in Milli-Q water.
6. Acetonitrile.
7. 0.1% formic acid in Milli-Q water.
8. 0.1% formic acid in acetonitrile.
9. SepPak cartridges, C_{18} (Waters, Milford, MA).
10. SpeedVac (Eppendorf, Cologne, Germany).
11. HPLC system model 1100 (Agilent Technologies, Waldbronn, Germany).
12. Ion trap mass spectrometer (HCTultra, Bruker, Bremen, Germany).

3. Methods

The methods described below outline (1) the sample preparation prior to IEF-FFE, (2) the separation of proteins using IEF-FFE, and (3) the enzymatic digestion of the IEF-FFE fractions including purification and concentration of the samples prior to RPLC-MS/MS.

3.1. Sample Preparation for IEF-FFE

A pooled urine sample was collected from 10 healthy donors. After collection, protease inhibitors (leupeptin, benzamidine, and phenylmethylsulfonyl fluoride) were added with a ratio of 1 to 100 (w/w) to avoid proteolysis in the sample. The sample was centrifuged at 3000 × *g* for 10 min at 4°C before the supernatant was collected and frozen at –80°C. The protein concentration was determined using the BCA Kit to 0.08 mg/mL. Prior to IEF-FFE the urine sample was concentrated to approximately 2.4 mg/mL using an ultrafiltration spin column with a molecular weight cutoff of 5 kDa. At the same time a buffer exchange to the IEF-FFE separation medium was performed (*see* **Notes 1–3**).

3.1.1. Preparation of the Urine Sample for IEF-FFE

1. Load the maximum volume (6 mL) of the collected urine sample to the ultrafiltration spin column.
2. Reduce the volume to approximately 1 mL using centrifugation at 4600 × *g* for approximately 20 min and empty the filtrate container.
3. Repeat the above steps until a volume of 30 mL of the urine sample has been loaded and concentrated.
4. Add IEF-FFE separation medium to fill the spin column and reduce the volume to approximately 1 mL using centrifugation at 4600 × *g* for approximately 20 min. Empty the filtrate container.
5. Repeat the buffer exchange step and collect the prepared urine sample in a microcentrifuge tube. Store the sample on ice until IEF-FFE separation.

3.2. IEF-FFE Separation

The process of performing an IEF-FFE separation of proteins is described in **Subheadings 3.2.1–3.2.4.** This includes (a) how to prepare the FFE instrument, (b) how the samples are introduced into the FFE instrument and how the generated fractions are collected, (c) how the FFE instrument is shut down after the separation has been finished, and (d) how the separated proteins are analyzed using SDS–PAGE.

3.2.1. Preparation of the FFE Instrument

1. Switch on the cooling circuit and set the appropriate experiment temperature (10°C).
2. Clean the interior surface of the chamber: move the separation chamber to an upright position and open the separation chamber by releasing the pairs of chamber clamps simultaneously. Use the following sequence for cleaning: Milli-Q water, isopropanol, petrolether, isopropanol, Milli-Q water (*see* **Note 4**).
3. Verify that the polypropylene chamber spacer (0.4 mm) is positioned correctly so as not to overlap the electrode and internal separation chamber seals or to cover the separation chamber media inlets.
4. Insert the anode and cathode membranes. Ensure that the smooth side of the membrane is facing the electrode seal without protruding over the electrode seal. Then place the filter paper strips on top of the anode and cathode membranes.
5. Close the separation chamber starting from the middle clips moving outward, which will squeeze any water that has been trapped under the spacer, leaving the spacer clear. Ensure that all media and counterflow inlets are free and are not covered by the spacer.
6. Open the (lift) valves at the top of the chamber to degas the chamber and fasten the wedge clamps of the media pump for tube I1–I7 (*see* **Fig. 1**). Immerse the medium inlet tubes and the counterflow tubes in a reservoir filled with Milli-Q water (*see* **Note 5**). Pump the water into the separation chamber to fill the separation chamber. Ensure that all air bubbles are displaced from the chamber. Remove any remaining bubbles in the chamber by reversing and forwarding the water flow through the chamber by using the pump control. When the chamber is filled with water and is free of bubbles start introducing the counterflow into the separation chamber by closing the counterflow wedge clamps on the media pump.
7. Ensure that there is no air in the counterflow tubes and then close the valves (to the down position) on the counterflow manifold. At this time, the water should begin dripping from all fractionation tubes into the fractionation housing (*see* **Note 6**). To verify that the system has been assembled correctly providing a consistent laminar flow over the length of the separation chamber, a performance test using a colored red dye (SPADNS) is recommended at this stage (*see* **Note 7**).
8. Stop the media pump. Attach the sample tube to the sample inlet S2 (*see* **Fig. 1**). Run the media pump forward at a flow rate of 400 mL/h. After the sample tube starts dripping, tighten the sample pump screw until the dripping stops.

9. Stop the media pump and immerse all media inlet tubes into a 0.2% HPMC solution in Milli-Q water. Move the separation chamber to a horizontal position. Run the media pump at 150 mL/h for 20 min to coat the surfaces of the separation chamber with HPMC (*see* **Note 8**).

3.2.2. Sample Loading and Collection

1. Prepare the IEF-FFE buffers and solutions for the pH 3–9 gradient as described in **Table 1**. Set up the separation and stabilization buffer tubes as follows: medium inlet tubes 1 and 2 in anodic stabilization medium, medium inlet tubes 3 and 4 in separation medium, medium inlet tubes 5–7 in cathodic stabilization medium, and counterflow inlets in counterflow medium.
2. Place the anode and cathode electrolyte solutions in the electrode buffer pump tray, immerse the electrode buffer tubes, and start the electrode pump.
3. Run the media pump at 200 mL/h for 15 min to fill the separation chamber with separation buffers and reduce the flow rate to 60 mL/h.
4. Adjust the voltage to 400 V and maximum current to 50 mA. Switch on the high voltage and wait 15 min for the current to stabilize (10–15 mA). Before the sample is loaded a determination of the pH gradient (*see* **Fig. 2**) and a p*I* marker performance test is recommended (*see* **Note 9**).

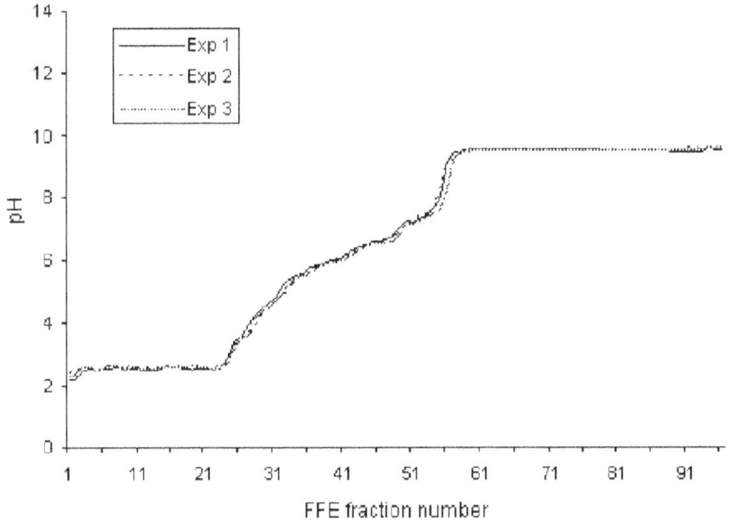

Fig. 2. IEF-FFE pH gradient from pH 3 to 9 used for the separation of urine proteins. Graphs from three independent experiments were overlaid. The flat regions observed below pH 3 and above pH 9 represent the pH values of the anodic and cathodic stabilization media, respectively.

5. Set the sample flow rate to 1 mL/h and start the sample flow.
6. Collect fractions of the separated proteins after approximately 30 min into a polypropylene 96-well plate (*see* **Note 10**).

3.2.3. Shutting Down the FFE Instrument

1. At the end of an IEF-FFE run, switch off the high voltage. Stop all pumps and place the media inlet tubes and counterflow tubes into a reservoir of Milli-Q water. Place the used sample tube (S2) into an empty microcentrifuge tube to collect reversed flush water and open the sample pump screw.
2. Run the media pump at 200 mL/h for 10 min and at 400 mL/h for an additional 30 min to rinse the separation chamber. During the rinsing procedure open the counterflow valve for 2 min to rinse the counterflow manifold.
3. Rinse the electrode circuit by removing all electrode tubes from the electrolyte solution and place them into an empty container. Empty the electrode circuit and place the electrode tubes in a reservoir of Milli-Q water. Run the electrode pump for 10 min and empty the electrode circuit.
4. Release the clamps of the separation chamber starting from both ends to the center and open the separation chamber in a raised position.
5. Take out the filter papers, wash with Milli-Q water, and store them dry. Remove the electrode membranes and rinse in a solution of glycerol/isopropanol (50:50 [v/v]).
6. To shut down the instrument, switch off the pumps, power supply, and external chiller. Leave the separation chamber unclamped in a raised position. The spacer may be left in the separation chamber for use on the next day.

3.2.4. SDS–PAGE Analysis of the IEF-FFE Fractions

The obtained protein-containing fractions are analyzed by SDS–PAGE using an XCell SureLock™ Mini-Cell in combination with precast NuPAGE® 4–12% Bis-Tris gels. Silver staining of the proteins is carried out using the SilverQuest kit (Invitrogen).

1. Apply 10 μL of the separated FFE fraction into a 500-μL sample vial.
2. Add 5 μL of SDS running buffer to the sample vial and mix the sample and buffer.
3. Pipette 10 μL of the protein sample on the gel.
4. Repeat these steps with all fractions to be analyzed (usually every second or third fraction).
5. Close the separation cell and adjust the power supply to 200 V and 100 mA.
6. Start electrophoresis and continue for 35 min.
7. After the run is finished take the gel out of the chamber, open the cassette, and place the gel in a glass tray.
8. Rinse the gel briefly with Milli-Q water.
9. Place the gel in 100 mL of Fixative and microwave at high power for 30 s.

10. Remove the gel from the microwave. Agitate for 5 min and decant the Fixative.
11. Add 100 mL of 30% ethanol and microwave at high power for 30 s.
12. Remove the gel from the microwave and agitate for 5 min at room temperature, then decant the ethanol.
13. Add 100 mL of Sensitizing solution to the washed gel and microwave at high power for 30 s. Remove the gel from the microwave and agitate for 2 min at room temperature, then decant the Sensitizing solution.
14. Wash the gel twice in 100 mL of Milli-Q water. Microwave at high power for 30 s. At each wash step, remove the gel from the microwave and agitate for 2 min at room temperature.
15. Add 100 mL of Staining solution. Microwave at high power for 30 s. Remove the gel from the microwave and agitate for 5 min at room temperature.
16. Decant the Staining solution and wash the gel with 100 mL of Milli-Q water for 20–60 s.
17. Add 100 mL of Developing solution and incubate for 5 min at room temperature with agitation. Do not microwave.
18. Once the desired band intensity is achieved, immediately add 10 mL of Stopper directly to the gel still immersed in Developing solution and agitate for 10 min. The color of the solution changes from pink to clear indicating the end of development.
19. Wash the gel with 100 mL Milli-Q water for 10 min (*see* **Fig. 3**).

3.3. Enzymatic Digestion of IEF-FFE Fractions

The process of identifying the proteins in the individual FFE fractions is described in **Subheadings 3.3.1–3.3.2.** This includes (1) the digestion and clean-

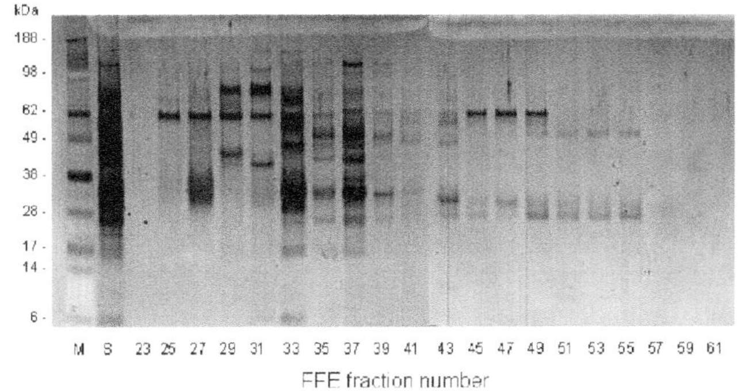

Fig. 3. Separation of proteins from human urine. Analytical SDS–PAGE analysis of IEF-FFE fractions obtained using the linear pH 3–9 gradient. Proteins were visualized by silver staining. A volume of 10 μL was loaded per FFE fraction. Labeling is as follows: M, protein marker; S, starting material.

up of the FFE fractions and *(2)* the following identification of the proteins using RPLC-MS/MS.

3.3.1. Enzymatic Digestion

1. Load 300 μL of the FFE fraction to be analyzed into a microcentrifuge tube and add 5 μL 200 m*M* TCEP solution. Incubate for 60 min at room temperature (*see* **Note 11**). Add 60 μL of 200 m*M* iodoacetamide and incubate for 60 min in the dark.
2. Adjust the pH to 7.8 with 550 μL of 100 m*M* ammonium bicarbonate.
3. Add 5 μg of trypsin and incubate a minimum of 4 h at 37°C. Add 400 μL of 0.1% trifluoroacetic acid (TFA) to terminate the digestion process.
4. Purify the generated peptides using a SepPak™ C_{18} reversed-phase cartridge. Equilibrate the cartridge two times with 1 mL of acetonitrile and an additional two times with 1 mL of 0.1% TFA. Load the sample. Wash two times using 1 mL of 0.1% TFA. Elute the peptides into a microcentrifuge tube using 400 μL of 70% acetonitrile.
5. Evaporate the sample to dryness using vacuum centrifugation and reconstitute in 250 μL of 0.1% TFA.

3.3.2. RPLC-MS/MS

ESI-based LC-MS/MS (HCTultra, Bruker, Bremen, Germany) analyses were carried out using an Agilent 1100 series NanoPump (Agilent Technologies, Waldbronn, Germany) on a 75-μm × 15-cm fused silica microcapillary reversed-phase column (Agilent). Sample volumes of 3 μL were loaded onto the precolumn (300-μm × 0.5-cm reversed-phase [C_{18}] column from Agilent) at a flow rate of 10 μL/min for 5 min using a microflow CapPump (Agilent). After sample loading, the sample was separated and analyzed at a flow rate of 200 nL/min with a gradient of 2% B to 40% B over 30 min. The column was directly coupled to the spray needle from New Objective (Woburn, MA). Mobile phase A was 0.1% formic acid and mobile phase B was 100% acetonitrile containing 0.1% formic acid. Peptides eluting from the capillary column were selected for collision-induced dissociation (CID) by the mass spectrometer using a protocol that alternated between one MS scan (300–1500 *m/z*) and three MS/MS scans. The three most abundant precursor ions in each survey scan were selected for CID, if the intensity of the precursor ion peak exceeded 10,000 ion counts. The electrospray voltage was set to 1.8 kV and the specific *m/z* value of the peptide fragmented by CID was excluded from reanalysis for 2 min.

Each MS/MS spectrum was searched against the IPI Human database, release no. 3.18, using the Mascot Software (Matrix Science Ltd., London, UK). The probability score calculated by the software was used as the criterion for correct identification. An expectation value of less than 0.05 was required for identification. The ion score of individual peptides was required to be higher than 15. In addition, peptides were required to have a minimum sequence length of seven amino acids and to be fully tryptic with one internal missed cleavage site allowed. Methionine oxidation was included as a variable modification and mass tolerances were 1.5 Da for MS and 0.5 Da for MS/MS. Proteins with at least one peptide passing these criteria were accepted as an identification (*see* **Table 2**).

4. Notes

1. A large number of membrane proteins have been identified in urine ***(12)***. The separation buffer used in IEF-FFE contained 6 *M* urea and 2 *M* thiourea (*see* **Table 1**) to increase the solubility of these proteins. In addition, tolerable additives for the IEF-FFE process include nonionic or zwitterionic detergents such as CHAPS, CHAPSO, ASB-14, digitonin, dodecyl-ß-D-maltoside, octyl-ß-D-glucoside, Triton X-100 and Triton X-114, and reducing agents such as up to 50 m*M* DTT or TCEP.
2. In general, the sample should be as similar as possible to the separation buffer concerning the chemical and physical properties such as density, conductivity, and viscosity. Consequently, a buffer exchange prior to separation of urine was performed to lower the salt concentration from the physiological 150 m*M* NaCl to below 25 m*M*, which is tolerated in the IEF-FFE separation.
3. Turbidity of the protein samples indicates insufficient protein solubility and/or protein precipitation. Turbid protein samples will compromise resolution of the IEF-FFE separation and the particulate matter should be removed by centrifugation prior to the separation. This does not apply to organelle preparations where samples are turbid by nature.
4. The system should be cleaned thoroughly prior to use in order to remove any organic and inorganic contaminations that may compromise the separation performance.
5. We strongly recommend the use of degassed Milli-Q water to avoid formation of air bubbles in the separation chamber. To degas water, fill the container and leave it open but covered overnight.
6. If one or more fractionation tubes do not drip, fast forward the media pump to flush air out of the fractionation tubing for at least 1 min.
7. To verify that the system has a consistent laminar flow over the length of the separation chamber a colored red dye (SPADNS) is used to obtain three stripes in the separation chamber from three of the separation buffer inlets. The appearance of the stripes correlates with the laminar flow. The stripe test is performed as follows:

   7.1. Prepare stripe medium by diluting 0.5 mL of 1% SPADNS stock solution into 50 mL of Milli-Q water.

Table 2
Proteins Identified in Fraction 39 (pH 6.05) from the IEF-FFE Separation of Human Urine

Accession no.	Protein name	Mascot score	Matched peptides no.	Sequence coverage (%)	Subcellular location	Theoretical value	
						M_r	pI
IPI00017601	Ceruloplasmin	221	5	7	Secreted	122,983	5.44
IPI00009950	Vesicular integral-membrane protein VIP36	215	4	16	Membrane	40,545	6.46
IPI00215894	Isoform LMW of kininogen-1	203	5	11	Secreted	48,936	6.29
IPI00396434	Prostatic acid phosphatase	200	5	10	Secreted	44,880	5.83
IPI00013179	Prostaglandin-H2 D-isomerase	200	3	21	Secreted	21,243	7.66
IPI00166729	α_2-Glycoprotein, zinc	172	4	15	Secreted	34,465	5.71
IPI00029260	Monocyte differentiation antigen CD14	115	3	10	Membrane	40,678	5.84
IPI00419424	IGKV1-5 protein	114	1	8	Secreted	26,503	6.30
IPI00747752	25-kDa protein	112	3	13	Secreted	25,563	5.64
IPI00022434	Serum albumin	109	3	5	Secreted	71,317	5.92
IPI00022426	AMBP protein	73	2	7	Secreted	39,886	5.95
IPI00291262	Clusterin	73	2	7	Secreted	53,031	5.89
IPI00218192	Isoform 2 of inter-α-trypsin inhibitor heavy chain H4	65	1	1	Secreted	101,521	6.21
IPI00022420	Plasma retinol-binding protein	56	2	8	Secreted	23,337	5.76
IPI00021000	Isoform A of osteopontin	53	2	8	Secreted	35,572	4.37
IPI00022895	α_{1B}-Glycoprotein	49	1	2	Secreted	54,809	5.58
IPI00160130	Cubilin	48	2	< 1	Membrane	407,195	5.14
IPI00061977	IGHA1 protein	46	1	3	Membrane	55,203	6.21

7.2. Stop the media pump and place inlet tubes 2, 4, and 6 into the SPADNS solution and the remaining inlet tubes 1, 3, 5, and 7 as well as counterflow inlet tubes into a reservoir of Milli-Q water.

7.3. Move the separation chamber to a horizontal position. Run the media pump at 250 mL/h and watch for the three stripes of SPADNS dye forming in the separation chamber. Distortion of the stripes indicates leaks, uneven plate geometry, or the presence of contaminants.

7.4. If distortions are observed try to adjust the clamp or if not successful open and repeat the cleaning procedure of the separation chamber.

8. Coating the surfaces of the separation chamber with HPMC will reduce electroendosmosis.
9. To verify the separation performance of the system the p*I* marker is prepared as mentioned in **Table 1**. The sample flow rate is set to 0.5 mL/h and the p*I* marker solution is pumped into the separation chamber using the sample inlet S2. Collect fractions after 30 min into a polypropylene 96-well plate. The colored p*I* markers should migrate as thin lines in the separation chamber and be focused into a maximum of three fractions in the collection plate. The pH gradient can be determined by measuring the pH of each of the collected fractions.
10. SPADNS may be added to the sample. In this way, the sample migration can be followed visually through the FFE system. Collection of fractions can be started when the red dye reaches the chamber end. Sample collection is stopped when the red dye is no longer detectable at the outlets of the collection tubes.
11. TCEP reduces even the most stable water-soluble alkyl disulfides over a wide pH range of pH 1.5–9.
12. The IEF-FFE buffers and solutions required for the pH 3–9 gradient are available as a kit from BD Diagnostics.

Acknowledgment

The urine sample was obtained in collaboration with Cecilia Sarto, Desio Hospital, Milan, Italy.

References

1. Anderson, N. L. and Anderson, N. G. (2002) The human plasma proteome: history, character, and diagnostic prospects. *Mol. Cell. Proteomics* **1**, 845–867.
2. Klose, J. (1975) Protein mapping by combined isoelectric focusing and electrophoresis of mouse tissues. A novel approach to testing for induced point mutations in mammals. *Humangenetik* **26**, 231–243.
3. O'Farrell, P. Z. and Goodman, H. M. (1976) Resolution of simian virus 40 proteins in whole cell extracts by two-dimensional electrophoresis: heterogeneity of the major capsid protein. *Cell* **9**, 289–298.

4. Washburn, M. P., Wolters, D., and Yates, J. R. (2001) Large-scale analysis of the yeast proteome by multidimensional protein identification technology. *Nat. Biotech.* **19**, 242–247.
5. Zhang, H., Li, X.-j., Martin, D. B., and Aebersold, R. (2003) Identification and quantification of N-linked glycoproteins using hydrazide chemistry, stable isotope labeling and mass spectrometry. *Nat. Biotech.* **21**, 660–666.
6. Zhang, H., Liu, A. Y., Loriaux, P., Wollscheid, B., Zhou, Y., Watts, J. D., and Aebersold, R. (2007) Mass spectrometric detection of tissue proteins in plasma. *Mol. Cell. Proteomics* **6**, 64–71.
7. Weber, G. and Bocek, P. (1998) Recent developments in preparative free flow isoelectric focusing. *Electrophoresis* **19**, 1649–1653.
8. Moritz, R. L., Ji, H., Schutz, F., Connolly, L. M., Kapp, E. A., Speed, T. P., and Simpson, R. J. (2004) A proteome strategy for fractionating proteins and peptides using continuous free-flow electrophoresis coupled off-line to reversed-phase high-performance liquid chromatography. *Anal. Chem.* **76**, 4811–4824.
9. Moritz, R. L., Clippingdale, A. B., Kapp, E. A., Eddes, J. S., Ji, H., Gilbert, S., Connolly, L. M., and Simpson, R. J. (2005) Application of 2-D free-flow electrophoresis/RP-HPLC for proteomic analysis of human plasma depleted of multi high-abundance proteins. *Proteomics* **5**, 3402–3413.
10. Malmstrom, J., Lee, H., Nesvizhskii, A. I., Shteynberg, D., Mohanty, S., Brunner, E., Ye, M., Weber, G., Eckerskorn, C., and Aebersold, R. (2006) Optimized peptide separation and identification for mass spectrometry based proteomics via free-flow electrophoresis. *J. Proteome Res.* **5**, 2241–2249.
11. Pieper, R., Gatlin, C. L., McGrath, A. M., Makusky, A. J., Mondal, M., Seonarain, M., Field, E., Schatz, C. R., Estock, M. A., Ahmed, N., Anderson, N. G., and Steiner, S. (2004) Characterization of the human urinary proteome: a method for high-resolution display of urinary proteins on two-dimensional electrophoresis gels with a yield of nearly 1400 distinct protein spots. *Proteomics* **4**, 1159–1174.
12. Adachi, J., Kumar, C., Zhang, Y., Olsen, J., and Mann, M. (2006) The human urinary proteome contains more than 1500 proteins, including a large proportion of membrane proteins. *Genome Biol.* **7**, R80.

10

Versatile Screening for Binary Protein–Protein Interactions by Yeast Two-Hybrid Mating

Stef J. F. Letteboer and Ronald Roepman

Summary

Identification of binary protein-protein interactions is a crucial step in determining the molecular context and functional pathways of proteins. State-of-the-art proteomics techniques provide high-throughput information on the content of proteomes and protein complexes, but give little information about transient interactions, about the binary protein pairs, or about the interacting epitopes. A powerful method to reveal this information is the yeast two-hybrid system. We have employed an optimized GAL4-based yeast two-hybrid system to dissect the photoreceptor cilium-associated protein complex around the retinitis pigmentosa GTPase regulator (RPGR) in mammalian photoreceptors. This enabled us to identify associating protein partners that, similar to RPGR, were also associated with a heterogeneous group of inherited retinal degenerations arising from ciliary defects. We describe how to generate high content pretransformed cDNA libraries, and perform an efficient yeast mating screen for protein-protein interactions with any bait protein of interest.

Key Words: Protein-protein interaction; yeast two-hybrid; interaction screening; yeast mating; binary interaction; protein network; bait; GAL4.

1. Introduction

Since the first draft of the human genome was published in 2001 *(1,2)*, much effort has been made to provide functional predictions of the genes and their encoded proteins. This switch of gears to functionally dissect the human proteome can only be made with knowledge about the interplay of proteins in complexes, pathways, and networks. Although excellent and valuable proteomics techniques have been developed for efficient and high-throughput identification of protein complexes, the basis of many of those methods is affinity

From: *Methods in Molecular Biology, vol. 484: Functional Proteomics: Methods and Protocols*
Edited by: J. D. Thompson et al., DOI: 10.1007/978-1-59745-398-1, © Humana Press, Totowa, NJ

purification of stable, steady-state complexes. As protein complexes most often also have very dynamic constituents, often involved in weak, transient interactions (with a high k_{off}) between complexes *(3)*, additional tools are required to identify these components. In addition, the size of some of the protein complexes, with double digit numbers of constituents, requires information about the proteins and protein domains that physically interact in order to dissect the functional pathways in which they participate.

When the yeast two-hybrid system was first published in 1989 *(4)*, it promised to be highly suitable to reveal such information. In this system, a "bait" protein is expressed as a fusion to the GAL4-DNA-binding domain, and this is used to screen a "prey" library of cDNA expressing the encoded proteins fused to the GAL4 activation domain, by coexpression in yeast (**Fig. 1**). Both bait and prey proteins are translocated to the yeast nucleus due to the presence of a nuclear localization signal. When a bait protein interacts with a prey protein, the DNA-binding domain that is attached to the promoter region of a reporter gene is brought in close vicinity to the transcription activation domain, thus restoring functional transcription factor activity and activating the reporter gene. As an alternative to the GAL4-based system, a combination of the DNA-binding *Escherichia coli* LexA repressor protein and the transcription activating *E. coli* acidic peptide B42 can be used in a similar manner *(5)*. The reporter gene commonly involves an auxotrophic marker of the yeast cells, e.g., biosynthesis of a specific amino acid, or a compound that can be detected by a simple colorimetric assay, e.g., the enzyme β-galactosidase. Alternative yeast two-hybrid methods have also been developed that employ, e.g., membrane recruitment in the cytoplasm to restore Ras signaling, rather than DNA binding in the nucleus *(6)*.

The combination of a sensitive (plus or minus) growth selection and expression in a eukaryotic cell, allowing many of the posttranslational modifications (such as phosphorylation, glycosylation, ubiquitination, farnesylation, and geranylgeranylation) to take place, has over the years proven to be one of the most powerful approaches for detecting and evaluating binary protein–protein interactions *(7)*. It has often provided the first functional insight into the role of an unknown gene product in relation to its environment, e.g., in case of the cloning of a novel disease gene. In the past decade, we have made particular use of the GAL4-based system in identifying novel protein–protein interactions in the retina that are related to inherited retinal degenerations *(8–13)*. We have optimized the screening procedure by introducing an efficient yeast mating protocol with frozen, pretransformed cDNA prey libraries in yeast for combining the bait and prey plasmids, instead of the more commonly used but less efficient and more laborious yeast (co)transformation procedures. In addition, we have included Gateway cloning technology (Invitrogen) to rapidly generate constructs

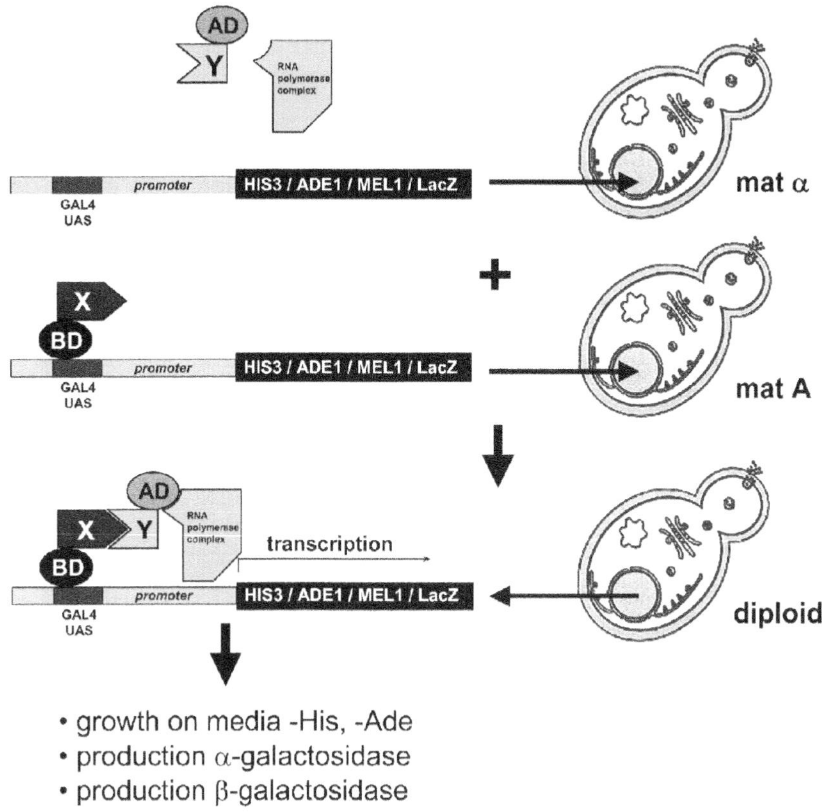

Fig. 1. Yeast two-hybrid mating strategy. A plasmid expressing a library prey protein (Y) fused to the GAL4 transcription activation domain (AD) is transformed into a yeast cell of mating type α. Upon combination (mating) with a yeast cell of the opposite mating type (A) expressing the bait protein (X) fused to the GAL4-DNA-binding domain (BD), a diploid yeast cell is formed, allowing the protein–protein interaction and consequently transcription of the reporter genes to take place. This is assessed by detecting growth on media lacking histidine and adenine, the development of a blue/green color by a plate assay detecting α-galactosidase, and detection of β-galactosidase by a LacZ filter-lift assay.

in different expression systems for essential validation of interactions. Using this setup, a single person is now able to perform four highly saturating yeast two-hybrid screens simultaneously in 1 week, requiring only the basic microbiology laboratory equipment.

Our studies exemplify the situation in which this method delivers the most biologically relevant results, namely when using tissue-specific two-hybrid cDNA libraries to screen for cytoplasmic interactors to specific motifs in the bait proteins. The outline of the system (*in trans* restoration of a functional transcription factor) limits the distance between the DNA- binding and transcription activation domains, making it more sensitive for smaller or compactly folded proteins or protein domains. In addition, steric hindrance of the folded proteins may disable detection of valid protein–protein interactions. Other drawbacks of the system include the requirement of N-terminal fusions of the bait and prey proteins with the DNA-binding and transcription activation domains (that could alter the biologically active conformation of the proteins), the requirement of the assayed proteins to translocate to the nucleus (which could be hindered in specific circumstances), and the ability of some bait protein constructs to transcriptionally activate the reporter genes in the absence of a prey protein construct (autoactivation), rendering them incompatible with the system. When taking these limitations into account, its unique properties ensures that yeast two-hybrid screening will continue to be a useful method to dissect protein complexes and networks in the years to come.

2. Materials

2.1. Generation of a Frozen Yeast Two-Hybrid Mating Library

The cDNA libraries that are used, either purchased or custom made, can be prepared from oligo(dT) or randomly primed cDNA. Optimally, each such library should be screened, as the former is more sensitive for detecting C-terminal protein domains of large proteins, as well as full-length smaller proteins, while the latter is more sensitive for detecting N-terminal domains of larger proteins, as well as different protein fragments that contain epitopes binding to the bait protein. The numbers of independent recombinants in the libraries used should be $>2 \times 10^6$ to get a good cDNA representation.

2.1.1. Preparation of Competent Yeast Cells

1. Quadruple reporter yeast strain PJ69-4α (genotype: MATα trp1-901 leu2-3, 112 ura3-52 his3Δ-200 gal4Δ gal80Δ LYS2::GAL1-HIS3 GAL2-ADE2 met2::GAL7-LacZ) ***(14)***.
2. YPAD: dissolve 50 g of Difco YPD Broth (BD Biosciences, San Jose, CA) in 1 L of Milli-Q water and add 40 mg adenine hemisulfate salt (Sigma-Aldrich, St. Louis, MO). Autoclave at 121°C for 15 min.
3. "YEASTMAKER," Yeast Transformation System 2 (BD Biosciences) (*see* **Note 1**).
4. 1.1× TE/LiAc (*see* **Note 2**): use the 10× stock solutions provided with the "YEASTMAKER," Yeast Transformation System 2 (BD Biosciences). Combine

1.1 mL of 10× TE buffer with 1.1 mL of 1 *M* LiAc (10×). Bring the total volume to 10 mL using sterile Milli-Q water.

2.1.2. Transformation of Competent Yeast Cells

1. PEG/LiAc (*see* **Note 2**): use the stock solutions provided with the "YEAST-MAKER," Yeast Transformation System 2 (BD Biosciences). Combine 8 mL of 50% PEG3350, 1 mL of 10× TE buffer, and 1 mL of 1 *M* LiAc (10×).
2. Herring testis carrier DNA, 10 μg/μL (BD Biosciences).
3. Dimethyl sulfoxide (Sigma-Aldrich) (DMSO); handle with care as this is very harmful.
4. YPD Plus liquid medium (BD Biosciences) (*see* **Note 3**).
5. 0.9% (w/v) NaCl solution: dissolve 0.9 g of sodium chloride (Merck, Darmstadt, Germany) in 100 mL Milli-Q water. Autoclave at 121°C for 15 min.
6. Library cDNA in pAD vector, 20 μg, at least 100 ng/μL (*see* **Notes 4** and **5**).

2.1.3. Plating the Transformation Mixture

1. SD –L plates: dissolve 6.7 g Difco yeast nitrogen base without amino acids (BD Biosciences), 182.2 g D-sorbitol (Sigma-Aldrich), and 0.64 g –Leu/DO (dropout) Supplement (Clontech, Mountain View, CA) in 960 mL Milli-Q water. Add 15–20 g Agar (MP Biomedicals, Solon, CA). Adjust the pH to 5.8 with NaOH. Autoclave at 121°C for 15 min. Cool down to 55°C in a waterbath. Add 40 mL sterile 50% D-(+)-glucose (Sigma-Aldrich) (*see* **Notes 6** and **7**).
2. Glass beads: 5 mm diameter (Omnilabo, Breda, The Netherlands).
3. 0.9% (w/v) NaCl solution: dissolve 0.9 g of sodium chloride (Merck) in 100 mL Milli-Q water. Autoclave at 121°C for 15 min.

2.1.4. Freezing the Cells

1. Plastic cell scraper (Corning, NY).
2. Cryogene vials (Corning).
3. Resuspension buffer 1: 5% v/v glycerol, 50 m*M* $MgSO_4$, 10 m*M* Tris, pH 8.0.
4. Resuspension buffer 2: 10% v/v glycerol, 50 m*M* $MgSO_4$, 10 m*M* Tris, pH 8.0.
5. Resuspension buffer 3: 65% v/v glycerol, 100 m*M* $MgSO_4$, 25 m*M* Tris, pH 8.0.

2.1.5. Titration

1. 0.9% (w/v) NaCl solution: dissolve 0.9 g of sodium chloride (Merck) in 100 mL Milli-Q water. Autoclave at 121°C for 15 min.
2. SD –L plates: dissolve 6.7 g Difco yeast nitrogen base without amino acids (BD Biosciences), 182.2 g D-sorbitol (Sigma-Aldrich), and 0.64 g –Leu/DO Supplement (Clontech) in 960 mL Milli-Q water. Add 15–20 g Agar (MP Biomedicals). Adjust the pH to 5.8 with NaOH. Autoclave at 121°C for 15 min. Cool down to 55°C in a waterbath. Add 40 mL sterile 50% D-(+)-glucose (Sigma-Aldrich) (*see* **Notes 6** and **7**).

2.2. Yeast Two-Hybrid Library Screening by Mating

2.2.1. Mating

1. PJ69-**4A** yeast strain (genotype: MAT**a** trp1-901 leu2-3, 112 ura3-52 his3-200 gal4 gal80 LYS2::GAL1-HIS3 GAL2-ADE2 met2::GAL7-LacZ).
2. Bait construct (gene of interest subcloned in pBD-GAL4 vector).
3. SD –W medium and plates: dissolve 6.7 g Difco yeast nitrogen base without amino acids (BD Biosciences), 182.2 g D-sorbitol (Sigma-Aldrich), and 0.74 g –Trp DO Supplement (Clontech) in 960 mL Milli-Q water. Add 15–20 g Agar (MP Biomedicals) for plates. Adjust the pH to 5.8 with NaOH. Autoclave at 121°C for 15 min. Cool down to 55°C in a waterbath. Add 40 mL sterile 50% D-(+)-glucose (Sigma-Aldrich) (*see* **Notes 6** and **7**).
4. SD –WH plates: replace –Trp by –Trp/–His DO Supplement (Clontech).
5. SD –WHA plates: replace –Trp by –Trp/–His/–Ade DO Supplement (Clontech).
6. YPAD: dissolve 50 g of Difco YPD broth (BD Biosciences) in 1 L of Milli-Q water and add 40 mg adenine hemisulfate salt (Sigma-Aldrich). For plates, add 15–20 g Agar (MP Biomedicals). Autoclave at 121°C for 15 min.
7. 0.9% (w/v) NaCl solution: dissolve 0.9 g of sodium chloride (Merck) in 100 mL Milli-Q water. Autoclave at 121°C for 15 min.

2.2.2. Screening

1. SD –LWHA plates: dissolve 6.7 g Difco yeast nitrogen base without amino acids (BD Biosciences), 182.2 g D-sorbitol (Sigma-Aldrich), and 0.60 g –Leu/–Trp/–His/–Ade DO Supplement (Clontech) in 960 mL Milli-Q water. Add 15–20 g Agar (MP Biomedicals). Adjust the pH to 5.8 with NaOH. Autoclave at 121°C for 15 min. Cool down to 55°C in a waterbath. Add 40 mL sterile 50% D-(+)-glucose (Sigma-Aldrich) (*see* **Notes 6** and **7**).

2.2.3. Determining Mating Efficiency

1. 0.9% NaCl solution: dissolve 0.9 g of sodium chloride (Merck) in 100 mL Milli-Q water. Autoclave at 121°C for 15 min.
2. SD –LW plates: dissolve 6.7 g Difco yeast nitrogen base without amino acids (BD Biosciences), 182.2 g D-sorbitol (Sigma-Aldrich), and 0.64 g –Leu/–Trp DO Supplement (Clontech) in 960 mL Milli-Q water. Add 15–20 g Agar (MP Biomedicals). Adjust the pH to 5.8 with NaOH. Autoclave at 121°C for 15 min. Cool down to 55°C in a waterbath. Add 40 mL sterile 50% D-(+)-glucose (Sigma-Aldrich) (*see* **Notes 6** and **7**).

2.2.4. Selection and Validation of Positive Clones

1. X-α-gal stock solution: dissolve X-α-gal (5-bromo-4-chloro-3-indolyl-α-D-galactopyranoside glycosynth) in *N,N*-dimethylformamide (DMF) at a concentration of 20 mg/mL. Store at –20°C.

2. SD –LWHA plates with and without 20 μg/mL X-α-gal (1:1000 dilution from stock solution).
3. X-gal stock solution: dissolve X-gal (5-bromo-4-chloro-3-indolyl-β-D-galactopyranoside, Invitrogen) in DMF at a concentration of 20 mg/mL. Store at –20°C.
4. Z-buffer with X-gal: prepare 100 mL of Z-buffer, 60 m*M* $Na_2HPO_4 \cdot 7\ H_2O$, 40 m*M* $NaH_2PO_4 \cdot H_2O$, 10 m*M* KCl, 1 m*M* $MgSO_4 \cdot 7\ H_2O$, pH 7.0; then add freshly prior to use, 0.27 mL of 2-mercaptoethanol and 1.67 mL of X-gal stock solution.
5. Liquid nitrogen.

3. Methods

It is important to use sterile techniques throughout the protocol as the yeast strains used do not possess antibiotic resistance; preferentially work in a laminar air flow (LAF) cabinet. When positive clones have been identified, activating all four reporter genes of the PJ69-4A/α diploid yeasts, the yeast two-hybrid system is also an excellent tool to map the interacting epitopes by cotransformation or mating of pAD and pBD constructs expressing the putatively interacting protein fragments. Similarly, when disease genes are the targets, the effects of mutations can be analyzed in a semiquantitative manner using ONPG as a substrate for β-galactosidase, producing a soluble orange compound that can be detected from a liquid assay by a spectrophotometer ***(11)***. However, in all cases the yeast data need to be confirmed and the biological relevance validated by independent biochemical and cell-biological methods such as GST pull-down analysis and coimmunoprecipitation and colocalization studies employing specific antibodies against the proteins of interest.

3.1. Generation of a Frozen Yeast Two-Hybrid Mating Library

3.1.1. Preparation of Competent Yeast Cells

1. Inoculate one colony (fresh, maximum 3 days old, 2–3 mm diameter) of PJ69-4α cells into 3 mL of YPAD medium in a sterile 15-mL round-bottom centrifuge tube. Vortex for 5 s to get a homogeneous single cell suspension.
2. Incubate at 30°C for 8 h in a shaking incubator (230–250 rpm).
3. Prepare three 250-mL flasks with 50 mL YPAD each. Each flask is inoculated with a different dilution of the culture from **step 2**. Prepare the following dilutions: 1:100, 1:250, and 1:1000 (*see* **Note 8**).
4. Incubate overnight at 30°C in a shaking incubator (230–250 rpm).
5. Measure the OD_{600} of the cultures against the blank (YPAD medium) (*see* **Note 9**). Use the culture in which the OD_{600} has reached 0.15–0.30 (*see* **Note 8**).
6. Centrifuge the cells at 900 × *g* for 5 min at room temperature (RT).
7. Discard the supernatant and resuspend the cell pellet in 100 mL of YPAD.
8. Incubate at 30°C for 3–5 h (final OD_{600} = 0.4–0.5).

9. Centrifuge the cells at 900 × *g* for 5 min at RT.
10. Discard the supernatant and resuspend the cells in 3 mL of 1.1× TE/LiAc solution.
11. Centrifuge the cells at 900 × *g* for 5 min at RT.
12. Discard the supernatant and resuspend the pellet in 1200 μL of 1.1× TE/LiAc solution (*see* **Note 10**).

3.1.2. Transformation of Competent Yeast Cells

1. Combine in a sterile microcentrifuge tube 1–10 μg plasmid DNA (cDNA library in pAD-Gal4) and 20 μL (200 μg) denatured herring testes carrier DNA (BD Bioscience) (*see* **Note 11**).
2. Add 600 μL TE/LiAc solution containing competent cells (from **Subheading 3.1.1, step 12**).
3. Gently mix by vortexing (vortex at half speed, for a few seconds).
4. Add 2.5 mL PEG/LiAc solution.
5. Mix thoroughly by gently vortexing (vortex at half speed, for a few seconds).
6. Incubate at 30°C for 45 min. Mix cells every 15 min by inverting and flicking the tube.
7. Add 160 μL DMSO and mix by inverting and flicking the tube. Place the tube in a 42°C water bath for 20 min. Vortex gently every 10 min.
8. Centrifuge the cells at 900 × *g* for 5 min at RT.
9. Remove the supernatant and resuspend in 3 mL of YPD plus liquid medium.
10. Incubate at 30°C with shaking (230–250 rpm) for 90 min.
11. Centrifuge the cells at 900 × *g* for 5 min at RT.
12. Discard the supernatant and resuspend in 40 mL of 0.9% (w/v) NaCl solution.

3.1.3. Plating the Transformation Mixture

1. Prepare dilutions in 0.9 % NaCl (e.g., 10^1–10^3) (*see* **Note 12**).
2. Plate (*see* **Note 13**) 100 μL of each dilution onto a 9-cm SD –L plate (*see* **Note 14**).
3. Incubate at 30°C for 2–4 days (*see* **Notes 15** and **16**).
4. Count colonies and calculate the transformation efficiency (number of transformants per microgram of DNA) and titer (number of yeast cells per milliliter) and total number of transformed yeast cells (*see* **Note 17**).
5. Plate (*see* **Note 13**) the rest of the yeast suspension onto 120 × 15-cm SD –L plates (333 μL per plate) (*see* **Note 14**).
6. Incubate at 30°C for 3–4 days (*see* **Notes 15, 18,** and **16**).

3.1.4. Freezing the Cells

1. Store the plates at 4°C for at least 1 h.
2. Scrape the cells in 10 mL Milli-Q water (at 4°C) using a cell scraper (*see* **Note 19**).

3. Rinse the plates with 10 mL Milli-Q water (at 4°C).
4. Pool all fractions and mix well.
5. Centrifuge the cells for 5 min at 900 × *g* at 4°C.
6. Resuspend the pellet in 500 mL Resuspension buffer 1 (at 4°C).
7. Centrifuge the cells for 5 min at 900 × *g* at 4°C.
8. Resuspend the pellet in 100 mL Resuspension buffer 2 (at 4°C).
9. Centrifuge the cells for 5 min at 900 × *g* at 4°C.
10. Resuspend the pellet in 1 pellet volume (~25 mL) Resuspension buffer 3 (at 4°C).
11. Pipette 0.2–0.5 mL of cell suspension in screw cap tubes (*see* **Note 20**).
12. Freeze at –80°C (*see* **Note 21**).

3.1.5. Titration

1. Thaw one tube on ice.
2. Prepare dilutions in 0.9 % NaCl (e.g., 10^5–10^9) (*see* **Note 12**).
3. Plate (*see* **Note 13**) 100 μL of each dilution onto a 9-cm SD –L plate (*see* **Note 14**).
4. Incubate at 30°C for 2–4 days (*see* **Notes 15** and **16**).
5. Count the colonies and calculate the titer (number of yeast cells per milliliter) (*see* **Note 22**).

3.2. Yeast Two-Hybrid Library Screening by Mating

3.2.1. Mating

1. Transform PJ69-**4A** yeast cells with the bait plasmid construct (cDNA of interest in pBD-GAL4). As high transformation efficiency is not important for this step, any small scale yeast transformation protocol can be used. A good method is to follow the steps described in **Subheadings 3.1.1–3.1.3**, but decreasing all culture, suspension, and sample volumes by a factor 10.
2. Plate (*see* **Note 13**) onto a SD –W plate and incubate for 2–3 days at 30°C (*see* **Notes 6, 14, 15,** and **16**).
3. Prescreen testing for intrinsic transcriptional activation (autoactivation) by the bait plasmid by restreaking three individual colonies on SD –W, SD –WH, and SD –WHA plates, followed by incubation for 2–3 days at 30°C. Growth on SD –WH and SD –WHA plates indicates autoactivation, disqualifying the bait for use in the screening procedure.
4. Inoculate 3 mL SD –W with one colony from the SD –W plate. Vortex for 5 s to disperse the cell clumps.
5. Incubate overnight at 30°C with shaking (230–250 rpm).
6. Inoculate 100 mL SD –W with overnight culture (1:50/1:100/1:250 dilutions) (*see* **Note 23**).
7. Incubate overnight at 30°C with shaking (230–250 rpm).
8. Determine OD_{600} (*see* **Note 9**) and use the culture with OD_{600} ~1 (=2.5 × 10^7cells/mL) (*see* **Note 23**).

9. Thaw a vial of frozen mating library on ice (*see* **Note 24**).
10. Incubate 4×10^8 library cells in 20 mL YPAD for 10 min at 30°C with shaking (230–250 rpm).
11. Add 8×10^8 bait cells and mix briefly.
12. Centrifuge the cell mixture for 5 min at $900 \times g$ at RT.
13. Resuspend the pellet in 2 mL YPAD.
14. Plate (*see* **Note 13**) onto four 15-cm YPAD plates (*see* **Note 14**).
15. Incubate for 4 h at 30°C (*see* **Note 15**).
16. Scrape cells in 10 mL 0.9% NaCl using a cell scraper.
17. Rinse plates with 10 mL 0.9% NaCl.
18. Pool all fractions and mix well.
19. Centrifuge the cells for 5 min at $900 \times g$.
20. Resuspend the pellet in 10 mL 0.9% NaCl.

3.2.2. Screening

1. Plate ~2.5×10^5 diploid clones (*see* **Notes 13, 14, 25**, and **26**) onto 15-cm SD –LWHA plates; prepare 10 plates.
2. Plate ~1×10^6 diploid clones (*see* **Notes 13, 14, 25**, and **26**) onto 15-cm SD –LWHA plates; prepare 10 plates.
3. Incubate at 30°C for 3–14 days, until colonies appear (*see* **Notes 15** and **26**).

3.2.3. Determining the Mating Efficiency

1. Prepare dilutions (from **Subheading 3.2.1, step 20**) in 0.9% NaCl (e.g., 10^1–10^6) (*see* **Note 12**).
2. Plate (*see* **Note 13**) each dilution onto a 9-cm SD –LW plate (*see* **Note 14**).
3. Incubate at 30°C for 2–4 days (*see* **Notes 15** and **16**).
4. Count the colonies and calculate the mating efficiency (*see* **Note 27**).

3.2.4. Selection and Validation of Positive Clones

1. Restreak colonies from **Subheading 3.2.2, step 3** onto fresh –LWHA plates, one with and two without X-α-gal, in a numbered, gridded pattern (see **Fig. 2A**), followed by incubation for 2–3 days at 30°C. Activation of the MEL1 reporter gene is detected directly on the X-α-gal plates by development of a blue-green color.
2. The LacZ reporter gene activation is detected by an X-gal filter lift assay using one plate without X-α-gal (see **Fig. 2B**). For this assay, a piece of paper (Whatman 3M or regular print paper) is cut in the right shape and placed onto the colony grid. To improve the transfer of the yeasts to the paper, it is softly but firmly pressed (using gloves) onto the surface of the plate for 1 min.
3. The paper is carefully lifted from the plate using forceps, and instantly frozen colony side up in liquid nitrogen for at least 30 s and then thawed on a clean filter paper, colony side up, for 2 min.

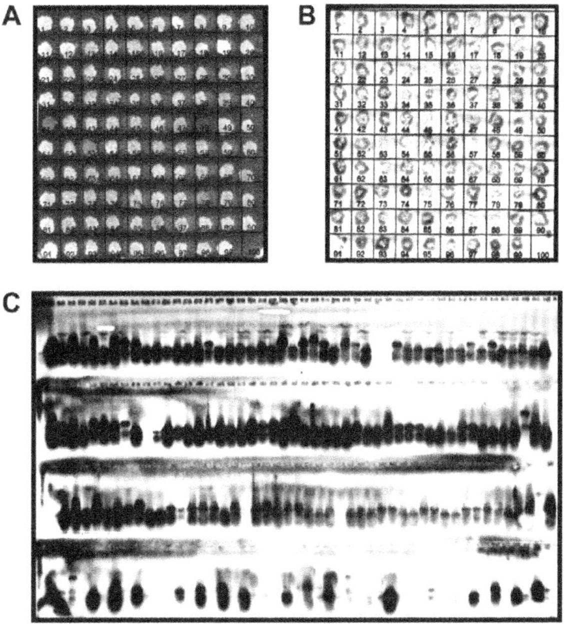

Fig. 2. Assessment of the results from yeast two-hybrid screening. (**A**) Detection of activation of the MEL1 reporter gene by a colorimetric X-α-gal plate assay, detecting α-galactosidase production and secretion by the growing yeast colonies. (**B**) Detection of activation of the LacZ reporter gene by a colorimetric X-β-gal plate assay, detecting β-galactosidase production in the yeast cytoplasm by a filter-lift assay. (**C**) Result from a validating interaction trap screen using RPGR as bait. The RPGRIP1 identity of 160 clones was determined by yeast colony PCR using vector-specific primers, followed by Southern blot analysis of the PCR products using RPGRIP1 as probe. Initial screens of RPGR using a yeast cotransformation protocol yielded on average eight times RPGRIP1, indicating a 20-fold increase in screening efficiency.

4. The freeze–thaw cycle is repeated once, and subsequently the paper is placed colony side up in a 15-cm Petri dish onto a filter paper and soaked in 4.5 mL of Z-buffer containing 2-mercaptoethanol and X-gal.
5. The Petri dishes are wrapped with parafilm and incubated at 30°C. They are monitored for the production of blue color indicating β-galactosidase activity, due to activation of the LacZ reporter gene, for up to 16 h. At that time, a faint blue color is produced by most of the yeast colonies due to native β-galactosidase activity.
6. Further downstream validation of the detected protein–protein interactions and their biological relevance can be performed in many ways and does not differ from the original yeast two-hybrid methods (*see* **Note 28**).

4. Notes

1. Any highly efficient yeast transformation protocol can be used. High efficiency is required to maintain an identical representation of all clones (small and large inserts). In our hands, the BD Biosciences YeastMaker 2 system was the most efficient one.
2. Prepare this solution fresh, just prior to transformation.
3. YPD Plus is specially formulated to promote transformation, increasing efficiency by 50–100%. Do NOT use standard YPAD medium for this step.
4. Many vectors/two-hybrid libraries can be purchased from different manufacturers. We have chosen to use the HybriZAP vectors from Stratagene (La Jolla, CA) as in this system inserts are packaged in phagemids independent of their size (up to 10 kb). Our classical use of this system has been described ***(11)***.
5. DNA purity is important and improves transformation efficiencies. Purification through common affinity columns, e.g., QIAprep plasmid purification kit (Qiagen, Venlo, The Netherlands), is sufficient.
6. Use about 25 mL of medium per 9-cm Petri dish. Do NOT make the plates too thin or they will become too dry, as they have to be incubated for at least 2 days. The plates can be sealed with parafilm or stored in a humidified incubator to protect them from desiccation. The plates can be stored for about a month at 4°C, but are preferably used fresh to get the best results (made 1–2 days in advance).
7. Use about 100 mL medium per 15-cm Petri dish (12 L total of medium for 120 plates). It is important NOT to make the plates too thin, as described in the previous note.
8. The 1:250 culture is usually the best. Take care not to overgrow the culture.
9. Make sure the yeast suspension is mixed well. Measure the OD_{600} immediately, because the cells will sediment really quickly.
10. Competent cells should be used for transformation immediately following preparation; however, if necessary they can be stored at room temperature for a few hours without significantly affecting the competency.
11. Although the herring testis carrier DNA has been denatured, we recommend denaturing the carrier DNA again upon receipt and prior to use. Transfer ~50 μL of herring DNA to a microcentrifuge tube and heat at 95°C for 5 min. Then, immediately chill the DNA by placing the tube in an ice bath.
12. Make serial dilutions. Add 100 μL of culture to 900 μL 0.9% NaCl (10^1 dilution). Add 100 μL of 10^1 dilution to 900 μL 0.9% NaCl (10^2 dilution), etc.
13. Use glass beads with a diameter of 5 mm. Use 5–10 beads per 9-cm Petri dish and 10–15 beads per 15-cm Petri dish. Shake the plates from side to side to evenly distribute the yeast suspension. Make sure that the beads do not start to swirl along the side of the Petri dish.
14. Make sure the plates are completely dry before transferring them to the incubator. Drying is enhanced by leaving the plates open in a laminar air flow (LAF) cabinet. Do NOT overdry the plates.
15. Use a humidified incubator to prevent the plates from drying out, or seal them with parafilm.

16. Incubate the plates until colonies are visible (2–3 mm in diameter). This is usually the case after 2 days of incubation.
17. The transformation efficiency can be calculated as follows: number of colonies counted × dilution factor (from **Subheading 3.1.3, step 1**) = cfu/100 µL; ×10 = cfu/mL; ×40 (resuspension volume from **Subheading 3.1.2, step 12**) = total number of cfu/amount of DNA (step 13) = transformation efficiency (cfu/µg). A confident lower limit is 10^5 cfu/µg; expected values are between 10^5 and 10^7 cfu/µg of DNA.
18. Inspect the plates daily. Make sure the plates do not dry out. If this is the case, seal the plates with parafilm. Discard the plates with fungus contamination immediately.
19. Only take out the plates that are used immediately.
20. 0.2 mL is better because only ~0.1 mL in needed per mating experiment.
21. Put the tubes directly in the –80°C freezer; do NOT use liquid nitrogen (slow freezing increases viability).
22. The titer can be calculated as follows: number of colonies counted × dilution factor (from **Subheading 3.1.5, step 2**) = cfu/100 µl; ×10 = cfu/mL. The expected titer is around 5×10^9 cfu/mL.
23. The 1:100 culture is usually the best. Take care not to overgrow the culture.
24. Approximately 100 µL of mating library cell suspension is needed per screen.
25. The higher the number of diploid clones plated, the longer it takes for positive clones to grow, the stronger the interactions need to be to withstand the selective pressure of the other yeasts, and the lower the amount of false positives commonly picked up. These properties can be used to increase or decrease the stringency of the screen. Be aware that when the stringency is decreased, the putative weaker interactions can be identified, but also the numbers of false positives isolated will rise. The latter can dramatically increase the downstream validation efforts.
26. The mating efficiency and thus the number of clones to be plated are not known in this step, and should be estimated. The mating efficiency is different per library and per experiment. Mating efficiency is usually between 5% and 50% (total $20–100 \times 10^6$ diploid cells).
27. The mating efficiency can be calculated as follows: number of colonies counted × dilution factor (from **Subheading 3.2.3, step 1**) = number of cfu/100 µl; × 10 = number of cfu/mL; × 10 (resuspension volume from **3.2.1, step 20**) = total number of cfu; 4×10^8 (input from **Subheading 3.2.1, step 10**) × 100% = mating efficiency. The expected mating efficiency is around 25% (total of 1×10^8 diploid cells).
28. When the screening is carried out in a reasonably high throughput, our first evaluation is commonly performed by polymerase chain reaction (PCR) on the positive yeast colonies using pAD-specific primers followed by sequencing of the PCR products. This makes it possible to determine if the inserts are in frame with the AD-encoding sequence, if multiple clones of the same gene have been detected, and if there is a relationship of certain clones with the (high) levels of activation that are preferably observed, indicating a strong interaction. If large numbers of

clones from a single gene are identified, Southern blotting of the PCR products could be helpful to discriminate the minority of clones that represents other genes (see **Fig. 2C**).

Acknowledgment

This work was supported by the European Commission IP "EVI-GenoRet" LSHG-CT-2005-512036.

References

1. McPherson, J. D., Marra, M., Hillier, L., Waterston, R. H., Chinwalla, A., Wallis, J., Sekhon, M., Wylie, K., Mardis, E. R., Wilson, R. K., *et al.* (2001) A physical map of the human genome. *Nature* **409**, 934–941.
2. Venter, J. C., Adams, M. D., Myers, E. W., Li, P. W., Mural, R. J., Sutton, G. G., Smith, H. O., Yandell, M., Evans, C. A., Holt, R. A., *et al.* (2001) The sequence of the human genome. *Science* **291**, 1304–1351.
3. Sprinzak, E., Altuvia, Y., and Margalit, H. (2006) Characterization and prediction of protein-protein interactions within and between complexes. *Proc. Natl. Acad. Sci. USA* **103**, 14718–14723.
4. Fields, S. and Song, O. (1989) A novel genetic system to detect protein-protein interactions. *Nature* **340**, 245–246.
5. Fashena, S. J., Serebriiskii, I. G., and Golemis, E. A. (2000) LexA-based two-hybrid systems. *Methods Enzymol.* **328**, 14–26.
6. Aronheim, A., Zandi, E., Hennemann, H., Elledge, S. J., and Karin, M. (1997) Isolation of an AP-1 repressor by a novel method for detecting protein-protein interactions. *Mol. Cell Biol.* **17**, 3094–3102.
7. Parrish, J. R., Gulyas, K. D., and Finley, R. L., Jr. (2006) Yeast two-hybrid contributions to interactome mapping. *Curr. Opin. Biotechnol.* **17**, 387–393.
8. Arts, H. H., Doherty, D., van Beersum, S. E., Parisi, M. A., Letteboer, S. J., Gorden, N. T., Peters, T. A., Marker, T., Voesenek, K., Kartono, A., Ozyurek, H., Farin, F. M., Kroes, H. Y., Wolfrum, U., Brunner, H. G., Cremers, F. P. M., Glass, I. A., Knoers, N. V. A. M., and Roepman, R. (2007) Mutations in the gene encoding the basal body protein RPGRIP1L, a nephrocystin-4 interactor, cause Joubert syndrome. *Nat. Genet.* **39**, 882–888.
9. Gosens, I., van Wijk, E., Kersten, F. F., Krieger, E., van der Zwaag, B., Marker, T., Letteboer, S. J. F., Dusseljee, S., Peters, T., Spierenburg, H. A., Punte, I. M., Wolfrum, U., Cremers, F. P. M., Kremer, H., and Roepman, R. (2007) MPP1 links the Usher protein network and the Crumbs protein complex in the retina. *Hum. Mol. Genet.* **16**, 1993–2003.
10. Kantardzhieva, A., Gosens, I., Alexeeva, S., Punte, I. M., Versteeg, I., Krieger, E., Neefjes-Mol, C. A., den Hollander, A. I., Letteboer, S. J. F., Klooster, J., Cremers, F. P. M., Roepman, R., and Wijnholds, J. (2005) MPP5 recruits MPP4 to the CRB1 complex in photoreceptors. *Invest. Ophthalmol. Vis. Sci.* **46**, 2192–2201.

11. Roepman, R., Schick, D., and Ferreira, P. A. (2000) Isolation of retinal proteins that interact with retinitis pigmentosa GTPase regulator by interaction trap screen in yeast. *Methods Enzymol.* **316**, 688–704.
12. Roepman, R., Bernoud-Hubac, N., Schick, D. E., Maugeri, A., Berger, W., Ropers, H. H., Cremers, F. P. M., and Ferreira, P. A. (2000) The retinitis pigmentosa GTPase regulator (RPGR) interacts with novel transport-like proteins in the outer segments of rod photoreceptors. *Hum. Mol. Genet.* **9**, 2095–2105.
13. Roepman, R., Letteboer, S. J. F., Arts, H. H., van Beersum, S. E. C., Lu,X., Krieger, E., Ferreira, P. A., and Cremers, F. P. M. (2005) Interaction of nephrocystin-4 and RPGRIP1 is disrupted by nephronophthisis or Leber congenital amaurosis-associated mutations. *Proc. Natl. Acad. Sci. USA* **102**, 18520–18525.
14. James, P., Halladay, J., and Craig, E. A. (1996) Genomic libraries and a host strain designed for highly efficient two-hybrid selection in yeast. *Genetics* **144**, 1425–1436.

11

Native Fractionation: Isolation of Native Membrane-Bound Protein Complexes from Porcine Rod Outer Segments Using Isopycnic Density Gradient Centrifugation

Magdalena Swiatek-de Lange, Bernd Müller, and Marius Ueffing

Summary

Networks of interacting protein control physiological processes in all living cells. Considerable effort has recently been invested in understanding protein interactions under normal and diseased conditions. One approach to elucidate the composition of protein complexes is native fractionation followed by immunological or MS-based identification of individual compounds. Native fractionation, in contrast to widespread affinity-based purification methods, allows analysis of protein interactions at the endogenous expression level and within a physiological context. In this chapter we describe a protocol for native fractionation of membrane-bound protein complexes from isolated porcine rod outer segments (ROSs). Protein complexes from isolated ROS membranes were solubilized using the nonionic detergent ß-dodecylmaltoside and fractionated by isopycnic sucrose density gradient centrifugation. Immunolabeling of individual sucrose gradient fractions demonstrated colocalization of proteins involved in the phototransduction pathway in photoreceptor outer segments.

Key Words: Membrane proteins; nonionic detergent; density gradient centrifugation.

1. Introduction

Protein–protein interactions are fundamental for almost every aspect of cellular function. Membrane-localized protein complexes are key players in most biological processes, as they regulate signaling pathways and intracellular processes and determine cellular responses to environmental stimuli.

From: *Methods in Molecular Biology, vol. 484: Functional Proteomics: Methods and Protocols*
Edited by: J. D. Thompson et al., DOI: 10.1007/978-1-59745-398-1, © Humana Press, Totowa, NJ

However, the analysis of protein complexes, especially when hydrophobic integral membrane proteins are at their core, has remained experimentally difficult. The analyses of protein complexes/interactions can be divided into high-throughput screenings, based on the yeast two-hybrid (Y2H) system *(1–3)*, split-ubiquitin technique *(4,5)*, and tandem-affinity *(6)* or immunoaffinity purification *(7)*. These experimental methods are often supported by computational interaction modeling, database systems, and analysis tools such as DIP, MINT, BIND, IntAct, and HPRD. The second experimental category aims to analyze the constituents of the protein complex in their physiological environment. Those analyses are often based on nondenaturing biochemical fractionation methods, e.g., Blue-Native PAGE *(8)* or isopycnic density gradient centrifugation *(9,10)*, and were successfully applied for purification of protein complexes from membrane fractions while retaining their native form.

The experimental procedure described here is based on density gradient centrifugation and involves four essential steps (*see also* **Fig. 1**):

1. Isolation of cellular membranes.
2. Nondenaturing solubilization of membrane-bound protein complexes.
3. Separation of protein complexes according to their density within a sucrose density gradient (**Fig. 2**).
4. Identification of individual proteins and their interaction partners by immunoblot analysis.

The isopycnic ("equilibrium") density gradient centrifugation, also known as density equilibration, resolves biological particles on the basis of varying intrinsic densities in a centrifugal field. Loaded into a linear density gradient and subjected to ultracentrifugation, the analyzed particles come to rest at points at which they are in density equilibrium with the surrounding solvent. The isopycnography has been utilized principally in the analysis of membranes, proteins, and nucleic acids.

Efficient solubilization is an essential prerequisite for native fractionation of membrane-bound protein complexes. The selection of the detergent is critical and directly affects the quality of subsequent analysis. Currently, nonionic detergents containing an uncharged, hydrophilic head and a hydrophobic, unbranched, and saturated alkyl tail are widely used for solubilization of membrane proteins. Alkyl glycosides, such as ß-dodecylmaltoside, are considered nondenaturing as they dissolve lipid–lipid or lipid–protein interactions rather than protein–protein interactions, and thus allow isolation of membrane protein complexes in their biologically active state. The alkyl glycosides are available in several forms containing variable alkyl chains attached to different polar sugar head groups (e.g., maltose, glucose) providing slightly different biochemical properties. We recommend testing different detergent

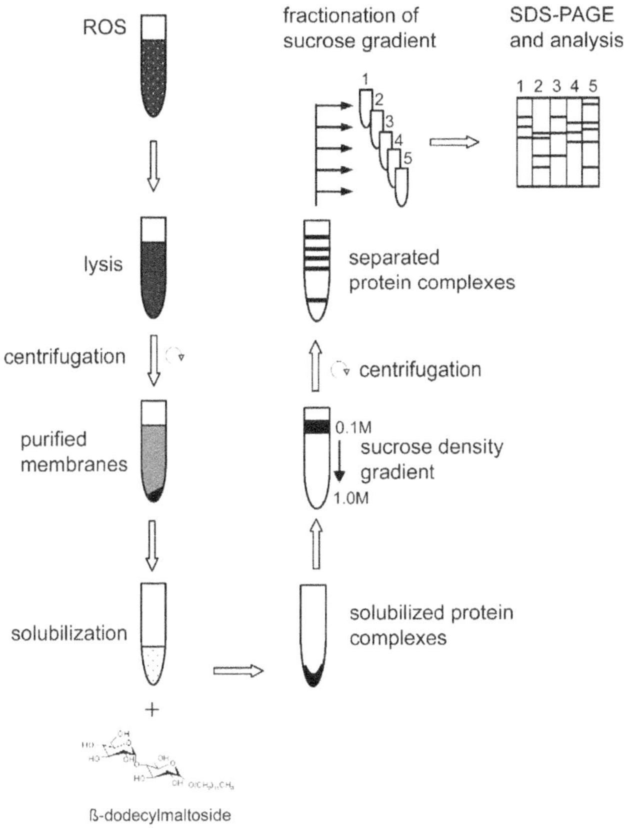

Fig. 1. Flow chart of native fractionation of membrane-bound protein complexes from porcine ROSs by isopycnic density gradient centrifugation.

types to optimize the solubilization procedure for each individual sample. Another critical parameter for successful solubilization of membrane proteins is the detergent-to-protein ratio, which correlates with the critical micelle concentration (CMC; defined as the maximum concentration of the detergent monomer above which aggregation into micelles occurs) of the chosen detergent. For detergents with high ($>1\%$) CMC the protein solubilization occurs at a concentration near the CMC, while for detergents with low CMC, more detergent has to be added to dissociate the lipid bilayer and form detergent–protein complexes. Comparative testing of ratios by varying both detergent and protein concentration is essential for determining the best suited detergent and optimal solubilization conditions.

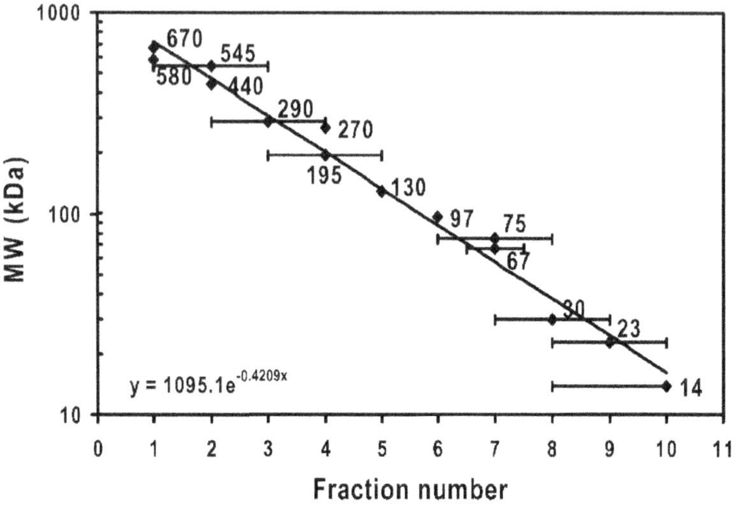

Fig. 2. Distribution of the molecular weights (MW) in 0.1–1.0 *M* sucrose density gradient. Marker proteins were solubilized in 1% (w/v) ß-DM, fractionated following the protocol described in **Subheading 3.2,** separated on SDS–PAGE following the protocol described in **Subheading 3.3,** and visualized by silver staining (**Subheading 3.6**). To create a standard curve of MW the positions of individual proteins were plotted on a half-logarithmic scale and an exponential trend line was added to the chart (*see* **Note 8**for details). The optimal separation was obtained for MWs between 620 and 16 kDa. The trend line equation is displayed below the chart.

The protocols for native fractionation of membrane-bound protein complexes presented here use isolated rod outer segments (ROSs) as starting material. Outer segments are the most peripheral subcellular structures of rod photoreceptors, connected to the inner segment by a specialized nonmotile cilium and protruding into the subretinal space toward the pigment epithelium. ROSs are filled with stacks of membranous disks containing the visual pigment rhodopsin densely arrayed in a phospholipid bilayer membrane. Rhodopsin, representing 70% of total protein in the outer segments *(11)*, may be considered a G-protein-coupled receptor of the highest endogenous expression, which offers a unique opportunity to study its interactions within a physiological context.

2. Materials

2.1. Subcellular Fractionation and Isolation of ROS Membranes

1. ROS isolated from 20–30 porcine eyes or other tissue or cells of interest.
2. Lysis buffer: 20 m*M* Tris–HCl, pH 7.2, stored at 0–4°C.
3. Bradford assay kit (Bio-Rad, Munich, Germany).

2.2. Isolation of Native Protein Complexes by Isopycnic Density Gradient Centrifugation

1. Solubilization buffer: 1% (w/v) ß-dodecylmaltoside (ß-DM, Sigma-Aldrich, Munich, Germany) in 20 m*M* Tris–HCl, pH 7.2. Prepare 10% (w/v) stock in 20 m*M* Tris–HCl, aliquot, and store at –20°C.
2. Sucrose solutions: 0.1 *M* sucrose in 20 m*M* Tris–HCl, pH 7.2, 0.06% ß-DM, and 1.0 *M* sucrose in 20 m*M* Tris–HCl, pH 7.2, 0.06% (w/v) ß-DM. Sucrose solutions are kept no longer than 1 week at 4°C. If desired, they can be aliquoted and frozen at –20°C.
3. Beckman Optima LE 80K centrifuge fitted with SP40Ti rotor, SP40Ti buckets, and ultraclear tubes (Beckman Coulter, Fullerton, CA).
4. Optional: gradient fractionator (e.g., Teledyne ISCO, Lincoln, NE).
5. Optional: peristaltic pump (e.g., Minipuls, Gilson, Middleton, WI).

2.3. Separation of Proteins by SDS–PAGE

1. 30% (w/v) acrylamide/bisacrylamide 37.5:1 solution (Bio-Rad, Munich, Germany) stored at 0–4°C. This solution is neurotoxic while unpolymerized and must be handled with extreme care.
2. 1 *M* Tris–HCl, pH 8.8. Stored at room temperature.
3. 0.5 *M* Tris–HCl, pH 6.8. Stored at room temperature.
4. 10% (w/v) SDS. Stored at room temperature.
5. TEMED (Bio-Rad, Munich, Germany, *see* **Note 1**).
6. 10% (w/v) ammonium persulfate. Small aliquots should be stored at –20°C.
7. Running buffer: (1×): 25 m*M* Tris, 192 m*M* glycine, 0.1% (w/v) SDS. Obtained as 10× TGS stock (Bio-Rad, Munich, Germany) and stored at room temperature.

2.4. Detection and Identification of Proteins

1. Semidry blotting system (Bio-Rad, Munich, Germany).
2. Transfer buffers: anode buffer I: 30 m*M* Tris, 20% methanol; anode buffer II: 300 m*M* Tris, 20% methanol; cathode buffer: 25 m*M* Tris, 40 m*M* 6-aminohexanoic acid, 20% methanol. Stored at room temperature.
3. BioTrace PVDF membrane (Pall, East Hills, NY).
4. Extra thick filter paper (Bio-Rad, Munich, Germany).
5. Tris-buffered saline: 50 m*M* Tris–HC, pH 8.0, 137 m*M* NaCl, 2.7 m*M* KCl. Routinely, TBS buffer is prepared as 10× stock and stored at room temperature.
6. TBS-T: Tris-buffered saline (1×) with 0.1% Tween. Stored at room temperature.
7. Blocking buffer: 5% (w/v) nonfat dry milk (Merck, Darmstadt, Germany) in TBS-T. Prepared fresh and stored not longer than 3 days at 0–4°C.
8. SuperSignal West Pico Chemiluminescent Substrate Kit (Pierce, Dreieich, Germany).
9. Hyperfilm ECL™ (GE Healthcare, Uppsala, Sweden).

10. Antibodies: antivisual arrestin, antitransducin α, and antirhodopsin (Affinity BioReagents, Golden, CO).

2.5. Stripping and Reprobing

1. Stripping buffer: 62.5 m*M* Tris–HCl, pH 6.8, 2% SDS (w/v), stored at room temperature. 100 m*M* 2-mercaptoethanol is added directly before use.

2.6. Silver Staining

1. Fixative solution: 50% methanol, 12% acetic acid, 0.0185% formaldehyde.
2. Washing solution: 50% ethanol.
3. Sensitizing solution: 0.8 m*M* $Na_2S_2O_3$.
4. Staining solution: 11.8 m*M* $AgNO_3$, 0.028% formaldehyde.
5. Developing solution: 0.57 *M* Na_2CO_3, 0.02 m*M* $Na_2S_2O_3$, 0.0185% formaldehyde.
6. Stopping solution: 50% methanol, 12% acetic acid.
7. Storage solution: 20% ethanol, 2% glycerol.

3. Methods

Isolated ROSs were the starting material for all downstream fractionations presented here. Isolation of ROSs from the retina is based on the method described by Molday and Molday ***(12)***. Briefly, ROSs are detached from the retinal tissue by gentle mechanical homogenization with a Potter-Elvehjem homogenizer and subsequently isolated from the homogenate by equilibrium density centrifugation in linear sucrose gradients (27–50%). The following protocol has also been successfully applied to isolate membrane-bound protein complexes from barley ***(9)*** and tobacco ***(10)*** thylakoid membranes.

3.1. Subcellular Fractionation and Isolation of ROS Membranes

1. Isolate 1–3 mg ROS from 20–30 porcine retinas.
2. Rupture ROS by hypoosmotic shock and separate intracellular membranes. Incubate 1 mg ROS in 100 μL of lysis buffer for 10 min on ice. If the material appears to aggregate add an additional 100 μL of lysis buffer (*see* **Note 2**).
3. Centrifuge samples at 16,000 × *g* for 5 min at 4°C. Collect and store the supernatant, containing cytosolic proteins if necessary. Process further with the membrane fraction (pellet).
4. Wash ROS membranes with 500 μL of lysis buffer, centrifuge at 16,000 × *g* for 5 min, and discard the supernatant.
5. Repeat **step 4**.
6. Resuspend membranes in 100 μL of lysis buffer and measure protein concentration using a Bradford assay (*see* **Note 3**).
7. Process directly to solubilization step or store isolated ROS membranes in lysis buffer at –80°C (*see* **Note 4**).

3.2. Isolation of Native Protein Complexes by Isopycnic Density Gradient Centrifugation

1. Prepare linear 0.1–1.0 *M* sucrose gradients. Use a gradient mixer with attached rubber tube whose outlet is inserted at the bottom of a centrifugation tube. Place the centrifugation tube upright in the rack below the gradient mixer.
2. Ensure that the mixer valve and stopper on the tubing are closed. Pipette 5 mL of 0.1 *M* sucrose solution into the first and 5 mL of 1.0 *M* sucrose solution into the second chamber of the gradient mixer. Place the stirring rod in each chamber and place the gradient mixer on the magnetic stirrer.
3. Start mixing and slowly open the valve to allow the solution to fill the connecting line between the two chambers. Avoid bubbles in this line.
4. Open the stopper on the tubing. Check if the liquid flowing through the first chamber is being mixed (*see* **Note 5**).
5. Prepare the sample. Spin down the ROS membrane equivalent of 1 mg protein at 16,000 × *g* for 5 min at 4°C.
6. Resuspend the ROS membranes in approx. 80 μL of lysis buffer. Add 10 μL of 10% (w/v) ß-dodecylmaltoside solution in 20 m*M* Tris–HCl to a final concentration of 1%. Fill with lysis buffer to the end volume of approx. 100 μL.
7. Solubilize membranes for 10 min on ice (*see* **Note 4**).
8. Remove unsolubilized material by centrifugation at 16,000 × *g* for 10 min at 4°C. Collect the supernatant.
9. Immediately overlay the supernatant on the sucrose gradients.
10. Carefully place the gradients in the rotor buckets and ultracentrifuge with a swing bucket rotor SW41Ti (Beckmann Coulter) for 16.5 h at 180,000 × *g* at 4°C.
11. Carefully remove the tubes from the buckets.
12. Carefully insert the centrifugation tube with separated protein complexes upright in a clamp stand.
13. Pierce the lowest point of the tube with a 20-gauge needle (*see* **Note 6**).
14. Collect the fractions of equal volume in reaction tubes (*see* **Note 7**).
15. Store individual gradient fractions at –80°C (*see* **Note 8**).

3.3. Separation of Proteins by SDS–PAGE

1. This protocol is optimized for a Bio-Rad PROTEAN II xi gel system fitted with 1.5-mm spacers and a 15-well comb. Before preparing a polyacrylamide (PAA) gel clean the glass plates well with a rinsable detergent (e.g., Deconex, Borer Chemie, Zuchwil, Swiss) and rinse extensively with distilled water and 70% ethanol.
2. Prepare a 9–15% gradient gel. For one gel, prepare 30 mL of 9% PAA solution by mixing 9 mL of acrylamide/bisacrylamide solution, 11.25 mL of 1 *M* Tris–HCl, pH 8.8, 0.3 mL of 10% SDS, and 9.45 mL of distilled water. Prepare 30 mL of 15% PAA solution by mixing 15 mL of acrylamide/bisacrylamide solution, 11.25 mL of 1 *M* Tris–HCl, pH 8.8, 0.3 mL of 10% SDS, and 3.45 mL of distilled water.

3. Degas both acrylamide solutions with constant stirring under vacuum pump for 5 min.
4. Assemble the gradient mixer as described in **Subheading 3.2, points 1–4**. Place the glass plates below the gradient mixer, insert the gradient mixer tubing between the glass plates, and attach well. Alternatively, use a needle connected to a gradient mixer tubing to obtain continuous flow.
5. Pour 25 mL 15% PAA solution into the first and 25 mL of 9% PAA solution into the second gradient mixer chamber. Start stirring (*see* **Note 9**).
6. Add 7.5 μL TEMED and 75 μL 10% APS into each chamber. Immediately open the stopper on the tubing and mixer valve and pour the gel, leaving enough space for a stacking gel.
7. Overlay the gel with water-saturated isobutanol and let polymerize for about 2 h.
8. After the gel has polymerized, remove the isobutanol and rinse with distilled water.
9. Prepare stacking gel by mixing 2.7 mL of acrylamide/bisacrylamide solution, 2.5 mL of 0.5 *M* Tris–HCl, pH 6.8, 0.1 mL of 10% SDS, and 4.7 mL of distilled water; add 6.5 μL of TEMED and 20 μL of 10% APS and mix well. Pour the stacking gel and insert the comb. The stacking gel should polymerize within 30 min. After polymerization is completed assemble the electrophoretic unit.
10. Prepare 1 L of running buffer by mixing 100 mL 10 × TGS stock with distilled water. Add running buffer to the upper and lower gel chambers and carefully remove the comb.
11. Prepare samples: mix 60 μL of the each gradient fraction with 20 μL 4× sample buffer (*see* **Note 10**).
12. Load the samples. Include one or more wells for prestained molecular weight markers.
13. Connect the electrophoretic unit to the power supply and start the run. Avoid overheating the gel: if possible perform the run in a cold chamber or under cooling (10°C). The gel can be run at 20 mA until the gel front reaches the separating gel and then at approx. 3 mA overnight. The dye fronts (bromophenol blue) can run off the gel, but the progress should be monitored by migration of prestained marker.

3.4. Immunoblotting for Rhodopsin-Associated Proteins

1. These instructions assume usage of a Bio-Rad semidry blotting system and BioTrace PVDF membranes (Pall) for protein transfer (*see* **Note 11**).
2. After completion of SDS–PAGE, disassemble the gel unit and measure and remove the separating part of the PAA gel from between the glass plates and place it in a clean tray filled with anode buffer I. Incubate under gentle rotation for 5–10 min.
3. Cut the membrane and filter paper (three sheets). The blot sandwich should be a few millimeters larger then the gel and membrane. Important: gloves must be worn at all times while handling the membrane to prevent cross-contamination.

4. Wet the membrane briefly in 100% methanol and incubate for 5 min in anode buffer I.
5. Equilibrate one extra thick filter paper in cathode buffer, one in anode buffer I, and one in anode buffer II.
6. Prepare the blot sandwich. Place the cathode buffer-equilibrated filter paper on the cathode plate and cover it with the equilibrated gel. Carefully place the preincubated PVDF membrane on the gel and cover with the filter paper equilibrated with anode I buffer followed by the last sheet of blotting paper wetted with anode II buffer. Depending on the orientation of the electrodes of the semidry blotter the blot sandwich can be inverted.
7. Remove all air bubbles between the gel and membrane. This can be done easily by rolling a Pasteur pipette across the surface of the gel/membrane sandwich. Close the system with the anode plate and activate the power supply. Blot for 1.5 h at 0.8 mA/cm^2 of gel.
8. After the transfer is completed disassemble the blotting unit. The prestained marker bands should be clearly visible on the membrane. Mark the position of the marker bands with a pencil as they tend to weaken during the blocking procedure.
9. Incubate the PVDF membrane with enough blocking buffer for at least 1 h. Blocking overnight is also possible.
10. Discard the blocking buffer and incubate the membrane in a 1:1000 dilution of antiarrestin antibodies (in blocking buffer; *see* **Notes 12** and **13**). Incubate the membrane on a rocking platform for 2 h at room temperature or overnight at 4°C.
11. Remove the primary antibody solution and wash the membrane four times for 10 min with 100 mL of TBS-T.
12. Incubate the membrane in a freshly prepared dilution of HRP-conjugated secondary antibodies in blocking buffer for 1 h at room temperature on the rocking platform.
13. Discard the secondary antibodies and wash the membrane four times for 10 min with 100 mL of TBS-T.
14. During the final wash mix equal volumes of component 1 and 2 of the chemiluminescent substrate kit (*see* **Note 14**).
15. Place the membrane in a new tray and cover with chemiluminescent substrate solution. Incubate in darkness for 5 min.
16. Discard the substrate solution, dry the membrane with Kim-Wipes, seal between two sheets of Saran wrap, and insert into the X-ray cassette.
17. Process in the dark room. Insert chemiluminescence film into the cassette with the membrane and expose it for suitable times.

3.5. Stripping and Reprobing Blots for Transducin and Rhodopsin

1. To determine the colocalization of proteins in the gradient fractions the blot membrane must be stripped of the previous signals and reprobed with another primary antibody. Alternatively, for the proteins with significant differences in MW, the blot membrane can be cut in fragments representing the MW of interest,

as monitored by the prestained marker. In such cases several antibodies can be tested simultaneously in one experiment.

2. Prepare 500 mL of stripping buffer and preheat to 50°C in a water bath (*see* **Note 15**).
3. Incubate the membrane at least four times for 30 min in 125 mL of stripping buffer at 50°C with agitation.
4. Wash the membrane at least four times for 10 min in 150 mL of TBS-T at room temperature with agitation.
5. Incubate the membrane for at least 1 h in blocking buffer and repeat the immunolabeling procedure (*see* **Note 16**) using antitransducin α antibodies (1:1000 in blocking buffer) and antirhodopsin antibodies (1:10,000 in blocking buffer).
6. Immunoblot demonstrating colocalization of rhodopsin with visual arrestin and transducin is shown in **Fig. 3**.

3.6. Silver Staining of the PAA Gel

1. As an alternative to the immunodetection, separated proteins can be visualized by silver staining. This staining procedure is based on the chemical reduction of silver ions to metallic silver on a protein band.
2. Prepare fresh fixative, washing, sensitizing, staining, and developing solutions as described in **Subheading 2.6** (*see* **Note 17**).

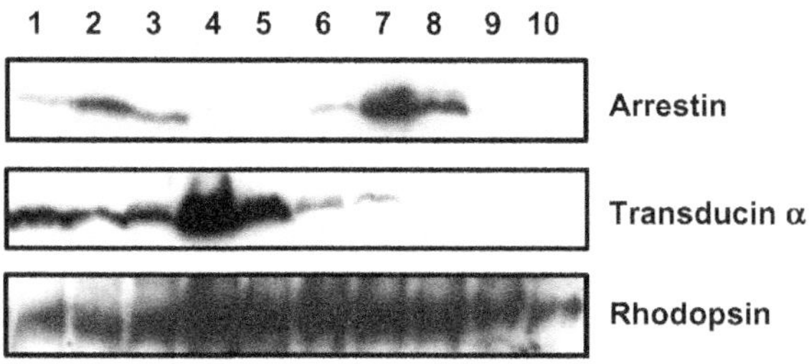

Fig. 3. Interactions of visual arrestin and transducin subunit α with rhodopsin in ROSs were confirmed by immunoblot analyses of sucrose gradient fractions. Transducin is a heterotrimeric G-protein activated by binding of photoexcited rhodopsin (metarhodopsin II). Once activated, transducin promotes the hydrolysis of cGMP by phosphodiesterase (PDE). A decrease of the intracellular cGMP level causes the closure of photoreceptor ion channels, leading to membrane hyperpolarization and, eventually, signal transmission (phototransduction cascade). Arrestin, in contrast, plays a key role in deactivation of the phototransduction cascade. Arrestin binds to the photolyzed, phosphorylated rhodopsin blocking its interaction with transducin. In agreement with their physiological roles, transducin and arrestin interact with two distinct pools of rhodopsin. Antibodies used are indicated on the right; the fraction number (from bottom to top) is indicated on the top of the panel.

3. Soak the gel in 500 mL of fixative solution for 30 min with gentle agitation.
4. Repeat the fixation step with the new 500 mL of fixative solution (*see* **Note 18**).
5. Decant the fixative and wash the gel three times for 20 min in 500 mL of washing solution with gentle agitation.
6. Soak the gel for 0.5 min in sensitizing solution.
7. Decant the sensitizer and wash the gel briefly in deionized water.
8. Soak the gel in 500 mL of staining solution for 20 min with gentle agitation. Be sure the gel is totally submerged in the solution.
9. Decant the staining solution. Rinse the gel shortly with deionized water (*see* **Note 19**).
10. Submerge the gel in developing solution until the protein bands appear.
11. When the appropriate staining intensity is reached, decant the developing solution and add stopping solution. Gently agitate the gel for 10 min.
12. Decant the stopping solution and add an appropriate volume of storage solution (*see* **Note 20**).

4. Notes

1. TEMED is a hazardous, flammable solution; store at +4°C, protected from light.
2. As isolated ROSs are open structures, the hypoosmotic shock is not used for cell disruption but rather to reopen ROSs that might seal on cilium breakage point after isolation, and to wash the preparation from contamination. Depending on the sample being analyzed, optimization of the cell rupture and membrane isolation method is necessary. The protocol for subcellular fractionation of animal tissue is described by Ryan ***(13)***.
3. We recommend the Bradford assay for determining protein concentration. The Bradford assay is based on the specific binding of Coomassie Brilliant blue G-250 to proteins and consequent stabilization of the anionic form of the dye, causing a shift of the absorbance maximum from 470 nm to 595 nm. The crucial step in this assay is preparation of the standard curve, selection of suitable protein standards (BSA or IgG), and establishing the zero point. Assay materials including dye, protein standard, and instruction book are available from Bio-Rad.
4. As isolated ROSs are open structures, the hypoosmotic shock is not used for cell disruption but rather to reopen ROSs that might seal on cilium breakage point after isolation, and to wash the preparation from contamination. Depending on the sample being analyzed, optimization of the cell rupture and membrane isolation method is necessary. The protocol for subcellular fractionation of animal tissue is described by Ryan ***(13)***.
5. The solubilization step is not only critical for disrupting the lipid bilayer but also for maintaining the protein complexes in their native form. While prolonged solubilization or highly concentrated detergents lead to protein denaturation, insufficient solubilization results in an accumulation of unsolubilized material. Therefore, the experiment must be carefully planned, as interrupting the procedure may risk loss of the sample.

6. Application of different sucrose concentrations will result in a separation of different densities. It is a matter of trial and error until precisely the right and reproducible conditions for separation of specific protein complexes are determined. As an alternative to the manual gradient preparation the peristaltic pump on low speed can be used. The sucrose gradients can be prepared the day before and stored at 0–4°C.
7. The hole should be sufficiently large to allow the sucrose solution to drip out at approx. 1 drop/s.
8. As an alternate to manual gradient fractionation, a mechanical gradient fractionator (e.g., Teledyne ISCO) may be used. Here, the fractions are collected in precise volumes by introducing a dense chase solution at the bottom of the centrifuge tube and then raising the gradient intact by bulk flow.
9. Estimation of molecular weight distribution within a sucrose density gradient may be done with native marker proteins of known molecular weight. Marker proteins are solubilized in 1% (w/v) ß-DM in 20 m*M* Tris–HCl, pH 7.2, and separated by ultracentrifugation in 0.1–1.0 *M* sucrose density gradients, following the protocol described in **Subheading 3.2.** We propose the following native protein mixtures as markers:

   9.1. HMW Electrophoresis calibration kit (Pharmacia): Thyroglobulin (669 kDa), ferritin (440 kDa), catalase (232 kDa), lactate dehydrogenase (140 kDa), and BSA (67 kDa).
   9.2. Kit for molecular weights 14,000–500,000 (Sigma-Aldrich): Urease (hexamer: 545 kDa; trimer: 272 kDa), BSA (dimer: 132 kDa; monomer: 66 kDa), albumin (45 kDa), carboanhydratase (29 kDa), and lactalbumin (14.2 kDa).
   9.3. Crosslinked phosphorylase *b* (Sigma-Aldrich): hexamer to monomer of phosphorylase *b*: 584.4, 487, 389.6, 292.2, 194.8, and 97.4 kDa, respectively.

   After gradient fractionation, individual proteins are separated by SDS–PAGE (*see* **Subheading 3.3**) and visualized by silver staining (*see* **Subheadings 3.6**). The positions of the individual proteins are then plotted on a half-logarithmic scale. To create a standard curve of molecular weight distribution in a sucrose gradient an exponential trend line is added to the chart and the corresponding equation is calculated and displayed. The molecular weight distribution in 0.1–1.0 *M* sucrose density gradient fractions is shown in **Fig. 2**.
10. In contrast to sucrose gradient, PAA gels are poured from the top, such that the heavy solution is loaded first. Alternatively, commercially available gradient-casting chambers (e.g., GE Healthcare, Bio-Rad) may be used.
11. For some proteins, e.g., rhodopsin, heating of the sample might cause protein aggregation and should be avoided.
12. The blotting system and membrane type should be optimized for the antibodies used. The advantage of a PVDF membrane is improved protein capture and retention, low background, and physical strength of the supporting membrane. The advantage of a nitrocellulose membrane is the ability to control transfer efficiency by Ponceau S staining.

13. For the first Western blot always use antibodies raised against the least abundant protein. As rhodopsin represents the most abundant ROS protein, immunodetection will be performed after stripping of the membrane. The amount of antibodies used can be reduced to 5–10 mL for a 20-cm membrane if the membrane and the primary antibody solution are sealed between two sheets of a plastic foil and incubated on rocking mixer.
14. The working solution is stable for a minimum of 24 h at room temperature. The solutions can be used in both light and dark conditions.
15. Temperatures higher than 50°C can damage the membrane. To avoid unpleasant smells in the laboratory work under an activated fume hood.
16. To control stripping efficiency, block the membrane in blocking solution for 1 h and reincubate with the secondary antibodies and substrate solution. Expose the membrane at least as long as the original exposure to show that primary antibodies are completely removed. If signals are still detected repeat the stripping procedure.
17. Silver nitrate will irreversibly stain the skin and fabric; it is also a severe skin and eye irritant and possible carcinogen. Always wear protective gloves and clothing during all steps of the staining procedure. Use clean containers and designate them for silver staining only.
18. The gel can be stored in fixative for up to 3 days, but longer fixation may affect staining efficiency.
19. Prolonged washing of the gel will remove silver ions from the polyacrylamide matrix and result in decreased sensitivity.
20. Stained gels can be stored up to 1 week without loss of staining quality. For more permanent storage, gels can be vacuum or air dried.

Acknowledgments

Work was funded by EU Grants PRO-AGE-RET QLK6-CT-2001-00385, RETNET MRTN-CT-2003-504003, EVI-GENORET: LSHG-CT-2005 512036, and INTERACTION PROTEOME LSHG-CT-2003-505520 and by funding from the German Federal Ministry of Education and Research: BMBF-Proteomics 031U108A/031U208A. We thank Dr. Ursula Olazabal for critical comments on the manuscript.

References

1. Uetz, P., Glot, L., Cagney, G., Mansfield, T. A., Judson, R. S., Knight, J. R., Lockshon, D., Narayan, V., Srinivasan, M., Pochart, P., Qureshi-Emlli, A., Li, Y., Godwin, B., Conover, D., Kalbfleisch, T., Vijayadamodar, G., Yang, M., Johnston, M., Fields, S., and Rothberg, J. M. (2000) A comprehensive analysis of protein-protein interactions in Saccharomyces cerevisiae. *Nature* **403**, 623–627.

2. Giot, L., Bader, J. S., Brouwer, C., Chaudhuri, A., Kuang, B., Li, Y., Hao Y. L., Ooi C. E., Godwin, B., Vitols, E., Vijayadamodar, G., Pochart, P., Machineni, H., Welsh, M., Kong, Y., Zerhusen, B., Malcolm, R., Varrone, Z., Collis, A., Minto, M., Burgess, S., McDaniel, L., Stimpson, E., Spriggs, F., Williams, J., Neurath, K., Ioime, N., Agee, M., Voss, E., Furtak, K., Renzulli, R., Aanensen, N., Carrolla, S., Bickelhaupt, E., Lazovatsky, Y., DaSilva, A., Zhong, J., Stanyon, C. A., Finley, R. L., Jr., White, K. P., Braverman, M., Jarvie, T., Gold, S., Leach, M., Knight, J., Shimkets, R. A., McKenna, M. P., Chant, J., and Rothberg J. M. (2003) A protein interaction map of *Drosophila melanogaster. Science* **302**, 1727–1736.
3. Parrish, J. R., Gulyas, K. D., and Finley, R. L. Jr. (2006) Yeast two-hybrid contributions to interactome mapping. *Curr. Opin. Biotechnol.* **17**, 387–393.
4. Miller, J. P., Lo, R. S., Ben-Hur, A., Desmarais, C., Stagljar, I., Noble, W. S., and Fields, S. (2005) Large-scale identification of yeast integral membrane protein interactions. *Proc. Natl. Acad. Sci. USA* **102, 12123–12128.**
5. Thaminy, S., Miller, J., and Stagljar, I. (2004) The split-ubiquitin membrane-based yeast two-hybrid system. In: *Methods in Molecular Biology* (Clifton, N. J. ed.), pp. **297–312.** Humana Press, Totowa, NJ.
6. Gavin, A.-C., Bösche, M., Krause, R., Grandi, P., Marzioch, M., Bauer, A., Schultz, J., Rick, J. M., Michon, A.-M., Cruciat, C.-M., Remor, M., Höfert, C., Schelder, M., Brajenovic, M., Ruffner, H., Merino, A., Klein, K., Hudak, M., Dickson, D., Rudi, T., Gnau, V., Bauch, A., Bastuck, S., Huhse, B., Leutwein, C., Heurtier, M.-A., Copley, R. R., Edelmann, A., Querfurth, E., Rybin, V., Drewes, G., Raida, M., Bouwmeester, T., Bork, P., Seraphin, B., Kuster, B., Neubauer, G., and Superti-Furga, G. (2002) Functional organization of the yeast proteome by systematic analysis of protein complexes. *Nature* **415**, 141–147.
7. Ho, Y., Gruhler, A., Heilbut, A., Bader, G. D., Moore, L., Adams, S.-L., Millar, A., Taylor, P., Bennett, K., Boutilier, K., Yang, L., Wolting, C., Donaldson, I., Schandorff, S., Shewnarane, J., Vo, M., Taggart, J., Goudreault, M., Muskat, B., Alfarano, C., Dewar, D., Lin, Z., Michalickova, K., Willems, A. R., Sassi, H., Nielsen, P. A., Rasmussen, K. J., Andersen, J. R., Johansen, L. E., Hansen, L. H., Jespersen, H., Podtelejnikov, A., Nielsen, E., Crawford, J., Poulsen, V., Sørensen, B. D., Matthiesen, J., Hendrickson, R. C., Gleeson, F., Pawson, T., Moran, M. F., Durocher, D., Mann, M., Hogue, C. W. V., Figeys, D., and Tyers, M. (2002) Systematic identification of protein complexes in Saccharomyces cerevisiae by mass spectrometry. *Nature* **415**, 180–183.
8. Schagger, H. and Von Jagow, G. (1991) Blue native electrophoresis for isolation of membrane protein complexes in enzymatically active form. *Anal. Biochem.* **199**, 223–231.
9. Müller, B. and Eichacker, L. A. (1999) Assembly of the D1 precursor in monomeric photosystem II reaction center precomplexes precedes chlorophyll a-triggered accumulation of reaction center II in barley etioplasts. *Plant Cell* **11**, 2365–2377.
10. Swiatek, M., Kuras, R., Sokolenko, A., Higgs, D., Olive, J., Cinque, G., Müller, B., Eichacker, L. A., Stern, D. B., Bassi, R., Herrmann, R. G., and Wollman, F. A. (2001) The chloroplast gene ycf9 encodes a photosystem II (PSII) core subunit,

PsbZ, that participates in PSII supramolecular architecture. *Plant Cell* **13**, 1347–1367.

11. Hamm, H. E. and Deric Bownds, M. (1986) Protein complement of rod outer segments of frog retina. *Biochemistry* **25**, 4512–4523.
12. Molday, R. S. and Molday, L. L. (1987) Differences in the protein composition of bovine retinal rod outer segment disk and plasma membranes isolated by a ricin-gold-dextran density perturbation method. *J. Cell Biol.* **105**, 2589–2601.
13. Ryan, N. M. (2004) Subcellular fractionation of animal tissues. In: *Methods in Molecular Biology* (Clifton, N. J. ed.), pp. 47–52. Humana Press, Totowa, NJ.

12

Mapping of Signaling Pathways by Functional Interaction Proteomics

Alex von Kriegsheim, Christian Preisinger, and Walter Kolch

Summary

Signaling pathways transduce extracellular stimuli from the membrane to the nucleus. Constitutive and thus inappropriate stimulation of these kinase cascades is associated with and observed in a majority of tumors. The transduction of signals in these pathways is achieved through protein–protein interactions regulated by changes in the phosphorylation status of key members. Therefore, the analysis of the interactions formed or broken in response to mitogenic stimulation is an important step toward understanding the molecular mechanisms of carcinogenesis. Today, mass spectrometry-based proteomics is one of the most widely used methods to unravel the molecular protein interaction networks that underlie these signaling cascades. This approach is powerful, but usually results in long lists of binding partners that may contain many false-positive hits and no information about the physiological role of the interacting proteins. Functional information can be derived by mapping changes in the interactome in response to specific stimuli or by comparing the interactome of related proteins with overlapping and different biological functions. As paradigms for these experimental approaches and the associated methodology, we describe here the functional proteomic analysis of the interactome of two distinct members of the mitogen-activated protein kinase (MAPK) cascade. The first is the analysis of interaction partners of the extracellular signal-regulated kinase (ERK) regulated by growth factor stimulation. The second is the differential analysis of binding partners of the C-terminal SH3 domain of the two small adaptor proteins Grb2 and GRAP.

Key Words: Functional interaction proteomics; signal transduction networks; protein interactions; MAP kinase; ERK; mitogen stimulation; adaptor proteins; SH3; protein domains; GST pulldowns; SILAC; proteomics; mass spectrometry.

From: *Methods in Molecular Biology, vol. 484: Functional Proteomics: Methods and Protocols*
Edited by: J. D. Thompson et al., DOI: 10.1007/978-1-59745-398-1, © Humana Press, Totowa, NJ

1. Introduction

Signal transduction typically starts with the binding of a ligand to its cognate receptor at the cell surface *(1–8)*. In the case of receptor tyrosine kinases (RTKs), ligand binding induces the phosphorylation of the cytoplasmic kinase domain of the receptors. These phosphate residues serve as docking sites for adaptor proteins, such as Grb2, and enable RTKs to recruit specific binding proteins and assemble multiprotein signaling complexes at the plasma membrane *(9)*. Adaptor proteins play an important role in the assembly of the signaling platforms that link the receptor complexes to downstream effectors. Adaptor proteins typically contain multiple functional binding regions such as SH2 and SH3 domains. Upon epidermal growth factor (EGF) stimulation Grb2 is one of the central adaptor proteins associated with assembling the functional receptor signaling complex *(10)*. Recruitment of Grb2 to the membrane occurs by its direct or indirect (via the Shc adaptor protein) association with the autophosphorylated intracellular domain of the EGF receptor. Grb2 is associated with SOS, a RasGEF bound to one of the SH3 domains of Grb2. Membrane-localized SOS is then able to interact with and activate Ras, which subsequently can activate the core kinase module of the extracellular signal-regulated kinase (ERK) pathway by recruiting and activating Raf kinases *(1,11)*. Raf then phosphorylates and activates MEK, which in turn activates ERKs by phosphorylating them in the activation loop. ERK, which can phosphorylate over 160 substrates *(1,12)*, is widely seen as a key effector that contributes to many fundamental biological processes including proliferation, differentiation, survival, transformation, and cell fate decisions to name a few. Localization, signal amplitude, and duration have all been shown to be crucial for ERK substrate selection *(1,12)*.

Adaptor proteins play a crucial role in mediating signaling events *(8)*. Many of these proteins contain several small modular domains that can interact with various regions in their respective binding partners. Examples are SH2 (or PTB) domains that specifically bind phosphorylated tyrosine residues and SH3 domains that interact with proline-rich sequence stretches, such as the PxxP motif.

We use mass spectrometry-based proteomics as it enables us to analyze and quantify the composition of protein complexes in cells in response to specific stimuli. This is a very efficient way to eliminate unspecific interactions, as they do not change in response to stimulation, and pinpoint those interactions that are functionally important by the fact that they change in correlation with a specific stimulus. In the first part of this chapter we describe the analysis and changes of ERK1 protein-binding partners upon induction of ERK1 phosphorylation by the EGF. We show the successful usage of stable isotopic labeling in cell culture (SILAC) for quantitative determination of these changes. The second part of this chapter describes the use of glutathione *S*-transferase (GST) tag-based pulldown experiments. Pulldowns using tagged proteins are usually the system of choice if there are no good immunoprecipitating antibodies against the endogenous protein available, or if a functional domain of a protein needs to be analyzed in isolation ***(13,14)***. As an example, we show the different binding properties of two SH3 domains in the two closely related adaptor proteins Grb2 and GRAP ***(7,8)***. There are many options for tags for pulldowns, and pulldowns can be performed either by expressing the bait protein in cells or by incubating cell lysates with the bait protein immobilized on a solid support. In our experience GST and flag-tags are satisfactory for proteomics experiments. The green fluorescent protein (GFP) that is commonly used to tag proteins for live cell imaging experiments conveniently also can be used as a tag for immunoprecipitation and subsequent mass spectrometry (MS) analysis ***(15)***. Tandem affinity purification (TAP) tags have been developed in various versions ***(16)***. They use a two-step purification procedure and thus permit the isolation of highly purified protein complexes, but because of the lengthy procedure are not well suited for the analysis of dynamic changes in protein interactions. In all pulldown methods that use antibodies it is crucial to covalently crosslink the antibodies to the solid support in order to avoid contamination of the samples with antibodies, which will hamper MS analysis.

2. Materials

2.1. Cell Culture, Lysis, and Immunoprecipitation from PC12 Cells

1. Dulbecco's modified Eagle's medium (DMEM; Gibco, 31885) supplemented with glycine, 10% heat-inactivated (1 h at 58°C in a water bath) horse serum, and 5% heat-inactivated (1 h at 58°C in a water bath) fetal calf serum.
2. Starvation medium; DMEM supplemented with glycine, 0.1% heat-inactivated (1 h at 58°C in a water bath) horse serum, and 0.05% heat-inactivated (1 h at 58°C in a water bath) fetal calf serum.
3. Phosphate-buffered saline (PBS): 8 g of NaCl, 0.2 g of KCl, 1.15 g of Na_2HPO_4, and 0.2 g of KH_2PO_4 per liter of water (*see* **Note 1**).
4. Rat tail collagen solution (Upstate, 08-115).
5. EGF (Roche, Cat. #1376454) at 20 μg/mL in DMEM.
6. HEPES lysis buffer: 20 m*M* HEPES-NaOH, pH 7.5, 150 m*M* NaCl, 1% NP-40, 2 m*M* EDTA, 1 m*M* phenylmethylsulfonylfluoride (PMSF), 2 m*M* sodium fluoride (NaF), 1 m*M* sodium vanadate (Na_3VO_4), 5 μg/mL leupeptin, 2.2 μg/mL aprotinin, 1 m*M* sodium pyrophosphate ($Na_3P_2O_7$), and 20 m*M* β-glycerophosphate.
7. HEPES wash buffer: 20 m*M* HEPES-NaOH, pH 7.5, 50 m*M* NaCl, 0.1% NP-40, 2 m*M* EDTA, 1 m*M* PMSF, 2 m*M* sodium fluoride, 1 m*M* sodium vanadate, 5 μg/mL leupeptin, 2.2 μg/mL aprotinin, and 20 m*M* β-glycerophosphate.
8. Spin columns: Micro Bio-Spin Chromatography Columns, empty (Bio-Rad 732-6204).
9. Glycine elution buffer: 200 m*M* glycine–HCl, pH 2.5, 500 m*M* NaCl, and 0.1% NP-40.

2.2. Cell Culture, Lysis, and GST Pulldowns from K562

1. Roswell Park Memorial Institute medium (RPMI 1640; Gibco, 21870) supplemented with glutamine (5 mL of 100× stock solution), 10% heat-inactivated (1 h at 58°C in a water bath) horse serum, and 5 mL of a 100× penicillin/streptomycin stock solution.
2. PBS: 8 g NaCl, 0.2 g KCl, 1.15 g Na_2HPO_4, and 0.2 g KH_2PO_4 per liter.
3. NP-40 lysis buffer: 20 m*M* HEPES-NaOH, pH 7.4, 150 m*M* NaCl, 0.5% NP-40, 2 m*M* EDTA, 1 m*M* PMSF, 2 m*M* sodium fluoride, 1 m*M* sodium vanadate, 5 μg/mL leupeptin, 2.2 μg/mL aprotinin, 1 m*M* sodium pyrophosphate, and 20 m*M* β-glycerophosphate.
4. Glutathione Sepharose 4B (GE-Healthcare).
5. Glycine elution buffer: 200 m*M* glycine–HCl, pH 2.5, 500 m*M* NaCl, and 0.1% NP-40.

2.3. Cross-Linking of Antibodies

1. Cross-linking buffer: 100 m*M* HEPES-NaOH, pH 8.5, and 10 mg/mL DMP (Pierce).
2. Cross-linking wash buffer: 100 m*M* HEPES-NaOH, pH 8.5.
3. HEPES lysis buffer: 20 m*M* HEPES-NaOH, pH 7.5, 150 m*M* NaCl, 1% NP-40, and 2 m*M* EDTA.
4. Protein A Sepharose (GE-Healthcare).
5. Rabbit anti-ERK1 antibody (Santa Cruz Biotechnology, sc-93).

2.4. Sodium Dodecyl Sulfate–Polyacrylamide Gel Electrophoresis (SDS-PAGE); Precasted Gels

1. Novex gel system.
2. NuPAGE® MOPS SDS Running Buffer.
3. 4x NuPAGE® LDS Sample Buffer with 100 m*M* DTT.
4. NuPAGE® 10% Bis-Tris Gel, 10 well, 1 mm thickness.

2.4.1. Colloidal Coomassie Solution

1. Dissolve 100 g $(NH_4)_2SO_4$ in 750 mL H_2O in a beaker, add 30 mL of H_3PO_4 and 1 g of Coomassie G-250, stir on a magnetic stirrer for 30 min, and then store in a light-proof bottle.
2. Prior to staining shake the bottle vigorously and pour 20 mL of the slurry into a 50-mL centrifuge tube. Add 5 mL methanol and vortex for 30 s. The slurry is now ready to use.

2.4.2. Destaining Solution

1. 25% methanol in H_2O (*see* **Note 2**).

2.5. Tryptic In-Gel Digest

1. 50% MeOH/50 m*M* NH_4HCO_3 in H_2O.
2. 50 m*M* NH_4HCO_3 in H_2O.
3. 100% acetonitrile.
4. 10 m*M* dithiothreitol (DTT) in 50 m*M* NH_4HCO_3 in H_2O.
5. 55 m*M* iodoacetamide in 50 m*M* NH_4HCO_3 in H_2O.
6. 125 ng/μL porcine modified trypsin (Promega) in 1 m*M* HCl in an H_2O stock solution. Dilute to 12.5 ng/μL with 50 m*M* NH_4HCO_3 in H_2O.
7. 1% trifluoroacetic acid (TFA) in 50% acetonitrile/H_2O.

3. Methods

To obtain accurate interaction partners of phosphorylated ERK1 it is of the utmost importance to proceed swiftly after EGF treatment. Since protein–protein interactions, especially those of activated ERK1 with its phosphorylated substrates, are very transient, it is important to limit the duration of the experiment and to keep the samples on ice at all times. We have indicated time points at which the experiment can be halted without any loss of detection and accuracy levels.

It is widely known that proteins with highly homologous protein domains (e.g., SH3 domains) can have completely different binding partners. These domains are usually rather small (less than 100 amino acids) and therefore require fusion to a protein to enable pulldown experiments. The usage of the GST tag, which itself is rather large (26 kDa), requires extensive preincubation of the samples with glutathione Sepharose beads and the GST protein before the actual pulldown experiment can be performed since the glutathione Sepharose beads and GST itself can bind and precipitate a multitude of proteins.

3.1. DMP Cross-Linking of Antibodies to Protein A

1. Pipette 100 μl of Protein A beads into a 1.5-mL microfuge tube and wash in 1 mL HEPES lysis buffer three times by sequentially mixing the beads with the buffer; centrifuge the beads to the bottom of the tube and remove the buffer with a 1-mL pipette (*see* **Note 4**).
2. Add 50 μg of antibody (200 μL) and 800 μL of HEPES lysis buffer to the beads and incubate on a roller at 4°C for 2 h.
3. Wash the beads (*see* 1) three times with 1 mL HEPES lysis buffer.
4. Wash the beads two times with 1 mL HEPES cross-linking wash buffer.
5. Incubate the beads with 1 mL cross-linking buffer containing DMP and shake on a rocker platform at room temperature for 1 h (*see* **Note 5**).
6. Wash the beads two times with 1 mL HEPES cross-linking wash buffer.
7. Quench the reaction by adding 1 mL 100 m*M* Tris–HCl, pH 7.5, and shake for 30 min at room temperature.

8. Wash with 1 mL HEPES lysis buffer twice followed by two washes with 1 mL elution buffer.
9. Wash with 1 mL HEPES lysis buffer twice and add 200 µL HEPES lysis buffer with sodium azide (0.02%) to the slurry. Keep the antibody beads at 4°C.

3.2. Preparation of Samples from PC12 Cells (see Notes 3 and 6)

1. PC12 cells are passaged when approaching 70% confluence and are split between one-half and one-quarter. The cells double every 48 h. They loosely attach to the surface and therefore splitting does not require trypsinization. To split remove 80% of medium and add 20% of fresh medium, shake the flask vigorously a couple of times to detach the cells from the surface, and split.
2. Seed the cells on collagen prior to EGF stimulation. Prepare the plates by diluting the collagen solution (Upstate Collagen Type I, rat tail 08-115) 1/200 in PBS. Then incubate 14-cm plates with 20 mL of the collagen/PBS solution for 30 min. Remove the collagen/PBS solution and plate the PC12 cells.
3. When the cells reach 50–70% confluence on the plates remove the DMEM, wash the cells with PBS, and serum starve the cells overnight in starvation medium.
4. The starved cells can then be stimulated with EGF (20 ng/mL) for desired periods of time.
5. After the treatment place the cells on ice and wash once with ice-cold PBS and lyse with 1 mL lysis buffer per 14-cm plate by scraping the cells off the plate into the lysis buffer.
6. Transfer the lysate in numbered 2-mL microfuge tubes and incubate on ice for 10 min with occasional vortexing.
7. Clear the lysates by centrifugation at 25,500 × *g* in a cooled (4°C) bench top centrifuge (Eppendorf 5127R) for 10 min.
8. Incubate the cleared lysates with the cross-linked antibody beads at 20 µg antibody per plate for 2 h.
9. Transfer the beads into a spin column and wash three times with ice-cold HEPES wash buffer by sequential mixing of the beads with the buffer and removing the buffer by centrifuging the buffer for a few seconds into a 2-mL microfuge tube at 1000 × *g*.
10. After the last wash incubate the dry beads with two bed volumes of the glycine elution buffer for 5 min on ice with occasional vortexing. Remove the eluate by centrifuging into a clean 2-mL microfuge tube. Repeat the elution once more.
11. Neutralize the pH of the combined eluates by adding 10% of the eluate volume of 2 *M* Tris–HCl, pH 9.
12. Concentrate the eluate by centrifugal filtration using a 3-kDa cutoff membrane (Eppendorf Microcon Ultracel YM-3) at 15°C, 14,000 × *g*, for 120 min.
13. Remove the concentrated sample by placing the sample reservoir upside down into a clean tube and centrifuge for 1 min at 1000 × *g*.
14. Determine the sample volume by pipetting and add one-fourth volume per volume of the LDS sample buffer. Denature the sample by heating it to 57°C for 15 min on a thermomixer. At this stage the samples can be frozen.

3.3. SDS–PAGE and Coomassie Staining

1. Wear gloves to avoid keratin contamination, and if possible do all manipulations in a dust-free environment, such as a laminar flow hood. Use precast gels to avoid contamination of samples by keratins, which is common with self-made gels. Open the gel pouch, rinse with water, and remove the adhesive tape from the bottom of the gel cassette
2. Insert a 4–12% NuPAGE Gradient gel (with 10 wells, 1 m*M* thickness) into the XCell SureLock mini-Cell with the comb facing the inside chamber and the plastic dam on the other side. Lock the gel and make sure that the electrodes are properly slotted.
3. Fill the inner chamber with 1 × MOPS running buffer and remove the comb.
4. Load 5 μL of the marker (Precision Plus Dual Colour Standard, Bio-Rad) in the first well and your sample in the third well. Load subsequent samples with one empty well between samples.
5. Fill the outer chamber with 1 × MOPS buffer, connect the electrodes to a power supply, and run the gel at constant 100 V until the dye front has reached the end of the gel.
6. Turn off the power and disconnect the electrodes. Remove the gel and open the cassette with the metal wedge provided by Invitrogen or a strong spatula or screwdriver.
7. After opening the cassette the gel will stick to one side; cut the gel with a wedge and drop the gel into a 14-cm cell culture dish with a lid.
8. Add 25 mL of the fixing solution and shake for 15 min at room temperature.
9. Replace the fixing solution with 25 mL of water and shake for 5 min at room temperature.
10. Remove the water and add the colloidal Coomassie staining solution and stain overnight.
11. Remove the stain and destain the gel with 25% methanol in water for 1 min and several washes of water until the background is clear.
12. Cut the gel with a scalpel into slices; try not to split the major protein bands.
13. Cut each slice into cubes of about 2 mm^3 and transfer them into clearly labeled 1.5-mL microfuge tubes. At this stage the samples can be frozen

3.4. GST Pulldowns from K562 Cell Lysates

3.4.1. GST Pulldowns for MS Analysis

This protocol assumes that you already have purified GST fusion proteins. There is a myriad of protocols available on either the WWW, in general laboratory method handbooks, and in vendors' manuals.

This protocol describes the use of GST pulldowns for proteomic analysis of interaction partners by mass spectrometry. The use of GST pulldowns for Western blot analysis is explained below.

1. K562 cells grow in suspension. They can thus be easily counted using a hemocytometer. Splitting does not require trypsinization, and should be done by diluting 5 mL of cells in 45 mL of fresh growth medium.
2. Grow K562 cells in 50 mL of RPMI growth medium in 175-cm^2 cell culture flasks to a cell density of 1–3 × 10^7 cells/mL. The amount required for cell lysate containing 100 mg of protein is approximately 10–12 flasks.
3. Harvest the cells by centrifugation in 50-mL cell culture tubes at 1000 × *g* for 2 min at room temperature. Use one 50-mL cell culture tube per 175 cm^2 flask.
4. Take off the cleared growth medium by using a Pasteur pipette attached to a vacuum pump.
5. Transfer the cells by adding 1 mL of ice-cold PBS into a 1.5-mL microfuge tube. Spin at 1000 × *g* for 2 min in a benchtop cooled (4°C) centrifuge.
6. Remove the PBS with a 1-mL pipette. Wash the cell pellet once more with ice-cold PBS.
7. Spin again at 1000 × *g* for 2 min at 4°C. Remove the PBS with a 1-mL pipette.
8. Immediately add 1 mL of ice cold NP-40 lysis buffer. Pipette up and down five times with a 1-mL pipette to resuspend the pellet.
9. Leave on ice for 15 min. Pipette up and down another five times and leave the tube on ice for another 15 min in order to permit efficient cell lysis.
10. Centrifuge the cell lysate in a benchtop centrifuge at 25,000 × *g* for 15 min at 4°C.
11. There will be a pellet of insoluble material at the bottom and a cleared supernatant and a lipid layer on top. Remove the supernatant (= cleared lysate) without interfering with the lipid phase and transfer to a new 1.5-mL microfuge tube.
12. Measure the protein concentration of the cleared lysate using a standard protein assay kit as available from various vendors.

GST pulldowns are performed from 20 mg of cleared lysate. These amounts refer to the usage of 20 mg of lysate in 4 mL of lysis buffer.

1. Take 100 μL (50 μl settled resin) slurry of glutathione Sepharose and spin at 1000 × *g* for 2 min at 4°C (*see* **Note 4**).
2. Take of supernatant and add 1 mL of lysis buffer, mix gently, and spin at 1000 × *g* for 2 min at 4°C. Repeat this wash three times.
3. Add the slurry to the 20 mg cell lysate in a 15-mL centrifuge tube and incubate on a roller for 2 h at 4°C. Spin at 1000 × *g* for 5 min at 4°C.
4. Transfer the cell lysate to a new 15-mL tube. Add 100 μL glutathione Sepharose slurry (washed with buffer as above) and 30 μg of GST protein. Incubate on a roller overnight at 4°C. Spin at 1000 × *g* for 5 min at 4°C.
5. Transfer the cell lysate to a new 15-mL tube. Add 100 μL glutathione Sepharose slurry (washed with buffer as above). Incubate on a roller for 2 h at 4°C.
6. Spin at 1000 × *g* for 5 min at 4°C. Transfer the now precleared cell lysate to a new 15-mL tube.

This preparation is done in the same way for all samples. For the actual pulldown experiments the following samples need to be done.

3.4.1.1. Blank

1. Pipette the corresponding amount of 20 mg cell lysate to a new 15-mL tube and add lysis buffer to a total volume of 4 mL.
2. Add 100 μL glutathione Sepharose slurry (washed with buffer as above).
3. Incubate on a roller for 2 h at 4°C.

3.4.1.2. GST Control

1. Pipette the corresponding amount of 20 mg cell lysate to a new 15-mL tube and add lysis buffer to a total volume of 4 mL.
2. Add 100 μL glutathione Sepharose slurry (washed with buffer as above) and 30 μg of GST protein.
3. Incubate on a roller for 2 h at 4°C.

3.4.1.3. Samples

1. Pipette the corresponding amount of 20 mg cell lysate to a new 15-mL tube and add lysis buffer to a total volume of 4 mL.
2. Add 100 μL glutathione Sepharose slurry (washed with buffer as above) and the appropriate amount (38 μg of the C-terminal SH3 domain of both Grb2 and GRAP fused to GST) of the GST fusion proteins.
3. Incubate on a roller for 2 h at 4°C.

All GST fusion proteins must be added at the same molar concentration (*see* **Note 7**)! After the pulldown is completed all samples (including blank and GST only) are treated as follows:

1. Spin at 1000 × *g* for 5 min at 4°C and take off the lysate with a 1-mL pipette.
2. Depending on future experiments it might be useful to snap-freeze the supernatant on dry ice and store at –70/–80°C.
3. Add 1 mL of lysis buffer and transfer the beads to a 1.5-mL microfuge tube. Spin at 1000 × *g* for 5 min.
4. Wash three times with lysis buffer.
5. Add 50 μL of 2 × SDS sample buffer and boil at 95°C for 5 min. The samples can now either be run on SDS–PAGE or frozen on dry ice.

SDS–PAGE and Coomassie staining are performed as described above (*see* **Subheading 3.3**).

3.4.2. GST Pulldowns for Western Blot Analysis

This protocol is similar to the one described above, but different amounts of the reagents are used.

1. Prepare the lysate as described above. Excess cell lysate that is not required for the pulldown can be aliquoted, frozen on dry ice, and stored at –70/–80°C.
2. GST pulldowns for Western blots are performed from 1–2 mg of cell lysate. These amounts refer to the usage of 1–2 mg of lysate in 500 μL of lysis buffer.

3. Preclear the required amount of cell lysate (1–2 mg of lysate per experiment, including blank and GST only) as described above.

For the pulldown experiments the following samples need to be done.

3.4.2.1. Blank

1. Pipette the corresponding amount of 1–2 mg cell lysate to a new 1.5-mL microfuge tube and add lysis buffer to a total volume of 500 μL.
2. Add 30 μl glutathione Sepharose slurry (washed with buffer as above).
3. Incubate on a roller for 2 h at 4°C.

3.4.2.2. GST Control

1. Pipette the corresponding amount of 1–2 mg cell lysate to a new 1.5-mL microfuge tube and add lysis buffer to a total volume of 500 μL.
2. Add 30 μL glutathione Sepharose slurry (washed with buffer as above) and 500 ng of GST protein.
3. Incubate on a roller for 2 h at 4°C.

3.4.2.3. Samples

1. Pipette the corresponding amount of 1–2 mg cell lysate to a new 1.5-mL microfuge tube and add lysis buffer to a total volume of 500 μL.
2. Add 30 μL glutathione Sepharose slurry (washed with buffer as above) and the appropriate amount (630 ng of the C-terminal SH3 domain of both Grb2 and GRAP fused to GST) of the GST fusion proteins.
3. Incubate on a roller for 2 h at 4°C.

All GST fusion proteins must be added at the same molar concentration! After the pulldown is completed all samples (including blank and GST only) are treated as follows:

1. Spin at 1000 × *g* for 5 min at 4°C and take off the lysate with a 1-mL pipette.
2. Depending on future experiments it might be useful to snap-freeze the supernatant on dry ice.
3. Add 700 μL of lysis buffer and wash the Sepharose beads three times with lysis buffer.
4. Add 30 μL of 2 × SDS sample buffer and boil at 95°C for 5 min. The samples can now either be run on SDS–PAGE or frozen on dry ice.

For Western blotting of the above mentioned samples please refer to general Western blotting protocols that are available in your laboratory, on the internet, or in the manuals of suppliers of antibodies or electrophoresis equipment. See **Figure 1** for IP and **Figure 2** for pulldown.

3.5. *Tryptic In-Gel Digest of Protein Bands* (see *Note 8*)

This protocol is a modified version of the original method published by the laboratory of Matthias Mann *(17)*.

Filter sterilize the ammonium bicarbonate (NH_4HCO_3) solution (50 m*M*) through a 0.22-μm filter prior to use. Take up 30 mL of NH_4HCO_3 into a 50-mL syringe, attach the filter, and press the solution into a fresh 50-mL tube.

All the steps described in this protocol must be carried out in a dust-free environment in order to reduce potential contaminations such as keratins.

If possible, filtered tips should be used.

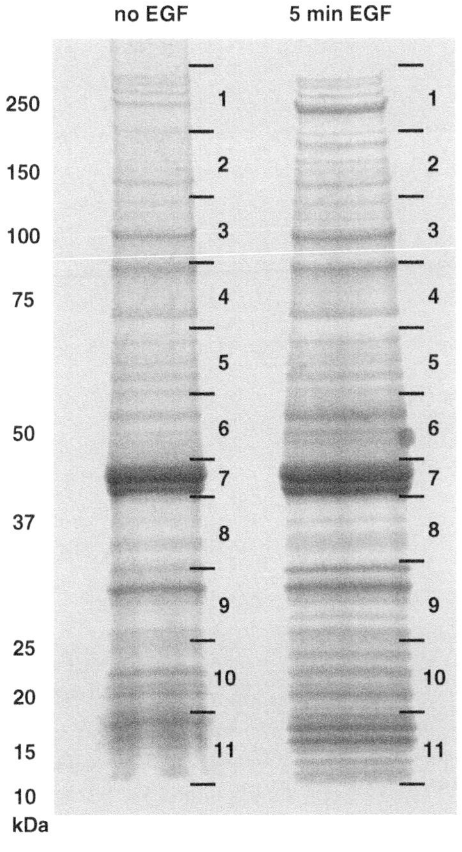

Fig. 1. Immunoprecipitation of ERK1 complexes. PC12 cells were either serum starved overnight and left untreated, or stimulated with EGF for 5 min. The cell lysates were subjected to immunoprecipitation with ERK1 antibody as described in the text and separated on a 4–12% gradient SDS gel. The gel was stained with Coomassie Brilliant blue. Eleven gel slices (labeled 1–11 in the picture) were excised, trypsin digested, and analyzed by mass spectrometry. Note the smaller gel slice number 7 (*see* **Note 11** for explanation). For example, gel slices 3 and 4 contained RSK1 to 4 with a decreased association upon EGF stimulation, slice 5 contained ERF with an increased association upon EGF stimulation, and slice 7 contained ERK1.

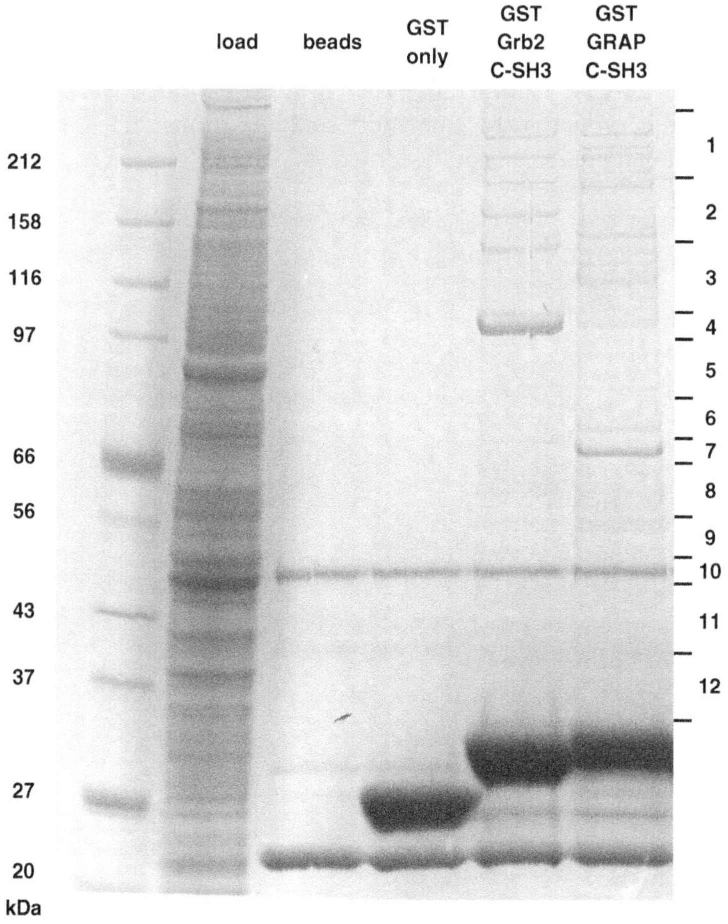

Fig. 2. GST pulldown of the C-terminal SH3 domains of Grb2 and GRAP. K562 cell lysate (20 μg loaded [lane 1]) was incubated with glutathione Sepharose beads alone (lane 2), glutathione Sepharose beads and GST (lane 3), glutathione Sepharose beads and GST-C-SH3 Grb2 (lane 4), and glutathione Sepharose beads and GST-C-SH3 GRAP (lane 5) and separated on a 4–12% gradient SDS gel. The gel was stained with Coomassie Brilliant blue. Twelve gel slices (labeled 1–10 in the picture) were excised from each lane, trypsin digested, and analyzed by mass spectrometry. For example, gel slices 4 contained dynamin 1 and 2 (Grb2) and small amounts of TRAP 150 (thyroid hormone receptor-associated protein 3) whereas gel slices 7 contained Hsp70 for both proteins (Grb2 low concentration; GRAP high concentration).

1. Add 500 µL of 50%MeOH/50 m*M* NH_4HCO_3 to the gel pieces and incubate on a thermoshaker at 22°C for up to 60 min under vigorous shaking to allow for destaining.
2. Remove the destaining solution and replace with 3 gel volumes of 50 m*M* NH_4HCO_3. Incubate on a thermoshaker at 22°C for 10 min under vigorous shaking.
3. Remove the supernatant and replace with acetonitrile. Incubate on a thermoshaker at 22°C for 10 min under vigorous shaking. The gel pieces should have shrunk.
4. Remove the supernatant and replace with 50 m*M* NH_4HCO_3. Incubate on a thermoshaker at 22°C for 10 min under vigorous shaking.
5. Remove the supernatant and replace with acetonitrile. Incubate on a thermoshaker at 22°C for 10 min under vigorous shaking. The gel pieces should have shrunk again.
6. Dry the shrunk gel pieces in a Speed-vac for approximately 5 min at 35°C.
7. Cover the gel pieces with 10 m*M* DTT (in 50 m*M* NH_4HCO_3) and incubate for 45 min at 56°C. Shaking is not required.
8. Remove any remaining supernatant and cover the gel pieces with 55 m*M* iodoacetamide (in 50 m*M* NH_4HCO_3) in order to acetylate the cysteine residues. Incubate for 30 min at room temperature in the dark.
9. Remove the supernatant and replace with 50 m*M* NH_4HCO_3. Incubate on a thermoshaker at 22°C for 10 min under vigorous shaking.
10. Dry the shrunk gel pieces in a Speed-vac for approximately 5 min at 35°C.
11. Cover the gel pieces with trypsin solution (final concentration: 12.5 ng/µL) and incubate on ice for approximately 30 min. Remove the remaining trypsin solution and cover the gel pieces with 50 m*M* NH_4HCO_3.
12. Incubate overnight at 37°C.
13. On the next day add 1–3 µL of 10% TFA to gain a final concentration of 1–2% TFA. The samples can now be either analyzed by MS or stored at –20°C.

3.6. RP-HPLC MS/MS and Sample Quantitation (see *Notes 9* and *10*)

The peptides were separated using nano-reversed-phase chromatography in the second dimension (UltiMate Nano LC System; LC Packings) and detected using a Q-Star Pulsar-i mass spectrometer (Applied Biosystems).

Then 10 µL of the digest was injected onto the LC system. The digest was run over a C_{18} reverse-phase (RP) cartridge (PepMap, 300 µm i.d. × 5 mm, LC Packings) functioning as a trap with a flow rate of 30 µL/min. The peptides bound to the C_{18} RP cartridge were then eluted from the trap and separated using a 75-µm i.d. C_{18} RP column (PepMap, 15 cm, LC Packings) with a 110 min gradient from 0–35% acetonitrile 0.1% formic acid with a flow rate of 200 nL/min. The eluted peptides were sprayed through a nano-LC needle (pico emitter, 20 µm i.d., 10-µm orifice, New Objectives) and analyzed using a data-dependent acquisition program on the Q-Star. Ions were excluded from MS/MS analysis for 120 s after analysis; collision energy was set automatically by the software. The four strongest ions that were multiply charged, not excluded, and had an ion count of greater than 30 were then selected by the software for further MS/MS analysis. The scan times were set as follows: MS, 1 s; first MS/MS; 1 s; second MS/MS 1.5 s; third MS/MS, 1.8 s; fourth MS/MS, 2 s.

The resulting MS/MS spectra were converted into Mascot readable files by an integrated script with the following settings: the ion charges were determined from the survey scan, ions with a charge higher than 5 were discarded, MS/MS scans were not grouped, peaks with an intensity lower than 0.1% of the maximum were removed, all centroid data were selected, and spectra with less than 10 peaks were discarded. Searching was done using a local copy of Mascot against the Rat-IPI database for PC12 cells or the human SwissProt database for K562 cells.

4. Notes

1. Unless otherwise stated, all solutions are prepared with Milli-Q water with a resistance of 18.2 MΩ/cm.
2. These protocols use some chemicals that are hazardous and toxic (such as methanol). It is strongly recommended that you make yourself familiar with the respective material safety sheets and obey the given recommendations.
3. These protocols have been optimized for PC12 and K562 cells, respectively, as examples for adherent and suspension cells from two different organisms (rat and human). K562 cells also have the advantage of rapid growth. These methods can be easily adapted for many other cell types. However, it is strongly recommended that test extractions be performed with a variety of different buffers in order to determine the optimal lysis buffer composition.
4. It is recommended that you cut off 5 mm of the pipette tip when pipetting viscous solutions and bead slurry, such as the Protein A or glutathione Sepharose 4B beads mentioned above.
5. The cross-linking agent DMP hydrolyzes over time and loses its activity. We therefore recommend storing the opened container in a desiccator.
6. The endogenous immunoprecipitation protocol has been optimized for the use of the ERK1 antibody. It can be easily adapted for many other antibodies from various suppliers. However, many antibodies are not suitable for immunoprecipitations, as will be stated on the data sheets of the antibody manufacturer. We would recommend a series of pilot experiments verified by Western blotting to determine the quality of the antibody, its suitability for immunoprecipitation, and the amount of antibody required to achieve maximal immunoprecipitation of the target protein.
7. In pulldown experiments proteins must always be added at the same molar concentration in order to permit a proper comparison between the analyzed protein fragments. For example, GST is 26 kDa. If the bait fragment of interest is

8 kDa, it will result in a 34-kDa fusion protein. Therefore 30 μg of GST equals (34/26) × 30 = 39 μg of the fusion protein.

8. Contamination of samples is a common problem in MS. Polymers, such as polyethylene glycol, most commonly are introduced into the sample by using cheap microfuge tubes and non-HPLC grade solvents and acids. We therefore suggest using glassware for buffer storage and replacement of these solutions on a regular basis. We also have found that polymer contamination can be reduced by limiting the amount of time the sample is stored in microfuge tubes, especially during and after the trypsin digest. Keratins are the most common protein contamination in MS samples. They are usually derived from skin flakes or hair. Special care must be taken to avoid these contaminations. It is also strongly recommended that you not wear garments made of wool when performing a protein digest since this will result in animal keratin as the major contaminants.
9. There is a variety of quantitative MS methods available that can be applied to determine changes in the composition of protein complexes, but an in depth description would go beyond the scope of this chapter.
10. We have only outlined our method of MS analysis. We strongly suggest talking through the project with your local MS facility manager or collaborating MS expert prior to starting the experiment.
11. Concentrated bands like number 7 in **Fig. 1**) should be cut out rather tightly without interfering with the adjacent parts of the gel. MS analysis will most likely show only peaks corresponding to the most prominent member of the particular gel piece (in this case this is Erk1). Any far lower concentrated proteins also present in a larger gel piece will not be detected.

Acknowledgments

We would like to thank A. Pitt, K. Burgess, R. Burchmore, and R. Goodwin at the Sir Henry Wellcome Functional Genomics Facility and W. Bienvenut and C. Ward at the Beatson Institute for Cancer Research for their continuous support with the mass spectrometry facilities and the members of the Kolch laboratory for many useful suggestions and discussions. This work has been supported by European Union FP6 grants "Interaction Proteome" contract LSHG-CT-2003-505520(AvK) and "Transnet" contract MRTN-CT-2004-512253 (CP).

References

1. Kolch, W. (2000) Meaningful relationships: The regulation of the Ras/Raf/MEK/ERK pathway by protein interactions. *Biochem. J.* **351(Pt. 2),** 289–305.
2. Vogelstein, B. and Kinzler, K. W. (2004) Cancer genes and the pathways they control. *Nat. Med.* **10(8),** 789–799.
3. Hahn, W. C. and Weinberg, R. A. (2002) Rules for making human tumor cells. *N. Engl. J. Med.* **347(20),** 1593–1603.

4. Blagoev, B., Kratchmarova, I., Ong, S. E., Nielsen, M., Foster, L. J., and Mann, M. (2003) A proteomics strategy to elucidate functional protein-protein interactions applied to EGF signaling. *Nat. Biotechnol.* **21(3),** 315–318.
5. Cho, S., Park, S. G., Lee, D. H., and Park, B. C. (2004) Protein-protein interaction networks: from interactions to networks. *J. Biochem. Mol. Biol.* **37(1),** 45–52.
6. von Kriegsheim, A., Pitt, A., Grindlay, G. J., Kolch, W., and Dhillon, A. S. (2006) Regulation of the Raf-MEK-ERK pathway by protein phosphatase 5. *Nat. Cell Biol.* **8(9),** 1011–1106.
7. Pawson, T. (1994) SH2 and SH3 domains in signal transduction. *Adv. Cancer Res.***64,** 87–110.
8. Pawson, T. and Nash, P. (2000) Protein-protein interactions define specificity in signal transduction. *Genes Dev.* **14(9),** 1027–1047.
9. Schlessinger, J. (2002) Ligand-induced, receptor-mediated dimerization and activation of EGF receptor. *Cell* **110(6),** 669–672.
10. Pawson, T. (2004) Specificity in signal transduction: from phosphotyrosine-SH2 domain interactions to complex cellular systems. *Cell* **116(2),** 191–203.
11. Wellbrock, C., Karasarides, M., and Marais, R. (2004) The RAF proteins take centre stage. *Nat. Rev. Mol. Cell Biol.* **5(11),** 875–885.
12. Yoon, S. and Seger, R. (2006) The extracellular signal-regulated kinase: multiple substrates regulate diverse cellular functions. *Growth Factors* **24(1),** 21–44.
13. Short, B., Preisinger, C., Schaletzky, J., Kopajtich, R., and Barr, F. A. (2002) The Rab6 GTPase regulates recruitment of the dynactin complex to Golgi membranes. *Curr. Biol.* **12(20),** 1792–1795.
14. Ren, S. Y., Bolton, E., Mohi, M. G., Morrione, A., Neel, B. G., and Skorski, T. (2005) Phosphatidylinositol 3-kinase p85{alpha} subunit-dependent interaction with BCR/ABL-related fusion tyrosine kinases: molecular mechanisms and biological consequences. *Mol. Cell. Biol.* **25(18),** 8001–8008.
15. Trinkle-Mulcahy, L., Andersen, J., Lam, Y. W., Moorhead, G., Mann, M., and Lamond, A. I. (2006) Repo-Man recruits PP1 gamma to chromatin and is essential for cell viability. *J. Cell Biol.* **172(5),** 679–692.
16. Puig, O., Caspary, F., Rigaut, G., *et al.* (2001) The tandem affinity purification (TAP) method: a general procedure of protein complex purification. *Methods* **24(3),** 218–229.
17. Shevchenko, A., Wilm, M., Vorm, O., and Mann, M. (1996) Mass spectrometric sequencing of proteins silver-stained polyacrylamide gels. *Anal. Chem.* **68(5),** 850–858.

13

Selection of Recombinant Antibodies by Eukaryotic Ribosome Display

Mingyue He and Michael J. Taussig

Summary

Ribosome display is a powerful method for selection of single-chain antibodies *in vitro*. It operates through the formation of libraries of antibody–ribosome–mRNA complexes that are selected on immobilized antigen, followed by recovery of the genetic information from the mRNA by RT-PCR. Both prokaryotic and eukaryotic versions are used. We describe our eukaryotic system, in which rabbit reticulocyte extracts are used for cell free transcription/translation and cDNA is recovered by *in situ* RT-PCR performed on the selected complexes.

Key Words: Single-chain antibody; library; selection; ribosome complex.

1. Introduction

Antibodies are the most widely used class of reagents for research, pharmaceutical, diagnostic, and therapeutic applications *(1,2)*. Protein display technologies offer an efficient and flexible route to the generation of recombinant antibodies, by selection from large libraries in which protein (phenotype) and encoding DNA (genotype) are coupled *(3)*. Ribosome display is a fully cell-free display method for the production and optimization of antibody-combining sites in which linkage of nascent, single-chain antibodies and their encoding mRNA is made as antibody–ribosome–mRNA (ARM) complexes in a cell-free system *(4)*. By interaction with an immobilized antigen, the formation of ribosome complexes allows coselection of specific antibodies together with their encoding mRNA, which is subsequently recovered as DNA via coupled

From: *Methods in Molecular Biology, vol. 484: Functional Proteomics: Methods and Protocols*
Edited by: J. D. Thompson et al., DOI: 10.1007/978-1-59745-398-1, © Humana Press, Totowa, NJ

reverse transcription-polymerase chain reaction (RT-PCR) amplification. This process can be repeated to enrich target (antibody) genes from a large population.

A major advantage of ribosome display over existing cell-dependent display methods is that it directly screens PCR-generated libraries without the need for bacterial cloning. The use of PCR libraries permits the display of larger populations as well as continuously searching for novel sequence diversity, providing a powerful tool for antibody evolution *in vitro*. In principle, all PCR-based mutagenesis methods, such as oligo-directed mutations, DNA shuffling, and "staggered" PCR, can be readily applied to create and diversify the DNA libraries ***(5)***.

Both prokaryotic and eukaryotic cell-free systems have been developed for ribosome display of antibodies ***(4,6)***, each with its own protocol and modifications. In this chapter, we describe our rabbit reticulocyte lysate method, originally termed "ARM" (antibody–ribosome–mRNA) display. A distinct feature of the ARM system is the use of an *in situ* RT-PCR procedure to recover DNA from ribosome complexes, which does not involve the prior dissociation of ribosome complexes ***(4)***. **Figure 1** shows the ARM display cycle.

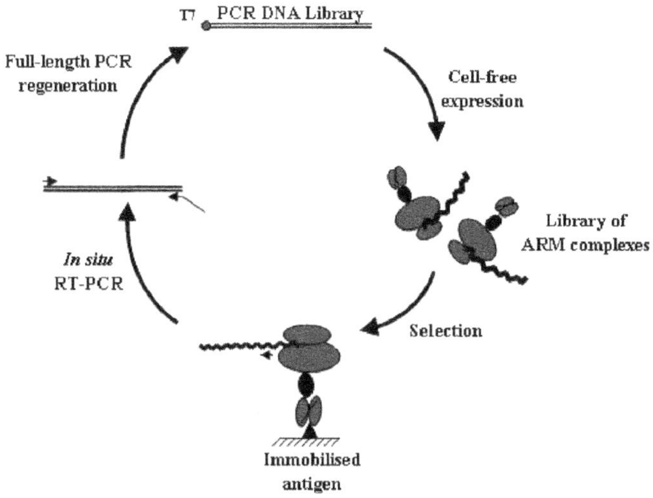

Fig. 1. The eukaryotic ribosome display cycle, showing steps of the PCR library, cell-free generation of AR*M* complexes, selection of AR*M* complexes, *in situ* RT-PCR recovery, and regeneration of full-length PCR construct. T7, T7 promoter.

2. Materials

All solutions, tubes, and tips used must be sterilized. Reagents should be nuclease free. Precautions should be taken to avoid DNA contamination. Primers, RT-PCR buffer, washing buffer, and dNTP solutions should be stored in aliquots.

2.1. Primers for DNA Recovery

2.1.1. Primers for Single-Tube RT-PCR Recovery

Primers are given in **Table 1**.

2.1.2. Primers for Single-Primer RT-PCR

Primers are given in **Table 2**.

Table 1
Primers for Single-Tube RT-PCR Recovery

Primer	Sequence
RT1	5′ -ACTTCGCAGGCGTAGAC-3′
T7Ab/back	GCAGC*TAATACGACTCACTATAGGAACAGA***CCACCATG**(C/G)AG GT(G/C)CA(G/C)*CTCGAG*(C/G)AGTCTGG
Ck/for	5′CTCTAGAACACTCTCCCCTGTTGAAGCTCTTTGTGACGGGCGA GCTCAGGCCCTGATGGGTGACTTCGCAGGCGTAGAC TTTG-3′

Table 2
Primers for Single-Primer RT-PCR[a]

Primer	Sequence
RTKz1	5′ -GAACAGACCACCATGACTTCGCAGGCGTAGAC-3′
Kz1	5′ -GAACAGACCACCATG-3′
T7Ab/back	GCAGC*TAATACGACTCACTATAGGAACAGA***CCACCATG**(C/G)AG GT(G/C)CA(G/C)*CTCGAG*(C/G)AGTCTGG
Ck/for	5′CTCTAGAACACTCTCCCCTGTTGAAGCTCTTTGTGACGGGCGA GCTCAGGCCCTGATGGGTGACTTCGCAGGCGTAGAC TTTG-3′

[a] Italics indicate the T7 promoter. Kozak sequence and initiation codon (ATG) are in bold. Underlined italics are restriction sites for cloning.

2.2. Molecular Biology Kit and Reagent

1. mRNA purification kit (Pharmacia Biotech Cat. #27-9255-01).
2. Titan™ one tube RT-PCR system (Boehringer Mannheim, Cat. #1888 382).
3. Qiagen QIAEX II gel extraction kit (Qiagen Cat. #20021).
4. Gel extraction kit (Sigma, Cat. #NA1111).
5. Rabbit reticulocyte TNT T7 quick for PCR DNA (Promega Cat. #L5540).
6. Taq DNA polymerase (Expand™ high fidelity PCR system: Boehringer Mannheim, Cat. #1732 641; Qiagen, Cat. #201203).
7. AMV reverse transcriptase (Promega Cat. #M5101).
8. 25 m*M* dNTPs: mix equal volumes of each 100 m*M* dNTP stock solution (Sigma, D 4788, D-4913, D5038, and T-9656).
9. 100 m*M* DTT (from Boehringer Mannheim Titan™ one tube RT-PCR system, *see* 2 above).
10. Dynabeads M-280 streptavidin (Dynal UK; 6.5 × 10^8/mL or 10 mg/mL; product #112.05/06).
11. RNase-free DNase I (Boehringer Mannheim Cat. #776 785 or Promega Cat. #M6101).
12. SUPERase In RNase inhibitor (Ambion, Cat. #2694).
13. SuperScript II reverse transcriptase (Invitrogen, Cat. #12236-022).
14. Agarose (Sigma, Cat. #A-9539).
15. 5 × gel loading buffer (40% w/v sucrose, 0.25% bromophenol blue).
16. TopYield Strips (NUNC, Cat. #248909).
17. 0.5-mL siliconized RNase-free microfuge tubes (Ambion, Cat. #12350).
18. Sterilized (DEPC-treated) distilled water: autoclaved Milli-Q water containing 0.1% (v/v) diethylpyrocarbonate.

2.3. Solutions

1. One-tube RT-PCR **Solution 1** (per 100 µL):

Dithiothreitol (DTT) (100 m*M* from Titan™ kit)	10 µL
dNTPs (10 m*M*)	4 µL
Upstream primer (16 µ*M*)	6 µL
Downstream primer (16 µ*M*)	6 µL
H_2O	74 µL
Store at –20°C	

2. One-tube RT-PCR **Solution 2** (per 96 µL):

5× RT-PCR buffer (from the Titan™ kit)	40 µL
H_2O	56 µL
Store at –20°C.	

3. Single-primer RT-PCR **Solution 3** (per 12 μL):

Primer RTKz1 (8 μ*M*)	1 μL
10 m*M* dNTP	2 μL
dH_2O	9 μL

4. Single-primer RT-PCR **Solution 4** (per 8μL):

5× first-strand buffer	4 μL
100 m*M* DTT	1 μL
SUPERasr In (20 U)	1 μL
SuperScript II (200 U)	1 μL
dH_2O 1 μL	

5. Buffer A: 0.1 *M* Na-phosphate buffer, pH 7.4.
6. Buffer D: Buffer A with 0.1% bovine serum albumin (BSA) (Sigma, Cat. #A-4503).
7. Buffer E: 0.2 *M* Tris–HCl, pH 8.5, with 0.1% BSA.
8. Phosphate-buffered saline (PBS), pH 7.4.
9. EZ-link™ sulfo-NHS-LC-LC-biotin (Pierce, Cat. #21338). Solution is made at a concentration of 1 mg/mL in water and stored at 4°C for at least 2 weeks.
10. Antigen solution (0.5–1 mg/mL) in PBS.
11. 50 m*M* magnesium acetate.
12. Washing buffer: PBS containing 0.01% Tween 20 and 5 m*M* Mg-acetate, stored at 4°C.
13. 10× DNase I digestion buffer: 400 m*M* Tris–HCl, pH 7.5, 60 m*M* $MgCl_2$, 100 m*M* NaCl. Autoclaved and stored at 4°C.
14. 10% Na-azide.

3. Methods

The method is described in the following steps: (1) construction of antibody library, (2) preparation of immobilized antigen, (3) ribosome display and antigen selection, and (4) *in situ* RT-PCR recovery.

3.1. Antibody Library Construction

Our single-chain human antibody libraries are constructed in the three-domain format of V_H/K and V_H/V_λ-C_κ (**Fig. 2**). V_H/K is generated by direct fusion of the heavy chain variable domain (V_H) to the complete κ light chain (**Fig. 2a**), while V_H/V_λ-C_κ is made by assembling V_H, V_λ and C_κ together (**Fig. 2b**). The heavy chain "elbow" region, a continuation of the V_H domain, is used as the peptide linker to join the V regions of heavy and light chains *(7)*. This design both simplifies the process of PCR construction and avoids

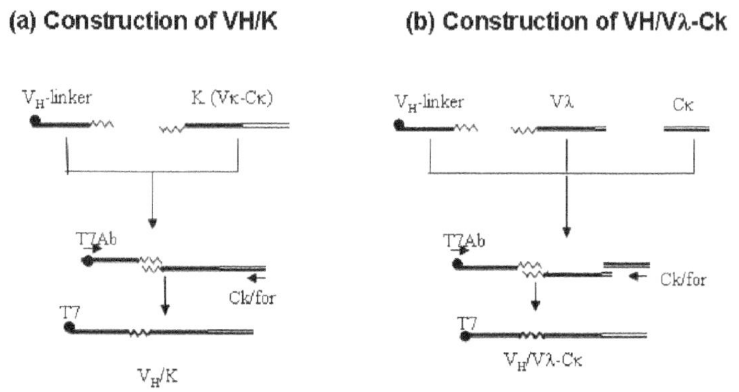

Fig. 2. PCR strategy for construction of PCR libraries. (**a**) Construction of V_H/K. (**b**) Construction of V_H/V_λ-C_κ The flexible linker is indicated by wavy lines. T7, T7 promoter.

the introduction of nonhuman sequences. The presence of the C_κ domain at the C-terminus provides a spacer to allow functional display of single-chain antibodies on the surface of ribosome, as well as providing a known priming site for RT-PCR recovery after selection. To produce stable ARM ribosome complexes in a rabbit reticulocyte lysate, a T7 promoter and Kozak sequence are required upstream to direct protein synthesis, while the stop codon at the 3′ end is removed to stall the ribosome with the translated mRNA (**Note 1**).

1. Isolate total mRNA from human peripheral blood lymphocytes (PBL) using the Pharmacia mRNA purification kit (instructions included with the kit).
2. Generate V_H-linker, K, V_λ and C_κ fragments by PCR. Individual fragments are generated by one-tube RT-PCR according to the manufacturer's instructions using the primers described *(7)*:
   One-tube RT-PCR mixture is set up as follows:

Solution 1	24 μL
Solution 2	24 μL
Enzyme Mix (from Titan^{TM} kit)	1 μL (*see* **Note 2**)
mRNA	1 μL (1–50 ng)

   A negative RT-PCR control (10 μL) is also set up without the mRNA (**Note 3**). Carry out RT-PCR thermal cycling: 1 cycle of 48°C for 45 min, followed by 94°C for 2 min; then 35 cycles of: 94°C for 30 s, 54°C for 1 min, 68°C for 2 min; finally, 1 cycle of 68°C for 7 min, then hold at 10°C.
3. Analyze RT-PCR products using 1% agarose gel and purify DNA fragments from the gel using the Sigma gel extraction kit.

4. Generate full-length construct by PCR assembly of different fragments. Individual PCR fragments in equal amounts are mixed to form pooled V_H-linker, V_λ and V_κ chain, separately. The C_κ domain is amplified separately using a plasmid template (**Fig. 2**). Full-length constructs for ribosome display are generated by assembly of DNA fragments. For example, V_H/K is constructed through assembly of V_H-linker and the complete κ chain through an overlapping sequence between the two fragments followed by PCR amplification of the assembled product using primers flanking the construct. Similarly, V_H/V_λ-C_κ is generated by PCR assembly of V_H-linker, V_λ and C_κ followed by PCR amplification with flanking primers (**Fig. 2**). PCR assembly reaction is set up as follows:

PCR fragment 1	5–25 ng
PCR fragment 2	5–25 ng
(or PCR fragment 3)	5–25 ng
10× PCR buffer (from Qiagen kit)	2.5 µL
5× Q solution (from Qiagen)	5 µL
2.5 m*M* dNTPs	1 µL
Taq DNA polymerase	1 U
dH_2O to final volume	25 µL

Carry out seven thermal cycles: 94°C for 30 s; 54°C for 1 min, and 72°C for 1.2 min, then extension at 72°C for 7 min.

Then set up second PCR to amplify the assembled product: Carry out 30 thermal

The assembly mixture (above)	2 µL
10× PCR buffer	5 µL
5× Q solution	10 µL
2.5 m*M* dNTPs	4 µL
16 µ*M* of T7Ab/back	1.5 µL
16 µ*M* of Hu-Cκ/for	1.5 µL
Taq DNA polymerase	2.5 U
dH_2O to final volume	50 µL

cycles: 94°C for 30 s, 54°C for 1 min, 72°C for 1.2 min; then, extension at 72°C for 7 min, finally hold at 10°C.

5. Analyze the PCR library by loading 5 µL of the sample onto a 1% agarose gel containing 0.5 µg/mL ethidium bromide.
6. Confirm the identity of the constructs by PCR mapping using primers annealing at various positions. The PCR libraries can be directly used or stored at –20°C (**Note 4**).

3.2. Preparation of Immobilized Antigens

Immobilized antigens for capturing specific ARM complexes can be prepared by either (1) antigen coupling to streptavidin Dynabeads through protein biotinylation or (2) antigen coating onto wells.

3.2.1. Coupling of Biotinylated Proteins to Streptavidin Dynabeads

1. Mix proteins in PBS (pH 7–8.5) with sulfo-NHS-biotin solution in proportions of 25 μg protein to 1 μg sulfo-NHS-biotin and incubate at room temperature (RT) for 30 min followed by dialysis against 2× 500 mL PBS overnight at 4°C. The biotinylated protein is ready for the next step or can be stored at 4°C.
2. Wash 50 μL of streptavidin Dynabeads M-280 3× with Buffer A and resuspend in 50 μL PBS.
3. Add 5 μg of biotinylated protein to the beads (ratio of biotinylated protein to beads of 10 μg to 1 mg) and incubate at room temperature for 30 min. After removing the supernatant, wash the beads three times with 50 μL PBS. Finally, resuspend in the original volume (50 μL) in Buffer D containing 0.02% Na-azide; beads may be stored at 4°C for 3–4 months.

3.2.2. Protein Coating onto Wells

1. Add 20 μL protein (at 0.5–1 mg/mL in PBS, pH 7–8.5) to each well of TopYield Strips and incubate at 4°C overnight.
2. Remove the solution and block the well with 100 μL 4% milk powder or 1% BSA in PBS for 1–2 h at RT.
3. Wash three times with PBS and store the strips at 4°C. Wash the wells briefly with ice-cold Washing Buffer before use.

3.3. Ribosome Display and Antibody Selection

To generate ARM complexes for selection, PCR libraries are directly expressed in a coupled rabbit reticulocyte lysate (TNT) system. Typically, 1 μg of PCR library is used in a standard 50 μL reaction. However, this system can be scaled up for of larger libraries (up to 10μg) in 250 μL of reaction reaction (*see* **Note 5**). For PCR DNA with the size of 1 Kb, 1 μg contains 9.1×10^{11} molecules.

1. Set up *in vitro* coupled transcription/translation to generate ribosome complexes:

TNT T7 Quick for PCR	40 μL (*see* **Note 5**)
PCR DNA	500 ng–1 μg
Methionine (1 m*M*) (from TNT kit)	1 μL

Mg-acetate (50 m*M*)	1 μL (*see* **Note 6**)
Distilled H_2O	to 50 μL

Incubate at 30°C for 60 min.

2. Remove the input PCR DNA fragment by adding 120 U RNase-free DNase I together with 7 μL 10× DNase I digestion buffer and H_2O to a final 70 μL. Incubate at 30°C for a further 20 min (*see* **Note 7**).
3. Dilute with 70–210 μL of cold PBS containing 5 m*M* magnesium acetate.
4. Add 100–150 μL of the TNT translation mixture, containing the generated ARM complexes to 2 μL antigen-coupled beads (or an antigen-coated well) (*see* **Subheading 3.2.2**) and incubate at 4°C for 2 h with gentle shaking or vibration.
5. Wash the beads (or wells) three times with 100 μL cold washing buffer, followed by two quick washes with 100 μL cold sterilized H_2O. Collect the beads after washes using a magnetic concentrator. The beads (or wells) carrying selected ARM complexes can be stored at –20°C or used directly for DNA recovery.

3.4. In Situ RT-PCR Recovery

After selection, *in situ* RT-PCR recovery is performed using one of the following procedures: (1) single-tube RT-PCR or (2) single-primer RT-PCR. While the former has advantages for use with beads, the latter can be applied to both beads and wells, with more appropriate application to wells, allowing flexible control of recovery according to downstream applications.

3.4.1. Single-Tube RT-PCR Recovery

Since the 3′ end of the selected mRNA is occupied by the stalled ribosome after translation, a downstream primer RT1 (**Table 1**), designed to hybridize at about 60 nt upstream of the 3′ end of the mRNA, is used in combination with the upstream primer T7Ab in a single-tube RT-PCR system (**Fig. 3a**). As the use of RT1 produces a shortened DNA fragment, a long primer Ck/for, which contains the missing 3′ end sequence, is used together with T7Ab to regenerate the full-length DNA for the subsequent cycle.

1. Set up a standard one-tube RT-PCR mixture as follows:

Solution 1 (*see* **Table 1**)	25 μL
Solution 2	24 μL
Enzyme mix	1 μL (*see* **Note 2**)

2. Resuspend the beads carrying bound ARMs in 10 μL H_2O. Add 2 μL of the bead suspension into 10–20 μL of the above RT-PCR solution and mix well.

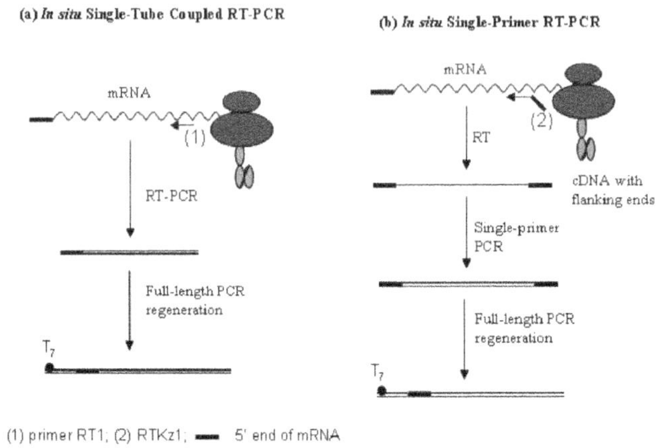

Fig. 3. *In situ* RT-PCR recovery. (**a**) Single-tube coupled RT-PCR. Reverse transcription (RT) is coupled with PCR in a single-tube reaction. (**b**) Single-primer RT-PCR reverse transcription is carried out first, followed by single-primer PCR amplification. The primers used are listed in **Tables 1** and **2**.

3. Carry out thermal cycling: one cycle of 48°C for 45 min, followed by 94°C for 2 min; then 30–40 cycles of 94°C for 30 s, 54°C for 1 min, and 68°C for 2 min; finally, 1 cycle of 68°C for 7 min, then hold at 10°C.
4. Analyze the PCR product by loading 5 μL of the sample onto a 1% agarose gel containing 0.5 μg/mL ethidium bromide.

3.4.2. Single-Primer RT-PCR Recovery

A single-primer RT-PCR procedure has also been developed for *in situ* recovery of DNA from ribosome complexes *(4)*. This procedure uses a novel sequence design of the RTKz1 primer (**Table 2**) to generate single-stranded cDNAs with complementary flanking 5′ and 3′ terminal sequences, so that the following PCR amplification can be performed using a single consensus primer (Kz1) (**Fig. 3b**). Again, the long primer Ck/for is required to pair with T7Ab for regeneration of the full-length DNA by PCR. This procedure works with a wide range of enzymes under standard conditions without the need for PCR optimization.

1. Set up the reverse transcription reaction by adding 12 μL **Solution 3** to each ARM-bound well. Incubate at 48°C for 5 min; then quickly place on ice for at least 30 s.
2. Add 8μL of **Solution 4** and incubate the mixture at 42°C for 45 min followed by 5 min at 85°C. Transfer the RT mixture to a fresh tube for subsequent single-primer PCR.
3. Set up the single-primer PCR mixture as follows:

10× PCR buffer	2.5 μL
5× Q solution	5 μL
2.5 m*M* dNTPs	2μL
Primer Kz1 (16 μ*M*)	1.5 μL
Taq DNA polymerase	1 U
dH_2O to final volume	25 μL

Carry out 30–35 cycles of thermal cycling as follows: 94°C for 30 s, 48°C for 1 min, 72°C for 1.2 min; then, extension at 72°C for 7 min, finally hold at 10°C.

4. Analyze the PCR by loading 5μL of the sample onto a 1% agarose gel containing 0.5 μg/mL ethidium bromide.

3.5. Regeneration of the Full-Length Construct

The use of an internal primer in the *in situ* RT-PCR recovery leads to shortening of the DNA fragment compared to the original fragment; therefore, a further PCR step is required to regenerate the full-length construct.

1. Set up the PCR mixture as follows:

10× PCR buffer	5 μL
5× Q solution	10 μL
2.5 m*M* dNTPs,	4 μL
16 μ*M* of T7Ab/back	1.5μL
16 μ*M* of Ck/for	1.5 μL
Taq DNA polymerase	2 U
PCR template from 3.4	1–10 ng
dH_2O to final volume	50 μL

Carry out 30 thermal cycles: 94°C for 30 s, 54°C for 1 min, 72°C for 1.2 min; then extension at 72°C for 7 min; finally hold at 10°C.

2. Analyze the PCR by loading 5μL of the sample onto a 1% agarose gel containing 0.5 μg/ml ethidium bromide. The full-length PCR can be used for either repeated cycles or protein expression (**Note 8**).

4. Notes

1. Although only the three-domain single-chain V_H/K and V_H/V_λ-C_κ is described in this chapter, the method is in principle equally applicable to other forms of single-chain or single-domain antibodies provided that a spacer is present at the C-terminus to allow the antibody combining site to be exposed on the surface of the ribosome. In addition to the C_κ domain used here, a number of different spacers

have been exploited, including gene III of filamentous phage M13, the C_H3 domain of human IgM, streptavidin, and GST *(4)*.

2. The one-tube RT-PCR can be carried out with comparable efficiency using AMV reverse transcriptase (Promega) and *Taq* DNA polymerase (Boehringer Mannheim) in combination with the Titan™ RT-PCR buffer. For example, to a 50 μL RT-PCR reaction, 0.5 μL (4–5 U) AMV and 0.5 μL (2 U) *Taq* are added to the mixture.
3. Negative controls lacking a template should be included in every RT-PCR or PCR experiment to assess DNA or mRNA contamination. The volume of PCR and RT-PCR can be scaled up to 100 μL or reduced to 5–10 μL according to applications.
4. The PCR libraries are usually stored in dH_2O at –20°C for routine use. Long-term storage should be at –20°C after ethanol precipitation and drying.
5. *In vitro* protein expression using Promega's TNT mixture can be scaled up to 100 μL or down to 20 μL without any significant reduction in recovery efficiency.
6. Mg-acetate concentration in the TNT mixture during translation affects ARM generation and recovery. We have shown that antibodies can be more efficiently recovered with Mg^{2+}concentration ranging from 0.5 to 2 m*M* *(7)*.
7. It is important to remove input DNA completely, as any contamination by the remaining DNA will cause a high background or DNA carryover in the DNA recovery step.
8. The number of cycles required to enrich for required antibodies depends on the nature of the antigen as well as the quality and diversity of the library used. Generally, three to five cycles should be sufficient to enrich a target demonstrably from a library (10^3–10^4-fold per cycle). Antibody enrichment can be estimated by comparing the ratios of input DNA and recovered DNA in each cycle.

Acknowledgments

We thank Hong Liu for technical assistance. Research at the Babraham Institute is supported by the Biotechnology and Biological Sciences Research Council (BBSRC), UK.

References

1. van Dijk, M. A. and van de Winkel, J. G. (2001) Human antibodies as next generation therapeutics. *Curr. Opin. Chem. Biol.* **5**, 368–374
2. Taussig, M. J., Stoevesandt, O., Borrebaeck, C. A. K., Bradbury, A. R., Cahill, D., *et al.* (2007) Proteome binders: planning a European resource of affinity reagents for analysis of the human proteome. *Nature Methods* **4**, 13–17.
3. Winter, G., Griffiths, A. D., Hawkins, R. E., and Hoogenboom, H. R. (1994) Making antibodies by phage display technology. *Annu. Rev. Immunol.* **12**, 433–455.
4. He, M. and Taussig, M. (2007) Eukaryotic ribosome display with *in situ* DNA recovery *Nature Methods* **4,** 281–288.

5. He, M. and Taussig, M. J. (2002) Ribosome display: cell-free protein display technology. *Briefings Funct Genomics Proteomics* **1**, 204–212.
6. Zahnd, C., Amstutz, P., and Pluckthun, A. (2007) Ribosome display: selecting and evolving proteins in vitro that specifically bind to a target. *Nat. Methods* **4,** 269–279.
7. He, M, Cooley, N., Jackson, A., and Taussig, M. (2004) Production of human single-chain antibodies by ribosome display. In: *Methods in Molecular Biology 248: Antibody Engineering Protocols*, 2nd ed. (Lo, B., ed.), pp. 177–189. Humana Press, Totowa, NJ.
8. He, M. and Taussig, M. J. (2005) Ribosome display of antibodies: expression, specificity and recovery in a eukaryotic system. *J. Immunol. Methods* **297**, 73–82.

14

Production of Protein Arrays by Cell-Free Systems

Mingyue He and Michael J. Taussig

Summary

Protein arrays make possible the functional screening of large numbers of immobilized proteins in parallel. To facilitate the supply of proteins and to avoid their deterioration on storage, we describe our protein *in situ* array (PISA) method for production of protein arrays in a single step directly from PCR DNA, using cell-free transcription and translation. In PISA, the *in vitro*-generated proteins are immobilized, as they are formed, on the surface of wells, beads, or slides coated with a protein-capturing reagent. In our preferred method, proteins are tagged with a double-hexahistidine sequence that binds strongly to Ni-NTA-coated surfaces. Advantages of PISA include avoiding bacterial expression and protein purification and making functional protein arrays available as required from genetic information.

Key Words: Protein array; protein immobilization; cell-free system; hexahistidine tag.

1. Introduction

Proteomics requires technologies for high-throughput, multiplexed analysis of protein function. Protein microarray is such a system. It simultaneously screens large numbers of proteins in a time- and cost-effective manner and has been applied increasingly for analysis of protein interactions, protein expression profiling, and biomarker discovery *(1)*. One of the bottlenecks is ensuring the supply of functional proteins. Cell-based expression methods suffer from limitations of production and functional maintenance of the huge diversity of proteins that could form the array elements. Moreover, recombinant protein production usually involves one of several *in vivo* expression systems followed by purification, which is a time-consuming process. Moreover, many

From: *Methods in Molecular Biology, vol. 484: Functional Proteomics: Methods and Protocols*
Edited by: J. D. Thompson et al., DOI: 10.1007/978-1-59745-398-1, © Humana Press, Totowa, NJ

proteins are either poorly expressed or not expressed as functional molecules in heterologous hosts *(2)*. Protein immobilization requires covalent or noncovalent attachment to a solid surface in such a way as to maintain long-term functionality (binding, enzymatic activity, etc.), which can often decline due to the denaturation and inherent instability of proteins on array surfaces. Cell-free protein synthesis may be exploited to overcome these problems *(3–5)*. It makes use of cell extracts to express proteins from polymerase chain reaction (PCR) DNA template(s), avoiding the need for bacterial cloning and enabling the rapid conversion of genetic information into functional proteins. In addition, the open and flexible systems permit addition of components and create defined environment(s) required for correct protein folding, modifications, or activity.

By coupling cell-free protein synthesis in parallel with *in situ* immobilization, it is possible to generate protein arrays from arrayed DNAs *(4)*. This novel strategy not only avoids the need for separate expression, purification, and printing of individual proteins, but also reduces the risk of deterioration in protein function during medium- or long-term storage. We have developed a cell-free protein array method, protein *in situ* arrays *(PISA)*, that generates protein arrays directly from PCR DNA by cell-free synthesis of tagged proteins on the tag-capturing surface, such that the newly synthesized proteins are immobilized *in situ* as they are synthesized *(3)* (**Fig. 1**). We have used this technology to make

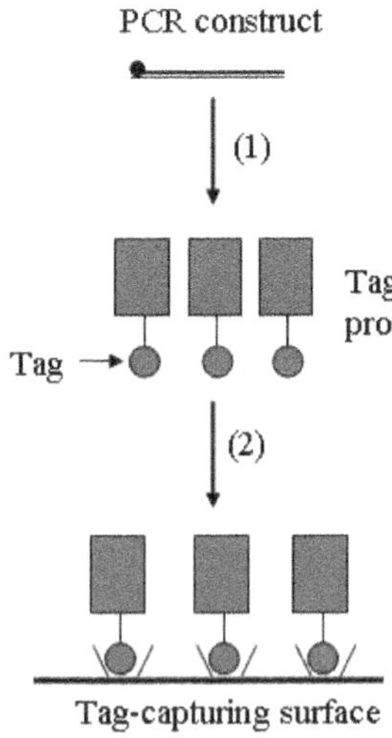

Fig. 1. Protein *in situ* array procedure showing cell-free synthesis of a tagged protein on the tag-binding surface and *in situ* immobilization. (1) Coupled *in vitro* transcription and translation. (2) *In situ* protein immobilization.

protein arrays for different applications **(5)**. Here, we describe the details of the *PISA* method for general utilization.

2. Materials

2.1. Primers

2.1.1. Primers for Making PCR Constructs Used in a Rabbit Reticulocyte Lysate System

1. T7/back(R):5′-GCAGC*TAATACGACTCACTATAG*GAACAGACCACC**ATG**-3′. An upstream primer containing T7 promoter (italics) and Kozak sequences (underlined) and the start codon ATG (bold).
2. G/back (R): 5′-TAGGAACAGACCACCATG(N)$_{15-25}$-3′. An upstream primer for PCR amplification of target genes. It contains a sequence overlapping with T7/back (R) (underlined) and 15–25 nucleotides from the 5′ sequence of the gene of interest. (N)$_{15-25}$ indicates the number of nucleotides.
3. G/for: 5′-CACCGCCTCTAGAGCG(N)$_{15-25}$-3′. A downstream primer for PCR amplification of target genes. It contains a sequence (underlined) overlapping a PCR fragment encoding a C-terminal region (*see* **Subheading 2.2**) and 15–25 nucleotides complementary to the 3′ region of a target gene.

2.1.2. Primers for Making the PCR Construct Used in Escherichia coli S30 Extracts

1. T7/back(E): 5′-GAAAT*TAATACGACTCACTATAG*GGAGACCACAACGGTTTCCCTCTAGAAATAATTTTGTTTAACTTTAAGAAGGAGATATACC**ATG**-3′. An upstream primer containing T7 promoter (italics) and ribosome-binding site (underlined) and the start codon ATG (bold).
2. G/back (E): 5′-CTTTAAGAAGGAGATATACCATG(N)$_{15-25}$-3′. An upstream primer for PCR amplification of target genes. It contains a sequence overlapping T7/back (E) (underlined) and 15–25 nucleotides from the 5′ sequence of the gene of interest. (N)$_{15-25}$ indicates the number of nucleotides.
3. G/for: 5′-CACCGCCTCTAGAGCG(N)$_{15-25}$-3′. A downstream primer for PCR amplification of a target gene. It contains a sequence (underlined) overlapping a PCR fragment encoding a T-domain (*see* **Subheading 2.2**) and 15–25 nucleotides complementary to the 3′ region of the target gene.

2.1.3. PCR Primers for PCR Amplification of a C-Terminal Region

1. Linker-tag/back: 5′-GCTCTAGAGGCGGTGGC-3′. An upstream primer for PCR generation of a termination region (*see* **Subheading 2.2**) in combination with T-term/for.
2. T-term/for: 5′-TCCGGATATAGTTCCTCC-3′. A downstream primer for PCR generation of either the termination region in combination with the Linker-tag/

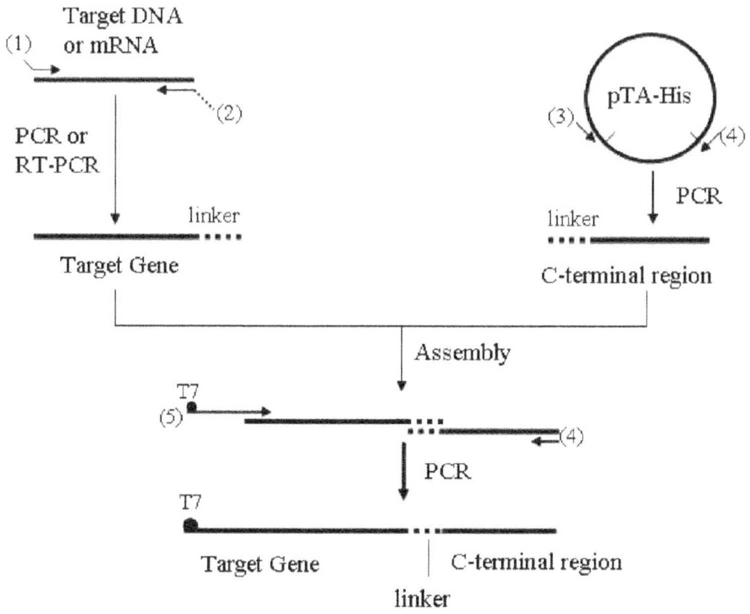

Fig. 2. A PCR construction strategy. The primers used are (1) G/back, (2) G/for, (3) Linker-tag/back, (4) T-term/for, and (5) T7/back. The broken line indicates the linker.

back or the full-length construct in combination with one of the T7 primers (*see* **Subheadings 2.1.1** and **2.1.2**; *also see* **Fig. 2**).

2.2. Plasmid Encoding a C-Terminal Region

A plasmid pTA-His has been created, containing a DNA insert encoding a C-terminal region, which is composed of (in order) a flexible linker, a double (His)$_6$ tag, two stop codons, a poly(A) tail, and a transcription termination region *(3)*. The detailed sequence is GCTCTAGAggcggtggctctggt ggcggttctggcggtgg-caccggtggcggttctggcggtggc AAACGGGCTGATGCTGCACATCACCATCAC-CATCACTCTAGAGCTTGGCGTCACCCGCAGTTCGGTGG TCACCACCA CCACCACCAC**TAATAA**(A)$_{28}$CCGCTGAGCAAT*AACTAGCATAACCCCT TGGGGCCTCTAAACGGGTCTTGAGGGGTTTTTTG*CTGAAAGGAGGAA CTATATCCGGA-3′. The lower case is a flexible linker encoding 19 amino acids. The underlined sequence encodes a novel double-(His)$_6$ tag sequence that has shown an order of magnitude or greater affinity for Ni-NTA modified surface than a conventional single-(His)$_6$ tag *(6)*. Two consecutive stop codons are in bold and (A)$_{28}$ is the poly(A) tail comprising 28xA. The transcription termination region is shown in italics.

2.3. Cell-Free Systems, Molecular Biology Reagents, and Kits

1. "TNT T7 Quick for PCR DNA" (Promega, UK).
2. RTS100 *E. coli* HY (Roche Molecular Biochemicals, UK).
3. Nucleotides (Sigma, UK).
4. Agarose (Sigma, UK).
5. Taq DNA polymerase (Qiagen, UK).
6. Gel elution kit QIAEX II (Qiagen, UK).
7. Ni-NTA-coated HisSorb strip/plates (Qiagen, UK).
8. Ni-NTA-coated magnetic agarose beads (Qiagen, UK).
9. Ni-NTA-coated microscope slide (Xenopore, USA)
10. HRP-linked anti-k antibody (The Binding Site, UK).
11. HRP-linked streptavidin (Amersham, UK).
12. 3,3′,5,5′-Tetramethylbenzidine (TMB) liquid substrate system for ELISA (Sigma, UK).
13. TSA™ Plus Fluorescence System (PerkinElmer, UK).

2.4. Solutions

1. 100 m*M* magnesium acetate.
2. Superblock (Pierce, UK).
3. Phosphate-buffered saline (PBS), pH 7.4.
4. Wash buffer 1: PBS containing 300 m*M* NaCl, 20 m*M* imidazole, pH 8.0.
5. Wash buffer 2: PBS containing 0.05% Tween 20.
6. Stripping buffer: 1 *M* $(NH_4)_2SO_4$, 1 *M* urea.

3. Methods

The method involves the following steps: (1) PCR construction, (2) PISA, and (3) detection of the arrayed proteins.

3.1. Generation of PCR Constructs for Cell-Free Expression

A PCR template is used for protein synthesis in a cell-free system. The PCR construct contains the essential elements for gene expression, including a promoter (usually T7), translation initiation site, and transcription and translation termination regions. The translation initiation site for eukaryotic systems is different from that for prokaryotic *E. coli* S30 extracts. To promote protein expression, the presence of a poly(A) tail is required after the stop codon. An affinity tag sequence is usually placed at either the N-or C-terminus of the target protein for *in situ* affinity immobilization on a surface (*see* **Note 1**). A flexible linker is also designed between the target protein and the tag sequence (**Fig. 2**).

To simplify the PCR construction, these essential elements can be cloned in order into a plasmid, which is then used as a template for a large amount of generation by PCR (**Fig. 2**). Here, we describe the use of a designed DNA fragment encoding a C-terminal region containing the required elements for cell-free protein synthesis (*see* **Subheading 2.2**). This fragment is linked to the C-terminus of the target protein (**Fig. 2**). At the N-terminus of the target protein, a T7 promoter and a translation initiation site are simply introduced using a long primer containing the corresponding sequences. **Figure 2** shows the PCR construction process.

3.1.1. Generation of a Target Gene and the C-Terminal Region

1. Set up a standard 50 μL PCR reaction using the Qiagen Taq system for amplifying (1) a target DNA using the primers G/back and G/for and (2) the C-terminal region from the plasmid pTA-His (*see* **Subheading 2.2**) using primer Linker-tag/back and T-term/for (**Fig. 2**) (*see* **Note 2**). Carry out thermal cycling for 30 cycles (94°C for 30 s, 54°C for 1 min, and 72°C for 1.2 min)
2. Analyze the resultant PCR products by 1% agarose gel electrophoresis and isolate the expected fragments using the Qiagen gel extraction kit.

3.1.2. Generation of the Construct by Assembly of the Gene and the C-Terminal Region

1. Set up a 25 μL PCR reaction by mixing the target gene and the C-terminal region in equimolar ratios (total DNA 50–100 ng). Carry out thermal cycling for eight cycles (94°C for 30 s, 54°C for 1 min, and 72°C for 1 min) to assemble the two fragments.
2. Amplify the assembled product by transferring 2 μL from step 1 above to a second PCR solution in a final volume of 50 μL for a further 30 cycles (94°C for 30 s, 54°C for 1 min, and 72°C for 1.2 min) using one of the T7/back primers and T-term/for.
3. Analyze the PCR product by 1% agarose gel electrophoresis and purify the DNA if required.
4. Confirm construct identity by PCR mapping using primers annealing at various positions along the desired sequence (*see* **Note 3**). The construct, either purified or unpurified, is ready for PISA (*see* below) or may be stored at –20°C for at least 6 months.

3.2. PISA on Nickel-Coated Wells, Magnetic Beads, and Glass Slides

The PISA procedure is carried out using a coupled cell-free system. We describe the use of either the rabbit reticulocyte lysate TNT system or RTS100 *E. coli* HY systems. Three different nickel-coated surfaces (i.e., Ni-NTA-coated microtiter plates, magnetic agarose beads, or glass slides) are used to capture His-tagged proteins.

1. Set up a translation mixture using either of the following cell-free systems:

   a. *Rabbit Reticulocyte Lysate TNT System*

TNT T7 Quick for PCR DNA	40 μL
1 m*M* methionine (from the kit)	1 μL
100 m*M* magnesium acetate	1 μL (*see* **Note 4**)
H_2O to	50 μL

   b. *RTS100 E. coli HY System*

E. coli lysate (from the kit)	12 μL
Reaction mix (from the kit)	10 μL
Amino acids (from the kit)	12 μL
Methionine (from the kit)	1 μL
Reconstitution buffer (from the kit)	5 μL
H_2O to	50 μL (*see* **Note 5**)

2. Add the translation mixture directly to either of the following surfaces:
   a. Add 10 μL translation mixture together with 0.1–0.25 μg PCR DNA (0.5–1 μL) into each Ni-NTA-coated well; or
   b. Mix 10 μL translation mixture containing 0.1–0.25 μg PCR DNA with 5–10 μL Ni-NTA-coated magnetic beads; or
   c. Spot 40 nL–2 μL translation mixture containing 50–100 ng PCR DNA per spot onto an Ni-NTA-coated glass slide.
3. Incubate the reaction at 30°C for 2 h.
4. Wash three times with Wash buffer 1 (*see* **Note 6**), followed by a final wash with 100 μL PBS, pH 7.4. Immobilized proteins are ready for functional assays (*see* below) or may be stored at 4°C.

3.3. Detection of Immobilized Proteins by Antibodies

1. Add horseradish peroxidase (HRP)-linked antibody (appropriately diluted with Superblock buffer) against the immobilized protein.
2. Incubate the mixture at room temperature for 1 h.
3. Wash three times with 100 μL Wash buffer 2, then a final wash with PBS.
4. Develop HRP activity using a TMB liquid substrate system for wells and beads and read at OD_{450} or using the tyramide signal amplification system on the glass slide, which is then scanned by an array scanner.

3.4. Reuse of Array Wells or Beads after Exposure to Detection Reagents

1. Wash the array wells or beads three times with 100 μL PBS containing 0.05% Tween.
2. Incubate with 50 μL freshly prepared stripping buffer at room temperature for 2 h.
3. Wash three times with 100 μL PBS containing 0.05% Tween, followed by a final wash with PBS, pH 7.4. The arrays are ready for reexposure to detection regents.

4. Notes

1. It has been reported that a tag may not be accessible when located at one or the other of the protein termini. In some circumstances, the location of a tag sequence may affect protein activity. In these cases, the tag should be tested at both the N-and C-termini.
2. The C-terminal region is usually produced in a large quantity by PCR and stored at –20°C for use as required.
3. PCR mapping is carried out by using a combination of various primers annealing at different positions in the construct. If all PCR reactions give the expected size, it strongly suggests the construction is correct.
4. Magnesium acetate added to the TNT mixture during translation has been found to improve protein expression. We have shown that single-chain antibodies and other proteins can be more efficiently produced with additional Mg concentrations ranging from 0.5 m*M* to 2 m*M*.
5. RTS100 *E. coli* HY can produce 3–25 μg proteins in a 50 μL reaction.
6. TNT lysate contains large amounts of hemoglobin, which sometimes sticks to Ni-coated magnetic beads. More washes are required to remove hemoglobin from the beads.

Acknowledgments

We thank Hong Liu for technical assistance. Research at the Babraham Institute is supported by Biotechnology and Biological Sciences Research Council (BBSRC), UK.

References

1. Bertone, P. and Snyder, M. (2005) Review: advances in functional protein microarray technology. *FEBS J.* **272,** 5400–5411.
2. Stevens, R. C. (2000). Design of high–throughput methods of protein production for structural biology. *Structure Fold. Des.* **8**, R177–185.
3. He, M. and Taussig, M. J. (2001) Single step generation of protein arrays from DNA by cell-free expression and *in situ* immobilization (PISA method). *Nucleic Acid. Res.* **29**, e73.

4. Ranachandran, N., Hainsworth, E., Bhullar, B., Eisenstein, S., Rosen, B., Lau, A. Y., Walter, J. C., and LaBaer, J. (2004) Self-assembling protein mircoarrays. *Science* **305**, 86–90.
5. He, M. and Taussig, M. J. (2003) DiscernArray™ technology: a cell-free method for the generation of protein arrays from PCR DNA. *J. Immunol. Methods* **274**, 265–270.
6. Khan, F., He, M., and Taussig, M. J. (2006) A double-His tag with high affinity binding for protein immobilisation, purification, and detection on Ni-NTA surfaces. *Anal. Chem.* **78**, 3072–3079.

15

Nondenaturing Mass Spectrometry to Study Noncovalent Protein/Protein and Protein/Ligand Complexes: Technical Aspects and Application to the Determination of Binding Stoichiometries

Sarah Sanglier, Cédric Atmanene, Guillaume Chevreux, and Alain Van Dorsselaer

Summary

In the present chapter we detail how mass spectrometry (MS) can be used to characterize noncovalent complexes, especially multimeric proteins and protein/ligand complexes. This original application of MS, also called "supramolecular MS" or "nondenaturing MS," appeared in the early 1990s and has continuously evolved since then. Nondenaturing MS is now fully integrated in structural biology programs and in drug discovery platforms. Indeed, appropriate sample preparation and fine tuning of the instrument make it possible to transfer weak assemblies without disruption from solution into the gas phase of the mass spectrometer. In this chapter we detail experimental conditions (sample preparation, optimization of instrumental parameters, etc.) required for the detection of noncovalent complexes by MS. We then focus on the type of information and accuracy that we get after interpreting electrospray ionization mass spectra obtained under nondenaturing conditions, with emphasis on the determination of the stoichiometry of protein/protein and protein/ligand complexes.

Key Words: Noncovalent interactions; nondenaturing mass spectrometry; multimeric protein; ligand binding stoichiometry.

1. Introduction

Since 1991 *(1)* electrospray ionization mass spectrometry (ESI-MS) has been the center of extensive research and development for a very specific application: the analysis of noncovalent complexes. The classical MS approach,

From: *Methods in Molecular Biology, vol. 484: Functional Proteomics: Methods and Protocols*
Edited by: J. D. Thompson et al., DOI: 10.1007/978-1-59745-398-1, © Humana Press, Totowa, NJ

so-called "molecular MS," analyzes, in the gas phase of the mass spectrometer, individual species initially present in solution after destruction of the noncovalent framework. On the other hand, "supramolecular MS" or "nondenaturing MS" aims at transferring intact noncovalent complexes that preexist in solution into the gas phase of the instrument.

The investigation of noncovalent complexes by MS is an original and unexpected application of MS in the biological field. At first, it may look inappropriate to use a technique that detects species in the gas phase to study assemblies maintained by weak interactions (such as electrostatic and van der Waals interactions, H-bounds, hydrophobic effect) because of their intrinsic fragility. Pioneering work performed by two American groups ***(1,2)*** in the early 1990s showed that specific protein/ligand interactions can survive the ESI process. Due to extensive work performed by several laboratories all over the world, experimental conditions making it possible to reproducibly perform such analyses have been established. Although nondenaturing MS remains the area of expertise of few laboratories, the number of publications relating the use of ESI-MS for noncovalent assemblies (protein/protein, protein/ligand, protein/metal, protein/RNA, protein/DNA, etc.) is exponentially growing (for recent reviews, *see* ***(3–8)***). Compared to more classical biophysical methods such as spectrophotometry, fluorescence techniques, crystallography, nuclear magnetic resonance (NMR), or surface plasmon resonance, nondenaturing MS is now well implemented as a complementary technique for characterizing protein/ligand or protein/protein interactions. The most interesting advantage of MS over other biophysical techniques consists of its ability to provide direct insight into all individual species present in solution through precise mass measurements. Finally, nondenaturing MS provides highly reliable and informative data including binding stoichiometry and specificity as well as an evaluation of relative binding affinity of complexes formed in solution.

In the present chapter we detail experimental conditions (sample preparation, optimization of instrumental parameters, etc.) required for the detection of noncovalent complexes. We also focus on the relevance of the information that can be deduced after interpretation of ESI mass spectra obtained under nondenaturing conditions, particularly on the determination of the stoichiometry of protein/protein and protein/ligand complexes.

2. Materials

2.1. Buffers

1. Milli-Q water.
2. Ammonium buffer: ammonium acetate ≥99.0% puriss. P. a. for mass spectroscopy (Fluka), ammonium bicarbonate or ammonium carbonate, triethylammonium bicarbonate, or pyridinium acetate.

3. Acetonitrile (Carlo Erba).
4. Formic acid.
5. Horse heart myoglobin (Sigma) for calibration of the MS instrument.

2.2. Desalting Procedure

1. Microconcentration on centrifugal filter units: centricon or microcon (Millipore), Vivaspin (Sartorius).
2. Gel filtration: NAP-5, NAP-10, and PD-10 gel filtration columns (GE Healthcare), Zeba (Perbio).
3. Equilibrium dialysis: Slide-A-Lyzer (Perbio).

2.3. Mass Spectrometry

1. Any electrospray-time-of-flight (ESI-TOF) or ESI-Q-TOF instrument.
2. Analysis under classical "denaturing conditions."
   a. Calibration of the mass spectrometer with horse heart myoglobin diluted to 2 μM in a H_2O/CH_3CN–1/1–solution acidified with 1% HCOOH.
   b. Dilute the sample to 2–5 μM in an H_2O/CH_3CN–1/1–solution + 1% HCOOH.
   c. Injection into the mass spectrometer.
   d. Record mass spectra on an appropriate *m/z* range (typically *m/z* 500–3000).

3. Analysis under "nondenaturing conditions."
   a. Calibration of the mass spectrometer with horse heart myoglobin diluted to 2 μM in a H_2O/CH_3CN–1/1–solution acidified with 1% HCOOH.
   b. Dilute the sample to 5–20 μM in ammonium buffer.
   c. Injection into the mass spectrometer.
   d. Record mass spectra on an appropriate *m/z* range (typically *m/z* 1000–5000).
   e. Adjust the pressure in the interface (Pi) and accelerating voltage (Vc) to obtain optimal transmission and desolvation without complex disruption (*see* **Subheading 3.2.3**).

2.4. Materials for the HPrK/P Example

1. The enzyme HPrK/P (Trx-His$_6$-S-tag) from *Bacillus subtilis* was expressed in *Escherichia coli* and purified as previously detailed ***(9)***.
2. ESI-MS measurements were performed on an electrospray quadrupole time-of-flight mass spectrometer Q-TOF-II fitted with a standard Z-spray source (Waters, Manchester, UK) and a *m/z* range extended to 25,000. Mass spectra were recorded at the exit of the TOF analyzer; the quadrupole was used in the "rf-only" mode.

2.5. Materials for the Aldose Reductase Example

1. The aldose reductase enzyme (ALR2) was expressed in *E. coli* and purified as previously detailed ***(10)***.
2. The inhibitors were prepared as highly concentrated solutions (5 m*M*) in ethanol. These solutions were then diluted to 100 μ*M* in 10 m*M* ammonium acetate (pH 7.0).
3. The coenzyme $NADP^+$ was purchased as a salt-free powder from Boehringer–Mannheim and dissolved to 1 m*M* in 10 m*M* ammonium acetate (pH 7.0).
4. The enzyme–inhibitor complexes were prepared by incubating the enzyme diluted to 10 μ*M* in 10 m*M* ammonium acetate with a 1 molar equivalent of $NADP^+$ and 2 molar equivalents of inhibitor. After a short incubation time at room temperature (10 min), the samples were continuously infused into the ESI ion source at a flow rate of 5 μL/min.
5. An electrospray time-of-flight mass spectrometer (ESI-TOF) equipped with a Z-spray ion source (LCT from Waters, UK) was used to perform the measurements. Electrospray ionization (ESI) conditions were optimized in order to keep the noncovalent specific interactions during ion desorption in the gas phase, while ensuring a good desolvation of the sprayed droplets. Calibration of the ESI-TOF instrument was performed with horse heart myoglobin diluted to 2 pmol/μL in a 1:1 water–acetonitrile mixture (v/v) acidified with 1% formic acid. Mass spectra were recorded in the positive ion mode on the mass range 500–4000 *m/z*.

3. Methods

3.1. Sample Preparation for Nondenaturing MS Analysis

Usually, buffers used for purifications or extractions of proteins or noncovalent complexes (phosphate buffers, Tris, HEPES, etc.) are nonvolatile salts that are not compatible with ESI-MS analysis, even at trace levels. Therefore, a prerequisite to perform noncovalent complex analysis by MS is to exchange the purification buffer, a procedure also called the "desalting step." The new buffer must fulfill two conditions: (1) being compatible with the ESI ionization process, i.e., volatile buffers are required and (2) integrity of the noncovalent assembly in solution must be preserved. Ammonium buffers best fulfill these requirements. Classical buffers usually used for nondenaturing MS analysis include volatile buffers such as ammonium acetate, ammonium carbonate or triethylammonium bicarbonate ***(11)***, pyridinium acetate, or water. Those buffers allow the pH of the solution to range from 5.0 to 8.5. Further pH adjustments toward more acidic or basic pHs can be achieved by adding small volumes of formic acid or ammonia, respectively. The ionic strength of the buffer can also range from 10 to 500 m*M* depending on the stability of the complex ***(12)***. In most studies, solutions between 10 and 200 m*M* ammonium buffers are used, ensuring optimal ESI mass spectra quality (*see* **Note 1**).

Classical methods used for small volume sample desalting include size exclusion chromatography (NAP-5™, NAP-10™, and PD-10™ gel filtration columns, GE Healthcare), microconcentration on centrifugal filter units (Centricon®, Microcon®, from Millipore; Vivaspin from Sartorius), and equilibrium dialysis (Slide-A-Lyzer, Perbio). These devices are all used according to supplier recommendations (*see* **Note 2**).

Figure 1 illustrates the importance of sample preparation for nondenaturing MS analysis. **Figure 1a** and **b** shows the need to have ESI compatible buffers for such kind of analysis. ESI mass spectra of the nucleocapsid protein NCp7 were recorded after two different sample preparation procedures: lyophilized NCp7 has been resuspended either in HEPES buffer (25 m*M*, pH 7.4) or in water prior to dilution to 20 μ*M* in a 50 m*M* ammonium acetate solution (pH 6.8). In the presence of HEPES buffer (**Fig. 1a**), no ion distribution corresponding to the NCp7 protein could be detected. The most intense ions present on the ESI mass spectrum correspond to HEPES and $[(HEPES)_n + Na]^+$ multimers, totally avoiding the detection of protein signals. However, when NCp7 is prepared in water and diluted to 20 μ*M* in $AcONH_4$ 50 m*M* (**Fig. 1b**), the only detected ion distribution can be attributed to the protein, allowing an accurate mass measurement of 5137.4 ± 0.2 Da corresponding to an $NCp7(Zn)_2$ complex. **Figure 1c** and **d** presents ESI mass spectra obtained for the recombinant human phosphatidylethanolamine binding protein (PEBP). After purification, the protein was lyophilized and resuspended in water. Nondenaturing MS analysis was performed directly on the sample in water (**Fig. 1c**) or after an additional desalting step (gel filtration using NAP-5™ columns, GE Healthcare) (**Fig. 1d**). Without desalting (**Fig. 1c**), ESI mass spectrum is very noisy with a low signal-to-noise ratio. Peaks are broad, sodium adducts are detected, and no accurate molecular mass can be measured. Such low-quality mass spectra are not compatible with the detection of ligand bound to the protein. After desalting (**Fig. 1d**), the signal-to-noise ratio is considerably improved. Ions distributions can be easily distinguished with narrow peak shapes, allowing unambiguous mass measurement. Sample preparation is now optimal for further nondenaturing MS analysis in the presence of different ligands, for instance.

To conclude, buffer exchange is an essential step in sample preparation (*see* **Notes 3** and **4**). It provides the protein sample free of nonvolatile salts, allowing acquisition of high-quality ESI mass spectra and accurate mass measurements.

3.2. Instrumental Conditions for Nondenaturing MS Analysis

3.2.1. Preferred Ionization Method

Matrix-assisted laser desorption/ionization (MALDI, ***13,14***) and ESI ***(15)*** are two "soft" ionization methods currently used for biomacromolecule analysis.

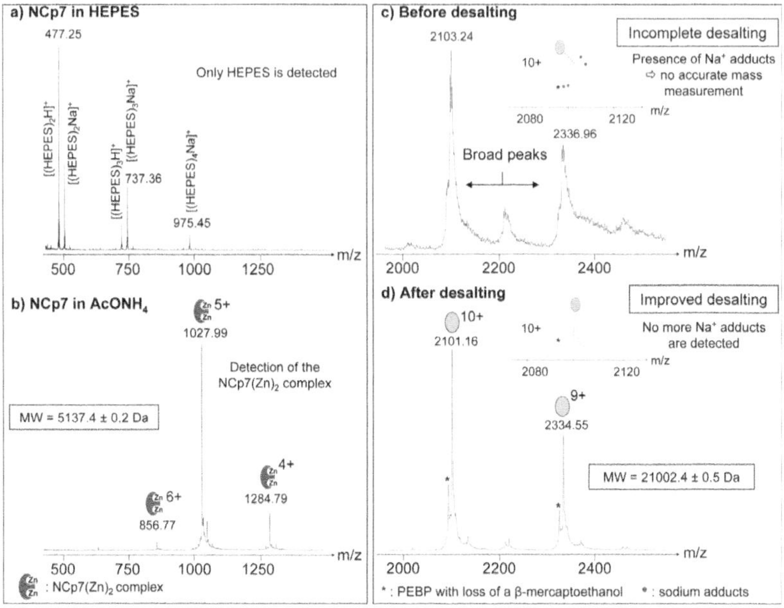

Fig. 1. Importance of sample preparation for MS analyses in nondenaturing conditions. Analysis of NCp7 after dilution to 20 μ*M* in $AcONH_4$ buffer (50 m*M*, pH 6.8) in the presence (**a**) and in the absence (**b**) of HEPES. (**a**) Nonvolatile buffer molecules (HEPES) lead to very intense peaks, which prevent the observation of NCp7 ions. (**b**) In the absence of these nonvolatile molecules, $NCp7(Zn)_2$ ions are easily detected and an accurate mass measurement is possible (5137.4 ± 0.2 Da). In the case of PEBP, analysis of the protein diluted to 15 μ*M* in $AcONH_4$ buffer (50 m*M*, pH 6.8) before (**c**) and after (**d**) desalting (gel filtration, NAP-5^{TM}, GE healthcare). Before desalting (**c**), the ESI mass spectrum shows that the presence of sodium traces induces peak broadening, preventing an accurate mass measurement. Removal of these salts by gel filtration makes it possible to obtain narrower peaks and subsequent accurate mass measurement (21,002.4 ± 0.5 Da), which is consistent with the theoretical mass (21,001.7 Da).

MALDI implies the use of a specific matrix, i.e., a small molecule that exhibits strong absorption at laser wavelength. Commonly used matrixes are derivatives of cinnamic acid or benzoic acid, which are rather acidic. Thus noncovalent interactions are mostly disrupted at the early stage of cocrystallization. Few studies, however, have reported MALDI detection of noncovalent complexes under specific conditions: it was observed that only spectra recorded from the upper layer of the samples show pronounced signals of noncovalent complexes: this phenomenon is called the "first shot phenomenon" ***(16–19)***.

With ESI, liquids are sprayed throughout a metallic capillary in the presence of a strong electric field forming small, multiply charged droplets. In case of noncovalent complex analysis, the best ionization method appears to be ESI since it requires liquid samples, and is therefore adapted to the use of ammonium buffers. Analytes can thus be transferred from solution into the gas phase in a very gentle manner allowing noncovalent bonds to be preserved. Miniaturization of the ESI technique, called nano-ESI, was achieved in 1994 by Wilm and Mann ***(20,21)***, who used capillaries (needles) with narrower diameters. Nano-ESI-generated droplets are about 10 times smaller than droplets obtained with pneumatically assisted ESI. As a result, nano-ESI is more efficient and hence has improved sensitivity. It also provides reduced flow rates thus affording longer analysis times and subsequent lower sample consumption ***(22,23)***. A commercial automated nano-ESI microchip system for noncovalent studies has been recently developed that combines the advantages of nanoflow electrospray MS with a high-throughput approach ***(24,25)***. The system shows a 10-fold increase in signal stability compared with nanoflow capillaries and a high level of nozzle-to-nozzle reproducibility ***(26)***.

3.2.2. Analyzers

When performing analysis under nondenaturing conditions (ammonium buffers with controlled pH and ionic strength), the native conformation of the protein is maintained. Consequently, less amino acids are accessible for protonation in a folded state than in an unfolded state. The effective charge of a protein in nondenaturing conditions is thus greatly decreased in comparison to the number of charges detected in the case of classical denaturing conditions (e.g., a mixture of water/acetonitrile acidified with formic acid, pH 3), resulting, on the ESI mass spectra, in detection of ions at higher *m/z* values, with less charges. Accordingly, analyzers with extended *m/z* ranges (over *m/z* 4000) should be preferred for noncovalent complex analysis. Many commercially available ESI instruments are coupled to quadrupole or ion trap mass analyzers, with a fairly limited *m/z* range, constituting a technical limitation for nondenaturing MS applications. Time-of-flight (TOF) instruments and hybrid quadrupole-TOF (Q-TOF) analyzers are particularly well adapted for nondenaturing MS experiments as they combine high sensitivity, high resolution, speed of acquisition, and extended mass range (theoretically unlimited) ***(27–29)***. Orthogonal hybrid instruments have additional potential for tandem MS measurements, providing supplementary structural information. In most commercially available instruments the *m/z* range of the quadrupole is limited to 4000, which restricts the ions selection to the analysis of ions with masses up to 60 kDa. The group of Robinson has recently reported the use of a quadrupole with *m/z* range extended to 32,000, allowing MS/MS experiments to be performed on large noncovalent assemblies ***(30)***.

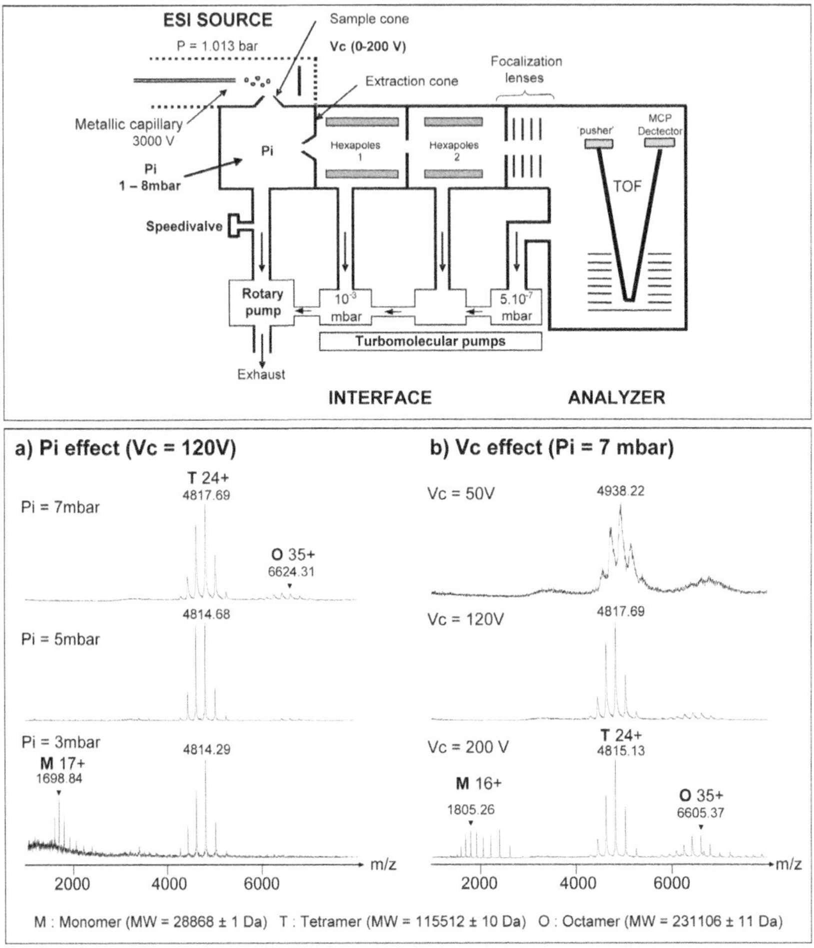

Fig. 2. Influence of interface parameters (Pi and Vc) optimization on TrmI oligomer detection. (**a**) Schematic view of the interface of the LCT instrument (Waters, Manchester, UK). The values of the pressures measured at different pumping stages are presented. The voltages applied on relevant lenses are also indicated. (**b** and **c**) The optimization of relevant interface parameters, Pi and Vc, respectively, for the detection of TrmI oligomers. ESI mass spectra were obtained with TrmI diluted to 80 μ*M* (monomer concentration) in 50 m*M* ammonium acetate buffer (pH 7.5). (**b**) Typical ESI mass spectra recorded at different pressures in the interface region (Pi) of the mass spectrometer (Vc was set to 120 V). At 7 mbar (upper spectrum), the most intense ion series corresponds to the TrmI tetramer, while a minor ion distribution can be attributed to the octameric form of TrmI. Decreasing the Pi to 5 (middle spectrum) and 3 mbar (lower spectrum) induces more efficient desolvation (narrow peaks) but also partial disruption of the tetramer into monomer and less efficient high *m/z* ion transmission (reduced TIC). (**c**) Typical ESI mass spectra recorded at different accelerating voltages (Vc) (Pi was set to 7 mbar). At low Vc values (50 V, upper spectrum),

3.2.3. Optimization of Interface Parameters of the Mass Spectrometer

A crucial point to maintain noncovalent interactions during the ionization/desorption process is the optimization of parameters of the mass spectrometer that control the energy communicated to the ions in the first pumping stage of the instrument. This is a key step for ensuring that the integrity of noncovalent complexes is preserved between the ion source of the instrument at atmospheric pressure and the high vacuum region of the analyzer. This region of intermediate pressure is called the interface and corresponds physically to the zone of the first hexapoles (see schematic representation on **Fig. 2a**). Two parameters are of utmost importance and need to be optimized for each new system to obtain optimum sensitivity and high-quality ESI mass spectra while preventing disruption of the complexes: (1) the pressure in the interface region (Pi), which affects the efficiency of the collisions [*see* **Note 5** ***(30–35)***] and (2) the accelerating voltage (Vc), which controls the kinetic energy communicated to the ions in the source of the instrument (*see* **Note 6**).

Figure 2b and **c** details the influence of Pi and Vc variations on the detection of the TrmI tetramer.

Vc and Pi are not independent parameters and should be optimized together to obtain the best compromise between sufficient ion desolvation and good transmission of high *m/z* ions without destruction of the noncovalent framework (**Fig. 3**). A careful optimization of Pi and Vc, different for each noncovalent assembly, is necessary to obtain the best results. Systematic control experiments, in which both Vc and Pi vary, are a prerequisite to unambiguously detect specific noncovalent complexes (*see* **Note 7**).

3.3. Observation of Noncovalent Complexes by MS and Information Deduced from Nondenaturing MS Experiments

3.3.1. MS-Based Strategy to Detect a Noncovalent Protein/Ligand (P/L) Complex

Observation of a noncovalent P/L complex (**Fig. 4**) by MS is a two-step strategy.

1. ESI-MS is performed in classical denaturing conditions: the noncovalent complex is diluted to 2–5 μ*M* in an H_2O/CH_3CN–1/1–mixture acidified with 1% HCOOH.

Fig. 2. (Continued) the signal-to-noise ratio is low leading to a low quality mass spectrum. Increasing Vc to higher values (Vc = 120 V, middle spectrum) considerably reduces peak broadening and enhances high *m/z* ion transmission. A further increase in Vc leads to partial dissociation of tetrameric into monomeric TrmI (Vc = 200 V, lower spectrum).

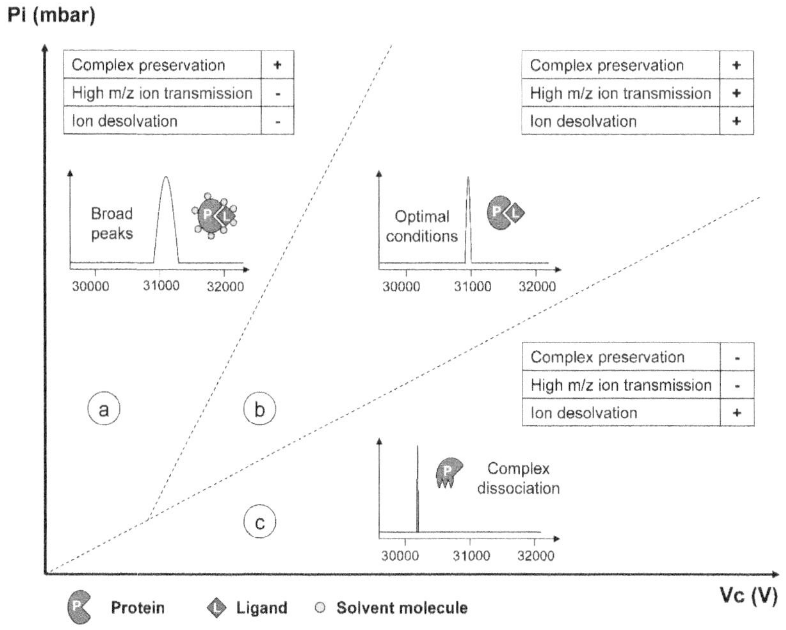

Fig. 3. Schematic representation of Pi (interface pressure) and Vc (accelerating voltage) optimization. (**a**) Region of incomplete desolvation, low high *m/z* ion transmission, and no disruption of noncovalent complexes (low Vc, high Pi).(**b**) Region of optimal tuning of Vc and Pi: region of best compromise between an efficient desolvation (narrow peaks), no dissociation of noncovalent complexes, and good high *m/z* ion focusing. (**c**) Region of disruption of noncovalent complexes and poor high *m/z* ion transmission (high Vc, low Pi) while desolvation is improved.

In such experimental conditions, noncovalent interactions between P and L are disrupted, proteins are denatured, and the molecular masses of individual species forming the complex are measured (M_P and M_L).

2. ESI-MS is performed in nondenaturing or "native" conditions: ESI-MS analysis is then performed in aqueous buffer at controlled pH and ionic strength as detailed above. Comparison of the molecular masses of the species measured under denaturing and nondenaturing conditions allows us to rapidly conclude that a noncovalent interaction between compound P and L exists.

3.3.2. Direct Determination of the Complex Stoichiometry

ESI-MS was shown to be a rapid and sensitive technique to unambiguously assess protein/ligand and protein/protein stoichiometries (**Fig. 4**). The comparison between the masses measured in native and denaturing conditions allows direct determination of complexes' stoichiometry. In case of multimeric

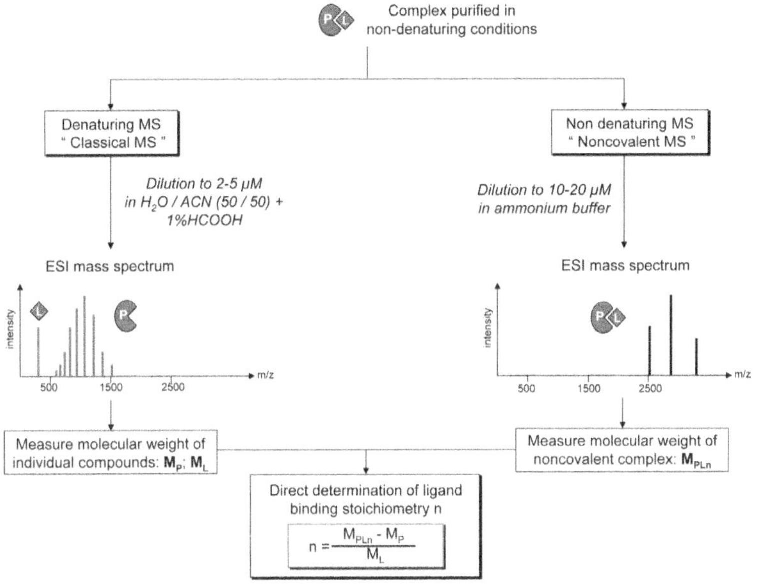

Fig. 4. MS-based strategy for noncovalent complex detection and determination of its binding stoichiometry. The purified complex is first analyzed in denaturing conditions. In such conditions molecular weights of individual species are determined (M_P and M_L). Then the same sample is analyzed under nondenaturing conditions, allowing mass measurement of the intact assembly (M_{PL}). Comparison of molecular masses obtained in denaturing and nondenaturing experiments makes it possible to evidence the existence of a noncovalent complex and to assess its binding stoichiometry.

proteins, the oligomeric state is directly given by the $MW_{native}/MW_{denaturing}$ ratio. For protein/ligand complexes, the stoichiometry of the bound ligand is given by $(MW_{native} - MW_{denaturing})/MW_{ligand}$. Examples that illustrate this point will be given in **Subheadings 3.5** and **3.6**.

3.3.3. Determination of the Complex Stability in Solution

Nondenaturing MS is perfectly suitable to perform in-depth characterization of the protein/ligand interaction to (1) analyze ligand selectivity for a protein, (2) study ligand-binding specificity for the protein-binding site, and (3) obtain valuable information about the relative binding affinity in solution for the protein/ligand system and subsequent ligand ranking according to their relative binding affinities. Thus, titration and competition experiments in solution can be set up and monitored by nondenaturing ESI-MS. An example is proposed in **Subheading 3.6**.

3.4. Validity of the Nondenaturing MS Approach: Do ESI Mass Spectra Give a Proper Image of Solution Equilibrium?

The essential prerequisite to the use of ESI-MS for the determination of binding stoichiometries of noncovalent complexes is that the peaks observed on mass spectra *in vacuo* are reliable to species effectively present in solution. Great care in the data acquisition as well as in the interpretation must be taken, since it is known that the solution-phase image might be distorted during ESI-MS analysis due to several factors, in particular during the evaporation of the ions in the gas phase, or during the transfer from the ion source to the analyzer through the interface region of the mass spectrometer [*see* **Note 8** *(36–42)*]. Thus, control experiments (involving different interacting partners or different experimental and instrumental conditions) should always be performed in order to avoid any misinterpretation.

3.5. Determination of the Oligomeric State of the Bifunctional Enzyme HPr Kinase/Phosphatase (HPrK/P) in B. Subtilis

3.5.1. The Biological Question

The HPr kinase/phosphatase enzyme is involved in the carbon catabolite repression mechanism observed in several low-GC (guanine, cytosine) Gram-positive bacteria. A high oligomerization state for HPrK/P is expected to play a key role in the regulation of its enzymatic activity.

At the time this study was undertaken, the data from the literature concerning the oligomerization state of HPrK/P from different bacteria were often approximate and confusing: oligomeric forms ranging from dimers ***(43)*** to octamers ***(44)*** and decamers ***(45)*** were reported depending on the bacteria and the analytical techniques used to assess the oligomeric form (gel filtration chromatography, ultracentrifugation). In this context, we evaluated the possibilities offered by nondenaturing ESI-MS to probe the oligomerization state of HPr kinase/phosphatase from *B. subtilis*.

3.5.2. Desalting Procedure

Sample desalting was achieved using centrifugal devices with a 10 kDa cutoff (Centricon YM10, Millipore). The final purification buffer (10 m*M* Tris buffer, pH 8.0) was exchanged against a 10 m*M* ammonium acetate (pH 6.8) solution. Six dilution/concentration steps were performed at 4°C and 6000 rpm (*see* **Note 9**). After desalting, the concentration of HPrK/P was determined spectrophotometrically using the Bio-Rad protein assay (Bio-Rad Laboratories, München, Germany) with Bio-Rad protein assay standard I lyophilized bovine plasma γ-globulin (Bio-Rad Laboratories, CA) as standard.

3.5.3. Analysis under "Classical" Denaturing Conditions

Calibration of the ESI-Q-TOF instrument was performed with horse heart myoglobin diluted to 2 pmol/μL in a 1:1 water–acetonitrile mixture (v/v) acidified with 1% formic acid. Mass spectra were recorded in the positive ion mode on the mass range *m/z* 500–4000. Accelerating voltage was set to 40 V and the pressure Pi in the interface region of the mass spectrometer was 2.5 mbar.

Desalted HPrK/P was first analyzed in classical denaturing conditions: the protein was diluted to 10 pmol/μL in a 1:1 water–acetonitrile mixture (v/v) acidified with 1% formic acid and directly infused into the mass spectrometer through a classical syringe pump at a flow rate of 5 μL/min. In these conditions the noncovalent interactions are disrupted in solution, which allows the molecular weight of the monomeric subunits to be measured with good precision (≥0.01%). This ESI-MS analysis revealed a highly pure and homogeneous protein preparation, as only one major ion series was detected. A molecular weight of 51,700 ± 1 Da was measured for the monomer (**Fig. 5a**), which is in good agreement with the molecular mass calculated from the expected amino acid sequence (51,699.3 Da).

3.5.4. Analysis under Nondenaturing Conditions

Calibration of the ESI-Q-TOF instrument on the extended mass range (*m/z* 2500-12,000) was achieved through a separate injection of a solution of 1 mg/mL CsI in 50% aqueous isopropanol (clusters of $Cs_{(n+1)}I_n$).

Desalted HPrK/P was then analyzed in nondenaturing conditions: the protein assembly was diluted to 20 pmol/μL in ammonium acetate (10 m*M*, pH 6.8) to preserve its native conformation in solution, before being continuously infused into the ESI ion source at a flow rate of 5 μL/min. Interface parameters (Pi and Vc) were optimized in order to obtain the best compromise between sufficient desolvation (narrow peaks), good ion transmission, and no destruction of the noncovalent assembly. Details about operating condition optimizations were previously described ***(34)***: the optimal values for Pi and Vc were found to be 6.5 mbar and 200 V, respectively. Both source and desolvation temperatures were 80°C. Mass spectra were acquired in the positive ion mode on the mass range *m/z* 2500–12,000 for 5 min and smoothed with the Savitzky Golay method.

ESI-MS analysis in nondenaturing conditions revealed three main ion series (**Fig. 5b**): (1) the major set of peaks with a charge state distribution ranging from 38+ to 44+ (the 41+ charge state being the most abundant) was observed in the mass range *m/z* 7000–8100 and led to a molecular weight of 310,337 ± 22 Da corresponding to the noncovalent association of six HPrK/P subunits; (2) a second minor ion series corresponding to the 103,404 ± 2 Da dimer with charge states ranging from 21+ to 25+ (centered on the 23+ charge state) was detected

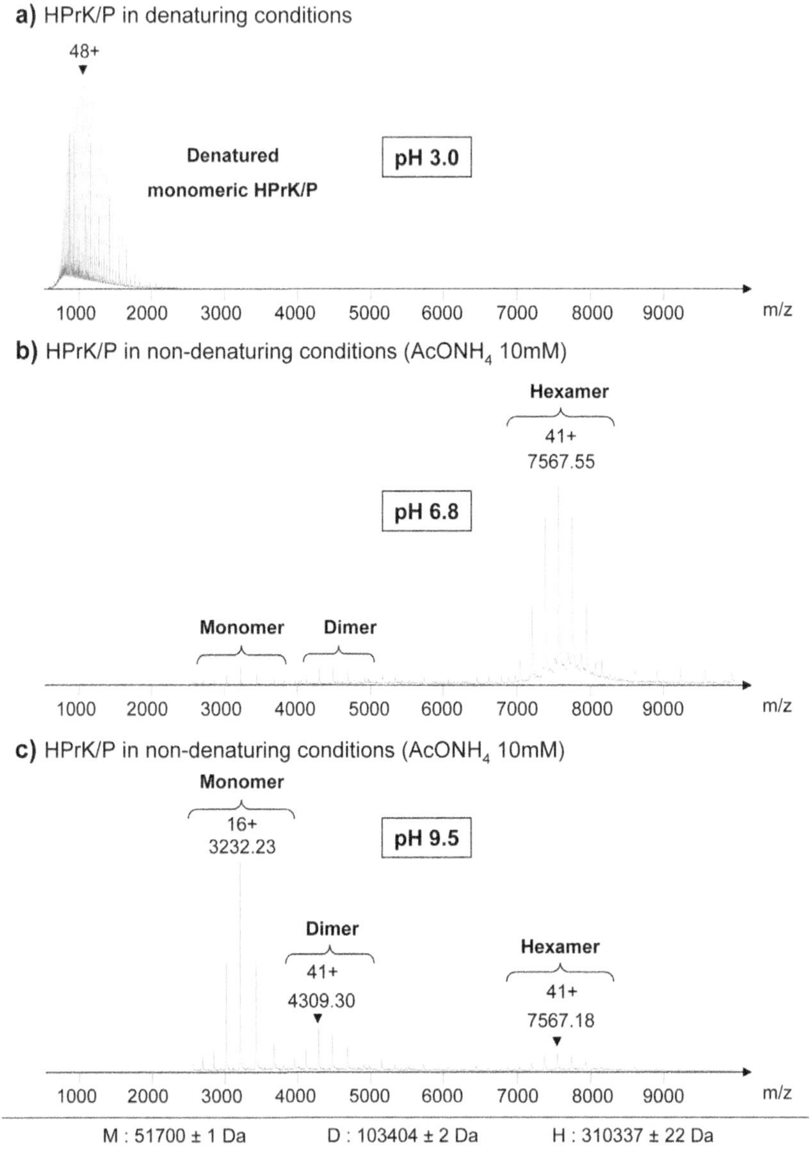

Fig. 5. ESI-MS analysis of the enzyme HPrK/P. (**a**) ESI mass spectra of HPrK/P from *B. subtilis* in "classical" denaturing conditions : HPrK/P was diluted to 10 μ*M* in a 1:1 water/acetonitrile mixture (v/v) acidified with 1% (v/v) formic acid, which enables an accurate mass measurement of the HPrK/P monomer (51,700 ± 1 Da). Vc = 40 V; Pi = 2.5 mbar. (**b** and **c**) ESI-MS analyses of HPrK/P in nondenaturing conditions

in the mass range *m/z* 4000–5000; (3) the third distribution with charge states ranging from 15+ to 17+ corresponded to monomeric subunits detected in the mass range *m/z* 3000–4000.

Since pH was known to play a key role in the switch between the kinase and phosphatase activity of the bifunctional HPrK/P, ESI-MS experiments were performed again under strictly identical experimental conditions except for pH (accelerating voltage set to 200 V and Pi to 6.5 mbar). As already mentioned, at pH 6.8 HPrK/P was mostly detected as a hexamer of 310,337 ± 22 Da. Increasing the pH of the ammonium acetate buffer to 9.5, by the addition of ammonium hydroxide, while keeping strictly identical conditions as at pH 6.8, resulted in a complete dissociation of the hexamer: the most intense signals on the ESI mass spectrum were those of the multiply charged monomer and dimer (**Fig. 5c**).

3.5.5. Data Interpretation and Conclusions

Comparison of molecular weights measured by ESI-MS in denaturing and native conditions demonstrated unambiguously that HPrK/P forms a specific noncovalent homohexamer of ~310 kDa at pH 6.8. The fact that pH variations induced strong changes on the ESI mass spectra provided a high level of confidence for a "structurally specific" hexamer and was correlated more closely with the phosphatase than the kinase activity of the bifunctional enzyme HPrK/P *(9)*.

ESI-MS analysis in nondenaturing conditions at different pHs revealed a direct correlation between pH dependence and oligomerization of the bifunctional enzyme, providing strong evidence for a structure–function relationship. ESI-MS measurements were consistent with the X-ray crystallography data obtained at the same period and that showed the existence of hexameric assembly ***(46–48)***.

Fig. 5. (Continued) (the protein was diluted to 20 μ*M*–hexamer concentration–in a 10 m*M* $AcONH_4$ buffer). Vc = 200 V; Pi = 6.5 mbar. (**b**) At pH 6.8 hexameric HPrK/P (310,337 ± 22 Da) is the major detected oligomerization state while dimeric (103,404 ± 2 Da) and monomeric forms are detected as minor components. (**c**) At pH 9.5 the oligomerization equilibrium is dramatically displaced toward monomeric and dimeric HPrK/P, which become the most abundant forms of the protein. Signals corresponding to the HPrK/P hexamer strongly decrease.

3.6. Determination of the Ligand-Binding Stoichiometries and Relative Solution Affinities for a Protein/Ligand System

3.6.1. The Biological Question

When considering protein/ligand interactions, several questions are of particular interest for biologists: (1) confirm the existence of a noncovalent interaction between the target protein and tested compounds, (2) determine ligand-binding stoichiometry, i.e., how many ligand molecules interact with the target protein, (3) evidence site specificity of the tested molecules, i.e., is the ligand binding "structurally" site-specific or does it bind nonspecifically anywhere at the surface of the protein?, and (4) the ability to gain insight into solution affinities from ESI-MS data.

In the following, possible use of ESI-MS for the characterization of protein/ligand interactions in terms of binding stoichiometry, binding specificity, and solution affinities will be described with the example of aldose reductase (ALR2). ALR2 is the first enzyme of the polyol pathway that converts glucose to sorbitol using $NADP^+$ as cofactor. ALR2 is implicated in the development of diabetic complications such as glaucoma, neuropathies, nephropathies, retinopathies, and cataracts. During diabetic hyperglycemia the increased flux of glucose through the polyol pathway results in biochemical imbalances in target tissues such as nerves, lenses, retina, and kidneys. Accordingly, inhibition of ALR2 represents an attractive strategy for preventing those diabetic-dependent complications.

3.6.2. Desalting Procedure

ALR2 was desalted by five dilution steps (5 × 60 min) in 10 m*M* ammonium acetate (pH 7.0) by using Centricon YM10 microconcentrators (Millipore). The final enzymatic concentration was spectrometrically measured (UV, 280 nm). The proteins were stored at 4°C in 10 m*M* ammonium acetate, pH 7.0, and used within a week after the end of their purification.

3.6.3. Determination of ALR2/$NADP^+$/Inhibitor Stoichiometry

The ternary complex formed between ALR2 (MW_{th} = 36,135 Da), its cofactor $NADP^+$ (MW_{th} = 744 Da), and an inhibitor (Fidarestat, I1, MW_{th} = 279 Da) was studied by ESI-MS.

3.6.3.1. Analysis under "Classical" Denaturing Conditions

Desalted ALR2 was diluted to 5 pmol/μL as explained in **Subheading 3.3.1.** Accelerating voltage was set to 20 V and the pressure Pi in the interface region of the mass spectrometer was 2.5 mbar. This ESI-MS analysis revealed highly pure and homogeneous protein preparation with a molecular weight of 36,138.9 ± 0.3 Da (**Fig. 6a**), which is in good agreement with the molecular mass calculated from the expected amino acid sequence (MW_{th} = 36,135 Da).

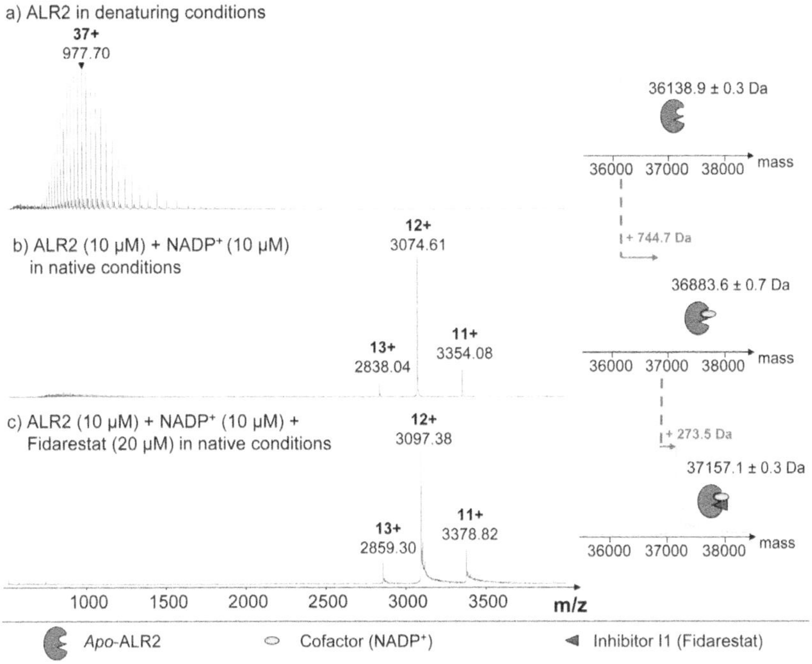

Fig. 6. ESI-MS analysis of an enzyme/cofactor/inhibitor complex (ALR2/NADP$^+$/Fidarestat). (**a**) ALR2 in denaturing conditions: analysis of ALR2 diluted to 5 μ*M* in a 1:1 water/acetonitrile mixture (v/v) acidified with 1% (v/v) formic acid allows an accurate mass measurement of the *apo*enzyme (MW = 36,138.9 ± 0.3 Da), which is in good agreement with the theoretical molecular weight (36,135 Da). Pi = 2.5 mbar and Vc = 20 V. (**b**) ALR2 in the presence of NADP$^+$: analysis of the ALR2 (10 μ*M*) in the presence of NADP$^+$ (10 μ*M*) in a 50 m*M* $AcONH_4$ buffer (pH 6.8) after 10 min incubation at room temperature. These nondenaturing conditions allow the detection of the *holo*enzyme, i.e., the 1:1 binary ALR2:NADP$^+$ complex (36,883.6 ± 0.7 Da). Pi = 5.0 mbar and Vc = 40 V. (**c**) ALR2 in the presence of both NADP$^+$ and Fidarestat (inhibitor I1): analysis of ALR2 (10 μ*M*) in the presence of NADP$^+$ (10 μ*M*) and Fidarestat (20 μ*M*) after a 10-min incubation in a 50 m*M* $AcONH_4$ buffer (pH 6.8) leads to the quantitative formation of the 1:1:1 ternary ALR2:NADP$^+$:I1 complex (37,157.1 ± 0.3 Da). Pi = 5.0 mbar and Vc = 40 V.

3.6.3.2. Analysis under Nondenaturing Conditions

Figure 6b and **c** shows the ESI mass spectra obtained for a preparation of ALR2 in the presence of its cofactor (**Fig. 6b**) or in the presence of both cofactor and inhibitor (**Fig. 6c**). The enzyme/cofactor/inhibitor complexes were prepared by incubating the enzyme diluted to 10 μ*M* in 10 m*M* ammonium acetate with 1 molar equivalent of $NADP^+$ and 2 molar equivalents of inhibitor (I1). After a 10-min incubation at room temperature, the samples were continuously infused into the ESI ion source at a flow rate of 5 μL/min (*see* **Note 10**). Interface parameters (Pi and Vc) were optimized in order to obtain the best compromise between sufficient desolvation (narrow peaks), good ion transmission, and no destruction of the noncovalent assembly. Details about operating condition optimizations were previously described ***(39)***. In the presence of the cofactor (**Fig. 6b**), ESI-MS analysis in nondenaturing conditions revealed a unique species with a molecular weight of 36,883.6 ± 0.7 Da, confirming that ALR2 forms a quantitative binary complex (1:1 stoichiometry) with $NADP^+$ (also called holo-ALR2). When both cofactor and inhibitor are present (**Fig. 6c**), a molecular mass of 37,157.1 ± 0.3 Da can be attributed to the quantitative formation of the ternary 1/1/1 ALR2/$NADP^+$/Fidarestat complex.

3.6.4. Determination of Ligand-Binding Specificity

In noncovalent interactions, the question of specificity of the interaction is an important issue. It is necessary to unambiguously distinguish "structurally specific" noncovalent complexes from nonspecific noncovalent complexes resulting from any gas-phase or in-solution artifactual association (*see* **Note 11**). For ALR2, we evaluated the interaction between ALR2 and inhibitors derived from sorbinil, a molecule that is currently used as a drug but that has medium affinity for ALR2. Analogue compounds of sorbinil comprising two asymmetric carbon atoms (four isomers) were evaluated in order to find the best stereochemistry and the best affinity. All four isomers (20 μ*M*) were individually incubated for 10 min at room temperature in a 10 μ*M* holo-ALR2 solution. Deconvoluted ESI mass spectra are presented in **Fig. 7**. Relative abundances of the different species are directly deduced from ESI mass spectra, assuming that ionization efficiencies of holo-ALR2 and holo-ALR2/inhibitors are similar ***(49)***. In strictly identical experimental MS conditions, different binding stoichiometries are observed. 4S isomers form 1/1 complexes with holo-ALR2. In case of 2S isomers binding of two or three inhibitor molecules is also observed. This statistical ligand multiaddition strongly suggests nonstructurally specific ligand binding of 2S isomers. Thus, ESI-MS was able to unambiguously determine that the 4S stereochemistry plays a central role in site-specific ligand binding.

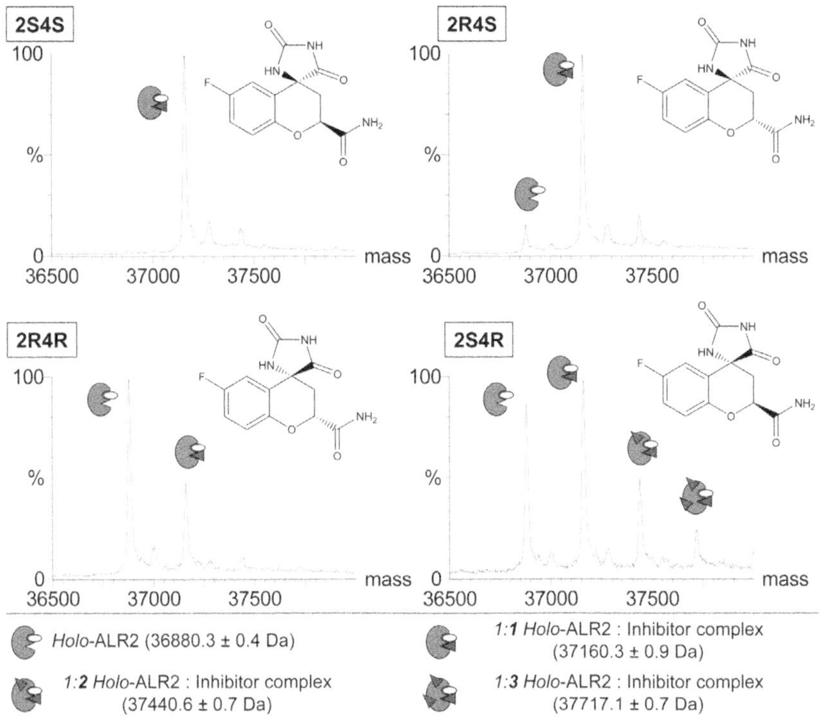

Fig. 7. Determination of inhibitor specificity by nondenaturing ESI-MS. ESI-MS analyses of ALR2 (10 μ*M*) in the presence of $NADP^+$ (10 μ*M*) and different stereoisomeric inhibitors (20 μ*M*) were recorded after a 10-min incubation in a 50 m*M* $AcONH_4$ buffer (pH 6.8). Pi and Vc were set to 5 mbar and 40 V, respectively. Stereoisometry of the two asymmetric carbons strongly influences the binding specificity. 2S4S (**a**), 2R4S (**b**), and 2R4R (**c**) compounds behave as specific binders: the only detected species corresponds to the 1:1:1 ternary ALR2:$NADP^+$:inhibitor complexes. Conversely, a statistical multiple binding of the 2S4R (**d**) inhibitor is observed, which indicates a nonstructurally specific interaction with the protein. Moreover, binding affinity is also affected by the stereochemistry of the inhibitor: 4S inhibitors show higher binding affinities than 4R ones.

3.6.5. Evaluation of Relative Solution Affinities of Different Inhibitors by Titration and Competition Experiments

Because of its unique advantage over other biophysical tools, ESI-MS provides direct insight into all individual species present in solution through precise mass measurements. In addition, the relative intensities of the different species observed on the mass spectrum can serve to estimate the relative abundances of the different compounds, providing important information about relative solution affinities *(49)*. The combination of these two pieces of information, accurate mass measurement and relative intensities of the peaks, can be used to rapidly determine which compounds from a mixture bind to which targets, and with what relative affinity. Molecular interactions with dissociation

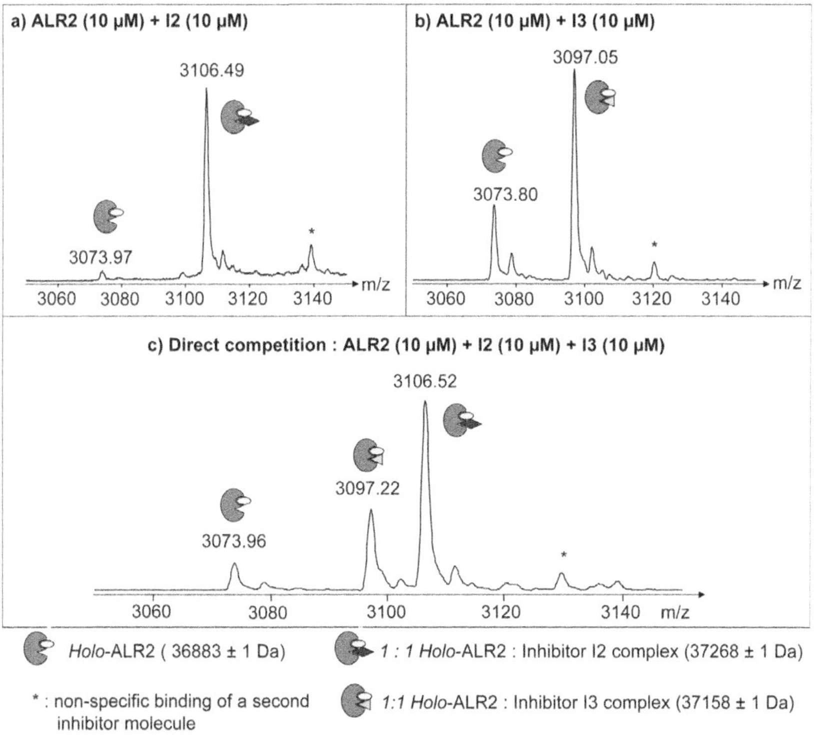

Fig. 8. Determination of relative ligand-binding affinities by nondenaturing ESI-MS. (**a** and **b**) Titration experiments performed in the presence of ALR2 (10 μ*M*), $NADP^+$ (10 μ*M*), and two different inhibitors I2 and I3 (10 μ*M*). In the presence of I2 (**a**), 98% of the detected species corresponds to the 1:1:1 ternary *holo*-ALR2:I2 complex, whereas only 75% of the detected compounds corresponds to the ternary *holo*-ALR2:I3 complex. This observation suggests a higher solution affinity for I2 compared to I3. (**c**) A direct competition experiment performed in the presence of ALR2 (10 μ*M*), $NADP^+$ (10 μ*M*), and a mixture of I2 and I3 (10 μ*M* each). The interpretation of the ESI mass spectrum reveals three compounds: the most intense one (63% of the detected ions) corresponds to the *holo*-ALR2/I2 complex while 28% and 9% of the detected signals can be attributed to the *holo*-ALR2/I3 and holo-ALR2 complexes, respectively. All analyses were performed after a 10-min incubation at room temperature in a 50 m*M* $AcONH_4$ buffer (pH 6.8). Again, these results confirm a better solution affinity for I2 than I3.

constants ranging from n*M* to m*M* have already been characterized using ESI-MS ***(50–52)***.

For the ALR2 project, titration experiments in the presence of increasing amounts of inhibitors and direct in-solution competition experiments in the presence of mixtures of inhibitors were monitored by ESI-MS in order to gain insight into inhibitor relative solution affinities. **Figure 8a** and **b** presents ESI mass spectra obtained for two different inhibitors, I2 and I3. When equimolar amounts of I2 were added to a 10 μ*M* ALR2 solution (**Fig. 8a**), almost all ESI-MS detected species (98%) correspond to the ternary holo-ALR2/I2 complex (MW = 37,268 ± 1 Da). On the contrary, in the presence of 10 μ*M* of I3 (**Fig. 8b**), only partial inhibitor binding was observed as about 75% of the detected species correspond to the ternary holo-ALR2/I3 complex (MW = 37,158 ± 1), while 25% have a molecular mass of 36,883 ± 1 Da, which could be attributed to holo-ALR2. From these titration experiments, it could be concluded that I2 seemed to have a higher solution affinity than I3. To confirm this hypothesis, a direct competition experiment was performed involving a mixture of equimolar amounts of I2 and I3 (10 μ*M* each). The resulting ESI mass spectrum (**Fig. 8c**) revealed three compounds: the most intense peak (63% of all the detected species) corresponds to the holo-ALR2/I2 complex; 28% of the compounds can be attributed to the holo-ALR2/I3 complex and 9% of the species are identified as holo-ALR2. This latest experiment enables a direct comparison of the two inhibitors: I2 has a higher solution affinity than I3. ESI-MS affinity ranking was in agreement with the data obtained in solution, as I2 and I3 have IC_{50} values of 108 n*M* and 580 nM, respectively.

3.7. Conclusions

ESI-MS is a powerful technique for the detailed characterization of protein/ligand interactions, providing reliable information such as binding stoichiometries, binding specificities, and evaluation of relative solution affinities of formed complex. Thus, ESI-MS can now be integrated in existing lead validation platforms and structural biology programs on the basis of the characterization of noncovalent target protein/ligand interactions. This approach offers several advantages compared to classical techniques used in drug discovery processes. Among them are the small quantities necessary to perform a complete MS validation, the rapidity of the technique, the direct visualization of ligand binding on ESI mass spectra, and the ability to work with unlabeled material.

4. Notes

1. The choice of the desalting buffer is sample dependent. In our laboratory, the standard desalting procedure consists of ammonium acetate, 50 m*M*, pH 6.8. Its ionic strength is increased to 100 m*M* or 200 m*M* when assemblies are stable only at high salt concentrations. pH can be adjusted using ammonia (no NaOH to avoid contamination with Na^+ ions) or acetic/formic acid.
2. The choice of the type of desalting procedure is sample dependent and cannot be predicted. In our laboratory, we first try desalting with gel filtration columns, which are less time consuming than microconcentration (often 4–10 dilution/concentration steps are required) or overnight dialysis on microdialysis units (precipitation of the protein may occur overnight). All the desalting devices are used according to supplier recommendations.
3. A relevant "trick" to perform MS analysis is to use fresh biological material. Freezing in ammonium acetate or even in the purification buffer should be avoided so as not to affect the stability of the complex. In our laboratory, samples are usually analyzed by mass spectrometry the day after purification and immediately after desalting.
4. After buffer exchange, it is highly advisable to check the activity of the protein complex in the ammonium buffer in order to ensure that conditions used for mass spectrometry analysis do not affect its biological activity.
5. Concerning the influence of the Pi of the instrument for detection of noncovalent complexes, several groups have reported that transmission of high *m/z* ions requires elevated pressures in the first vacuum stages of mass spectrometers ***(30–35)***. On our Q-TOF and LCT instruments (Waters, Manchester, UK), the first vacuum stage of the instrument is located between the sample cone and the extraction cone (see schematic representation in **Fig. 2a**). The pressure in this region (Pi) is regulated with a speedivalve, which throttles pumping by the rotary pump and allows the Pi to be adjusted between 1 and 8 mbar. Pi is directly linked to the internal energy communicated to the ions via collisions with residual gaseous molecules present in this part of the mass spectrometer. As the distance between two consecutive collisions (mean free path) with ambient gaseous molecules is inversely proportional to Pi, lower pressures in the interface (1–3 mbar) imply longer distances between two successive collisions. Consequently, gas phase ions have enough time to be "warmed up" and to accumulate energy, which further results in "destructive" collisions. Although ion desolvation is improved, such energetic collisions may lead to the dissociation of labile noncovalent subassemblies. Inversely, increasing the pressure in the interface results in more frequent but lower energy and less "destructive" collisions after which the "thermalized" ions (corresponding to large macromolecules) are transferred without any damage to the analyzer. Elevating pressure is also associated with less efficient ion desolvation, which is observed on ESI mass spectra by significant peak broadening. Higher pressures also substantially improve ion transmission at high *m/z*. In summary, increased Pi values permit improved collisional cooling and focusing of large ions in the quadrupole guides and, therefore, better transmission through the quadrupoles and TOF.

6. The Vc is important in the detection of noncovalent complexes. Varying the Vc induces changes in the initial kinetic energy communicated to the ions in the electrospray source (*see* **Fig. 2**). At high accelerating voltages, ions have higher initial kinetic energies that cause strong energetic collisions and possibly dissociation of weak interactions. Decreasing the Vc leads to a considerable loss in sensitivity due to nonoptimal transmission of high *m/z* ions and much less efficient desolvation, resulting in dramatically reduced mass accuracy. Better desolvation and focalization of high *m/z* ions at high accelerating voltages make interpretation of the recorded mass spectra much easier. Accordingly, the peak broadening effect previously mentioned for high interface pressures can be reduced by increasing the Vc. Fine tuning of the instrument in order to obtain the best compromise between sufficient desolvation, optimal transmission of intact high *m/z* ions, and nondestructive gas-phase collisions needs to be achieved to detect specific noncovalent edifices of high molecular weights.
7. In practice, for each studied complex, several ESI mass spectra are recorded for different (Pi, Vc) couples.
8. Possible reasons for discrepancies between MS data and solution data have been described in the literature ***(36–38)***; the stability of noncovalent complexes during the ESI-MS process strongly depends on the type of interaction (electrostatic contacts, hydrogen bonds, van der Waals interactions) involved in the formation of the complex. During ion transfer from the solution to the gas phase both electrostatic interactions and hydrogen bonds are strengthened. In contrast, complexes that are stabilized in solution by hydrophobic effects appear to be weakened ***(39,40)***. To understand those effects, it is necessary to remember that water molecules evaporate when passing into the gas phase of the mass spectrometer. Without water molecules around the complex, it is reasonable to assume that hydrophobic interactions do not contribute significantly to any complex stabilization in the gas phase of the mass spectrometer. This assumption has been verified by several groups comparing X-ray crystallography and ESI-MS results ***(3,4,10,41,42)***.
9. Desalting on microconcentrators is often a tedious job. Proteins can stick on the ultrafiltration membrane, which necessitates changing the device regularly (it is common to use at least two devices per desalting). However, this type of desalting affords very high quality ESI-MS spectra.
10. An additional centrifugation step (11,000 rpm for 2 min) can be performed before injection of the incubated mixture in the mass spectrometer in order to separate any precipitate and to avoid capillary plugging.
11. As precisely detailed by Smith and Light-Wahl ***(36)***, several control experiments can be performed to provide evidence for structurally specific interactions: (1) adjustment of interface conditions does not modify the detection of a preferred stoichiometry, (2) complex dissociation due to modification of the conditions in solution (pH, temperature, buffer, etc.) and subsequent change on the ESI mass spectrum, (3) complex dissociation upon variations in the interface conditions (more harsh interface conditions should disrupt labile complexes), and (4) sensitivity of the complex formation to modifications in the complex components.

Acknowledgments

The authors would like to thank Valérie Vivat-Hannah for critical reading of the manuscript. Guillaume Chevreux thanks the CNRS and Sanofi-Aventis for financial support. We also thank all our collaborators for providing us with starting material, especially Jacques Haiech, Hugues de Rocquigny, Yannick Goumon, Carine Tisné, and Alberto Podjarny.

References

1. Ganem, B., Li, Y. T., and Henion, J. D. (1991) Detection of noncovalent receptor-ligand complexes by mass spectrometry. *J. Am. Chem. Soc.* **113**, 6294–6296.
2. Katta, V. and Chait, B. T. (1991) Observation of the heme-globin complex in native myoglobin by electrospray-ionization mass spectrometry *J. Am. Chem. Soc.* **113**, 8534–8535.
3. Loo, J. A. (1997) Studying noncovalent protein complexes by electrospray ionization mass spectrometry. *Mass Spectrom. Rev.* **16**, 1–23.
4. Loo, J. A. (2000) Electrospray ionization mass spectrometry: a technology for studying noncovalent macromolecular complexes. *Int. J. Mass Spectrom.* **200**, 175–186.
5. Heck, A. J. and Van Den Heuvel, R. H. (2004) Investigation of intact protein complexes by mass spectrometry. *Mass Spectrom. Rev.* **23**, 368–389.
6. van den Heuvel, R. H. and Heck, A. J. (2004) Native protein mass spectrometry: from intact oligomers to functional machineries. *Curr. Opin. Chem. Biol.* **8**, 519–526.
7. Potier, N., Rogniaux, H., Chevreux, G., and Van Dorsselaer, A. (2005) Ligand-metal ion binding to proteins: investigation by ESI mass spectrometry. *Methods Enzymol.* **402**, 361–389.
8. Sharon, M. and Robinson, C. V. (2007) The role of mass spectrometry in structure elucidation of dynamic protein complexes. *Annu. Rev. Biochem.* **76**, 167–193.
9. Ramstrom, H., Sanglier, S., Leize-Wagner, E., Philippe, C., Van Dorsselaer, A., and Haiech, J. (2003) Properties and regulation of the bifunctional enzyme HPr kinase/phosphatase in Bacillus subtilis. *J. Biol. Chem.* **278**, 1174–1185.
10. Darmanin, C., Chevreux, G., Potier, N., Van Dorsselaer, A., Hazemann, I., Podjarny, A., and El-Kabbani, O. (2004) Probing the ultra-high resolution structure of aldose reductase with molecular modelling and noncovalent mass spectrometry. *Bioorg. Med. Chem.* **12**, 3797–3806.
11. Lemaire, D., Marie, G., Serani, L., and Laprevote, O. (2001) Stabilization of gas-phase noncovalent macromolecular complexes in electrospray mass spectrometry using aqueous triethylammonium bicarbonate buffer. *Anal. Chem.* **73**, 1699–1706.
12. Vis, H., Dobson, C. M., and Robinson, C. V. (1999) Selective association of protein molecules followed by mass spectrometry. *Protein Sci.* **8**, 1368–1370.
13. Tanaka, K., Waki, H., Ido, Y., Akita, S., Yoshida, Y., and Yoshida, T. (1988) Protein and polymer analyses up to m/z 100,000 by laser ionization time-of-flight mass spectrometry. *Rapid Commun. Mass Spectrom.* **2**, 151–153.

14. Karas, M. and Hillenkamp, F. (1988) Laser desorption ionization of proteins with molecular masses exceeding 10,000 daltons. *Anal. Chem.* **60**, 2299–2301.
15. Fenn, J. B., Mann, M., Meng, C. K., Wong, S. F., and Whitehouse, C. M. (1989) Electrospray ionization for mass spectrometry of large biomolecules. *Science* **246**, 64–71.
16. Kiselar, J. G. and Downard, K. M. (2000) Preservation and detection of specific antibody–peptide complexes by matrix-assisted laser desorption ionization mass spectrometry. *J. Am. Soc. Mass Spectrom.* **11**, 746–750.
17. Strupat, K., Rogniaux, H., Van Dorsselaer, A., Roth, J., and Vogl, T. (2000) Calcium-induced noncovalently linked tetramers of MRP8 and MRP14 are confirmed by electrospray ionization-mass analysis. *J. Am. Soc. Mass Spectrom.* **11**, 780–788.
18. Horneffer, V., Forsmann, A., Strupat, K., Hillenkamp, F., and Kubitscheck, U. (2001) Localization of analyte molecules in MALDI preparations by confocal laser scanning microscopy. *Anal. Chem.* **73**, 1016–1022.
19. Wattenberg, A., Sobott, F., Barth, H.-D., and Brutschy, B. (2000) Studying noncovalent protein complexes in aqueous solution with laser desorption mass spectrometry. *Int. J. Mass Spectrom.* **203**, 49–57.
20. Wilm, M. S. and Mann, M. (1994) Electrospray and Taylor-cone theory, Dole's beam of macromolecules at last? *Int. J. Mass Spectrom. Ion Processes* **136**, 167–180.
21. Wilm, M. and Mann, M. (1996) Analytical properties of the nanoelectrospray ion source. *Anal. Chem.* **68**, 1–8.
22. Benesch, J. L., Sobott, F., and Robinson, C. V. (2003) Thermal dissociation of multimeric protein complexes by using nanoelectrospray mass spectrometry. *Anal. Chem.* **75**, 2208–2214.
23. Fandrich, M., Tito, M. A., Leroux, M. R., Rostom, A. A., Hartl, F. U., Dobson, C. M., and Robinson, C. V. (2000) Observation of the noncovalent assembly and disassembly pathways of the chaperone complex MtGimC by mass spectrometry. *Proc. Natl. Acad. Sci. USA* **97**, 14151–14155.
24. Benkestock, K., Van Pelt, C. K., Akerud, T., Sterling, A., Edlund, P. O., and Roeraade, J. (2003) Automated nano-electrospray mass spectrometry for protein-ligand screening by noncovalent interaction applied to human H-FABP and A-FABP. *J. Biomol. Screen.* **8**, 247–256.
25. Schultz, G. A., Corso, T. N., Prosser, S. J., and Zhang, S. (2000) A fully integrated monolithic microchip electrospray device for mass spectrometry. *Anal. Chem.* **72**, 4058–4063.
26. Keetch, C. A., Hernanndez, H., Sterling, A., Baumert, M., Allen, M. H., and Robinson, C. V. (2003) Use of a microchip device coupled with mass spectrometry for ligand screening of a multi-protein target. *Anal. Chem.* **75**, 4937–4941.
27. Ayed, A., Krutchinsky, A. N., Ens, W., Standing, K. G., and Duckworth, H. W. (1998) Quantitative evaluation of protein-protein and ligand-protein equilibria of a large allosteric enzyme by electrospray ionization time-of-flight mass spectrometry. *Rapid Commun. Mass Spectrom.* **12**, 339–344.

28. Fitzgerald, M. C., Chernushevich, I., Standing, K. G., Whitman, C. P., and Kent, S. B. (1996) Probing the oligomeric structure of an enzyme by electrospray ionization time-of-flight mass spectrometry. *Proc. Natl. Acad. Sci. USA* **93**, 6851–6856.
29. Rostom, A. A. and Robinson, C. V. (1999) Disassembly of intact multiprotein complexes in the gas phase. *Curr. Opin. Struct. Biol.* **9**, 135–141.
30. Sobott, F., Hernandez, H., McCammon, M. G., Tito, M. A., and Robinson, C. V. (2002) A tandem mass spectrometer for improved transmission and analysis of large macromolecular assemblies. *Anal. Chem.* **74**, 1402–1407.
31. Tahallah, N., Pinkse, M., Maier, C. S., and Heck, A. J. (2001) The effect of the source pressure on the abundance of ions of noncovalent protein assemblies in an electrospray ionization orthogonal time-of-flight instrument. *Rapid Commun. Mass Spectrom.* **15**, 596–601.
32. Krutchinsky, A. N., Chernushevich, I. V., Spicer, V. L., Ens, W., and Standing, K. G. (1998) Collisional damping interface for an electrospray ionization time-of-flight mass spectrometer. *J. Am. Soc. Mass Spectrom.* **9**, 569–579.
33. Chernushevich, I. V. and Thomson, B. A. (2004) Collisional cooling of large ions in electrospray mass spectrometry. *Anal. Chem.* **76**, 1754–1760.
34. Sanglier, S., Ramstrom, H., Haiech, J., Leize, E., and Van Dorsselaer, A. (2002) Electrospray ionization mass spectrometry analysis revealed a 310 kDa noncovalent hexamer of HPr kinase/phosphatase from Bacillus subtilis. *Int. J. Mass Spectrom.* **219**, 681–696.
35. Schmidt, A., Bahr, U., and Karas, M. (2001) Influence of pressure in the first pumping stage on analyte desolvation and fragmentation in nano-ESI MS. *Anal. Chem.* **73**, 6040–6046.
36. Smith, R. D. and Light-Wahl, K. J. (1993) The observation of noncovalent interactions in solution by electrospray ionization mass spectrometry: promise, pitfalls and prognosis. *Biol. Mass Spectrom.* **22**, 493–501.
37. Robinson, C. V., Chung, E. W., Kragelund, B. B., Knudsen, J., Aplin, R. T., Poulsen, F. M., and Dobson, C. M. (1996) Probing the nature of noncovalent interactions by mass spectrometry. A study of protein-CoA ligand binding and assembly. *J. Am. Chem. Soc.* **118**, 8646–8653.
38. Hernandez, H., Hewitson, K. S., Roach, P., Shaw, N. M., Baldwin, J. E., and Robinson, C. V. (2001) Observation of the iron-sulfur cluster in Escherichia coli biotin synthase by nanoflow electrospray mass spectrometry. *Anal. Chem.* **73**, 4154–4161.
39. Li, Y. T., Hsieh, Y. L., Henion, J. D., Senko, M. W., McLafferty, F. W., and Ganem, B. (1993) Mass spectrometric studies on noncovalent dimers of leucine zipper peptides. *J. Am. Chem. Soc.* **115**, 8409–8413.
40. Li, Y. T., Hsieh, Y. L., Henion, J. D., Ocain, T. D., Schiehser, G. A., and Ganem, B. (1994) Analysis of the energetics of gas-phase immunophilin-ligand complexes by ion spray mass spectrometry. *J. Am. Chem. Soc.* **116**, 7487–7493.
41. Rogniaux, H., Van Dorsselaer, A., Barth, P., Biellmann, J.-F., Barbanton, J., van Zandt, M., Chevrier, B., Howard, E., Mitschler, A., Potier, N., Urzhumtseva, L., Moras, D., and Podjarny, A. (1999) Binding of aldose reductase inhibitors: corre-

lation of crystallographic and mass spectrometric studies. *J. Am. Soc. Mass Spectrom.* **10**, 635–647.
42. El-Kabbani, O., Rogniaux, H., Barth, P., Chung, R. P., Fletcher, E. V., Van Dorsselaer, A., and Podjarny, A. (2000) Aldose and aldehyde reductases: correlation of molecular modeling and mass spectrometric studies on the binding of inhibitors to the active site. *Proteins* **41**, 407–414.
43. Kravanja, M., Engelmann, R., Dossonnet, V., Bluggel, M., Meyer, H. E., Frank, R., Galinier, A., Deutscher, J., Schnell, N., and Hengstenberg, W. (1999) The hprK gene of Enterococcus faecalis encodes a novel bifunctional enzyme: the HPr kinase/phosphatase. *Mol. Microbiol.* **31**, 59–66.
44. Jault, J. M., Fieulaine, S., Nessler, S., Gonzalo, P., Di Pietro, A., Deutscher, J., and Galinier, A. (2000) The HPr kinase from Bacillus subtilis is a homo-oligomeric enzyme which exhibits strong positive cooperativity for nucleotide and fructose 1,6-bisphosphate binding. *J. Biol. Chem.* **275**, 1773–1780.
45. Brochu, D. and Vadeboncoeur, C. (1999) The HPr(Ser) kinase of Streptococcus salivarius: purification, properties, and cloning of the hprK gene. *J. Bacteriol.* **181**, 709–717.
46. Fieulaine, S., Morera, S., Poncet, S., Monedero, V., Gueguen-Chaignon, V., Galinier, A., Janin, J., Deutscher, J., and Nessler, S. (2001) X-ray structure of HPr kinase: a bacterial protein kinase with a P-loop nucleotide-binding domain. *EMBO J.* **20**, 3917–3927.
47. Marquez, J. A., Hasenbein, S., Koch, B., Fieulaine, S., Nessler, S., Russell, R. B., Hengstenberg, W., and Scheffzek, K. (2002) Structure of the full-length HPr kinase/phosphatase from Staphylococcus xylosus at 1.95 A resolution: mimicking the product/substrate of the phospho transfer reactions. *Proc. Natl. Acad. Sci. USA* **99**, 3458–3463.
48. Steinhauer, K., Allen, G. S., Hillen, W., Stulke, J., and Brennan, R. G. (2002) Crystallization, preliminary X-ray analysis and biophysical characterization of HPr kinase/phosphatase of Mycoplasma pneumoniae. *Acta Crystallogr. D Biol. Crystallogr.* **58**, 515–518.
49. Peschke, M., Verkerk, U. H., and Kebarle, P. (2004) Features of the ESI mechanism that affect the observation of multiply charged noncovalent protein complexes and the determination of the association constant by the titration method. *J. Am. Soc. Mass Spectrom.* **15**, 1424–1434.
50. Griffey, R. H., Hofstadler, S. A., Sannes-Lowery, K. A., Ecker, D. J., and Crooke, S. T. (1999) Determinants of aminoglycoside-binding specificity for rRNA by using mass spectrometry. *Proc. Natl. Acad. Sci. USA* **96**, 10129–10133.
51. Griffey, R. H., Sannes-Lowery, K. A., Drader, J. J., Mohan, V., Swayze, E. E., and Hofstadler, S. A. (2000) Characterization of low-affinity complexes between RNA and small molecules using electrospray ionization mass spectrometry. *J. Am. Chem. Soc.* **122**, 9933–9938.
52. Sannes-Lowery, K. A., Griffey, R. H., and Hofstadler, S. A. (2000) Measuring dissociation constants of RNA and aminoglycoside antibiotics by electrospray ionization mass spectrometry. *Anal. Biochem.* **280**, 264–271.

16

Protein Processing Characterized by a Gel-Free Proteomics Approach

Petra Van Damme, Francis Impens, Joël Vandekerckhove, and Kris Gevaert

Summary

We describe a method for the specific isolation of representative N-terminal peptides of proteins and their proteolytic fragments. Their isolation is based on a gel-free, peptide-centric proteomics approach using the principle of diagonal chromatography. We will indicate that the introduction of an altered chemical property to internal peptides holding a free α-N-terminus results in altered column retention of these peptides, thereby enabling the isolation and further characterization by mass spectrometry of N-terminal peptides. Besides pointing to changes in protein expression levels when performing such proteome surveys in a differential modus, protease specificity and substrate repertoires can be allocated since both are specified by neo-N-termini generated after a protease cleavage event. As such, our gel-free proteomics technology is widely applicable and amenable for a variety of proteome-driven protease degradomics research.

Key Words: Gel-free proteomics; N-terminal COFRADIC; protein processing; proteases; substrates.

1. Introduction

There are several advantages of gel-free proteomics following selection and identification of protein N-terminal peptides *(1)*. First, the greatest reduction in sample complexity prior to mass spectrometry (MS)/MS analysis is achieved without any loss of information since every protein is represented only by its N-terminal peptide. Second, as many protein isoforms diverge mainly at their N-terminal extremities it is possible to distinguish them. As an example, so-called xenoproteomics experiments, i.e., simultaneous analysis of proteomes

From: *Methods in Molecular Biology, vol. 484: Functional Proteomics: Methods and Protocols*
Edited by: J. D. Thompson et al., DOI: 10.1007/978-1-59745-398-1, © Humana Press, Totowa, NJ

from different species as present in xenografs, have been performed successfully using N-terminal peptides *(2)*. Third, newly generated N-termini are indicative for protein cleavage by proteases, allowing screening for their substrates in a differential proteomics setup *(3)*. In the protocol to select N-terminal peptides by COmbined FRActional DIagonal Chromatography (COFRADIC) *(4)* outlined below, we focus on this latter application. The commercial rights for this and other COFRADIC applications belong to pronota (www.pronota.com).

Only a few techniques were reported for analyzing protein N-terminal sequences in a gel-free, high-throughput manner *(5)*. Two methods somewhat related to N-terminal COFRADIC were recently reported: the use of protein sequence tags *(6)* and positional proteomics *(7)*. However, both methods have not been applied for proteome-wide characterization of protein processing until now.

In our approach, following their extraction from cells or tissues, proteins are reduced, cysteines are alkylated, and free α- and ϵ-amines are blocked by trideuteroacetylation, making it possible to later characterize the *in vivo* nature (blocked or free, see below) of protein N-termini. Following protein cleavage this modification is an extra confirmation for the identification of newly formed N-termini since these should be trideuteroacetylated. As a consequence of this acetylation step, digestion by trypsin results in peptides ending on an arginine residue. The N-terminal COFRADIC procedure then serves to separate internal and C-terminal peptides from N-terminal ones. The modification reaction between the two sequential and identical chromatographic separation steps uses 2,4,6-trinitrobenzenesulfonic acid (TNBS). This bulky, hydrophobic reagent now reacts only with free α-amines of internal peptides, hereby inducing a hydrophobic shift during the secondary separation. In this way, nonshifted N-terminal peptides (blocked by acetylation) are sorted for further MS/MS analysis.

To distinguish between different proteomes, stable isotope labeling is necessary, introducing known measurable peptide mass differences (**Subheading 3.3**). Using N-terminal COFRADIC in a differential way, the dynamics and status of N-terminal modifications on proteins are characterized. Furthermore, when screening for protease substrates, typically, samples with and without protease activity are compared. Peptides from newly generated N-termini will be present only in one proteome sample and will therefore be present as a peptide with a single isotopic envelope distribution in a mass spectrum *(3)*. This, together with the trideuteroacetylation step mentioned above, makes proteome-wide identification and characterization of protease substrates very straightforward.

Just like other enzymatic systems proteases almost never work alone. They tend to work in networks in which one protease sequentially activates other proteases (e.g., the caspase cascade and during blood clotting), or where several, different proteases become active at the same time (e.g., release of proteases by lysosomal membrane permeabilization). Together with unwanted protease activity induced by cell or tissue lysis, this often complicates the *in vivo* study of protease substrates. When used in a differential way, unwanted protein processing is evident following differential N-terminal COFRADIC since N-terminal peptides formed by this "unwanted activity" will be equally present in treated and control samples. Compensating for protease networking is more difficult and highly challenging, since often there is interest in categorizing the substrates of only one particular protease working in its normal *in vivo* environment or network. Therefore, we suggest performing two types of screens. First, we identify substrates by adding a purified or recombinant protease to a relevant lysate (further referred to as the *in vitro* screen) containing substrates in their native state. The generated list of substrates not only allows the assessment of cleavage site specificity, but can also be used to validate the results obtained from the second screen (further referred to as the *in vivo* screen) where the protease is active in its biological context. Based on the results of the *in vitro* screen those cleavage events in the *in vivo* screen that are due to activity of the protease of interest can be assigned.

2. Materials

2.1. Protein Extraction (Subheading 3.1)

1. Jurkat cell line (ATCC, Manassas, VA, #CRL-1658) and RPMI 1640 medium (Invitrogen, Carlsbad, CA, #61870-010) or adapted arginine-free RPMI medium (*see* **Subheading 3.3.2**).
2. Complete EDTA-free protease inhibitor cocktail tablet (Roche Diagnostics, Mannheim, Germany, #11873580001).
3. Complete protease inhibitor cocktail tablet (Roche Diagnostics, #11697498001).
4. Lysis buffer 1: 50 m*M* morpholinoethanesulfonic acid (MES), 50 m*M* sodium phosphate, pH 7.4, 150 m*M* NaCl, 1 m*M* dithiothreitol (DTT), 1 m*M* EDTA-free protease inhibitors (1 tablet per 100 mL of lysis buffer, see **Notes 1** and **2**).
5. Lysis buffer 2: 50 m*M* HEPES, pH 7.4, 100 m*M* NaCl, 0.8% CHAPS, protease inhibitors (Roche, 1 tablet per 100 mL).
6. Bio-Rad DC Protein Assay Kit (Bio-Rad, München, Germany #500-0006).
7. Recombinant HIV-1 protease (ProteinOne, Bethesda, MD, #P5102).
8. Disposable desalting columns packed with Sephadex™ G-25 (GE Healthcare Bio-Sciences, Uppsala, Sweden, #17-0853-01, #17-0854-01, or #17-0851-01).

2.2. N-Terminal COFRADIC (Subheading 3.2)

1. Tris(2-carboxyethyl)phosphine (TCEP, Pierce, Rockford, IL, #20490).
2. Iodoacetamide (Fluka BioChemica, Buchs, Switzerland, #57670).

3. Sulfo-*N*-hydroxysuccinimide acetate (s-NHS-acetate, Pierce, #26777).
4. Trideutero-*N*-hydroxysuccinimide acetate *(8)*.
5. Hydroxylamine (Fluka BioChemica, #55458).
6. Hydrogen peroxide (30% [w/w] in H_2O, Sigma-Aldrich, St. Louis, MO,#H1009).
7. 2,4,6-Trinitrobenzenesulfonic acid (TNBS, Fluka BioChemika; 1 *M* solution in water, #92822).
8. Disposable desalting columns packed with Sephadex™ G-25 (GE Healthcare).
9. Sequencing grade modified trypsin (Promega, Madison, WI, #V5111).
10. Analytical reverse-phase high-performance liquid chromatography (RP-HPLC) column: 2.1 mm internal diameter (i.d.) × 150 mm (length) 300SB-C_{18} column, Zorbax® (Agilent, Waldbronn, Germany).
11. Agilent 1100 Series HPLC system.
12. HPLC grade water (e.g., Baker HPLC analyzed, Mallinckrodt Baker B.V., Deventer, the Netherlands).
13. HPLC grade acetonitrile (e.g., Baker HPLC analyzed, Mallinckrodt Baker B.V.).
14. HPLC solvent A: 10 m*M* ammonium acetate (pH 5.5) or 0.1% trifluoroacetic acid (TFA) in water/acetonitrile, 98/2 (v/v) (see **Note 3**).
15. HPLC solvent B: 10 m*M* ammonium acetate (pH 5.5) or 0.1% TFA in water/acetonitrile, 30/70 (v/v) (see **Note 3**).
16. TFA (Rathburn, Walkerburn, UK).

2.3. Protein Isotopic Labeling (Subheading 3.3)

1. ^{18}O-rich water (93.7% $H_2{}^{18}O$ [w/w] pure, ARC Laboratories, Apeldoorn, The Netherlands, #OLM-240).
2. TCEP (Pierce, #OLM-240): prepare a 10 m*M* stock solution in water.
3. Iodoacetamide (Fluka BioChemica, #57670): prepare a 100 m*M* stock solution in water.
4. Guanidinium hydrochloride (Fluka BioChemica, #50939): prepare a 6 *M* stock solution in water.
5. Amino acids.

   a. $^{13}C_6$-L-Arginine hydrochloride (Cambridge Isotope Laboratories, Andover, MA, #CLM-2265).
   b. $^{13}C_6{}^{15}N_4$-L-Arginine hydrochloride (Cambridge Isotope Laboratories, #CNLM-539).
   c. L-Arginine (Sigma-Aldrich, #A-8094).

6. Cell culture:

   a. Dialyzed fetal bovine serum (Invitrogen, #26400-044).
   b. Dulbecco's modified Eagle's medium (DMEM), F-12K or RPMI 1640 without L-arginine (Invitrogen). Note: the compositions of these media are available from Invitrogen as a custom service. The custom-synthesized media have exactly the same composition as the regular media (DMEM, #21885-108; RPMI 1640, #61870-010; F-12K, #21127-022 all from Invitrogen), except that they are deficient of the specified amino acid.

c. Penicillin–streptomycin (10,000 U of penicillin and 10,000 μg/mL of streptomycin) (Invitrogen, #15070-063).
d. HEK 293T cell line (ATCC, #CRL-11268).
e. Jurkat cell line (ATCC, #CRL-1658).
f. K-562 cell line (ATCC, #CCL-243).
g. A-549 cell line (ATCC, #CCL-185).
h. NK-92 cell line (ATCC, #CRL-2407).
i. NK-92MI cell line (ATCC, #CRL-2408).
j. SH-SY5Y cell line (ATCC, #CRL-2266).

7. Prepare concentrated stocks (400 m*M*) of $^{13}C_6$, $^{13}C_6{}^{15}N_4$, and $^{12}C_6$ L-arginine hydrochloride in phosphate-buffered saline (PBS) (f.c. [final concentration] for RPMI, #61870; 200 μg/mL or 1.15 m*M* L-arginine or 1.15 m*M*, f.c.. for F-12K, #21127; 422 μg/mL or 2 m*M* and f.c. for DMEM, #21885; 84 μg/mL or 0.398 m*M*) to make complete RPMI 1640 (containing $^{12}C_6$ L-arginine) and RPMI 1640 with $^{13}C_6$ or $^{13}C_6{}^{15}N_4$ L-arginine. Dissolve and divide in small aliquots to avoid multiple freeze–thaw cycles. Add the optimized amount of stock $^{13}C_6$, $^{13}C_6$ $^{15}N_4$, or $^{12}C_6$ L-arginine hydrochloride to the reconstituted arginine-deficient RPMI 1640 media (containing 10% dialyzed fetal bovine serum [free of amino acids], 1% penicillin-streptomycin, and other components whenever required), as to prepare the heavy and light forms of the media. respectively. Subsequently, filter the medium through a 0.22-μm filter and store it at 4°C until use.

3. Methods

3.1. Extraction Procedures

Efficient protein extraction yielding soluble proteins after disruption of biological membranes is required prior to N-terminal COFRADIC. Since our main focus here is the identification of protease substrates, the major differences between the lysis methods described below depend on whether *in vitro* or *in vivo* substrate catalogues will be constructed. We describe three different protein extraction procedures preceding differential N-terminal COFRADIC approaches. **Subsection 3.1.1** outlines procedures for *in vitro* protease substrate screening whereas **Subsection 3.1.2** is recommended for protease-unrelated studies or studies in which postlytic *in vitro* enzymatic activity is unwanted. Both protocols use cells in culture. When starting from dissected animal tissue, **Subsection 3.1.3** must be applied.

For *in vitro* screens, as many potential substrates as possible must be extracted, preferably in their native form. In addition, the extraction conditions should be compatible with subsequent activity of the protease of interest. Therefore, we suggest extracting proteins by multiple freeze–thaw cycles on the cells of interest in a buffer optimal for protease activity or adaptable to achieve such conditions. Detergent-based cell lysis is to be avoided since most detergents are inefficiently removed and interfere with mass spectrometric analyses. Furthermore, detergents may lead to protein denaturation and thus to protease access to epitopes in irrelevant substrates. Also, detergents might influence protease activity.

As a major drawback, some proteins might be missed since their extraction needs detergents. To avoid contaminating downstream protease activity, broad-spectrum protease inhibitors against the three classes of proteases other than the one under investigation should be included, although many proteases are not well targeted by these inhibitors and behave as exceptions in their class. For reasons of general protein solubility the pH of the extraction buffer should be around 7. When studying proteases displaying an acid pH optimum, adjust the pH of the lysate after its extraction. Ionic strength, chelators, and other buffer components can best be optimized for each individual protease to reach its optimal activity.

Since the relevant "library" of possible substrates and specific conditions for activity differ considerably between proteases, in contrast to an *in vivo* screen, we cannot supply an optimal protocol well suited for every protease. As an example we describe the protein extraction steps and procedures to screen for substrates of the recombinant HIV-1 protease in a representative lysate of cultured human Jurkat T cells.

3.1.1. Protein Extraction from Cultured Cells for Subsequent Protease Incubation

1. a. In the case of metabolic labeling of proteins by $^{13}C_6$-Arg SILAC: culture Jurkat cells separately in adapted RPMI 1640 medium in the presence of $^{12}C_6$ or $^{13}C_6$-arginine as described in **Subsection 3.3.2**. Harvest equal numbers of light and heavy labeled cells and wash them two times with PBS to remove residual media components.
   b. In the case of postmetabolic, enzymatic labeling of peptides by $H_2^{18}O$ after protein extraction and digestion: harvest the cells cultured in normal RPMI 1640 medium and wash them two times with PBS. Divide the sample in two aliquots of equal cell numbers. Details with reference to the labeling procedure are outlined in **Subsection 3.3.1**.
2. Resuspend individual cell pellets in lysis buffer 1.
3. Freeze both samples by putting them on dry ice for 15 min followed by thawing on ice at 4°C for 15 min. Repeat this step three times.
4. Centrifuge the samples for 15 min at 16,000 × *g* (4°C) and recover the supernatant.
5. Measure the protein concentration using the Bio-Rad DC Protein Assay Kit according to the manufacturer's instructions. Equalize small differences in protein concentration by diluting the most concentrated sample with the appropriate volume of lysis buffer 1.

6. Acidify both samples with 2 *N* HCl to pH 5.5 and increase the salt concentration to 300 m*M* NaCl using a 5 *M* stock since the HIV-1 protease has a slightly acid pH optimum and cleaves more efficiently at higher salt concentrations ***(9)*** (*see* **Note 4**).
7. Add to one sample the recombinant HIV-1 protease to a final concentration of 200 n*M* and incubate for 75 min at 37°C (treated sample, *see* **Note 5**). Add no protease or, alternatively, an inactive protease variant to the other sample and incubate under conditions identical to the treated sample (control sample, *see* **Note 6**).
8. After incubation, the protease activity can be blocked by adding an excess of a potent protease inhibitor to both samples; however, as is the case for the HIV-1 protease, such (often patented) inhibitors are not always available. In that case, immediately inhibit any remaining protease activity by adding chaotropes (e.g., guanidinium hydrochloride) in sufficiently high concentrations (4– 6 *M*) combined with cysteine alkylation (*see* below).
9. The pH of both samples is increased to 7.5 using 2 *M* NaOH and guanidinium hydrochloride is added dry to a final concentration of 4 *M* (*see* **Note 4**).
10. Proceed directly to *step* 2 of **Subsection 3.2.1**. Mixing of both samples is discussed in **Subsection 3.4**.

In screens where extraction conditions do not need to be tuned for monitoring specific protease activity and the integrity of the three-dimensional structure of the substrate is unnecessary, postlysis effects due to remaining protease activity should be avoided during extraction. Below, we describe a general protocol for protein extraction for *in vivo* screens starting from cultured cells or dissected tissue.

3.1.2. Protein Extraction from Cultured Cells

1. In the case of metabolic labeling of proteins by $^{13}C_6$-Arg SILAC: culture cells separately in the appropriate medium and in the presence of $^{12}C_6$ or $^{13}C_6$-arginine according to labeling conditions described in **Subsection 3.3.2**. Perform treatment of cells during culture (i.e., stimulate cells to evoke protease activity or use as control) and harvest numbers of light and heavy labeled cells such that equal amounts of proteins (*see* **Notes 6** and **7**) for treated and control cells are obtained. Wash the cells thoroughly with PBS.
2. In the case of postmetabolic, enzymatic labeling of peptides by $H_2^{18}O$ after protein extraction and digestion: culture the cells in their normal medium, perform appropriate treatment of the cells during culture, and harvest numbers of light and heavy labeled cells to obtain equal amounts of protein (*see* **Notes 6** and **7**) for treated and control sample. Wash the cells thoroughly with PBS.
3. Resuspend each cell pellet in lysis buffer 2 and lyse the cells on ice for 15 min (*see* **Note 2**). More specific protease inhibitors can be added to this lysis buffer if required.
4. Centrifuge the samples for 15 min at 16,000 × *g* (4°C) and recover the supernatant.

5. Measure the protein concentration using the Bio-Rad DC Protein Assay Kit according to the manufacturer's instructions. Equalize small differences in concentration by diluting with an appropriate volume of lysis buffer 2.
6. Desalt the protein mixture using disposable desalting columns according to the manufacturer's instructions with the appropriate volume of guanidinium hydrochloride in sodium phosphate (pH 7.5). The final concentration of guanidinium hydrochloride should be 4 *M* after drying down the protein mixture to its original starting volume.
7. Proceed directly to *step 2* of **Subsection 3.2.1**. Mixing of both samples is discussed in **Subsection 3.4**.

3.1.3. Protein Extraction from Dissected Animal Tissue

1. During dissection, wash the tissue samples several times thoroughly with PBS and remove residual body fluid components as completely as possible. Snap-freeze the samples in liquid nitrogen and store at –80°C until further processing.
2. Subject the frozen tissue to mechanical dissociation by a pestle in a liquid nitrogen-cooled mortar.
3. Suspend the powder in 4 *M* guanidinium hydrochloride and 50 m*M* sodium phosphate buffer at pH 7.5 (*see* **Note 2**).
4. Extract proteins by incubating this suspension on an orbital shaker for 1 h at 4°C.
5. Centrifuge the protein sample for 60 min at 90,000 × *g* and at 4°C and recover the supernatant.
6. Measure the protein concentration using the Bio-Rad DC Protein Assay Kit according to the manufacturer's instructions. Equalize small differences in concentration by adding lysis buffer.
7. Proceed directly to *step 2* of **Subsection 3.2.1**. Mixing of both samples is discussed in **Subsection 3.4**.

3.2. N-Terminal COFRADIC

3.2.1. Sorting of N-Terminal Peptides

1. Prepare proteomes from treated and control samples as described in **Subheading 3.1.**
2. Desalt the protein mixtures on a disposable desalting column according to the manufacturer's instructions with the appropriate amount of guanidinium hydrochloride in sodium phosphate (pH 7.5) to generate a final concentration of 4 *M* guanidinium hydrochloride in 50 m*M* sodium phosphate (pH 7.5) after vacuum drying the desalted protein mixtures to their original volume.
3. Add freshly prepared TCEP·HCl (1 m*M* f.c.) and iodoacetamide (2 m*M* f.c.) solutions. Let the reduction/alkylation reaction proceed in the dark for 1 h at 37°C.
4. Desalt the protein mixtures on a desalting column in 2 *M* guanidinium hydrochloride in 50 m*M* sodium phosphate (pH 8.0) after drying down to its original volume.

5. Add freshly prepared 5 m*M* sulfo-*N*-hydroxysuccinimide acetate or 10 m*M* trideutero-*N*-hydroxysuccinimide acetate (prepare a fresh 500 m*M* stock in 1% DMSO). Incubate for 90 min at 30°C.
6. Revert partial acetylation of hydroxyl groups by adding 2 μL of hydroxylamine and incubate for an additional 15 min at 30°C.
7. Desalt the mixtures of modified proteins in 20 m*M* NH_4HCO_3 (pH 7.6).
8. Reduce the overall volume of each sample to 1 mL by vacuum drying.
9. Boil the protein mixtures for 10 min at 95°C and then transfer for 10 min to an ice bath.
10. Add sequence grade modified trypsin (the enzyme/substrate ratio should be about 1/50) and incubate overnight at 37°C.
11. Proceed to *step 2* of **Subsection 3.3.1** when using differential ^{18}O labeling.
12. Acidify the modified primary fractions by adding 2 μL of TFA or 4 μl of 100% acetic acid (*see* **Note 8**) and centrifuge the peptide mixtures for 10 min at 10,000 × *g* to remove insoluble material. Transfer the supernatant to an HPLC sample vial.
13. Add the appropriate volume of 30% (w/v) H_2O_2 solution to reach a final concentration of 0.5% and incubate for 30 min at 30°C (*see* **Note 9**).
14. Load the sample on the reverse-phase column (*see* **Subheading 3.2.2**) for the primary COFRADIC separation and fractionate in 12–15 consecutive fractions of 4 min each starting 20 min following sample injection (about 7% of acetonitrile concentration), as very few peptides elute earlier in the gradient.
15. Dry these primary fractions to complete dryness and redissolve each primary fraction in 50 μL sodium borate buffer (pH 9.5).
16. Add 10 μL of a 15 m*M* TNBS solution and incubate for 1 h at 37°C.
17. Repeat the previous step three times to ensure near quantitative TNBS modification of free α-amino groups.
18. Load the TNBS-treated fraction onto the reverse-phase column, starting with the most hydrophobic primary fraction, and subsequently fractionate using the same solvent gradient as during the primary run. Collect the N-terminal peptides (*see* **Note 10**) in 16 equal-volume secondary fractions in an 8-min-long time interval starting 2 min prior to and ending 2 min after the primary collection interval (*see* **Note 11**). An example of COFRADIC sorting N-terminal peptides is depicted in **Fig. 1**.
19. Dry the collected N-terminal peptides and store at –20°C until further LC-MS/MS analysis (*see* **Note 12**).

3.2.2. Setting Up the Reverse-Phase Diagonal Chromatographic System for Sorting N-Terminal Peptides

1. Apply the following binary solvent gradient for separating the peptide mixture:

   a. Following injection of the sample onto the column, apply a 10 min isocratic run with 100% of solvent A at a constant flow rate of 80 μL/min (*see* **Note 13**).

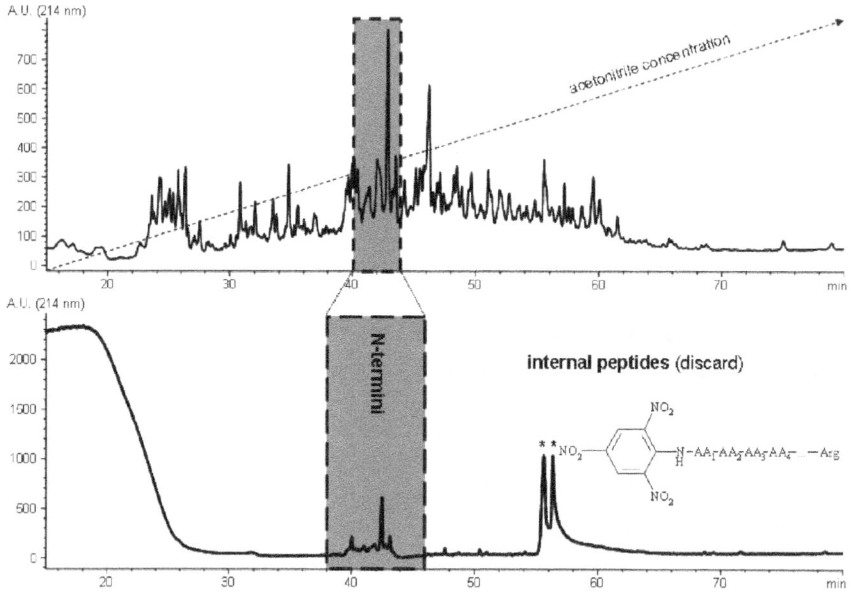

Fig. 1. Sorting of N-terminal peptides. Cultured human Jurkat cells were subjected to three freeze–thaw cycles to extract proteins and subsequently processed as indicated under the method in **Subsection 3.1.1**, *step 4*. The upper panel shows the RP-HPLC chromatogram (UV absorbance measured at 214 nm) of the separation of the tryptic digest of this protein mixture (i.e., the primary COFRADIC run). This peptide mixture was fractionated into 13 primary fractions of 4 min each (from 20 to 72 min). Shown in the lower panel is the RP-HPLC chromatogram of secondary fraction 6 after treatment of the peptide mixture with TNBS (i.e., the secondary COFRADIC run). Unaltered N-terminal peptides are collected in 16 equal-volume secondary fractions in an 8-min-wide time window starting 2 min prior to the original, primary elution interval of fraction 6 (indicated in a gray background with a dashed line). TNBS-modified peptides (i.e., internal peptides that carried a free α-amino group) now obtained a hydrophobic trinitrophenyl group and are thus shifted to later elution times. Note that background peaks due to impurities in TNBS are indicated with an asterisk.

b. Apply a linear, binary gradient over 100 min to 100% of solvent B.
c. Apply a 10 min isocratic wash with 100% of solvent B, followed by a linear gradient over 5 min to 0% of solvent B (100% of solvent A).
d. Reequilibrate the column for another 20 min with 100% of solvent A before injection of another sample.

2. Depending upon the type of peptide isolated and thus the preceding protein preparation steps we observed that peptides typically elute between 20 and 100 min of gradient time, corresponding to acetonitrile concentrations of 7% and 63%, respectively. Collect the primary fractions as indicated in *step 13* of **Subsection 3.2.1**.

3.3. Differential Quantitative Proteomic Labeling Approaches Exploited for N-Terminal COFRADIC

When performing large-scale, differential proteomics surveys, labeling methods incorporating stable, heavy isotopes into proteins or peptides are typically used. By determining the ratio of the intensities originating from the isotopically "light" and "heavy" ion signals of a peptide in a mass spectrum, the relative abundance of the peptide (and protein) in the two represented varieties can be assessed. Isotope labeling can be done on two different levels: either through physiological incorporation (metabolic labeling) or by introduction of a specific enzymatic or chemical derivatization step on the peptide or protein level (postmetabolic labeling) ***(10,12,17)***. Here, we focus on the strategies that we routinely follow to introduce stable heavy isotopic label(s) when performing N-terminal COFRADIC, the selection of which mainly depends on the sample's origin.

We recently introduced an acetylation step on the protein level introducing a trideutero-acetyl group ***(8)*** on every free α-and ϵ-amino group. As mentioned above, cleavage event(s) will now appear as single trideuteroactelyted neo-N-termini (*see* **Note 14**).

Representative for postmetabolic peptide labeling is proteolytic ^{18}O labeling by trypsin ***(13)***. Trypsin catalyzes the exchange of oxygen atoms at the C-terminal carboxyl groups of tryptic peptides and produces in this way labelled peptides that carry two oxygen-18 isotopes at their C-termini. This labeling is introduced following proteome digestion and before chromatographic and mass spectrometric analyses to identify and quantify (relatively) peptides. The primary advantage of this labeling approach is that it is applicable to every proteolytic digest independent of its origin of sampling, whether tissue extractions, body fluids, or cell culture lysates.

Routinely, we also use SILAC (stable isotopic labeling of amino acids in cell cultures; *see* **Note 15**). SILAC was developed as a simple and accurate approach for MS-based quantitative proteomics ***(14)*** and relies on the incorporation of essential amino acids with substituted stable isotopic nuclei (D, ^{13}C, and ^{15}N). During the N-terminal COFRADIC protocol, except for the majority of the C-terminal peptides, all peptides end on arginine. Accordingly, heavy form(s) of arginine are the SILAC amino acids to be used since these will introduce (at least) one label per peptide. Interestingly, there are at least three benefits when using $^{13}C_6$ or $^{13}C_6{}^{15}N_4$ L-arginine. First, the spacing between the light and heavy isotopes is increased (6 to 10 Da) as compared to oxygen-16/18 labeling making the determination of abundance ratios straightforward, since peaks are more easily declustered. Second, SILAC labels are very stable during COFRADIC and MS experiments in contrast to the oxygen-16/18 labeling where back-exchange can occur in acidic environments. Finally, triplex experiments may be performed since $^{12}C_6$, $^{13}C_6$, or $^{13}C_6{}^{15}N_4$ arginine forms can be used. The flow path for both labeling strategies is illustrated in **Fig. 2**. One possible flaw is the arginine-to-proline conversion, which can occur in mammalian cells. This results in label dilution in two different peptide forms both representing the heavy form of the peptide (*see* **Fig. 3**). Thus far, in our hands, in all cell lines tested (including primary cell lines), proline conversion occurs but can be reduced to background levels by reducing the L-arginine concentration to 5–20% of the concentration suggested by manufacturers of cell media, and this without notably affecting cell growth and morphological appearances (*see* **Note 16**).

3.3.1. Peptide Labeling with Oxygen-18 Atoms

1. *Step 2* of this protocol is preceded by *step 10* of **Subsection 3.2.1**.
2. Following digestion in 10 m*M* ammonium bicarbonate (pH 7.6), vacuum dry peptide mixtures.
3. Redissolve the peptides in 25 μL of 0.1 *M* KH_2PO_4 (pH 4.5) and redry.
4. Add 100 μL of ^{18}O-rich water ("heavy peptides") or 100 μL of natural water ("light peptides") and incubate overnight at 37°C.
5. Transfer 10 μL of the 10 m*M* TCEP solution to an Eppendorf tube and vacuum dry. Add 10 μL of the 100 m*M* iodoacetamide solution to 75 μL of 6 *M* guanidinium

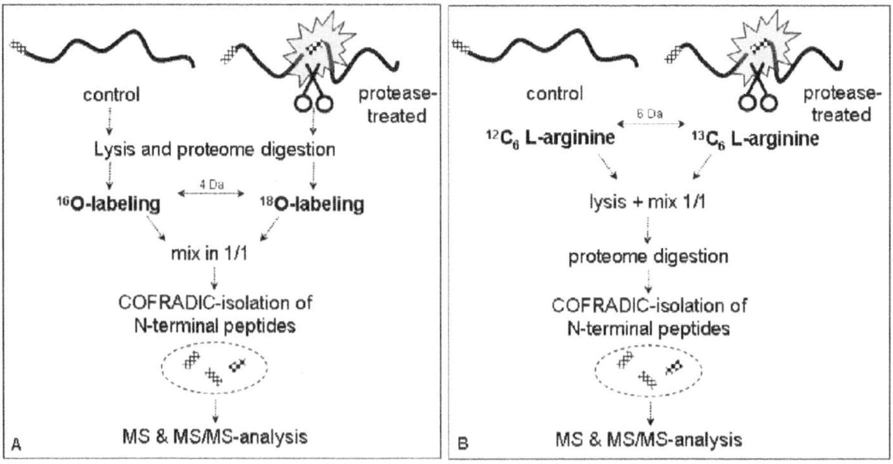

Fig. 2. Schematic strategic experimental outline when making use of diverse quantitative proteomic labeling approaches. As outlined, the flow path for sample processing differs when making use of either oxygen-18 (**A**) or SILAC labeling (**B**). When using SILAC, samples can be processed simultaneously, ruling out potential artifacts introduced by parallel processing of samples with postmetabolic oxygen-18 labeling. For oxygen-18 labeling, samples are mixed at the peptide level.

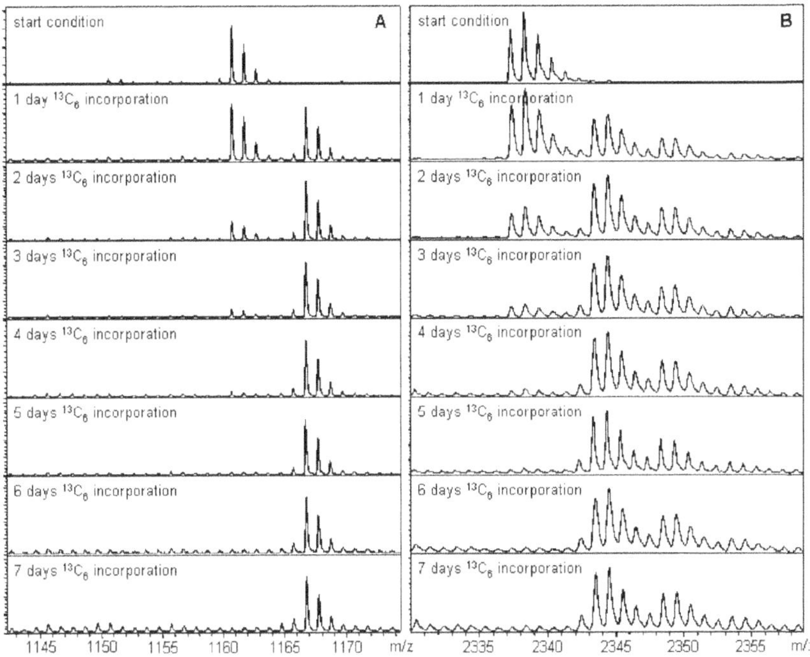

Fig. 3. SILAC labeling strategy in combination with N-terminal COFRADIC. (**A**) The SILAC labeling with $^{13}C_6$ L-arginine at various points in time. Jurkat cells were switched to $^{13}C_6$ L-arginine-containing RPMI medium on day 0 and samples were obtained on days 0, 1, 2, 3, 4, 5, 6, and 7 during the labeling process. After acetylation, lysates were digested with sequencing-grade modified trypsin and separated on an RP-HPLC. Corresponding fractions in time in the different setups were analyzed by MALDI-MS. The panels show the extent of incorporation of $^{13}C_6$ L-arginine into the peptide at the indicated time points. Complete incorporation of $^{13}C_6$ L-arginine into proteins was observed in digests obtained from cell lysates harvested on day 5. (**B**) Jurkat cells readily convert $^{13}C_6$ L-arginine to $^{13}C_5$ -proline. This results in the formation of two clusters of heavy peptides differing by 5 Da for all proline-containing peptides. The correct weight of the heavy peptides is thus the sum of the $^{13}C_6$ L-arginine and the $^{13}C_6$ L-arginine + $^{13}C_5$ -proline peak. By reducing the amount of $^{13}C_6$ L-arginine, proline conversion was no longer observed.

hydrochloride (this is for 100 μL of ^{18}O-rich water or natural water to achieve an f.c. for guanidinium hydrochloride of 4 *M*) in a second Eppendorf tube and dry.

6. Transfer the peptide mixture to the "TCEP vial," mix thoroughly, and incubate at 37°C for 1 h.
7. Transfer the reduced peptide mixture to the "iodoacetamide + Gu.HCl vial" and incubate again for 1 h at 37°C in the dark. At this time point, samples can be stored at –20°C (*see* **Note 17**).

8. Mix both samples in a 1/1 ratio.
9. Continue with *step 12* of **Subsection 3.2.1**.

3.3.2. SILAC Labeling with Heavy Arginine

1. Label the cell population with $^{13}C_6$, $^{13}C_6{}^{15}N_4$, or $^{12}C_6$ L-arginine hydrochloride during cell culture at 37°C, 5% CO_2 for at least five population doublings (usually complete incorporation is achieved after six doublings).
2. Harvest cells from each population and extract proteins as outlined in **Subsection 3.1.**

3.4. Sample Mixing

For differential proteomics, mixing of peptide samples in a near 1/1 ratio is favored for meaningful quantification information. As this ratio is based on the total protein amount present in both samples it is important to start with equal sample amounts. Also, small differences in total protein concentration after lysis can be accounted for as indicated in the protocols of **Subsection 3.1** to obtain unavoidable but similar losses of protein material in the following desalting steps.

For ^{18}O-labeled peptides, the point of sample mixing is fixed in the procedure (after labeling and before the primary COFRADIC run) as described in **Subsection 3.3.1**. As for SILAC-labeled proteins, samples can basically be mixed as early as possible in the protocol (directly after lysis or even before) guaranteeing like treatment of samples. However, to avoid postlysis effects when studying protease substrates it is beneficial to mix samples at a later time point in the procedure when high chaotrope concentration, alkylation (e.g., in the case of cysteine proteases or proteases depending on disulfide bridges for their activity), and acetylation blocked any protease activity. Since most proteases will have lost their activity after one of the modification reactions, samples can then be mixed rather safely before subsequent desalting steps.

In any case, a precise measurement of protein concentration should precede the mixing step. By mixing different sample volumes it is possible to adjust for small differences in protein concentration. However, for both ^{18}O-labeled peptides and SILAC peptides it is preferable to first mix a small part of the samples in a 1/1 ratio and to use this mixture for a "preprimary" COFRADIC run. Collected primary fractions of this separation are then measured in MS mode. Based on the observed average ratio of the peptide peaks, the mixing volumes of the rest of the samples can be adjusted to obtain a 1/1 ratio.

4. Notes

1. Besides EDTA, these tablets also lack pepstatin A, a generally used inhibitor for aspartic proteases.
2. Since the total protein amount after extraction depends on the cell type and the number of lysed cells or amount of tissue, it is necessary to determine the amount of protein material harvested. A total protein amount of at least 2 mg when using a total extract should be obtained. Correct the buffer volume to get a total protein concentration between 2 and 4 mg/mL taking into account the volume of sample that needs to be loaded on desalting columns. Several types of columns are available, differing in the sample volume applied. The same type of columns should be used during the whole procedure.
3. The elution profile of a peptide depends on the ion pairing agent of the HPLC solvents. In ammonium acetate systems peptides tend to elute at lower concentrations of organic solvent than in TFA systems. For cataloguing proteomes we suggest using TFA as this ion-pairing agent produces extremely sharp peaks and as such a high resolution can be obtained when sorting amino terminal peptides.
4. High concentrated stock solutions are used to avoid large decreases in total protein concentration by volume increase of the sample. The volume of stock solutions added to a given buffer solution should be tested in advance.
5. The conditions of concentration and time for protease incubation should be optimized using alternative techniques (e.g., Western analysis of known substrates to follow their processing in function of time/protease concentration). During optimization and for the final analysis a constant protein (substrate) concentration should be respected (*see also* **Note 2**).
6. The analysis should be repeated with label swapping between samples. Besides accounting for an extra validation of substrates from a single experiment, repeating the analysis will partly overcome the undersampling problem, which is an intrinsic drawback of mass spectrometers working in automated MS/MS mode due to random selection of peptide ions for fragmentation.
7. Cell treatment can influence both the amount and nature of proteins extracted (e.g., lysis of cells in different phases of cell death). Therefore, it is necessary to determine the extracted protein amount upon stimulation and correct for differences between treated and control samples.
8. When peptides are labeled with oxygen-18, TFA cannot be used as an ion pairing ion in HPLC solvents as this may lead to acid-catalyzed exchange of oxygen atoms in carboxyl groups ***(13)***. Typically, acetic acid is first added to lower the pH to 5 before injecting peptides onto the RP column, which is run in an ammonium acetate system.
9. The use of hydrogen peroxide to uniformly oxidize methionines to their sulfoxide form is recommended since this prevents accidental hydrophilic shifts of methionyl peptides between chromatographic runs. When performing methionine oxidation prior to the primary RP-HPLC separation (*step 9* of **Subsection 3.2.1**) it is important to respect the oxidation time (30 min) and temperature (30°C) since prolonged incubation leads to unwanted and uncon-

trolled oxidation of methionine to methionine sulfone, and the side chain of other amino acids such as cysteine and tryptophan is also oxidized. This implies that following the oxidation step it is necessary to proceed immediately with the RP-HPLC separation of the peptide mixture.

10. Besides N-terminal peptides, other types of peptides are unavoidably cosorted by COFRADIC. Peptides carrying (or acquiring) a blocked, nonacetylated N-terminal amino acid such as a pyrrolidone carboxylic acid or a cyclic *S*-carbamoylmethylcysteine are cosorted since they do not react with TNBS. Although they appear to "pollute" the mixture of sorted peptides, for differential proteomics purposes their presence can be beneficial as several peptides per protein can be quantified, thus increasing the accuracy of the abundance ratio of their proteins.
11. In theory, N-terminal peptides should elute in the same time frame during the primary and secondary runs. In practice, given the fact that HPLC is not absolutely reproducible, the elution window tends to enlarge and especially abundant N-terminal peptides tend to smear over larger intervals. Therefore, peptides are collected both before (2 min) and after (2 min) their primary collection interval. Since the number of peptides collected in these intervals is much lower than those collected in the expected elution window, such a secondary fraction may be pooled reducing the number of LC-MS/MS analyses.
12. To link MS/MS spectra of COFRADIC-sorted peptide ions efficiently to peptide/protein sequences in databases, search engines such as Mascot ***(15)*** need to consider the (potential) presence of several modifications on the analyzed peptides. An overview of both the fixed modifications (due to the protein preparation method) and potential (variable) modifications (modifications that are likely to be present in [a part of] the sorted peptides) is presented in **Table 1**. Furthermore, the sequence of a sorted peptide indicating irreversible protein processing is often not exactly predicted by search engines as they do not consider *in vivo* "processing and ragging" of protein (termini). Hence, identification of such peptides may be missed. To overcome such flaws, we constructed DBToolkit (freely available via http://www.proteomics.be), an algorithm that uses protein databases as input, imitates protein processing, and creates FASTA-formatted, peptide databases ***(16)***. Using such peptide-centric databases, we noted an increase of at least 30% of identified MS/MS spectra of N-terminal peptides using Mascot ***(3)***.
13. In the overall COFRADIC setup the reproducibility of peptide separation is critical. Adequate HPLC instrumentation is now available creating highly reproducible solvent gradients and thus equally reproducible peptide separations. We use Agilent's electronic flow controller for maintaining a constant solvent flow through the column independent of the backpressure and we thermo-control as many parts of the system as possible (e.g., the column compartment as well as the tubing delivering the solvent to the column and the fraction collector). Taking care of these issues, we generally observe a standard deviation of only a few seconds on the retention time of peptides in a complex peptide mixture over a gradient of nearly 2 h.

Table 1
Recommended Parameters for Searching Databases with MS/MS Spectra of Peptides Sorted by N-Terminal COFRADIC [a]

Fixed modifications	Variable modifications
Trideutero-acetylation (K)	Acetylation (N-terminus)
Carbamidomethyl (C)	Trideutero-acetylation (N-terminus)
Oxidation (M)	Deamidation (NQ)
	Oxidation (M)
	Pyrocarbamidomethyl cysteine (C)
	Pyroglutamic acid (N-terminal Q)
Optional fixed modifications	**Optional variable modifications**
^{18}O Labeling	
^{18}O C-term (double)	
SILAC labeling	
$^{13}C_6$ L-arginine	$^{13}C_5$ proline*

[a] Since the COFRADIC sorting chemistries lead to additional modifications on sorted peptides we here provide an overview of recommended and essential settings of amino acid modifications when searching databases with engines such as Mascot or SEQUEST.
*Only when proline conversion occurs.

14. Protease substrates are often characterized by only one identified MS/MS spectrum ("single-hits"). The presence of a trideutero-acetyl group at the α-amino group of peptides, being present in single isotopic forms, the searched peptide/protein database indicative of the cleavage specificity of the protease of interest, the internal start position and manual validation of identified MS/MS spectra that strictly met the criterium of being ranked one, and scoring above Mascot's 95% confidence interval score are all making the identification more confident.
15. SILAC cannot be applied for labeling harvested tissue samples, although metabolic labeling of intact species ***(17,18)*** has been performed.
16. As for some cell lines, the propagation in media containing dialyzed serum (devoid of all substances less than about 10 kDa) may require some optimization, meaning supplementing extra growth factors to the serum.
17. To obtain complete trypsin inactivation the combined action of reductive alkylation under strong denaturing conditions is required.

Acknowledgments

F.I. is a Research Assistant of the Fund for Scientific Research–Flanders (Belgium) F.W.O.–Vlaanderen).

References

1. 1. Gevaert, K., Goethals, M., Martens, L., Van Damme, J., Staes, A., Thomas, G. R., and Vandekerckhove, J. (2003) *Nat. Biotechnol.* **21,** 566–569.
2. Meuleman, P., Libbrecht, L., De Vos, R., de Hemptinne, B., Gevaert, K., Vandekerckhove, J., Roskams, T., and Leroux-Roels, G. (2005) *Hepatology* **41,** 847–856.
3. Van Damme, P., Martens, L., Van Damme, J., Hugelier, K., Staes, A., Vandekerckhove, J., and Gevaert, K. (2005) *Nat Methods* **2,** 771–777.
4. Gevaert, K., Van Damme, J., Goethals, M., Thomas, G. R., Hoorelbeke, B., Demol, H., Martens, L., Puype, M., Staes, A., and Vandekerckhove, J. (2002) *Mol. Cell. Proteomics* **1,** 896–903.
5. Gevaert, K., Van Damme, P., Ghesquiere, B., and Vandekerckhove, J. (2006) *Biochim. Biophys. Acta* **1764**, 1801–1810.
6. Kuhn, K., Thompson, A., Prinz, T., Muller, J., Baumann, C., Schmidt, G., Neumann, T., and Hamon, C. (2003) *J. Proteome Res.* **2,** 598–609.
7. McDonald, L., Robertson, D. H., Hurst, J. L., and Beynon, R. J. (2005) *Nat. Methods* **2,** 955–957.
8. Ji, J., Chakraborty, A., Geng, M., Zhang, X., Amini, A., Bina, M., and Regnier, F. (2000) *J. Chromatogr. B Biomed. Sci. Appl.* **745,** 197–210.
9. Szeltner, Z. and Polgar, L. (1996) *J. Biol. Chem.* **271,** 5458–5463.
10. Beynon, R. J. and Pratt, J. M. (2005) *Mol. Cell. Proteomics* **4,** 857–872.
11. Mann, M. (2006) *Nat. Rev. Mol. Cell. Biol.* **7,** 952–958.
12. Miyagi, M. and Rao, K. C. (2007) *Mass Spectrom. Rev.* **26,** 121–136.
13. Staes, A., Demol, H., Van Damme, J., Martens, L., Vandekerckhove, J., and Gevaert, K. (2004) *J. Proteome Res.* **3,** 786–791.
14. Ong, S. E., Blagoev, B., Kratchmarova, I., Kristensen, D. B., Steen, H., Pandey, A., and Mann, M. (2002) *Mol. Cell. Proteomics* **1,** 376–386.
15. Krijgsveld, J., Ketting, R. F., Mahmoudi, T., Johansen, J., Artal-Sanz, M., Verrijzer, C. P., Plasterk, R. H., and Heck, A. J. (2003) *Nat. Biotechnol.* **21,** 927–931.
16. Wu, C. C., MacCoss, M. J., Howell, K. E., Matthews, D. E., and Yates, J. R., 3rd (2004) *Anal. Chem.* **76,** 4951–4959.
17. Perkins, D. N., Pappin, D. J., Creasy, D. M., and Cottrell, J. S. (1999) *Electrophoresis* **20,** 3551–3567.
18. Martens, L., Vandekerckhove, J., and Gevaert, K. (2005) *Bioinformatics* **21,** 3584–3585.

17

Identification and Characterization of N-Glycosylated Proteins Using Proteomics

David S. Selby, Martin R. Larsen, Cosima Damiana Calvano, and Ole Nørregaard Jensen

Summary

Glycoproteins constitute a large fraction of the proteome. The fundamental role of protein glycosylation in cellular development, growth, and differentiation, tissue development, and in host–pathogen interactions is by now widely accepted. Proteome-wide characterization of glycoproteins is a complex task and is currently achieved by mass spectrometry-based methods that enable identification of glycoproteins and localization, classification, and analysis of individual glycan structures on proteins. In this chapter we briefly introduce a range of analytical technologies for recovery and analysis of glycoproteins and glycopeptides. Combinations of affinity-enrichment techniques, chemical and biochemical protocols, and advanced mass spectrometry facilitate detailed glycoprotein analysis in proteomics, from fundamental biological studies to biomarker discovery in biomedicine.

Key Words: Glycoprotein; glycopeptide; affinity chromatography; lectins; HILIC; titanium dioxide; tandem mass spectrometry.

1. Introduction

Posttranslational modifications (PTMs) have a significant effect on protein function and their characterization is receiving much attention in current proteomics research *(1)*. The literature contains reports of hundreds of different types of posttranslational modifications, ranging from comparatively straightforward modifications such as enzymatic processing and methylation to more complex modifications such as glycosylation. Glycosylation is known to be one

From: *Methods in Molecular Biology, vol. 484: Functional Proteomics: Methods and Protocols*
Edited by: J. D. Thompson et al., DOI: 10.1007/978-1-59745-398-1, © Humana Press, Totowa, NJ

of the most common and complex types of modification *(2)*, with many different biological roles *(3,4)*. Some of the functions of glycosylation relate to the general effect that the size and shape of the glycan have on the behavior of the peptide backbone, for instance, the use of glycosylation to influence protein folding and the assembly of protein complexes *(5)*. Other roles are likely to be more closely related to the specific configuration of branched glycan structures, such as cell recognition, cell–cell interaction, and immune responses *(6)*. A simple form of glycosylation, O-GlcNAcylation, is believed to be competitive with phosphorylation and probably has a role in intracellular signal transduction *(7)*.

Eukaryotic organisms have three types of common glycosylation *(4)*:

1. N-linked glycosylation, where the glycan is attached to an asparagine via an amide bond, with the initial glycan attachment performed by a glycosyltransferase during peptide synthesis. The asparagine has to be in an Asp-Xxx-(Ser, Thr or Cys) motif, where Xxx is any amino acid other than proline.
2. O-linked glycosylation, where the glycan structure is attached to a serine or threonine as a true posttranslational modification after initial peptide synthesis.
3. The carbohydrate portion of glycosylphosphoinositol lipid anchors, where the glycan acts as a linker unit between the C-terminus of the protein and the lipid anchor that is embedded in a membrane.

There are also other types of glycosylation, such as the simple O-GlcNAcylation mentioned above *(7)* and C-glycosylation of tryptophan *(8)*. Of all these types of glycosylation, the most frequently observed, at least in plant and mammalian systems, is N- and O-linkage *(9)*. The N- and O-linked glycans have a branched carbohydrate structure, which can be contrasted with the linear primary structure of DNA, RNA, and proteins.

Each of the different types of glycosylation is investigated with different experimental protocols and it is not possible to describe methods for all of them in this chapter. Thus the remainder of this chapter will mainly focus on N-linked glycosylation, which is very widely studied (for instance, over 140 references published in 2006 were found in a Pubmed search of "N-linked glycosylation" and "N-linked glycan"), in part because it combines a well-known consensus sequence, common core structure, and an (almost) universally active enzyme glycosidase *(10)*, features not shared with the other types of glycosylation.

Full characterization of protein glycosylation involves a number of different levels of analysis, to find the protein sites that are glycosylated, the level of glycan occupancy at each site, and the actual glycan structures at each site.

It is often quite difficult to perform such a complete analysis, due to the effects of substoichiometric levels of glycan attachment with the highly heterogeneous structures, with some studies revealing in excess of 20 structures on a single attachment site *(11)*. This combination of low levels of individual structures and complexity has meant that, at least to date, there is no single technique capable of providing a full qualitative or quantitative structural analysis of glycoproteins or glycopeptides, although nuclear magnetic resonance (NMR) is perhaps the most powerful structural tool. NMR, is, however, significantly less sensitive than mass spectrometry (MS) and requires rather homogeneous glycoprotein samples, which means that NMR is less suitable than MS for most proteomic studies, where the amount of sample is a limiting factor. Thus most glycoproteomic studies use MS for detection (*see* **Note 1**), typically combined with some type of glycan-specific enrichment or derivatization/tagging.

Intact glycoproteins are often enriched by using lectins *(12)* for affinity purification. Lectins are proteins that recognize glycan epitopes and that are easily immobilized on solid supports to allow batch-wise enrichment of glycoproteins. A range of lectins is commercially available, which differ in their specificity and selectivity. Serial lectin affinity chromatography (SLAC) takes advantage of the different selectivity to recover various subsets of glycoproteins from complex samples *(13,14)*. A selection of lectins and their characteristics are listed in **Table 1**.

There are examples of antibody-based glycoprotein probing strategies. Antibodies raised against specific glycan structures are useful for the detection of glycoproteins that carry a particular epitope *(15)*.

It is always advantageous and often necessary to enrich for glycopeptides prior to MS analysis. This is because nonmodified peptides will compete efficiently for charges in the ionization process, leading to ionization bias and discrimination against glycopeptides. We have previously shown that solid-phase extraction using graphite powder facilitates recovery of very hydrophilic peptides, such as glycopeptides *(16)*. MALDI MS analysis of the recovered glycopeptides is a very sensitive and rapid mean to generate a glycan profile for individual glycosylation sites in proteins *(11)*. Hydrophilic interaction chromatography (HILIC) is another useful method for purification of hydrophilic species, such as glycosylated peptides. Samples are loaded in organic solvents and recovered by increasing the aqueous content of the mobile phase. N-linked, O-linked, and GPI-anchored peptides can be purified in this way and subsequently characterized by MALDI or electrospray ionization (ESI) tandem mass spectrometry *(17–19)*. More recently, we found that sialic acid-containing glycopeptides can be recovered for MS analysis by using TiO_2 affinity enrichment *(20,21)*. A detailed overview of several of these methods and their application to the analysis of glycoproteins in body fluids was recently published *(22)*.

N-linked glycoproteins and glycopeptides can be chemically derivatized and immobilized using periodate oxidation and hydrazide chemistries via their *cis* vicinal diols (23). The immobilized proteins/peptides are subsequently released by treatment with N-glycosidase enzymes and identified by MS. This approach

Table 1
Lectins and Synthetic Materials Suitable for Enrichment of Glycoproteins or Glycopeptides from Protein/Peptide Mixtures[a]

	Saccharide specificity	Application
Lectin		
Concanavalin A (Con A)	Man/Glc	Many N-linked glycans
Wheat germ agglutinin (WGA)	$(GlcNAc)_{1-3}$, sialic acid	Sialylated and GlcNAc terminated N- and O-linked glycans
Pisum sativum (PSA) and *Lens culinaris* (LCA)	Man/Glc	Similar to Con A, but binding enhanced by core fucosylation
Jacalin	Gal (Man)	Isolation of IgA, mucins, and many O-linked glycans
Sambucus nigra agglutinin (SNA)	Siaa6Gal/GalNAc, (Gal/GalNAc)	Sialylated glycoconjugates
Ulex europaeus agglutinin (UEA I)	Fuc	Fucosylated glycoconjugates
Synthetic material		
ZIC-HILIC	General glycan residues	General N-linked glycans
Titanium dioxide	Charged terminal groups (e.g., sialic acid, phosphorylated glycans)	Glycans or glycopeptides containing acidic groups

[a]More extensive lists of lectins are available ***(29,30)***.

was used to recover glycoproteins from complex samples, such as blood ***(23,24)***. Inclusion of stable isotope labeling in such capture strategies facilitates relative quantitation of glycopeptides ***(25)***. These chemistry-based methods frequently require rather large amounts of starting material and issues relating to the generation of side products need to be addressed.

An overview of the strategy and analytical approaches that will be described in the remainder of this chapter is shown in **Fig. 1**. The basic method involves the use of some form of proteolytic digestion, combined with the use of affinity enrichment material in microcolumns and sensitive MS detection and characterization. First, the glycoprotein-containing samples are digested with protease to generate peptides and glycopeptides. The glycopeptides are then enriched using selective, miniaturized chromatographic methods, viz. immobilized lectins, HILIC, graphite, or TiO_2. The recovered glycopeptides are then analyzed using either MALDI MS/MS or LC-ESI-MS/MS. We recommend using high-resolution mass spectrometers, such as quadrupole time of flight

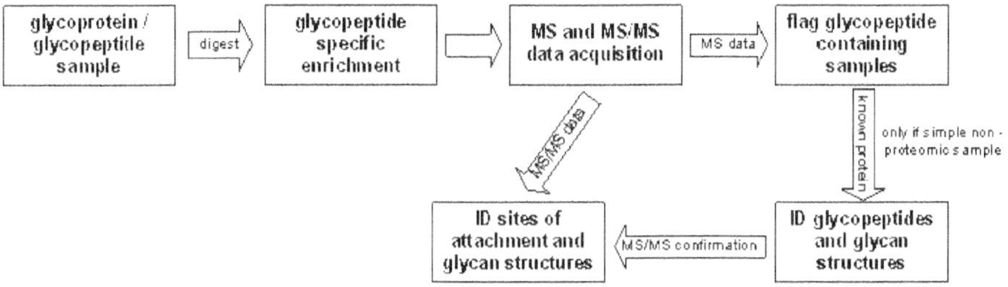

Fig. 1. A flowchart illustrating a general approach for N-linked glycan enrichment.

(Q-TOF) and/or ion trap Fourier transform ion cyclotron resonance (FTICR) or ion trap Orbitrap instruments, to achieve high mass accuracy in the MS and MS/MS mode. This ensures more straightforward assignment of glycan species and structure elucidation *(26)*. The ion trap type instruments provide MS^n capabilities, which is sometimes useful for detailed analysis of glycans. It is also advantageous to probe or release the glycan structures using endo- and exoglycosidase enzymes. Monitoring the reaction by MS often allows assignment of structural features based on the known enzyme specificity and mass determination of the glycan.

2. Materials

2.1. Lectin Enrichment

1. Various lectins (agarose concanavalin A [Con A], agarose wheat germ agglutinin [WGA], agarose *Sambucus nigra* bark lectin [SNA]), suspended in 10 m*M* HEPES, pH 7.5, 0.15 *M* NaCl, 0.1 m*M* $CaCl_2$, 0.01 m*M* $MnCl_2$, 20 m*M* glucose, 0.08% sodium azide as product specifications (Vector Laboratories Inc.).
2. Various sugars (methyl-α-D-mannopyranoside, *N*-acetyl-D-glucosamine, and lactose) purchased from Sigma-Aldrich.
3. GELoader tips (Eppendorf).
4. Disposable 1-mL syringe, fitted to the GELoader with a cut down 200-μL tip (see **Fig. 2**).
5. Lectin load solution: 20 m*M* Tris–HCl, pH 7.4, 0.15 *M* NaCl, 1 m*M* $MnCl_2$, 1 m*M* $CaCl_2$.
6. Lectin wash solution: 20 m*M* Tris–HCl, pH 7.4, 0.5 *M* NaCl, 1 m*M*$MnCl_2$, 1 m*M*$CaCl_2$.
7. Lectin elute solution: appropriate sugar solution for each lectin. In particular:

   a. Con A: 200 m*M* ethyl-α-D-mannopyranoside in loading buffer.
   b. WGA: 500 m*M* *N*-acetyl-β-D-glucosamine in loading buffer.
   c. SNA: 500 m*M* lactose in loading buffer.

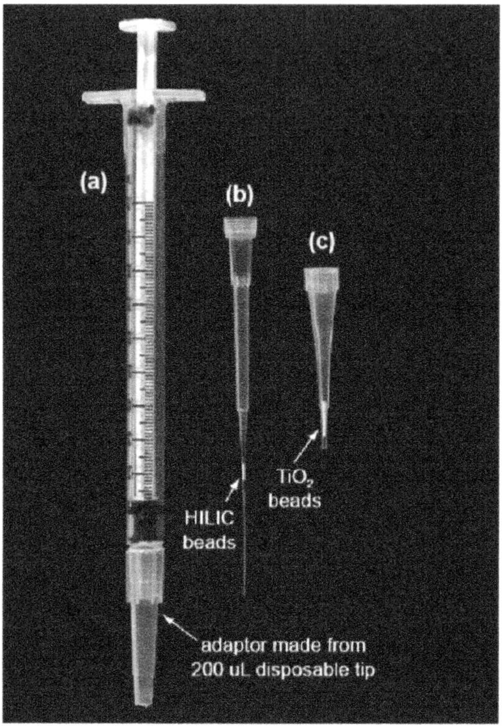

Fig. 2. A 1-mL syringe with an adaptor designed to fit the top of a GELoader tip or 10-μL tip microcolumn (**a**), shown to scale with the GELoader tip (**b**) and 10-μL tip (**c**) microcolumns.

2.2. HILIC Enrichment

1. ZIC-HILIC chromatographic media (ZIC-HILIC, silica 10 μm, 200 Å, Sequant AB, Umeå, Sweden), suspended in acetonitrile or methanol (*see* **Notes 2** and **3**).
2. C-8 StageTips *(27)* made from either GELoader Tips or 10-μL disposable syringe tips (Proxeon Biosystems, Odense, Denmark).
3. Disposable 1-mL syringe, fitted to the GELoader with a cut down 200-μL tip (*see* **Fig. 2**).
4. HILIC wash: 80% acetonitrile, 19.5% water, 0.5% formic acid; can be stored at 4°C for up to 1 week.
5. HILIC elute: 0.5% aqueous formic acid; can be stored at 4°C for up to 1 week.

2.3. TiO_2 Enrichment

1. Titansphere TiO_2 5 μm chromatographic material (GL Sciences, Tokyo, Japan) suspended in acetonitrile.

2. C-8 StageTips made from either GELoader Tips or 10-μL disposable syringe tips (Proxeon Biosystems, Odense, Denmark).
3. Disposable 1-mL syringe, fitted to the GELoader with a cut down 200-μL tip (*see* **Fig. 2**).
4. TiO_2 loading buffer: 100 mg/mL 2,5-dihydroxybenzoic acid (DHB) in 70% acetonitrile, 25% water, 5% trifluoroacetic acid; prepare fresh daily.
5. TiO_2 wash: 80% acetonitrile, 19% water, 1% trifluoroacetic acid; can be stored at 4°C for up to 1 week.
6. TiO_2 elute: 20 μL 25% ammonia in 980 μL water; add more ammonia solution if required to adjust the pH to approximately 10.5; can be stored at 4°C for up to 1 week.
7. Dephosphorylation buffer: 50 m*M* aqueous ammonium bicarbonate.
8. Alkaline phosphatase from calf intestine.

2.4. Deglycosylation

1. *N*-Glycosidase F (PNGase F) in glycerol containing solution (Roche Diagnostics, Mannheim, Germany).
2. Deglycosylation buffer: 50 m*M* aqueous ammonium bicarbonate (*see* **Note 4**).

2.5. Mass Spectrometric Analysis of N-Linked Glycopeptides and Deglycosylated Peptides

1. Poros R2 and OLIGO R3 chromatographic media, 20 μm (Applied Biosystems, CA), suspended in 70% acetonitrile, 30% water.
2. Reverse phase wash: 0.5% aqueous formic acid; can be stored at 4°C for up to 1 week.
3. MALDI elute: 10 mg/mL DHB in 50% acetonitrile, 49.9% water, 0.1% trifluoroacetic acid; prepare fresh daily.
4. ESI elute: 60% acetonitrile, 39.5% water, 0.5% formic acid; can be stored at 4°C for up to 1 week.
5. MALDI mass spectrometer and/or ESI mass spectrometer, at least one of which should be able to perform tandem mass spectrometry (MS/MS).

3. Methods

The general procedures described rely upon the deposition of material with an affinity for N-linked glycan groups into a pipette tip to form a microcolumn. A 1-mL syringe with an adaptor cut down from a 200-μL disposable pipette tip (*see* **Fig. 2**) is used to provide gentle air pressure to force solutions through the column. All of these methods assume, unless otherwise mentioned, that you are starting with a chemical or enzymatic digest that contains both glycopeptides and nonglycosylated peptides. The samples can come either from a complex proteomic type sample or from a simpler sample, such as a purified protein or 1-D gel band.

3.1. Lectin Enrichment

1. Add 15 μL of Con A slurry in an Eppendorf tube and wash three times with 80 μL of lectin loading buffer. Pipette up and down to mix; do not vortex the solution to avoid the breakage of lectin-agarose beads bond (*see* **Notes 5** and **6**).
2. Add 30 μg of glycoprotein digest diluted in 200 μL of lectin load buffer to the lectin solution.
3. Gently shake for 2 h at 4°C.
4. Make a partially constricted GELoader pipette tip by squeezing the end.
5. Centrifuge the sample for 15 min at 5000 × *g*.
6. Collect the agarose beads and load into the StageTip. The lectin beads are packed by applying gentle air pressure with the plastic syringe.
7. Wash unbound material from the column with 20 μL of lectin washing solution (three times).
8. Elute the glycosylated peptides from lectin material with 20 μL of elution solution (200 mM methyl-α-D-mannopyranoside in loading buffer). Retain the elute.
9. Dry the glycopeptides down in a vacuum centrifuge and store until required for mass spectrometric analysis.

3.2. HILIC Enrichment

1. 1–20 pmol of the digested sample is made up in 10–20 μL of HILIC wash solution (*see* **Note 7**).
2. Prepare the HILIC microcolumns. This is done by vortexing the ZIC-HILIC bead-containing solution and depositing a few microliters of the resulting slurry into 10 μL of HILIC wash solution that was loaded into a StageTip. The HILIC beads are packed on top of the C8 plug by applying gentle air pressure with the plastic syringe. The length of the column is dependent on the amount of peptides you wish to analyze, with a 3- to 5-mm column sufficient for up to 20 pmol.
3. Clean the microcolumn with 15 μL of HILIC elution solution, flushing the solution through with gentle pressure from the syringe.
4. Condition the column by flushing with 30 μL of HILIC wash solution.
5. Load the sample containing glycopeptides onto the column using the syringe. Ensure the volume loaded is at least 10 μL.
6. Wash unbound material from the column with 20–40 μL of HILIC wash solution.
7. Elute the glycosylated peptides from the HILIC material with 7–15 μL of HILIC elute solution. *Retain the eluate.*
8. Glycopeptides with relatively hydrophobic peptides may still be bound to the C8 plug. Elute these glycopeptides from the plug with 3μL of HILIC; wash Solution—*pool with eluate from step 8.*

9. Dry the glycopeptides down in a vacuum centrifuge and store until required for MS analysis (*see* **Note 8**).

3.3. TiO_2 Enrichment

1. 1–20 pmol of the digested sample is made up in 10 μL of the dephosphorylation buffer and 0.2 U alkaline phosphatase is added.
2. Incubate overnight at 37°C to remove any phosphate groups (*see* **Note 9**).
3. Prepare TiO_2 microcolumns. This is done in a manner similar to the preparation of HILIC microcolumns (*see* **Subheading 3.2,** *step 2*), but a TiO_2 bead slurry instead of an HILIC slurry should be used.
4. Dilute the dephosphorylated peptide solution with the TiO_2 loading buffer, from a ratio of 1:3 to 1:5, with the higher ratio used for more complex samples. The sample is loaded onto the column and run dry using the syringe.
5. Wash the sample on the column with 5–10 μL of TiO_2 loading buffer.
6. Wash the column with 20 μL TiO_2 wash.
7. Elute the sample with a minimum of 20 μL TiO_2 elute.
8. Glycopeptides with relatively hydrophobic peptides may still be bound to the C8 plug. Elute these glycopeptides from the plug with 3 μL of TiO_2 wash; *pool with the eluate from step 8*.
9. Dry the glycopeptides down in a vacuum centrifuge and store until required for mass spectrometric analysis (*see* **Note 8**).

3.4. Deglycosylation of N-Linked Glycopeptides

1. Prepare enriched glycopeptides using the methods given in **Subheadings 3.1–3.3**. Redissolve these in 10 μL 50 m*M* ammonium bicarbonate solution and add 0.2 U of PNGase F (*see* **Note 4**).
2. Incubate at 37°C from 3 h to overnight. This should remove all N-linked glycans from the peptides, other than those containing α(1–3) core fucosylation (*see* **Note 10**).
3. Store in the freezer until required for MS analysis.

3.5. Mass Spectrometric Analysis of N-Linked Glycopeptides and Deglycosylated Peptides

1. Treat samples prepared according to **Subheadings 3.1–3.4** as follows:

    a. Lectin-enriched glycopeptides (**Subheading 3.1**); resuspend in 10 μL reverse phase wash and go to *step 2*.
    b. HILIC-enriched glycopeptides (**Subheading 3.2**); resuspend in 10 μL reverse phase wash and go to step 3 for MALDI or *step 4* for ESI.
    c. TiO_2-enriched glycopeptides (**Subheading 3.3**); resuspend in 10 μL reverse phase wash and go to *step 2*.
    d. Deglycosylated peptides (**Subheading 3.4**); thaw and go to *step 2*.

2. Desalt the sample with R2 and R3 microcolumns (*see* **Note 11**):

   a. Prepare the microcolumns. This is done in a manner similar to the preparation of HILIC microcolumns (**Subheading 3.2**, *step 2*), except that an R2 or R3 slurry should be used.
   b. Condition the microcolumn by flushing with 20 μL of reverse-phase wash, using the syringe to apply gentle air pressure.
   c. Load the sample containing glycopeptides or deglycosylated peptides onto the column. If you load an aliquot of less than 10 μL, add sufficient reverse-phase wash solution to increase the volume to at least 10 μL (*see* **Note 12**).
   d. Wash unbound material with 20 μL of reverse-phase wash (*see* **Note 12**).
   e. Elute the peptides from the microcolumn with up to 10 μL of ESI elute solution; *retain the eluate* (*see* **Note 13**). Go to *step 3* for MALDI samples and *step 4* for ESI samples.

3. MALDI: Deposit an aliquot of up to 1 μL of sample on a MALDI plate, followed by 0.5 μL of MALDI matrix solution. Wait for the spots to dry and acquire data with a MALDI mass spectrometer. See **Fig. 3** for an example of the type of MALDI-time of flight mass spectrometry (TOFMS) results that can be expected when using the HILIC and TiO_2 methods described in this protocol to enrich glycopeptides from fetuin, a glycoprotein.
4. ESI: The solvent composition for ESI samples should be adjusted until appropriate for the type of analysis required, for instance, 50% acetonitrile (ACN)/49.5% water/0.5% formic acid for direct infusion, or 0.5% aqueous formic acid for reverse-phase LC/MS/MS.

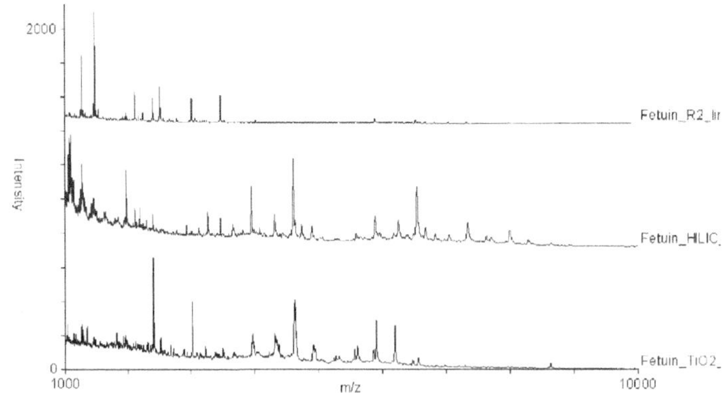

Fig. 3. MALDI-TOFMS of fetuin, illustrating the use of ZIC-HILIC and titanium dioxide microcolumns for the enrichment of glycopeptides, when compared to reverse-phase desalting. (**top**) Shows 1 pmol of tryptic-digested fetuin after desalting with R2 reverse-phase material, (**middle**) 1 pmol (of 10 pmol total) purified with HILIC, and (**bottom**) 2 pmol after titanium dioxide purification.

Table 2
Common Glycan Residues, Masses, and Related Oxonium Ions

Residue	Nominal mass	Related oxonium ions
Hexose (Glc, Man, etc.)	162	163, 366 (+ *N*-acetylhexosamine)
Deoxyhexose (Fuc)	146	147 (low)
N-Acetylhexosamine (GlcNAc, GalNAc)	203	204, 366 (+ hexose)
Sialic acid (Sia)	291	292, 274 (–H_2O)

5. Analysis of results:
   a. Deglycosylated peptides**:** formerly glycosylated peptides can be identified by looking for deamidated peptides containing the N-linked glycan consensus sequence (*see* **Note 4**).
   b. Glycopeptides: glycopeptide spectra can be analyzed by reference to the mass differences relating to the different glycan residues and the appearance of related oxonium fragment ions at low mass to charge in MS/MS (*see* **Table 2**).

4. Notes

1. It is important to remember that mass spectrometers cannot separate isobars (species of the same mass). This means that in glycan structure analysis MS alone cannot readily differentiate isomeric sugar species, such as the different hexose sugars (e.g., mannose, glucose, galactose). It is sometimes possible to use high-energy collision-induced dissociation to resolve some isobaric glycan structures by generating cross ring cleavages, or alternatively, a knowledge of biology or use of specific glycosidases can allow assignment of specific glycan structures in combination with MS results.
2. High-purity solvents should be used throughout. This means HPLC grade or similar for organic solvents and 18 MΩ water.
3. The ZIC-HILIC used here is made from silica beads with zwitterionic sulfobetain groups, which provides superior enrichment when compared to bare silica HILIC materials.
4. Enzymatic treatment will deglycosylate and deamidate the asparagine to aspartate (R-NH-glycan to R-OH). Optional use of 50% ^{18}O water for the bicarbonate buffer provides for doublet (+1 and +3 Da) deamidation peaks for the formerly glycosylated peptides, ensuring that they will not be confused with other deamidated peptides.
5. The protocol described involves the use of a single lectin microcolumn (Con A) for enrichment of a class of N-linked glycopeptides (*see* **Table 1** for specificity details). This protocol can also be used with WGA and SNA (*see* **Subheading**

2.1 for required materials) by substituting for the preferred lectin at *step 1* and its corresponding elution solution at *step 8*.

6. If you want to enrich for multiple classes of glycoproteins it is possible to prepare a multilectin column ***(28)***, but in that case you should take into account that each lectin has a different binding capacity. For instance, SNA binds 1.5 mg of protein/mL of gel, Con A binds 4 mg of protein/mL of gel, and WGA binds 8 mg/mL. Thus, a multilectin column with equal binding capacities from each of these lectins would be prepared in the ratio of 3:2:1, by volume ***(29,30)***.
7. HILIC purification works best with samples that are not too complicated, for instance, 1-D gel bands or pools of glycoproteins that were obtained by lectin enrichment.
8. When the solution volume is reduced to 10 μL or less, analyze a small aliquot by MALDI-TOFMS, using a glycopeptide compatible matrix, such as 2,5-dihydroxybenzoic acid.
9. Enzymatic dephosphorylation is necessary to prevent enrichment of phosphopeptides in addition to glycopeptides, since the TiO_2 material has a high affinity for phosphopeptides.
10. N-linked glycans containing fucose that is α(1–3) linked to the asparagine can be removed with *N*-glycosidase A (PNGase A) from almond meal, instead of PNGase F. Use the same protocol as for PNGase F, but substitute 0.2 U of PNGase A for PNGase F.
11. Either R2 or R3 media can be used for desalting peptides and glycopeptides. R3 is able to bind more hydrophilic species than R2, but R3 may not efficiently elute some more hydrophobic species. Thus the most suitable medium is sample dependent.
12. Optional step: If you are using an R2 column, rather than discarding the eluate at steps c and d, load it onto an R3 column, which may catch some peptides/glycopeptides that were not retained on the R2 column. Then the sample can be eluted off both microcolumns and analyzed further.
13. Optional step: If you want to analyze only samples by MALDI, the sample can be carefully eluted directly onto the MALDI plate with 0.5–1.0 μL of MALDI elute solution and the use of *very gentle* air pressure from the syringe.

References

1. Jensen, O. N. (2006) Interpreting the protein language using proteomics. *Nat. Rev. Mol. Cell. Biol*, **7**, 391–403.
2. Sharon, N. and Lis, H. (1997) Glycoproteins: structure and function. In *Glycosciences: Status and Perspectives* (Gabius, H.-J., Gabius, S., eds.). Chapman & Hall, Wienheim, Germany, pp. 133–162.
3. Varki, A. (1993) Biological roles of oligosaccharides–all of the theories are correct. *Glycobiology* **3**, 97–130.
4. Varki, A., Cummings, R., Esko, J., Freeze, H., Hart, G., and Marth, J. (eds.). (1999) *Essentials of Glycobiology*. Cold Spring Harbor Press, Cold Spring Harbor, NY.

5. Helenius, A. and Aebi, M. (2001) Intracellular functions of N-linked glycans. *Science* **291**, 2364–2369.
6. Rudd, P. M., Elliott, T., Cresswell, P., Wilson, I. A., and Dwek, R. A. (2001) Glycosylation and the immune system. *Science* **291**, 2370–2376.
7. Wells, L., Vosseller, K., and Hart, G. W. (2001) Glycosylation of nucleocytoplasmic proteins: signal transduction and O-GlcNAc. *Science* **291**, 2376–2378.
8. Hofsteenge, J., Muller, D. R., Debeer, T., Loffler, A., Richter, W. J., and Vliegenthart, J. F. G. (1994) New-type of linkage between a carbohydrate and a protein—-C-glycosylation of a specific tryptophan residue in human Rnase U-S. *Biochemistry* **33**, 13524–13530.
9. Harvey, D. J. (1999) Matrix-assisted laser desorption/ionization mass spectrometry of carbohydrates. *Mass Spectrom. Rev.* **18**, 349–450.
10. Medzihradszky, K. F. (2005) Characterization of protein N-glycosylation. *Methods Enzymol.* **405**, 116–138.
11. Mortz, E., Sareneva, T., Julkunen, I., and Roepstorff, P. (1996) Does matrix-assisted laser desorption/ionization mass spectrometry allow analysis of carbohydrate heterogeneity in glycoproteins? A study of natural human interferon-gamma. *J. Mass Spectrom.* **31**, 1109–1118.
12. Gabius, H. J., Andre, S., Kaltner, H., and Siebert, H.C. (2002) The sugar code: functional lectinomics. *Biochim. Biophys. Acta* **1572**, 165–177.
13. Wang, Y., Wu, S. L., and Hancock, W. S. (2006) Monitoring of glycoprotein products in cell culture lysates using lectin affinity chromatography and capillary HPLC coupled to electrospray linear ion trap-Fourier transform mass spectrometry (LTQ/FTMS). *Biotechnol. Prog.* **22**, 873–880.
14. Drake, R. R., Schwegler, E. E., Malik, G., Diaz, J. I., Block, T., Mehta, A., and Semmes, O. J. (2006) Lectin capture strategies combined with mass spectrometry for the discovery of serum glycoprotein biomarkers. *Mol. Cell. Proteomics* **5**, 1957–1967.
15. Peracaula, R., Royle, L., Tabares, G., Mallorqui-Fernandez, G., Barrabes, S., Harvey, D. J., Dwek, R. A., Rudd, P. M., and de Llorens, R. (2003) Glycosylation of human pancreatic ribonuclease: differences between normal and tumor states. *Glycobiology* **13**, 227–244.
16. Larsen, M. R., Cordwell, S. J., and Roepstorff, P. (2002) Graphite powder as an alternative or supplement to reversed-phase material for desalting and concentration of peptide mixtures prior to matrix-assisted laser desorption/ionization-mass spectrometry. *Proteomics* **2**, 1277–1287.
17. Hagglund, P., Bunkenborg, J., Elortza, F., Jensen, O. N., and Roepstorff, P. (2004) A new strategy for identification of N-glycosylated proteins and unambiguous assignment of their glycosylation sites using HILIC enrichment and partial deglycosylation. *J. Proteome Res.* **3**, 556–566.
18. Omaetxebarria, M. J., Hagglund, P., Elortza, F., Hooper, N. M., Arizmendi, J. M., and Jensen, O. N. (2006) Isolation and characterization of glycosylphosphatidylinositol-anchored peptides by hydrophilic interaction chromatography and MALDI tandem mass spectrometry. *Anal. Chem.* **78**, 3335–3341.

19. Hagglund, P., Matthiesen, R., Elortza, F., Hojrup, P., Roepstorff, P., Jensen, O. N., and Bunkenborg, J. (2007) An enzymatic deglycosylation scheme enabling identification of core fucosylated N-glycans and O-glycosylation site mapping of human plasma proteins. *J. Proteome Res.* **6**, 3021–3031.
20. Larsen, M. R., Thingholm, T. E., Jensen, O. N., Roepstorff, P., and Jorgensen, T. J. D. (2005) Highly selective enrichment of phosphorylated peptides from peptide mixtures using titanium dioxide microcolumns. *Mol. Cell.Proteomics* **4**, 873–886.
21. Larsen, M. R., Jensen, S. S., Jakobsen, L. A., and Heegaard, N. H. (2007) Exploring the sialiome using titanium dioxide chromatography and mass spectrometry. *Mol. Cell. Proteomics* **6**, 1778–1787.
22. Bunkenborg, J., Hägglund, P., and Jensen, O. N. (2007) Modification-specific proteomic analysis of glycoproteins in human body fluids by mass spectrometry. In *Proteomics of Human Body Fluids: Principles Methods, and Applications* (Thongboonkerd, V., ed.). Humana Press, Totowa, NJ.
23. Zhang, H., Li, X. J., Martin, D. B., and Aebersold, R. (2003) Identification and quantification of N-linked glycoproteins using hydrazide chemistry, stable isotope labeling and mass spectrometry. *Nat. Biotechnol.* **21**, 660–666.
24. Zhang, H. and Aebersold, R. (2006) Isolation of glycoproteins and identification of their N-linked glycosylation sites. In *New and Emerging Proteomic Techniques* (Nedelkov, D., Nelson, R. W., eds.), Vol. 328, pp. 177–185. Humana Press, Totowa, NJ,
25. Kaji, H., Saito, H., Yamauchi, Y., Shinkawa, T., Taoka, M., Hirabayashi, J., Kasai, K., Takahashi, N., and Isobe, T. (2003) Lectin affinity capture, isotope-coded tagging and mass spectrometry to identify N-linked glycoproteins. *Nat. Biotechnol.* **21**, 667–672.
26. Harvey, D. J. (2005) Structural determination of N-linked glycans by matrix-assisted laser desorption/ionization and electrospray ionization mass spectrometry. *Proteomics* **5**, 1774–1786.
27. Rappsilber, J., Ishihama, Y., and Mann, M. (2003) Stop and go extraction tips for matrix-assisted laser desorption/ionization, nanoelectrospray, and LC/MS sample pretreatment in proteomics. *Anal Chem*, **75**, 663–670.
28. Yang, Z. P. and Hancock, W. S. (2004) Approach to the comprehensive analysis of glycoproteins isolated from human serum using a multi-lectin affinity column. *J. Chromatogr. A* **1053**, 79–88.
29. Cummings, R. D. (1997) Lectins as tools for glycoconjugate purification and characterization. In *Glycosciences: Status and Perspectives* (Gabius, H.-J., Gabius, S., eds.), pp. 191–199. Chapman & Hall, Wienheim, Germany.
30. Gabius, H. J., Siebert, H. C., Andre, S., Jimenez-Barbero, J., and Rudiger, H. (2004) Chemical biology of the sugar code. *Chembiochemistry* **5**, 741–764.

IV

Protein Analysis

18

Data Standards and Controlled Vocabularies for Proteomics

Lennart Martens, Luisa Montecchi Palazzi, and Henning Hermjakob

Summary

Proteomics data can be diverse and complex, and are typically produced on a large scale. To allow sharing and centralized storage and dissemination of such results, the Human Proteome Organization (HUPO) Proteomics Standards Initiative (PSI) has created a set of community standards for the exchange of mass spectrometry and protein interaction data. We describe the origins and overall concepts behind these standards, as well as the individual efforts that are ongoing in the field of mass spectrometry proteomics and protein interactions.

Key Words: Proteomics; standards; ontologies; protein interactions; mass spectrometry; HUPO-PSI; mzData; mzXML; PSI-MI; mzML; protein identification; peptide identification.

1. Introduction

Science relies heavily on the publication, and therefore the sharing of, findings with others. This concept was elegantly expressed by Sir Isaac Newton when he paraphrased the twelfth-century French philosopher Bernard of Chartres by stating that: "*If I have seen further it is by standing on the shoulders of Giants*." Because many different and at least partially complementary techniques are available to proteomics researchers today, the ability to combine results from diverse sources is especially appealing as it holds the promise of increased research efficiency and can thereby substantially aid the assembly of more complete data sets for subsequent in-depth analysis. The

From: *Methods in Molecular Biology, vol. 484: Functional Proteomics: Methods and Protocols*
Edited by: J. D. Thompson et al., DOI: 10.1007/978-1-59745-398-1, © Humana Press, Totowa, NJ

same factors that make data sharing and integration so desirable however, effectively conspire against achieving this goal. The many different platforms for discovery impose their own data formats and work flows, rendering assembly of the diverse results impractical and sometimes even impossible. Widely adopted, standardized interchange formats can alleviate much of this problem, yet in order to enable effective data sharing, the following two essential requirements should minimally be fulfilled by such a standard: it has to make the data readily *accessible*, and it should provide *sufficient* data to allow correct interpretation and potentially also replication.

To make data accessible, both the *format* in which the data are retrieved from various sources and the *wording* used to annotate these formats should be consistent. Achieving this latter goal requires the use of a *controlled vocabulary* (CV; a limiting list of clearly defined terms, with optional relationships between the terms) or an *ontology* (which moves beyond a mere CV by actually attempting to model a part of the real world). Finally, the presence of sufficient data can be guaranteed by defining (and enforcing adherence to) minimal reporting requirements. Interestingly, the field of micro arrays has already established its standards according to these overall schemes ***(1)***. To ensure that these requirements are also met for the proteomics community, standards development efforts have been initiated, most notably by the Human Proteome Organization (HUPO) Proteomics Standards Initiative (PSI). This chapter introduces the PSI, its use of controlled vocabularies, and the individual standards developed for mass spectrometry and molecular interactions, as these are the most mature and are already in active use.

2. Methods

2.1. The Human Proteome Organization Proteomics Standards Initiative

2.1.1. Goals and Organizational Structure

The Human Proteome Organization (HUPO) Proteomics Standards Initiative (PSI) was founded at the HUPO meeting hosted by the National Institues of Health (NIH) in Bethesda, Maryland, April 28–29, 2002, to define community standards for data representation in proteomics to facilitate data comparison, exchange, and verification ***(2)***. Organizationally, the PSI is divided into several working groups that each focuses on a particular domain or topic (gel electrophoresis, mass spectrometry, molecular interactions, protein modifications, proteomics informatics, and sample processing), as well as three intergroup activities overseeing integrative activities (controlled vocabularies, minimal information about a proteomics experiment [MIAPE], and steering group). Membership in the PSI working groups is open to anyone interested in actively contributing to the standards, and the PSI coordinates its community mainly via its website (http:///www.psidev.info) and two PSI meetings each year, one in Spring and one at the yearly HUPO World Congress in Autumn.

2.1.2. HUPO PSI Standards Development

Standards development by the HUPO PSI is largely based on voluntary contributions by participating members of the community, with membership of the working groups open to anyone interested. To organize the development of the standards by the different working groups, the PSI has defined the four documents that make up a standard:

1. a formal requirements specification,
2. minimal reporting requirements for the standard,
3. a data exchange format definition, and
4. a domain-specific controlled vocabulary or ontology.

In addition to formalizing the aspects of a standard, review processes for the different types of documents have also been elaborated by the HUPO PSI *(3)*. Most notably, these encompass a public review stage during which interested members of the community can provide feedback on the proposed standards, along with a more formal, invited peer review of the documents or specifications.

2.2. Controlled Vocabularies in the PSI

PSI CVs are sets of terms recommended as reference lexicon in order to standardize the meaning and syntax of the terminologies used while exchanging proteomics data. An example of why this is necessary is given by the molecular interaction (MI) format, in which the yeast two hybrid method (term MI:0018) can be written by various authors in myriad ways (e.g., 2 hybrid, 2-hybrid, 2H, two-hybrid), and that is excluding spelling errors.

All PSI CVs are encoded in Open Biomedical Ontology (OBO; http://obo.sourceforge.net/main.html) format in which each term must have a preferred reference name and an unambiguous consensus definition stating its meaning in the context of proteomics. Moreover, each term is coupled with a unique identifier and can be associated with number of synonyms or alternative spellings different from the preferred name. In an OBO file terms are structured in a hierarchy or a graph (where each term can have multiple parents) through semantic binary relationships of type "is_a" (e.g., rose *is_a* flower) or "part_of" (e.g., petal is *part_of* flower). Each PSI workgroup creates and maintains a CV as part of the proposed standard by collecting terms required to support the data exchange format, the formal requirements specification, and the minimal information reporting guidelines. In both mass spectrometry and molecular interaction workgroups the hierarchy of the CV reflects the exchange format structure and each top level term is associated with a location in the exchange format where its child terms can be used. This strategy makes it easier for the users to create data files, since the exchange format can be used as a template that needs to be filled in with the appropriate CV terms. Furthermore, a CV hierarchy adapted to the format facilitates the development of automatic semantic validation tools that check whether a data file is compliant with the minimal information reporting guidelines.

The PSI CVs are dynamically maintained via dedicated mailing lists that allow any user to request new terms in agreement with the community involved. Once a consensus is reached the new terms are added within a few days. This is a key mechanism to keep good coverage of novel proteomic technologies represented in the CVs and to ensure the flexibility of the exchange standard in reporting emerging data types with the existing format but associated with dedicated new CV terms.

Although PSI CVs largely cover the terminology of the proteomic domain, they are not intended to be a standalone ontology reference (like the Gene Ontology) modeling the reality of any proteomic experiment. As a matter of fact the PSI CVs are fragmented in different open biomedical investigation (OBO) files, closely related to specific exchange formats. However, the PSI participates in the ongoing effort of developing an OBI ontology (http://obi.sourceforge.net/) *(4)* by providing sets of well-defined proteomics vocabularies to be located in a comprehensive representation of biological experimental observation.

2.3. Standards for Mass Spectrometer-Based Proteomics

Continuous improvements to the instruments, the increasing availability of (protein) sequence databases, and the development of powerful separation methods have all contributed to the crucial role that mass spectrometers play in current high-throughput proteomics approaches. The raw data output of these instruments is typically captured in a vendor-specific (and sometimes even instrument model-specific) binary output format, however. The only way to gain access to these formats (apart from outright reverse engineering) is to use appropriate vendor-supplied software libraries. Although these libraries are often included free of charge when buying an instrument, access to these files remains restricted to researchers who have actually purchased that instrument. The inherent limitations of such proprietary data formats and their impact on science have been clearly described *(5)*. Since these raw files usually include much more detailed information than is required for protein or peptide identification, most researchers prefer to rely on heavily processed peak lists instead *(6)*. These peak lists are much smaller, text-based files that essentially capture only mass-over-charge (*m/z*) and intensity information for centroided peaks. Although a number of different formats for peak lists exist, these are so simple that transformations between them are typically straightforward. Despite their apparent convenience, however, peak lists are suboptimal formats for sharing mass spectrometry data; they make parts of the data *accessible*, but fail to capture *sufficient* data. More and more researchers are also making use of additional information not captured in the peak lists *(6)*. To provide an instrument-independent data format that filled the gap between the proprietary raw formats from the vendors and the minimalist peak lists, the Institute for Systems Biology (ISB) in Seattle, WA and the HUPO PSI independently designed new mass spectrometry output formats based on XML. The mzXML format of the ISB *(7)* is already extensively used as the common input format for mass spectrometry data processing tools, and several conversion programs have been made available to extract mzXML files from the proprietary formats of different vendors (http://sashimi.sf.net). An independent analysis of the strengths and shortcomings of the mzXML format is available *(8)*. The mzData format of the PSI *(9)* was developed as a community standard with strong participation from the instrument vendors. By actively soliciting this vendor involvement, the PSI ensured built-in support for mzData in the actual instrument software, an important step toward widespread adoption of the format. The many vendors and software tools that have implemented mzData can be found on the PSI website (http://www.psidev.info/index.php?q=node/95).

Since the presence of two independent mass spectrometry standards was correctly perceived to be an unfavorable situation by the ISB, the PSI, and the community at large, the two development teams decided to join forces under the PSI banner to develop a single successor to both mzXML and mzData *(10)*. The objective of this ongoing collaboration, which continues to receive support from the instrument vendors, is to integrate the specific strengths of each format, while simultaneously eliminating any remaining problems.

2.4. Standards for Protein–Protein Interactions

The understanding of protein interactions is a key to the understanding of biology at the molecular level, and many experiments aim to determine them, from small-scale enzymatic essays to large-scale technologies such as tandem affinity purification *(11)*. However, the results of these experiments are not yet systematically captured in databases, as authors are not obliged to submit the data to a public database prior to publication, as for instance DNA sequence data. The published data are often accessible only in the form of PDF tables or proprietary formats on authors' and journals' web sites, or not at all. The value of published protein interaction data was recognized by projects and funding agencies, leading to the creation of several independent databases for protein interactions, for example BioGRID (http://www.thebiogrid.org/), DIP (http://dip.doe-mbi.ucla.edu/), HPRD http://www.hprd.org/), IntAct (http://www.ebi.ac.uk/intact), MINT (http://mint.bio.uniroma2.it/mint/), and MPact (http://mips.gsf.de/genre/proj/mpact). These projects collect interaction data abstracted from the literature or directly submitted to the databases. However, no single database can possibly capture all the published interaction data, and even the data captured by the databases were previously offered in different, incompatible formats.

In 2004, the HUPO Proteomics Standards Initiative published the PSI MI XML 1.0 standard, jointly developed by a broad range of both academic and commercial organizations ***(12)***. This standard is now widely implemented; data in PSI MI format is available, among others, from BioGRID, DIP, HPRD, IntAct, MINT, and MPact. This allows users to easily download and combine the data from multiple sources for their own analysis. Tools supporting the PSI MI standard include the Cytoscape network visualization system ***(13)***, XSLT scripts for the conversion of PSI MI XML files into HTML, and a validator allowing semantic validation in addition to standard XML syntax validation. Given a PSI MI file, the validator will check the correct use of controlled vocabularies as well as a set of data consistency rules (http://www.ebi.ac.uk/intact/validator/). Building on the successful implementation of the 1.0 PSI MI standard, version 2.5 has been released in December 2005 (http://www.psidev.info/ index.php?q=node/60). Version 2.5 extends the scope of the standard from protein–protein interactions to molecular interactions in general, providing additional interactor types such as DNA, RNA, and chemical entities. It also provides a more detailed modeling of quantitative parameters, for example, dissociation constants. Overall, the PSI MI 2.5 format provides a comprehensive framework for the exchange and validation of detailed molecular interaction data.

While a detailed representation of molecular interactions is essential for high-quality database curation and detailed data analysis, many applications require less detailed data, and the PSI received frequent requests for a standardized tabular data format providing interactor pairs and a minimal set of additional parameters. Thus, the PSI 2.5 format also provides a minimalist tabular description of binary molecular interactions, derived from the BioGRID format. Data in this MITAB format are currently available from the DIP, IntAct, and MINT databases.

While the standardized data representation facilitates the exchange of molecular interaction data, it does not in itself solve the problem of redundant data curation by independent databases. In the International Molecular Exchange Consortium (IMEx) (http://imex.sf.net), based on the PSI MI format, several molecular interaction databases, currently DIP, IntAct, and MINT, with BioGRID and BindingDB as observers, aim to coordinate their curation efforts and to exchange all curated data, similar to the well-established exchange of DNA sequence data by the International Nucleotide Sequence Database Collaboration (http://www.insdc.org). IMEx members are already coordinating both their curation standards and their curation topics, and are currently implementing regular data exchange. In collaboration with journal editors, in particular from *PROTEOMICS* and *Nature Biotechnology* ***(14)***, the IMEx partners are encouraging direct deposition of molecular interaction data in the IMEx databases as part of the publication process, to overcome the fragmentation of published molecular interaction data and to provide a network of comprehensive, stable, high-quality molecular interaction data resources.

Acknowledgments

Development of the PSI standards is funded in part by the EU ProDaC, Grant LSHG-CT-2006-036814. The authors would like to thank Rolf Apweiler for his support and the HUPO PSI community for their contributions toward the development of the standards.

References

1. Ball, C. A. and Brazma, A. (2006) MGED standards: work in progress. *OMICS* **10(2),** 138–144.
2. Kaiser, J. (2002) Proteomics. Public-private group maps out initiatives. *Science* **296(5569),** 827.
3. Vizcaíno, J. A., Martens, L., Hermjakob, H., Julian, R. K., and Paton, N. W. (2007) The PSI formal document process and its implementation on the PSI website. *Proteomics* **7(14),** 2355–2357.
4. Whetzel, P. L., Brinkman, R. R., Causton, H. C., *et al.* (2006) Development of FuGO: an ontology for functional genomics investigations. *OMICS* **10(2),** 199–204.
5. Wiley, H. S. and Michaels, G. S. (2004) Should software hold data hostage? *Nat. Biotechnol.* **22(8),** 1037–1038.
6. Martens, L., Nesvizhskii, A. I., Hermjakob, H., *et al.* (2005) Do we want our data raw? Including binary mass spectrometry data in public proteomics data repositories. *Proteomics* **5(13),** 3501–3505.
7. Pedrioli, P. G. A., Eng, J. K., Hubley, R., *et al.* (2004) A common open representation of mass spectrometry data and its application to proteomics research. *Nat. Biotechnol.* **22(11),** 1459–1466.
8. Lin, S. M., Zhu, L., Winter, A. Q., Sasinowski, M., and Kibbe, W. A. (2005) What is mzXML good for? *Expert Rev. Proteomics* **2(6),** 839–845.

9. Orchard, S., Hermjakob, H., Julian, R. K., *et al.* (2004) Common interchange standards for proteomics data: public availability of tools and schema. *Proteomics* **4(2),** 490–491.
10. Orchard, S., Jones, A. R., Stephan, C., and Binz, P.-A. (2007) The HUPO pre-congress Proteomics Standards Initiative workshop. HUPO 5th annual World Congress. Long Beach, CA, 28 October–1 November 2006. *Proteomics* **7(7),** 1006–1008.
11. Puig, O., Caspary, F., Rigaut, G., *et al.* (2001) The tandem affinity purification (TAP) method: a general procedure of protein complex purification. *Methods* **24(3),** 218–229.
12. Hermjakob, H., Montecchi-Palazzi, L., Bader, G., *et al.* (2004) The HUPO PSI's molecular interaction format—a community standard for the representation of protein interaction data. *Nat. Biotechnol.* **22(2),** 177–183.
13. Shannon, P., Markiel, A., Ozier, O., *et al.* (2003) Cytoscape: a software environment for integrated models of biomolecular interaction networks. *Genome Res.* **13(11),** 2498–2504.
14. Editors. (2007) Democratizing proteomics data. *Nat. Biotechnol.* **25(3),** 262.

19

The PRIDE Proteomics Identifications Database: Data Submission, Query, and Dataset Comparison

Philip Jones and Richard Côté

Summary

The PRIDE database has been developed to allow the proteomics community to share publicly, or within private collaborations, the vast volume of data generated by proteomics laboratories across the globe. These data are being generated at an expanding rate as increasingly sophisticated technologies become available. Compounding this problem, the infrastructure and techniques used to generate these data vary in terms of the instrumentation used, the protein sequence databases searched, the search engines employed, and the automatic or manual filtering of identifications following the initial automated search. The PRIDE project provides an infrastructure to solve these problems, including a generic, standards-based format that can be annotated to capture data generated using any proteomics pipeline, a protein accession mapping service to overcome the problem of disparate protein sequence databases being searched, and tools for query, comparison, and analysis of proteomics data. This chapter describes the main practical considerations in making use of PRIDE, including the available resources: the PRIDE database, the Ontology Lookup Service (OLS), the protein identifier cross-referencing service (PICR), the Proteome Harvest PRIDE submission spreadsheet, and the PRIDE BioMart.PRIDE can be accessed at http://www.ebi.ac.uk/pride.

Key Words: PRIDE; proteomics; mass spectrometry; public data repository; BioMart; HUPO-PSI; mzData; XML; protein identification; peptide identification; proteome harvest.

1. Introduction

A vast amount of data is being generated by proteomics laboratories across the world, with several high-impact journals publishing a large number of articles describing the identification, quantitation, and distribution of proteins,

From: *Methods in Molecular Biology, vol. 484: Functional Proteomics: Methods and Protocols*
Edited by: J. D. Thompson et al., DOI: 10.1007/978-1-59745-398-1, © Humana Press, Totowa, NJ

peptides, and posttranslational modifications. These experiments are normally in different tissues, under different disease conditions, at various developmental stages and under a variety of environmental conditions. Journal author guidelines often encourage or even mandate the publication of the experimental data to accompany the submission of a manuscript. Very often this is achieved by the generation of supplementary material in the form of a spreadsheet, PDF document, or other printable format. Unfortunately, the use of such techniques for disseminating data is not conducive to allowing comparison and further analysis of separate sets of experimental data.

The PRIDE project ***(1,2)*** was initiated to provide a solution to this problem. The core of PRIDE is a relational database designed to store experimental proteomics data, together with an XML schema developed for the exchange (i.e., submission and retrieval) of complete experimental data sets. This is all made available to the community through a web interface (http://www.ebi.ac.uk/pride) at the EMBL-European Bioinformatics Institute (EBI) in Cambridge, United Kingdom.

This interface contains forms for data query and submission as well as forms to allow the management of this data.

Submitting data to a repository such as PRIDE that utilizes a complex schema is not a trivial undertaking. To mitigate this, the PRIDE team has developed a submission tool implemented as a Microsoft Excel workbook. This tool allows the potential submitter to create a PRIDE XML file by populating the spreadsheets included in the workbook. The workbook also provides direct access to controlled vocabularies and ontologies for annotation of the data. For laboratories that wish to submit data on a regular basis, populating a spreadsheet may not prove to be the most efficient means for generating PRIDE XML files. In this case, a Java API is available that can be used to build a software pipeline for submission to PRIDE.

2. Materials

2.1. The PRIDE Web Interface: Navigation and Query

This section focuses on the web interface elements that are used to extract information from the PRIDE database as well as additional tools that are part of the PRIDE interface. For details of the XML validation and submission forms, *see* **Section 2.3** and **3.3**.

The PRIDE database has been designed to allow installation at multiple sites as well as the central/main PRIDE service at EBI.

2.2. The BioMart Query Interface

The PRIDE BioMart interface allows the user to build complex queries. The user is able to create multiple filters based upon different criteria and can specify precisely which attributes (equivalent to columns in a spreadsheet) are included in the search output. It is also possible to select from a number of different output formats, including an HTML table to be viewed in an internet browser, column separated or tab separated values, or a Microsoft Excel spreadsheet. Additionally, BioMart provides a web service interface for programmatic access to data. Further documentation of the BioMart *(3)* project can be found at http://www.biomart.org/.

The PRIDE BioMart is accessible from the left-hand menu on the PRIDE web site (*see* **Section 2.1**) or can be accessed directly at http://www.ebi.ac.uk/pride/biomart/martview.

2.3. Submitting Data to PRIDE

Data are submitted to PRIDE in the form of a PRIDE 2.1 XML file. The XML schema for this format can be found at http://www.ebi.ac.uk/pride/help_resources/pride.xsd. The schema is also documented using Altova XMLSpy. This documentation can be viewed at http://www.ebi.ac.uk/pride/schemaXmlspyDocumentation.do.

The PRIDE XML format makes direct use of the HUPO PSI mzData XML format, version 1.05 *(4)*, as an embedded element to support the submission of mass spectra to PRIDE. The mzData 1.05 XML schema can be found at http://psidev.info/docstore/mzdata.xsd.

The Proteome Harvest PRIDE submission spreadsheet has been developed to allow laboratories with limited bioinformatics support to create valid PRIDE XML files, simply by populating an Excel spreadsheet. This resource is documented in detail at http://www.ebi.ac.uk/pride/proteomeharvest/ where links are included to download the latest version of the spreadsheet. This page also includes "e-learning" tutorial movies that can be run in a browser that has the latest Adobe Flash plug-in installed.

An alternative submission tool, Pride Wizard *(5)*, has been developed by the University of Manchester and is available from http://www.mcisb.org/software/PrideWizard/. This tool includes the facility to add iTRAQ™ *(6)* labels, allowing quantitation data to be encoded in PRIDE XML.

For laboratories that wish to create their own data pipeline using the Java programming language, a compiled PRIDE core jar file is available from http://sourceforge.net/project/showfiles.php?group_id=122040. ("PRIDE Compiled API"). This API includes infrastructure to allow a complete PRIDE java object model to be constructed that can then be used to generate a valid PRIDE XML file. Using the API is outside the scope of this chapter and will not be described further.

Once a valid PRIDE XML or mzData file has been created, it is possible to submit this directly to PRIDE. The submission process and associated infrastructure are described in **Section 3.3**.

2.4. The Ontology Lookup Service

The Ontology Lookup Service (OLS) *(7)* was developed as a spin-off of the PRIDE project and is used extensively in PRIDE to provide ontology and controlled vocabulary queries. This service provides functionality that goes beyond PRIDE and proteomics data, however. It is possible to both search and browse this service through the web interface available at http://www.ebi.ac.uk/ontology-lookup. The use of the OLS web application will be described in detail in **Section 3.4**.

The OLS also provides a rich, programmatic web service implemented using SOAP (Simple Object Access Protocol) version 1.1 (http://www.w3.org/TR/soap). The WSDL (Web Service Definition Language) documentation for the OLS SOAP service is described at http://www.ebi.ac.uk/ontology-lookup/WSDLDocumentation.do, including a hyperlink to the WSDL itself. The use of this web service is outside the scope of this chapter and will not be described further.

2.5. The Protein Identifier Cross-Referencing Service

The Protein Identifier Cross-Referencing Service (PICR) was developed by the PRIDE team and is used extensively in PRIDE. This service provides functionality that goes beyond PRIDE and proteomics data and provides a mechanism to resolve protein identifiers across multiple source databases. It is possible to search and browse this service through the web interface available at http://www.ebi.ac.uk/Tools/picr/. The use of the PICR web application will be described in detail in **Section 3.5**.

The PICR service also provides a rich, programmatic web service implemented using SOAP, as described above. The WSDL documentation for the PICR SOAP service is described at http://www.ebi.ac.uk/Tools/picr/WSDLDocumentation.do, including a hyperlink to the WSDL itself. The use of this web service is outside the scope of this chapter and will not be described further.

3. Methods

3.1. The PRIDE Web Interface: Navigation and Query

The PRIDE web interface includes pages and forms to provide the user with access to the core functionality of PRIDE, together with documentation of this functionality and documentation of the data submission process. Documentation and guidance for software engineers and bioinformaticians wishing to deploy local installations of PRIDE are also provided. The use of these pages and forms is described in this section. Details of the submission process are described in detail in **Section 3.3**.

3.1.1. Searching PRIDE

For queries that involve building complex filters with control over the individual data items included in the output, the user is referred to **Section 3.2** describing the BioMart query interface to PRIDE. The core PRIDE web application includes some basic query mechanisms, however.

3.1.1.1. PRIDE "Simple Query"

It is possible to perform a simple query using the "Search PRIDE" text box:

1. Navigate using an Internet browser to the PRIDE home page located at http://www.ebi.ac.uk/pride.
2. Locate the "Search PRIDE" text box at the top left-hand corner of the page and enter your search term. Possible search term types are listed in ***Note 1***.
3. You will then be taken to the "Search Results: Summary and Format Selection" page. If no results match your search, you will be informed. Otherwise you will be presented with a summary of the matching results as described in **Section 3.1.1.4**.

3.1.1.2. Using the Advanced Search Interface

1. You will find a menu on the left-hand side of the majority of the PRIDE web pages (with the exception of the mass spectrum viewer and the BioMart page). Click on the link "Advanced Search" on this menu.
2. Enter your search term into the appropriate search box on this form. You can enter experiment accession numbers, protein accession numbers, peptide sequences, and parts of reference lines or select items from controlled vocabularies (describing the sample) to perform a search. Note that in all cases, you can enter only a single search term. To conduct a more complex query, use the BioMart interface described in **Section 3.2**.
3. You will then be taken to the "Search Results: Summary and Format Selection" page. If no results match your search, you will be informed. Otherwise you will be presented with a summary of the matching results as described in **Section 3.1.1.4**.

3.1.1.3. Browsing PRIDE Experiments

The "Browse Experiments" page provides a direct entry point to experiments categorized by project or various sample parameters, described in ***Note 2***. Following a link on this page takes the user to the search summary page for all of the experiments matching the search.

1. Click on the "Browse Experiments" link on the PRIDE menu.
2. You will be presented with a form composed of several tables with different search categories. Near the top you will find the "Browse By Project" section. Below this you will find the various sample parameter search sections.
3. It is possible to sort the columns in this view by clicking on the heading of the column. Repeated clicking reverses the direction of the sort.
4. Click on the project name or term of interest.
5. You will then be taken to the "Search Results: Summary and Format Selection" page. If no results match your search, you will be informed. Otherwise you will be presented with a summary of the matching results as described in **Section 3.1.1.4**.

3.1.1.4. Functionality of the Experiment Summary View

Following a query of PRIDE using either the simple or advanced search form, you will be taken to the search summary view illustrated in **Fig. 1**. This form provides several options available for investigating the individual experiments in more detail.

It is possible to compare the protein identifications found in up to 10 experiments, with the results being displayed as a Venn diagram or histogram as appropriate. This is achieved by checking the check boxes of the experiments in which you are interested on the "Compare Protein Identification Sets" column. If two

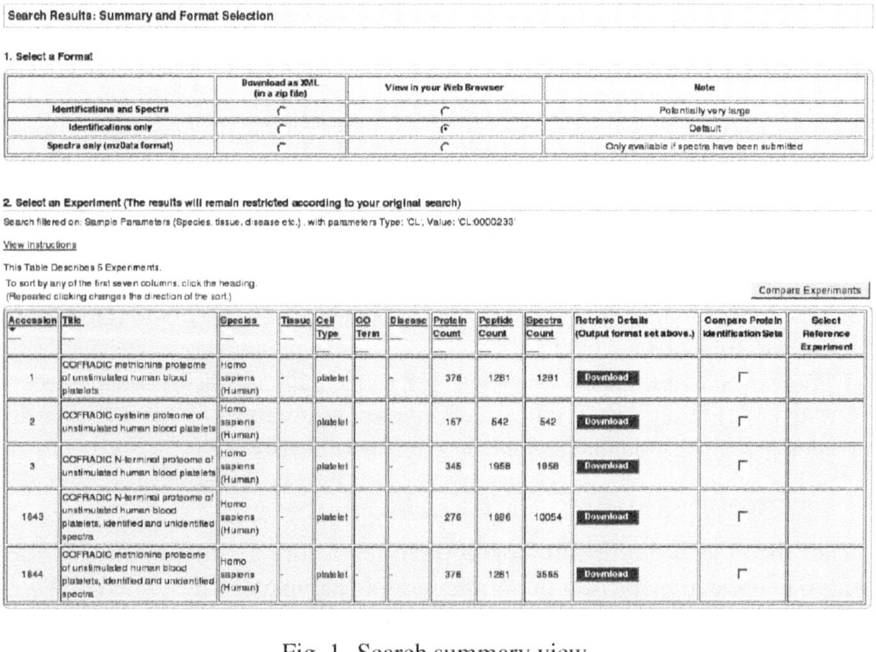

Search Results: Summary and Format Selection

1. Select a Format

	Download as XML (in a zip file)	View in your Web Browser	Note
Identifications and Spectra			Potentially very large
Identifications only			Default
Spectra only (mzData format)			Only available if spectra have been submitted

2. Select an Experiment (The results will remain restricted according to your original search)

Search filtered on: Sample Parameters (Species, tissue, disease etc.), with parameters Type: 'CL', Value: 'CL:0000233'

View instructions

This Table Describes 5 Experiments.

To sort by any of the first seven columns, click the heading.
(Repeated clicking changes the direction of the sort.)

Compare Experiments

Accession	Title	Species	Tissue	Cell Type	GO Term	Disease	Protein Count	Peptide Count	Spectra Count	Retrieve Details (Output format set above.)	Compare Protein Identification Sets	Select Reference Experiment
1	COFRADIC methionine proteome of unstimulated human blood platelets	Homo sapiens (Human)	-	platelet	-	-	378	1281	1281	Download		
2	COFRADIC cysteine proteome of unstimulated human blood platelets	Homo sapiens (Human)	-	platelet	-	-	157	542	542	Download		
3	COFRADIC N-terminal proteome of unstimulated human blood platelets	Homo sapiens (Human)	-	platelet	-	-	345	1958	1958	Download		
1643	COFRADIC N-terminal proteome of unstimulated human blood platelets, identified and unidentified spectra	Homo sapiens (Human)	-	platelet	-	-	276	1696	10054	Download		
1644	COFRADIC methionine proteome of unstimulated human blood platelets, identified and unidentified spectra	Homo sapiens (Human)	-	platelet	-	-	376	1281	3585	Download		

Fig. 1. Search summary view.

or three experiments are selected, the results will be displayed as a standard Venn diagram. If between 4 and 10 experiments are selected, you should select a single "reference experiment" in the right-hand column of the summary view, against which all of the other selected experiments will be compared. The resulting comparison will then be displayed as a histogram.

There are several available options for downloading the details of each experiment. These options are presented at the top of the result summary page under the heading "1. Select a Format," illustrated in **Fig. 1**. These options include the choice to view the results as HTML or to retrieve them as a compressed ("zipped") PRIDE XML file. The user can also select the portion of the data in the experiment to be returned, with the following options: "Identifications and Spectra," "Identifications only," and "Spectra only." Once a selection has been made, the user can then click on the "Download" button adjacent to the experiment in which they are interested.

3.2. The BioMart Query Interface

The PRIDE BioMart is embedded in and accessible from the left-hand menu on the PRIDE web site (*see* **Section 2.1**) or can be accessed directly at http://www.ebi.ac.uk/pride/biomart/martview.

Whichever method you use to access the service, you will be presented with the form illustrated in **Fig. 2**. This interface is used to build your query. Generally speaking there are three main steps involved in query building: the creation of filters to restrict the data included in your results (i.e., restricting the number of

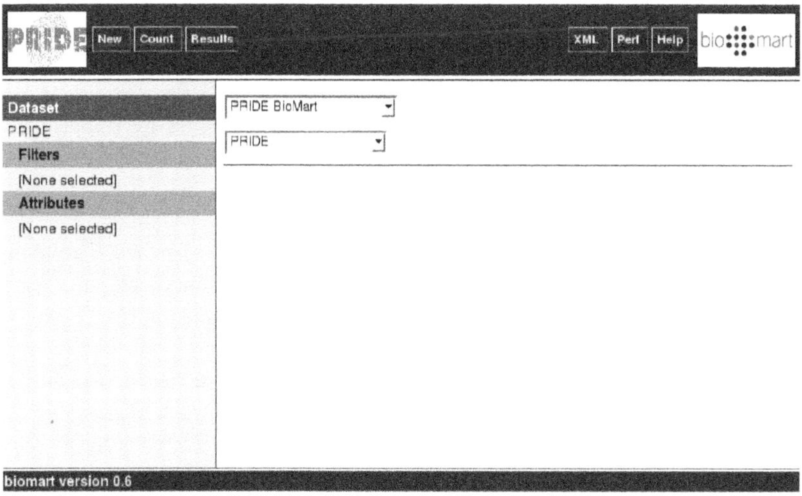

Fig. 2. Form presented when accessing the service.

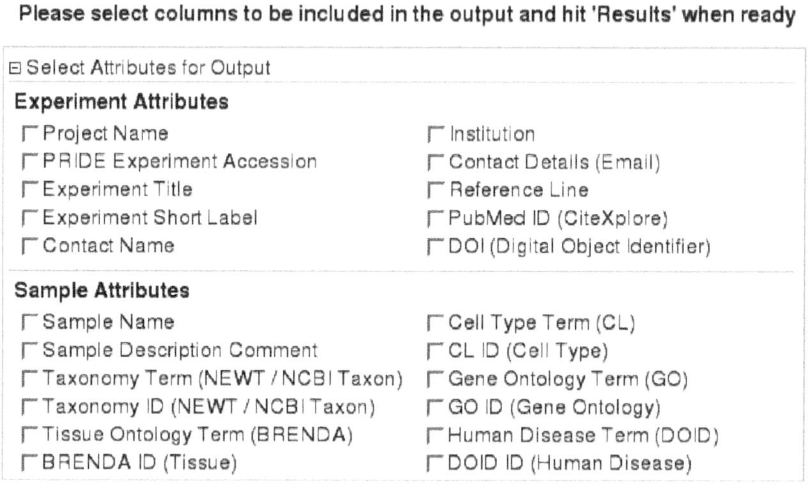

Fig. 3. Selection of attributes.

rows of data returned), the selection of attributes (i.e., the selection of columns of data to include), and finally the selection of a format for the results (i.e., HTML table, tab separated values, comma separated values, or a Microsoft Excel spreadsheet).

1. Selection of Attributes: Click the "**Attributes**" link in the left panel of the BioMart user interface. The right-hand panel will change as illustrated in **Fig. 3**. Select the attributes for inclusion as columns of data by clicking on the check boxes to the left of the attribute descriptions. In this example, click on "PRIDE Experiment Accession," "Submitted Protein Accession," and "Peptide Sequence."
2. Creation of Filters: Click on the "**Filters**" link in the left panel of the BioMart user interface and click the + symbol to the left of "Filter by Experiment" that will appear on the right-hand panel. The right hand panel should then appear as illustrated in **Fig. 4**. Click in the text area to the right of the "Filter by Experiment Accession" label and enter the number 2 into this field (*see* **Note 13**).
3. Click on the "Count" button at the top of the BioMart interface. The number of PRIDE experiments that match your filter criteria will be displayed (one in this case). Note that this is *not* the same as the number of rows of results that will be returned to you, which may be considerably greater.
4. Click on the "Results" button at the top of the BioMart interface. You will now be presented with the first 10 results that match your query as a representative set as illustrated in **Fig. 5**. The purpose of this step is to allow you to modify your query before accessing all of the available results.
5. In the right-hand panel, click on the select pull-down labeled "rows as (HTML)" and select "TSV" to allow you to retrieve the results as a tab separated values file.

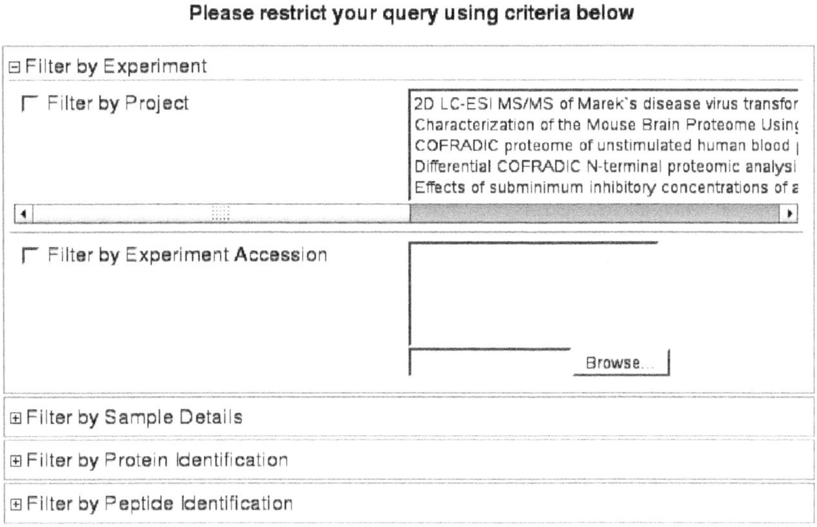

Fig. 4. Creation of Filters.

6. In the right-hand panel, click on the select pull-down labeled "Export all results to (File)" and select "Browser." Then click on GO. The complete set of results matching your filter will be displayed in a new browser window.

3.3. Submitting Data to PRIDE

Data can be submitted to PRIDE in the form of a valid PRIDE 2.1 XML file or an mzData 1.05 XML file (the latter if you wish to submit spectra only

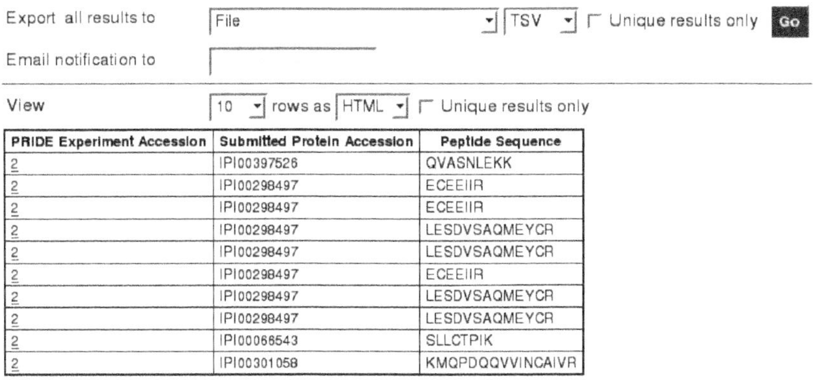

PRIDE Experiment Accession	Submitted Protein Accession	Peptide Sequence
2	IPI00397526	QVASNLEKK
2	IPI00298497	ECEEIIR
2	IPI00298497	ECEEIIR
2	IPI00298497	LESDVSAQMEYCR
2	IPI00298497	LESDVSAQMEYCR
2	IPI00298497	ECEEIIR
2	IPI00298497	LESDVSAQMEYCR
2	IPI00298497	LESDVSAQMEYCR
2	IPI00068543	SLLCTPIK
2	IPI00301058	KMQPDQQVVINCAIVR

Fig. 5. First 10 results that match your query.

to PRIDE). For details of mechanisms for generating PRIDE XML files, *see* **Section 2.3**.

Note that for a new submission, you should not include the <**ExperimentAccession/**> element in the XML file. An experiment accession number will be assigned automatically following a successful data submission.

Once an XML file has been generated, submitters may make use of the XML validation tool built into PRIDE to check that their XML file validates correctly against the schema:

1. On the left-hand menu on the PRIDE home page, click on "Validate XML." You will now be presented with a form as illustrated in **Fig. 6**
2. Click on the "Browse" button on this form and browse to the XML file that you wish to validate. Alternatively, paste the fully qualified path and file name into the text box adjacent to the Browse button.
3. Select the appropriate "File Type" (PRIDE 2.1 XML or mzData 1.05 XML) and then click "Validate File."
4. After a few seconds delay, a report will be returned to you indicating that the file is valid, or if there is an error you will be given details of the position and nature of the problem.

Once you are satisfied that you have created a valid XML file, you can then proceed to submitting the file to the PRIDE database.

Submission requires that you log in to the PRIDE system with a valid username and password. If you do not have a user account on PRIDE, you can register for an account (for free, of course) by clicking on the "Register" link in the left-hand menu. Otherwise you can log in to PRIDE by clicking on the "Log in" link on the left-hand menu. You can then begin the submission process:

1. Log in to PRIDE by clicking on the "Log in" menu item on the left-hand menu. (Or register on the PRIDE system if you are a new user, as described above.)

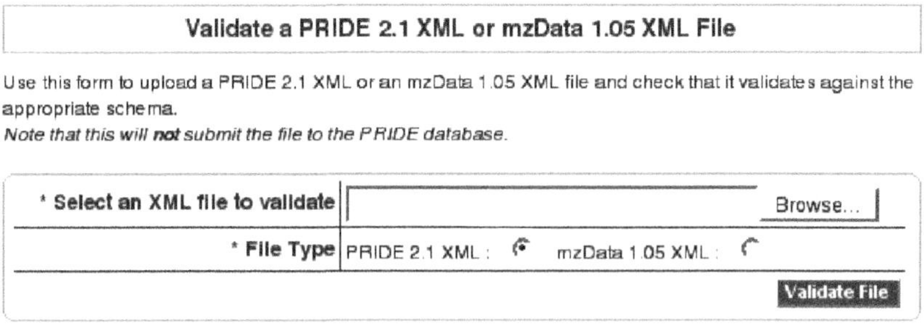

Fig. 6. Validate a PRIDE.

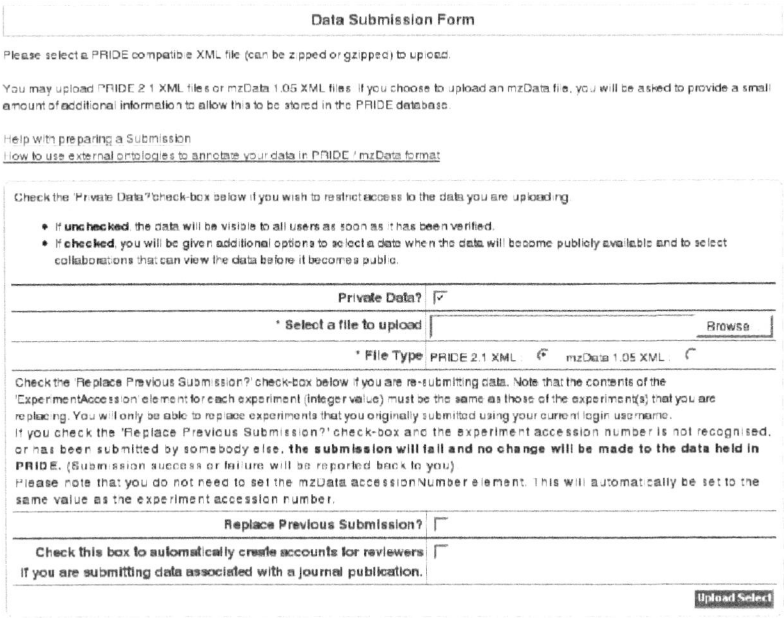

Fig. 7. Data submission form.

2. The left-hand menu will now extend slightly, with the addition of a "Submit data" menu item that you should now click. You will be presented with the submission form illustrated in **Fig. 7**.
3. Click on the "Browse" button on this form and browse to the XML file that you wish to validate. Alternatively, paste the fully qualified path and file name into the text box adjacent to the Browse button.
4. Note that by default, the "Private Data?" check box is checked. If you wish to submit data publicly, uncheck this box by clicking it once.
5. Select the appropriate "File Type" (PRIDE 2.1 XML or mzData 1.05 XML).
6. If this is a new submission, leave the "Replace Previous Submission?" checkbox unchecked (*see* **Note 3**).
7. If you are submitting data privately and you wish to create a reviewer account, check the box labeled "Check this box to automatically create accounts for reviewers if you are submitting data associated with a journal publication." You will be sent an email following submission, with details of an anonymous login account that you can send to your reviewers to allow them access to the private data set that you have submitted.
8. Click on the "Upload" button at the foot of the form.
9. If you have selected to submit your data privately, you will now be presented with a second form that you can use to specify a future date when the data should (automatically) become public. Leave this field blank if you do not wish this to occur.

10. After submission a progress bar will be displayed followed by a feedback page that indicates whether or not your submission has been successful. This feedback page includes the PRIDE accession numbers that have been assigned to the experiments that you have submitted. If you entered a valid email address when you registered on the PRIDE system, you will also receive an email containing the details of the submission outcome.

3.4. The Ontology Lookup Service

3.4.1. Searching for Ontology Terms

1. Navigate using an Internet browser to the OLS home page located at http://www.ebi.ac.uk/ols.
2. Select the ontology or controlled vocabulary that you want to search from the "Search Ontology" pull-down menu (*see* **Note 4**). If you wish to browse the selected ontology click on the "Browse" button (*see* **Section 3.4.2**).
3. Type the term you wish to search in the "Term Name" text box. As you type, a list of suggested terms will appear. The list will be updated as you type, refining the search results (*see* **Note 5**). You can use the arrow keys or your mouse cursor to select the appropriate term. If more than 20 results are returned for a search, the last entry in the result box will be "… and more." If you select this value, you will be redirected to a result page where all the search values are listed in tabular form.
4. Once a search result is selected, the unique identifier for this term will be displayed in the "Term ID" text box. Additional information will also be retrieved from the OLS for this term and can include definitions, comments, synonyms, and cross-references to other databases or ontologies.
5. It is now possible to browse the ontology containing the newly found term as a root for the ontology browser, as described in **Section 3.4.2**, by clicking on the "Browse" button.

3.4.2. Browsing an Ontology

The ontology browser web page is divided into multiple sections **Fig. 8**. The main section is the ontology tree browser on the left of the page. On the right of the page, several information boxes are present. The uppermost is a brief description on how to use the browser. The "Relations" box will indicate the relationship between a term and its immediate parent. The "Term Information" box will indicate the unique ID and name for a selected term. If available, a link to a specific term at the authoritative website for the ontology being browsed will be displayed as an "external link." The "Zoom" button will allow the user to reroot the tree browser, using the selected term as a root. The "Associated Information" box will contain any additional information available for the selected term, which can include definitions, comments, synonyms, and cross-references. Finally, the "Term Hierarchy" box contains a graphic illustration of all possible paths from the selected term to the root(s) term(s) of the ontology.

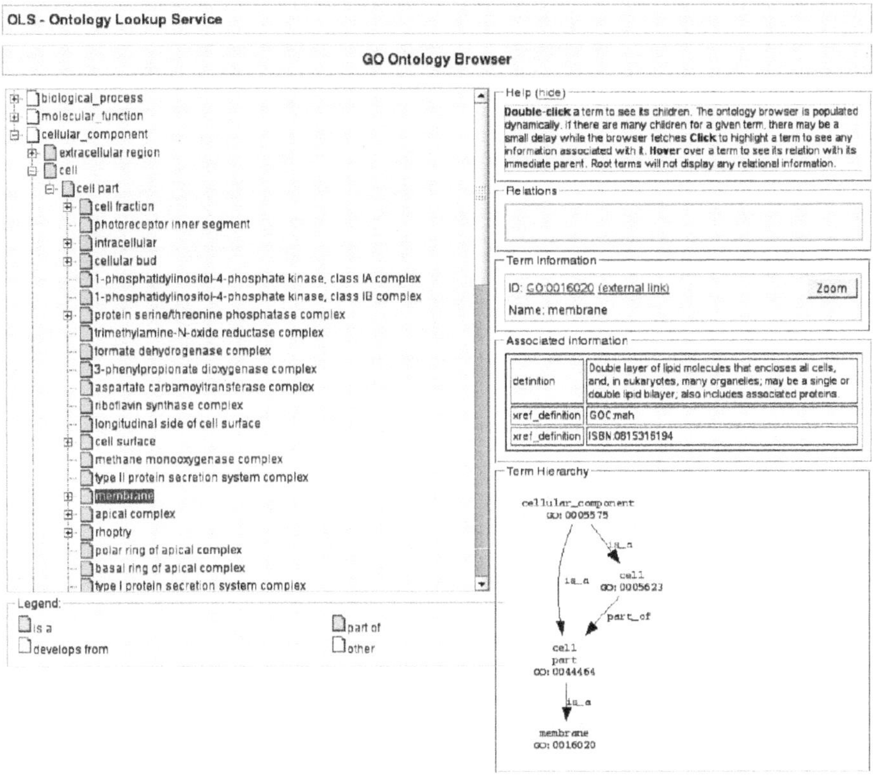

Fig. 8. Ontology browser web page.

1. Unless a specific term has been preselected as a browsing root, the default root terms of the ontology are shown in the browsing pane. Double-clicking on the term will load any child terms, if any. Once the child terms have been loaded, double-clicking on a term will expand/collapse the display (*see* **Note 6**).
2. Relationships between terms are color-coded in the browsing pane. The colored symbol next to a term name indicates its relationship with its parent (i*s a, part of, develops from* or *other; see* **Note 7**).
3. Clicking once on a term will highlight it and update the "Term Information," "Associated Information," and "Term Hierarchy" boxes.
4. Hovering over a term will update the "Relations" box.

3.5. The Protein Identifier Cross-Referencing Service

1. Navigate using an Internet browser to the PICR service home page (**Fig. 9**) located at http://www.ebi.ac.uk/Tools/picr/WSDLDocumentation.do.

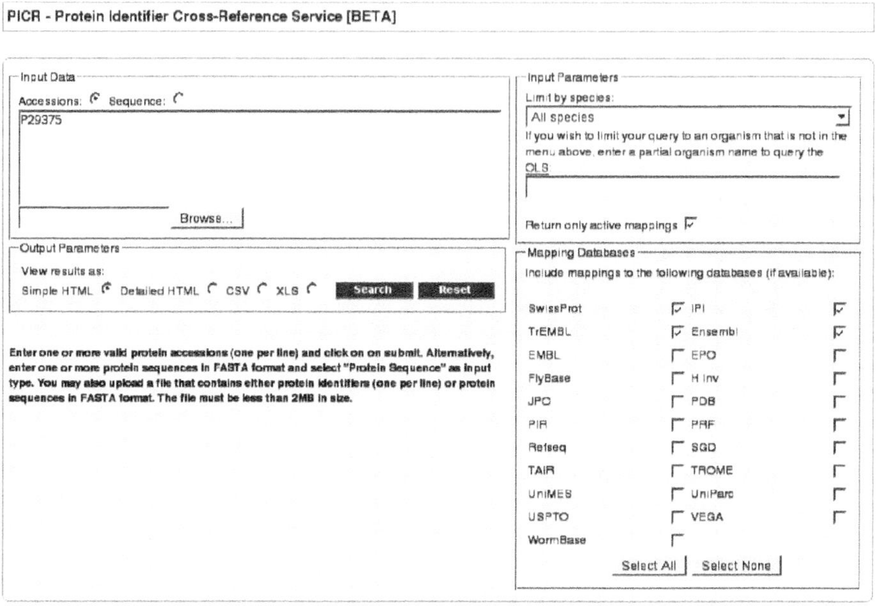

Fig. 9. PICR service home page.

2. Paste a list of protein identifiers in the "Input Data" text box, one identifier per line. You can only submit a maximum of 100 protein identifiers at one time. Alternatively, you can click on the "Browse" button and select a text file to upload. The file should contain one identifier per line. You can also search for identifier mappings using sequences in FASTA format. Sequences can be entered in the "Input Data" text box or a properly formatted text file can be uploaded as described above. The same limit of 100 protein sequences applies. If you are mapping sequences, you need to update the "Input Parameter" box and select "Sequence" as the input data type.
3. By default, the PICR service will return all available protein mappings, but it is possible to limit them by taxonomy and by active status. To retrieve only active mappings (*see* **Note 8**), check the "Return only active mappings" box. To limit the mappings to a particular taxonomy, select the desired option from the "Limit by species" menu (*see* **Note 9**).
4. Select which databases you wish to map to from the "Mapping Databases" option box (*see* **Note 10**).
5. Select how you wish to view the results. The default option is the "Simple HTML" table where each row represents a submitted protein identifier or sequence and each column represents a selected mapping database (*see* **Note 11**). The "Detailed HTML" option will give a full description of each UniParc entry corresponding to the submitted protein accession or sequence, including the entry time stamp and

a full description of the mappings (database, accession and version, active status, taxonomy, gi number, date added, date modified or deleted). The "CSV" option will produce a comma-separated file to download whose layout is identical to that of the "Simple HTML" view (*see* **Note 12**).

Click on the "Search" button. A search progress bar will be displayed on the screen as your search is processed. Once done, the search results will be displayed on screen or a file download dialog box will appear, depending on the selected options above.

4. Notes

1. It is possible to use any of the following identifier types to search using the "simple search" box on the home page:
   - PRIDE Experiment accession number. These values are plain integers.
   - PRIDE controlled vocabulary term (e.g., PRIDE:0000018).
   - GO (Gene ontology) term: GO:0000176.
   - Protein accession (e.g., IPI00295313).
2. The "Browse PRIDE Experiments" page includes five sections for browsing the experiments in PRIDE by sample. These sections include the following:
   - Taxonomy (using the NCBI taxonomy or NEWT at the EBI).
   - BRENDA tissue ontology term.
   - Cell Type ontology term.
   - Gene ontology term, used to annotate the subcellular location of the sample.
   - Disease ontology term.
3. There are several safeguards in place to prevent accidental overwriting of data in PRIDE. If you wish to resubmit an experiment to PRIDE, you must ensure the following:
   - The experiment accession number in the new XML file is the same as the experiment accession number in the XML you are replacing.
   - You attempt to resubmit under the same login account as the original submission.
   - You must check the "Replace Previous Submission?" check-box on the submission form.
4. By default, when the search page is loaded, the Gene Ontology *(8)* is selected. To search across all the ontologies and CVs, select the "Search in all ontologies" option at the top of the menu. If this option is selected, the search results will be prefixed with the short label of the ontology in the result box.
5. An example search would be to type "mitochondria" in the search box while the GO ontology is selected. The list updates itself as the search string is updated. If

nothing seems to be happening, hit the spacebar to add an empty space character to your search query.

6. A term might have a plus (+) or minus (–) symbol next to it in the browsing pane. A + next to a term indicates that the term has child terms that are not currently shown in the tree. A – next to a term indicates that it is possible to collapse a portion of the tree and hide some terms from the display.
7. *Is a, part of,* and *develops from* are the major relationship types between terms, though others are less widely used. To simplify the display, only the three major types have been color coded.
8. The UniProt Archive ***(9)*** contains all current and historical protein sequences and mappings. When mappings are deleted from the source database, for various reasons, they are retained in UniParc but are labeled as inactive.
9. Although we have tried to get the maximum taxonomic coverage for the mappings, some source databases do not provide taxonomy information and, as such, those mappings cannot be properly identified and will be excluded from any search that is limited by taxonomy.
10. Some mapping options actually refer to more than one database. For example, selecting *Ensembl* will query all the organism-specific Ensembl releases, as is the case for *RefSeq, Vega,* and *Trome*. Selecting *SwissProt* and *TrEMBL* will also include the respective spice variant databases.
11. Some mappings might be highlighted in red. These mappings are historical and inactive, as the referenced entries have been removed or renamed from the current release of the mapped databases. Some mappings might be highlighted in blue. These mappings, while valid, are not based on 100% sequence identify and may include splice variants and sequence variants.
12. The CSV version will not have the highlighted information as described in **Note 9**.
13. Complex filters can be created involving any number of filter elements. For example, it is possible to create a filter based upon characteristics of the sample, together with details of the protein search database and the search engine used.

Acknowledgments

PRIDE is supported through BBSRC iSPIDER and HUPO Plasma Proteome Project funding as well as an EU Marie Curie fellowship. The Proteome Harvest data submission spreadsheet is funded through the BBSRC Proteome Harvest grant.

References

1. Jones, P., Côté, R. G., Martens, L., Quinn, A. F., Taylor, C. F., Derache, W., *et al.* (2006) PRIDE: a public repository of protein and peptide identifications for the proteomics community. *Nucleic Acids Res.* **34(Database issue),** D659–663.

2. Martens, L., Hermjakob, H., Jones, P., Adamski, M., Taylor, C., States, D., *et al.* (2005) PRIDE: the proteomics identifications database. *Proteomics* **5(13),** 3537–3545.
3. Durinck, S., Moreau, Y., Kasprzyk, A., Davis, S., De Moor, B., Brazma, A., *et al.* (2005) BioMart and Bioconductor: a powerful link between biological databases and microarray data analysis. *Bioinformatics* **21(16),** 3439–3440.
4. Orchard, S., Jones, P., Taylor, C., Zhu, W., Julian, R. K., Hermjakob, H., *et al.* (2006) Proteomic data exchange and storage: the need for common standards and public repositories. *Methods Mol. Biol.* **367,** 261–270.
5. Siepen, J. A., Swainston, N., Jones, A. R., Hart, S. R., Hermjakob, H., Jones, P., *et al.* (2007) An informatic pipeline for the data capture and submission of quantitative proteomic data using iTRAQTM. *Proteome Sci.* **5,** 4.
6. Wiese, S., Reidegeld, K. A., Meyer, H. E., and Warscheid, B. (2007) Protein labeling by iTRAQ: a new tool for quantitative mass spectrometry in proteome research. *Proteomics* **7(3),** 340–350.
7. Côté, R. G., Jones, P., Apweiler, R., and Hermjakob, H. (2006) The Ontology Lookup Service, a lightweight cross-platform tool for controlled vocabulary queries. *BMC Bioinformatics* **7,** 97.
8. Ashburner, M., Ball, C. A., Blake, J. A., Botstein, D., Butler, H., Cherry, J. M., *et al.* (2000) Gene ontology: tool for the unification of biology. The Gene Ontology Consortium. *Nat. Genet.* **25(1),** 25–29.
9. Leinonen, R., Diez, F. G., Binns, D., Fleischmann, W., Lopez, R., and Apweiler, R. (2004) UniProt archive. Bioiniformatics **20(17),** 3236–3237.

20

Searching the Protein Interaction Space Through the MINT Database

Andrew Chatr-aryamontri, Andreas Zanzoni, Arnaud Ceol, and Gianni Cesareni

Summary

Many fundamental processes involve protein–protein interactions. Recent advances in technology make it possible to perform large-scale, genome-wide interaction mapping experiments that result in an always increasing amount of data. Protein–protein interaction databases are thus becoming a major resource for investigating biological networks and pathways. In this chapter we describe the Molecular INTeraction database (MINT). The MINT database aims at storing, in a structured format, information about protein–protein interactions (PPIs) by extracting experimental details from work published in peer-reviewed journals.

Key Words: Protein–protein interaction; database; protein networks.

1. Introduction

The Molecular INTeraction Database (MINT, http://mint.bio.uniroma2.it/mint/) is a relational database designed to collect experimentally verified protein–protein interactions.

Created in 2002 *(1)*, MINT has now undergone a profound reorganization of both data model and database structure that resulted in the adoption of the IntAct relational model *(2)*. Furthermore, the number of stored interactions has dramatically increased, with more than 100,000 entries and up to 63,000 unique interactions as of January 2007 *(3)*.

With the new database structure MINT is now able to represent both binary and *n*-ary interactions (i.e., complexes) and molecule types other than protein as

From: *Methods in Molecular Biology, vol. 484: Functional Proteomics: Methods and Protocols*
Edited by: J. D. Thompson et al., DOI: 10.1007/978-1-59745-398-1, © Humana Press, Totowa, NJ

interaction participants. In addition, MINT will be compatible with toolkits for data storage, representation, and analysis developed by the IntAct consortium.

The whole interaction dataset stored in MINT is freely available at the database website (http://mint.bio.uniroma2.it/mint/download.do) in several formats: XML documents according to Proteomics Standards Initiative-Molecular Interation (PSI-MI) Level 1 and 2.5 standards *(4)*, MITAB formatted files (a tab-delimited format defined by the PSI-MI group where all complexes are represented as binary interactions; *see* **Note 4**), and a simplified tab-delimited file where all participants of an interaction are represented in a single line.

Methods are provided here for searching MINT over the Internet, exploring the interaction network using the MINT Viewer, submitting interaction data to MINT, and downloading the interaction dataset.

2. Materials

The latest version of the MINT database released in January 2007 is described here. The database can be accessed by any workstation connected to the Internet. Recent versions of common browsers supporting Java version 1.4 (or above) are recommended in order to properly visualize the protein interaction networks through the MINT Viewer. Note that the Java Virtual Machine (JVM) provided by Microsoft is not fully compatible, therefore another JVM (for instance the one provided by SUN at http://java.sun.com/j2se/downloads/index.html) should be installed on Windows machines. Mac users are strongly encouraged to use the Safari browser.

3. Methods

3.1. Searching MINT over the Internet

To access the database open the browser and connect to the MINT homepage address (http://mint.bio.uniroma2.it/mint). Then click the "Search" link in the top panel of the homepage (**Fig. 1**).

3.1.1. The Search Page

From the Search page the database can be queried using different criteria (**Fig. 2**).

1. Protein text search: users can search the database for their favorite protein by providing protein or gene names (i.e., *TP53)*, accession number (**Note 1**), or

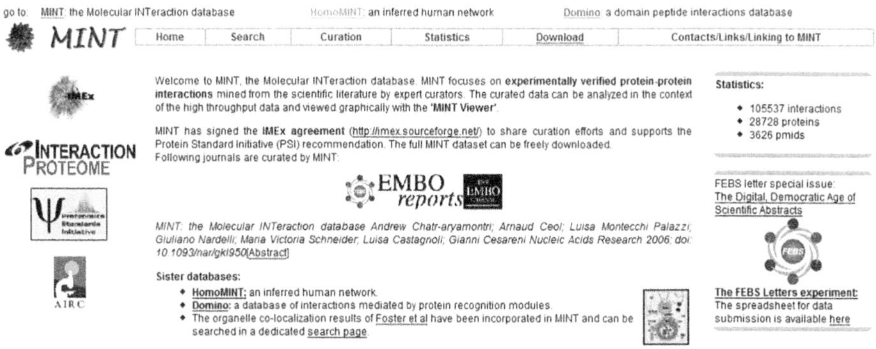

Fig. 1. The MINT database home page.

keywords (i.e., *phosphorylation* or *apoptosis*) in the corresponding text boxes. The search can be carried out on the full MINT dataset or on a given subset of the database (i.e., only mammalian proteins). This search leads to a list of interaction partners and finally to a list of experiment descriptions.

2. Interaction search by publication: it is possible to directly retrieve the list of interactions described on a given publication by entering its PubMed ID (PMID) in the corresponding text box.

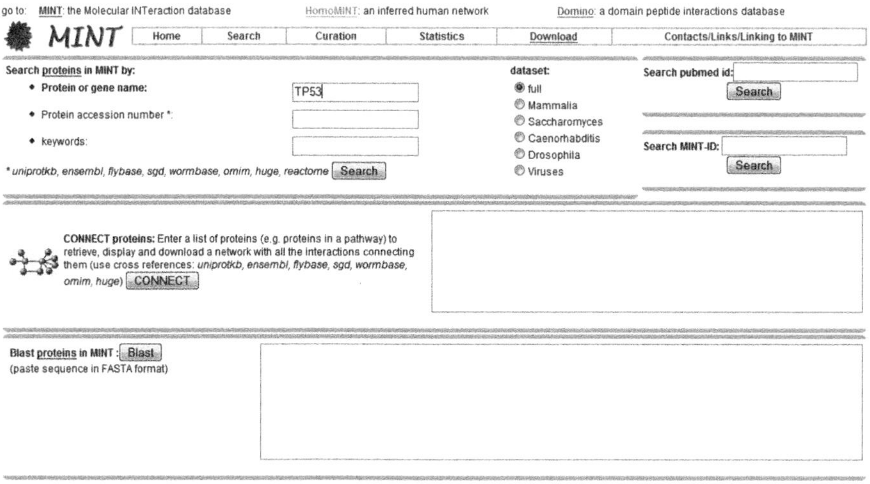

Fig. 2. The MINT search page.

3. Similarity search: the user can also use a protein sequence of interest in FASTA format to perform a BLAST search *(5)* against all the protein sequences stored in MINT. The query is performed by clicking the BLAST button.

3.1.2. The Result Page

A list of database entries matching the search criteria is returned to the user.

1. A protein search will lead to a list of interaction partners (**Fig. 3**). In case of ambiguity for a query protein, for instance where multiple proteins share a gene name or the same protein exists in different organisms, the user may select a protein of interest from a list in which molecules are briefly described by a short identification label, the source organism, their description, gene names, and domain composition. The protein of interest is selected by clicking the protein short-label.
2. A two-panel view is presented to the user (**Fig. 4**). In the left panel, a summary of the protein of interest is shown, comprising protein annotation extracted from the UniProt resource *(6)* along with cross-references to other relevant databases. Those references provide information about, for instance, diseases associated with the gene (OMIM) or the domain composition of the protein. In the right panel a list of interacting partners for the protein of interest is provided in a tabular format. The first column displays the short label, the organism, and a UniProt cross-reference for the partner. The second column reports the number of experiments documenting the interaction. The third column provides a confidence score for the interaction (**Note 2**). By clicking on the interaction number (*see* **Subheading**

Fig. 3. The results of a protein text search using as a keyword the "TP53" gene name.

Fig. 4. The two-panel view. In the left panel there is a brief summary of protein features. In the right panel all the interacting partners stored in MINT are reported.

3.1.2.2), in the left panel a short description of the interaction is provided along with the MINT interaction accession number (**Fig. 5**). Interactions are described here in their full complex composition (*see* **Note 4**).

3. Clicking on the MINT interaction accession number allows retrieval of detailed information about the experiment supporting the interaction (*see* **Step 2** in **Subheading 3.1.2**). A graph view is loaded by clicking the MINT Viewer link (or the interaction button) in the upper part of both panels. The MINT Viewer allows

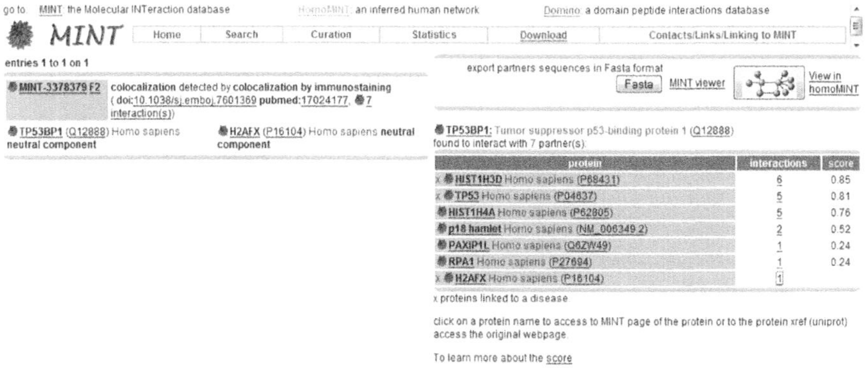

Fig. 5. By clicking on the interaction number (**Fig. 4, right panel**) in the left a short description of the interaction is provided.

Fig. 6. The results of an interaction search.

the interactive exploration of the interaction network of the protein of interest (**Subheading 3.2**).

4. In case of an interaction search (or as a result of **Subheading 3.1.2.1**), a list of interactions is presented (**Fig. 6**). The MINT interaction accession number is linked to the detailed description (**Fig. 7**) of the experiment supporting the given

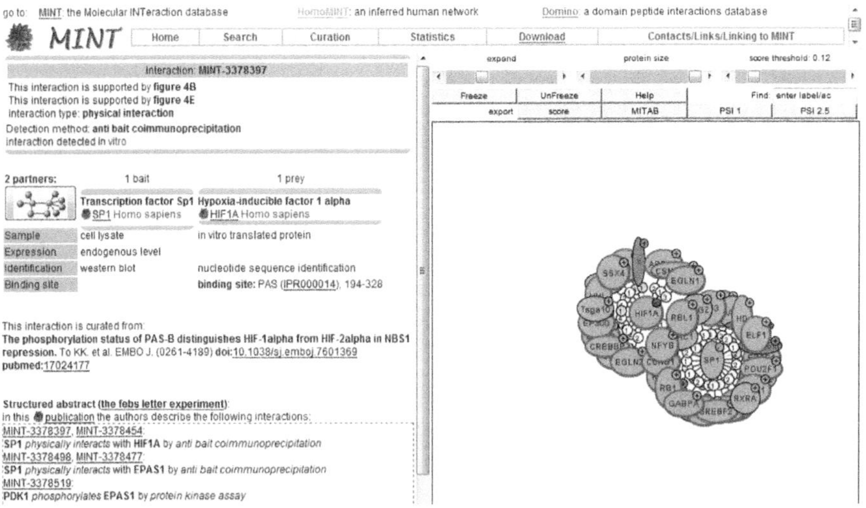

Fig. 7. A detailed description of the experiment supporting a MINT interaction.

interaction. It consists of the PubMed ID of the publication, the experimental technique used to assay the interaction, and the condition in which the experiment was carried out. Moreover, each partner is further annotated with the experimental description (experimental role, sampling process, the identification method) and biological form of the proteins (the binding site and its associated domains, the biological role, mutations and post-translation modifications).

5. In case of a similarity search, a list of proteins producing a significant alignment is returned. For each protein a short label and source organism are provided along with the BLAST bit-score and E-value. By clicking on the protein short label, the user retrieves the two-panels protein view described earlier (**Subheading 3.1.2.1**).

3.2. Visualizing Interactions with the MINT Viewer

The interactions involving a given protein are displayed graphically in the MINT Viewer, a Java applet derived from the applet Graph (http://java.sun.com). The nodes, which represent proteins, are assigned a size proportional to the protein's molecular weight and a color that depends on the species. They are linked by edges (**Fig. 8A**) that represent the interactions, and that are weighted (number on the line) according to the number and type of supporting experiments. The graph can be expanded (**Fig. 8B**) at nodes of interest (left click on "+") and edited interactively by moving or deleting nodes (right click). Proteins linked to diseases according to the OMIM database are highlighted in red. It is also possible to filter out of the network proteins with a confidence score too low by scrolling the bar named confidence score (**Fig. 8C**).

Nodes and edges are linked to the description page of the protein and the interactions they represent, respectively (described in *Steps 2* and *3* in **Subheading 3.1.2**). The resulting network can be captured in different formats: PSI-MI XML documents, MITAB (PSI-MI tab-delimited standard), and Osprey (**Note 3**).

3.3. Submitting Interaction Data

To maintain high-quality annotation of the data stored in MINT only specifically trained MINT curators are allowed to access the curation page and thus the process of submitting information into the database. Nevertheless, experimentalists are encouraged to submit their interactions to the database, by providing the results of large screening experiments in their own custom formats or by using standardized forms developed in the PSI-MI project. These forms are provided as Excel files and for each field a window menu suggests the most appropriate term.

Syntax and semantics for data representation are provided by the PSI-MI standards. The PSI-MI workgroup develops and maintains a common data standard, allowing users to retrieve all relevant data from different data providers and to perform comparative analysis.

The minimal information required to submit an interaction includes the UniProt accession numbers of the interaction partners and the PubMed ID of the article reporting the experiment that supports the interaction (**Fig. 9**). The following steps permit full description of the interactors' features and the experimental conditions.

In the interactor page (**Fig. 10**) it is possible to describe valuable information such as the protein range involved in the binding, and mutations or posttranslational modifications affecting the strength of the interaction. It is also possible to specify the expression level of the protein, whether it is tagged, and which method was used to identify the interactor.

The experiment page (**Fig. 11**) contains descriptions of the interaction detection method, the interaction type, and the model organism in which the interaction occurs.

(A)

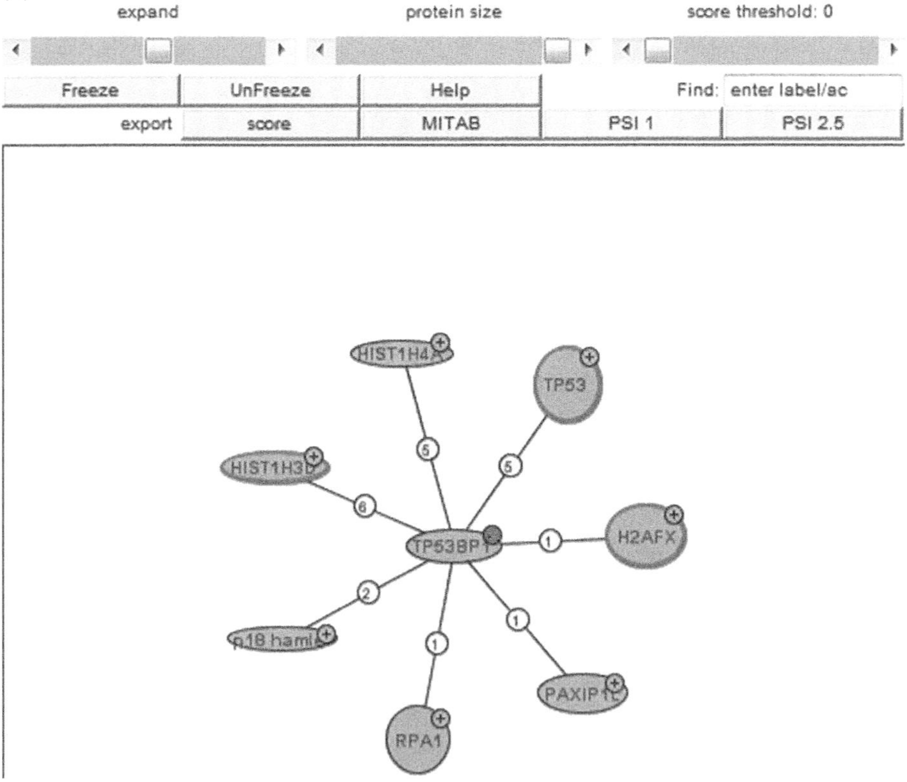

Fig. 8. The MINT Viewer allows visualizing graphically the interaction network of a given protein (**A**). The graph can be expanded (**B**) at nodes of interest. Interactions below a defined confidence score threshold can be filtered out (**C**).

(B)

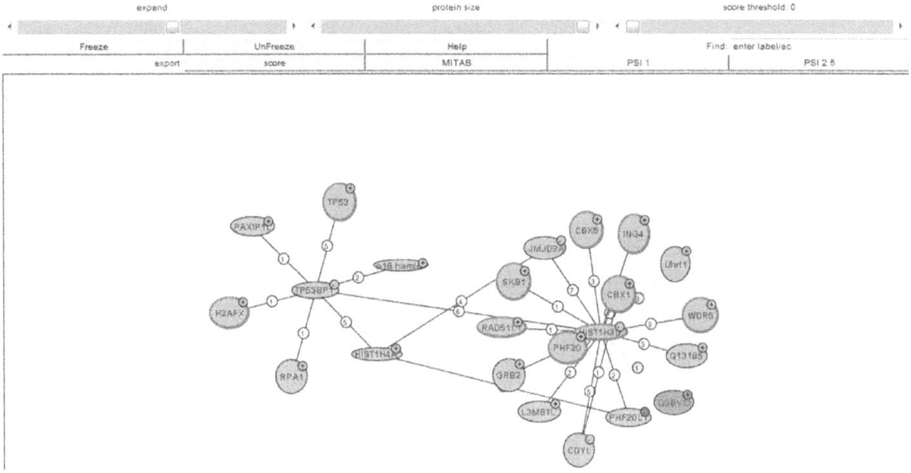

(C)

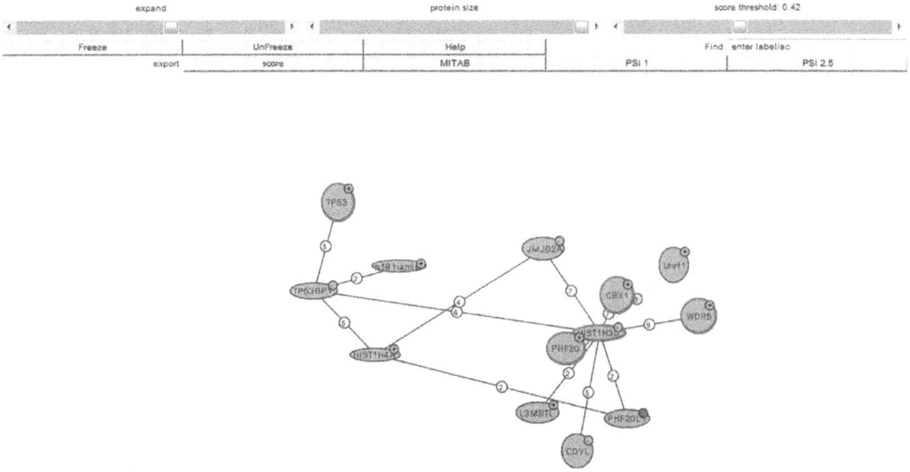

Fig. 8. (Continued)

3.4. Downloading the Interaction Dataset

Although the web-interface provides essential access to interactions for users who focus on a few proteins, MINT also makes the full dataset available for download, for further or orthogonal analyses in different formats.

The PSI-MI files are structured XML documents that aim at providing a complete representation of an experiment. Those files are not human friendly and are used either as an exchange format between databases or for being loaded in independent tools such as visualization software developed by the IntAct

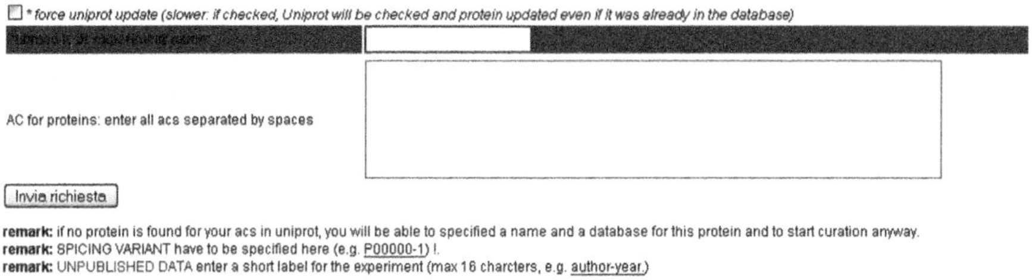

Fig. 9. The first step of the submission procedure. The curator is asked to fill a form with the minimal information required (PubMed ID and the accession numbers of the interaction partners).

consortium. Moreover, PSI-MI files use a controlled vocabulary that permits the classification and the comparison of experimental results.

The MITAB is a simple tab-delimited format that can be edited in a spreadsheet program, developed by the PSI-MI group. Since the file format is standardized, the user knows that wherever the file comes from, all columns will be on the same position and the vocabulary used will be the same. In a MITAB file, all entries are exploded into binary interactions (**Note 4**).

The MINT text file is a simplification of the MITAB format with a less detailed description of the experiment; all complexes are represented on a single line: the bait is shown in the first column and all preys in the second.

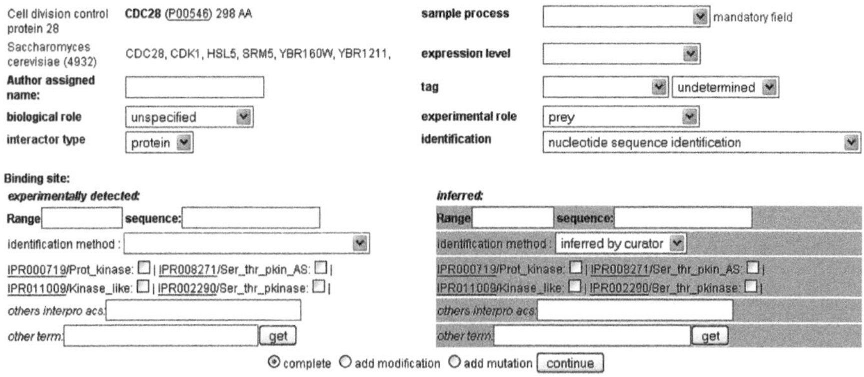

Fig. 10. The second step of the submission procedure, the interactor page, allows the curator to insert valuable information regarding the interaction partners.

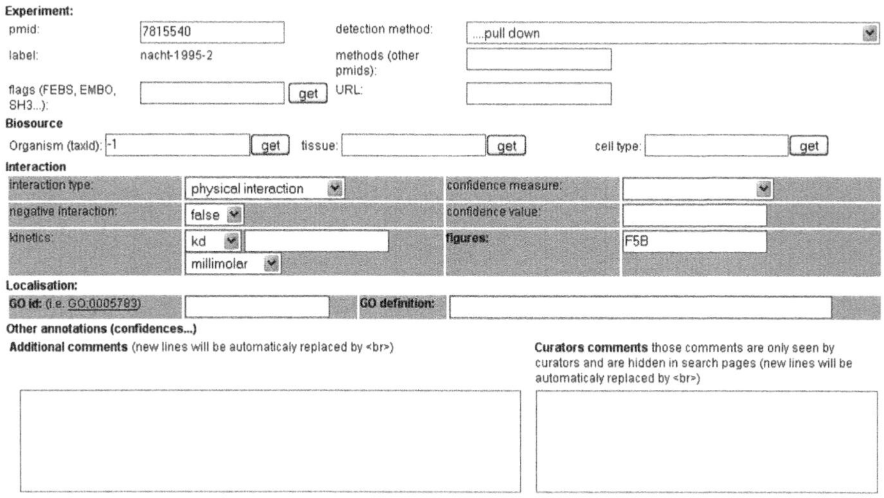

Fig. 11. The experiment page collects information regarding the interaction itself. The curator can also provide kinetics data.

4. Notes

1. MINT supports protein accession numbers from several databases such as UniProt *(6)*, ENSEMBL *(7)*, FlyBase *(8)*, SGD *(9)*, Wormbase *(10)*, HUGE *(11)*, Reactome *(12)*, and OMIM *(13)*.
2. To attribute a reliability index to the reported interactions, a confidence level has been assigned to each interaction, based on the full interaction network in MINT and on the experimental detection method and experimental conditions *(14)*. No single experimental approach has maximum sensitivity (no false negative) and specificity (no false positive), thus confidence can only be built on the integration of orthogonal experimental evidence. The score is calculated as a function of the cumulative evidence (x) according to the formula:

   $$S = 1 - a^x$$

   The Cumulative Evidence is a function of: (1) Size of the experiment. Experiments are defined large scale if the article reporting them describes more than 50 interactions otherwise they are defined small scale. (2) Interaction type. It depends on the type of experiment supporting the interaction and emphasizes evidences of direct interaction with respect to experimental support that does not provide unequivocal evidence of direct interaction, i.e co-ip, pull down etc. (3) Number of different publications (n) supporting the interaction.
3. Osprey *(15)* is a software platform for the visualization of protein networks that can be downloaded at the following URL: http://biodata.mshri.on.ca/osprey/. Osprey is available for Windows, Mac OS X, and Linux.

4. Two binary representations of a complex are used in MINT, according to the experimental role of the proteins ***(16)***. (a) In the spoke model the experiment involves one bait and many preys (for instance, tandem affinity purification); the complex is represented as all possible protein pairs involving the bait and one prey. (b) In the matrix model the role of each partner is neutral (e.g., cosedimentation); all possible pairs of protein are shown.

References

1. Zanzoni, A., Montecchi-Palazzi, L., Quondam, M., Ausiello, G., Helmer-Citterich, M., and Cesareni, G. (2002) MINT: a Molecular INTeraction database. *FEBS Lett.* **513**, 135–140.
2. Hermjakob, H., Montecchi-Palazzi, L., Lewington, C., Mudali, S., Kerrien, S., Orchard, S., Vingron, M., Roechert, B., Roepstorff, P., Valencia, A., *et al.* (2004) IntAct: an open source molecular interaction database. *Nucleic Acids Res.* **32**, D452–D455.
3. Chatr-aryamontri, A., Ceol, A., Palazzi, L. M., Lardelli, G., Schneider, M. V., Castagnoli, L., and Cesareni G. (2007) MINT: the Molecular INTeraction database. *Nucleic Acids Res.* **35**, D572–D574.
4. Hermjakob, H., Montecchi-Palazzi, L., Bader, G., Wojcik, J., Salwinski, L., Ceol, A., Moore, S., Orchard, S., Sarkans, U., von Mering, C., *et al.* (2004) The HUPO PSI's molecular interaction format—-a community standard for the representation of protein interaction data. *Nat. Biotechnol.* **22**, 177–183.
5. Altschul, S. F., Madden, T. L., Schaffer, A. A., Zhang, J., Zhang, Z., Miller, W., and Lipman, D. J. (1997) Gapped BLAST and PSI-BLAST: a new generation of protein database search programs. *Nucleic Acids Res.* **25**, 3389–3402.
6. Bairoch, A., Apweiler, R., Wu, C. H., Barker, W. C., Boeckmann, B., Ferro, S., Gasteiger, E., Huang, H., Lopez, R., Magrane, M., *et al.* (2005) The Universal Protein Resource (UniProt). *Nucleic Acids Res.* **33**, D154–D159.
7. Birney, E., Andrews, D., Caccamo, M., Chen, Y., Clarke, L., Coates, G., Cox, T., Cunningham, F., Curwen, V., Cutts, T., *et al.* (2006) Ensembl 2006. *Nucleic Acids Res.* **34**, D556–D561.
8. Grumbling, G. and Strelets, V. (2006) FlyBase: anatomical data, images and queries. *Nucleic Acids Res.* **34**, D484–D488.
9. Hirschman, J. E., Balakrishnan, R., Christie, K. R., Costanzo, M. C., Dwight, S. S., Engel, S. R., Fisk, D. G., Hong, E. L., Livstone, M. S., Nash, R., *et al.* (2006) Genome Snapshot: a new resource at the Saccharomyces Genome Database (SGD) presenting an overview of the Saccharomyces cerevisiae genome. *Nucleic Acids Res.* **34**, D442–D445.
10. Schwarz, E. M., Antoshechkin, I., Bastiani, C., Bieri, T., Blasiar, D., Canaran, P., Chan, J., Chen, N., Chen, W. J., Davis, P., *et al.* (2006) WormBase: better software, richer content. *Nucleic Acids Res.* **34**, D475–D478.

11. Kikuno, R., Nagase, T., Nakayama, M., Koga, H., Okazaki, N., Nakajima, D., and Ohara, O. (2004) HUGE: a database for human KIAA proteins, a 2004 update integrating HUGEppi and ROUGE. *Nucleic Acids Res.* **32**, D502–D504.
12. Joshi-Tope, G., Gillespie, M., Vastrik, I., D'Eustachio, P., Schmidt, E., de Bono, B., Jassal, B., Gopinath, G. R., Wu, G. R., Matthews, L., *et al.* (2005) Reactome: a knowledgebase of biological pathways. *Nucleic Acids Res.* **33**, D428–D432.
13. McKusick, V. A. (1998) *Mendelian Inheritance in Man. A Catalog of Human Genes and Genetic Disorders.* Johns Hopkins University Press, Baltimore, MD.
14. Chatr-Aryamontri, A., Ceol, A., Licata, L., and Cesareni, G. (2008) Protein interactions: integration leads to belief. *Trends Biochem Sci.* May 8, 2008.
15. Breitkreutz, B. J., Stark, C., and Tyers, M. (2003) Osprey: a network visualization system. *Genome Biol.* **4**, R22.
16. Bader, G. D. and Hogue, C. W. (2002) Analyzing yeast protein-protein interaction data obtained from different sources. *Nat. Biotechnol.* **20**, 991–997.

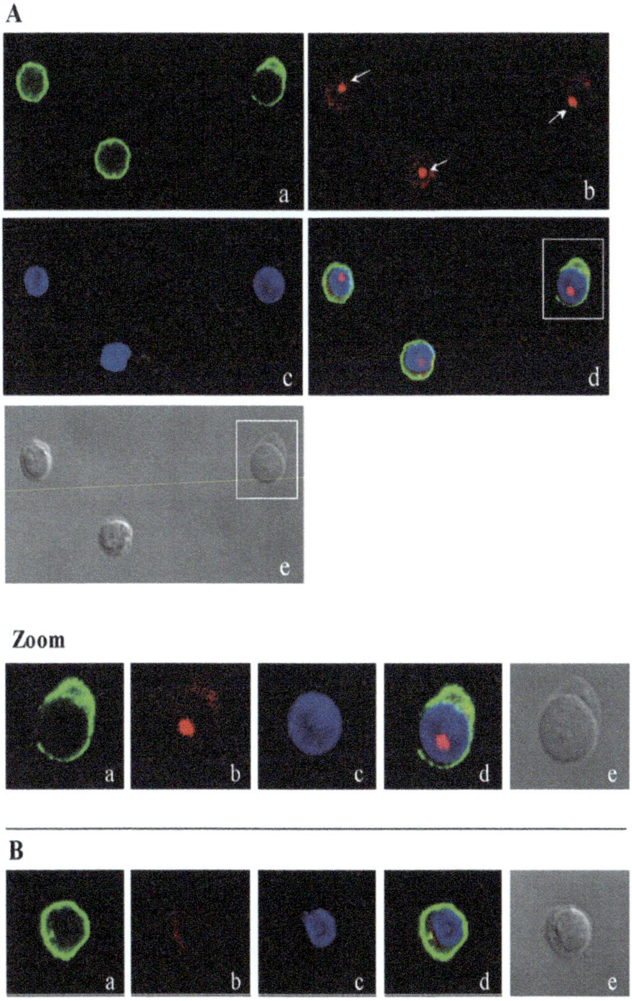

Color Plate 1. Confocal microscopy of $CD8^+$ T cells: (**A**) membrane is stained in green with α-CD45 (**a**), nucleolus in red with α-nucleolin (**b**, arrows), and nucleus in blue with DAPI (**c**). (**d**) The merged and (**e**) the differential interference contrast (DIC) images are shown. A zoom of the cell inside the white square is also shown. (**B**) Membrane of another $CD8^+$ T cell is stained in green with α-CD45 (**a**), actin in red with α-actin (**b**), and nucleus in blue with DAPI (**c**). (**d**) The merged and (**e**) the differential interference contrast (DIC) images are shown. (Fig. 5, Chapter 4; *see* discussion on p. 64).

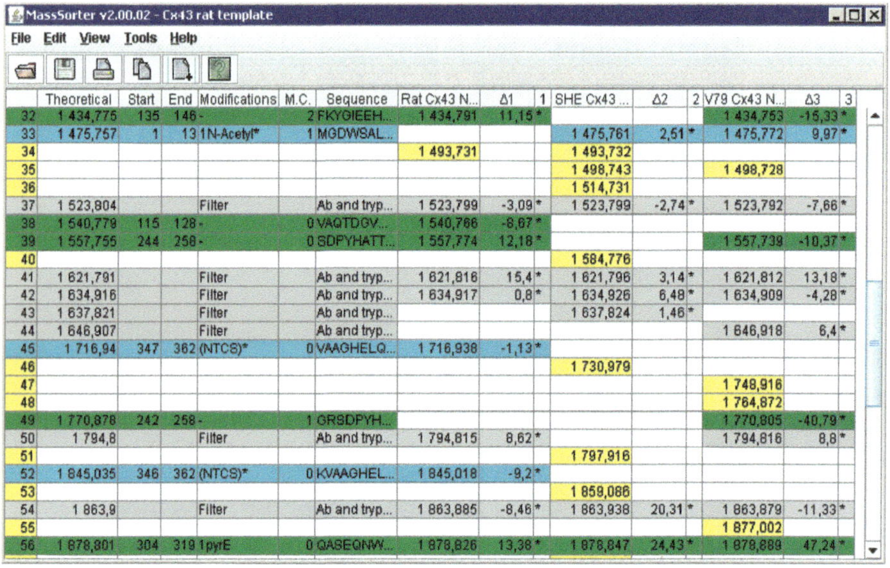

	Theoretical	Start	End	Modifications	M.C.	Sequence	Rat Cx43 N...	Δ1	1	SHE Cx43 ...	Δ2	2	V79 Cx43 N...	Δ3	3
32	1 434,775	135	146	-	2	FKYGIEEH...	1 434,791	11,15	*				1 434,753	-15,33	*
33	1 475,757	1	13	1N-Acetyl*	1	MGDWSAL...				1 475,761	2,51	*	1 475,772	9,97	*
34							1 493,731			1 493,732					
35										1 498,743			1 498,728		
36										1 514,731					
37	1 523,804			Filter		Ab and tryp...	1 523,799	-3,09	*	1 523,799	-2,74	*	1 523,792	-7,66	*
38	1 540,779	115	128	-	0	VAQTDGV...	1 540,766	-8,67	*						
39	1 557,755	244	258	-	0	SDPYHATT...	1 557,774	12,18	*				1 557,739	-10,37	*
40										1 584,776					
41	1 621,791			Filter		Ab and tryp...	1 621,816	15,4	*	1 621,796	3,14	*	1 621,812	13,18	*
42	1 634,916			Filter		Ab and tryp...	1 634,917	0,8	*	1 634,926	6,48	*	1 634,909	-4,28	*
43	1 637,821			Filter		Ab and tryp...				1 637,824	1,46	*			
44	1 646,907			Filter		Ab and tryp...							1 646,918	6,4	*
45	1 716,94	347	362	(NTCS)*	0	VAAGHELQ...	1 716,938	-1,13	*						
46										1 730,979					
47													1 748,916		
48													1 764,872		
49	1 770,878	242	258	-	1	GRSDPYH...							1 770,805	-40,79	*
50	1 794,8			Filter		Ab and tryp...	1 794,815	8,62	*				1 794,816	8,8	*
51										1 797,916					
52	1 845,035	346	362	(NTCS)*	0	KVAAGHEL...	1 845,018	-9,2	*						
53										1 859,086					
54	1 863,9			Filter		Ab and tryp...	1 863,885	-8,46	*	1 863,938	20,31	*	1 863,879	-11,33	*
55													1 877,002		
56	1 878,801	304	319	1pyrE	0	QASEQNW...	1 878,826	13,38	*	1 878,847	24,43	*	1 878,889	47,24	*

Color Plate 2. A fraction of the DST comparing Cx43 from rat, Syrian hamster, and Chinese hamster. For simplicity, only one sample is shown for each species. The rows 33, 45, 46, 47, 52, 53, and 55 are specifically mentioned in the text. Rows 32 and 38 are examples of unmodified matches. Row 56 is an example of a modified match and row 37 corresponds to a filter peak. Row 35 is an example of two experimental masses that are unmatched, but identical to each other within the chosen accuracy (in this case 50 ppm). (Fig. 2, Chapter 23; *see* discussion on p. 352).

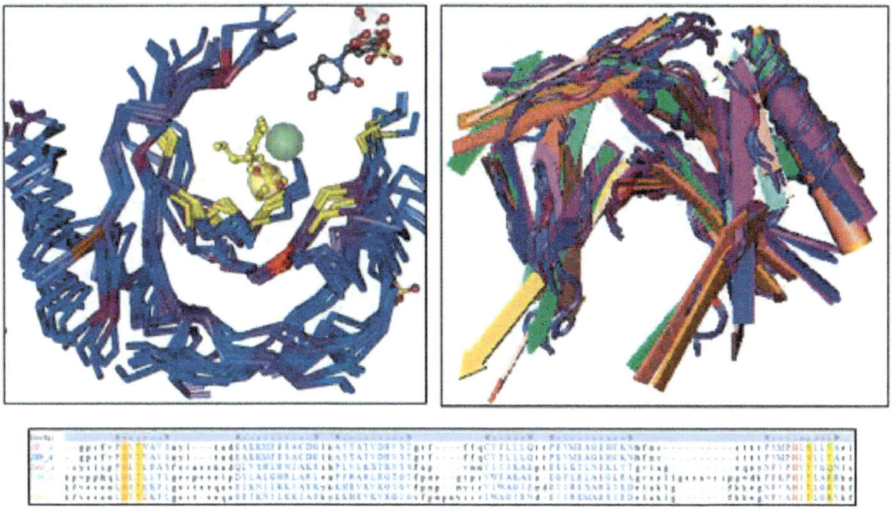

Color Plate 3. Structure-guided alignment and superposition using Cn3d showing the conserved regions and conserved binding residues. (Fig. 4, Chapter 28; *see* discussion on p. 483).

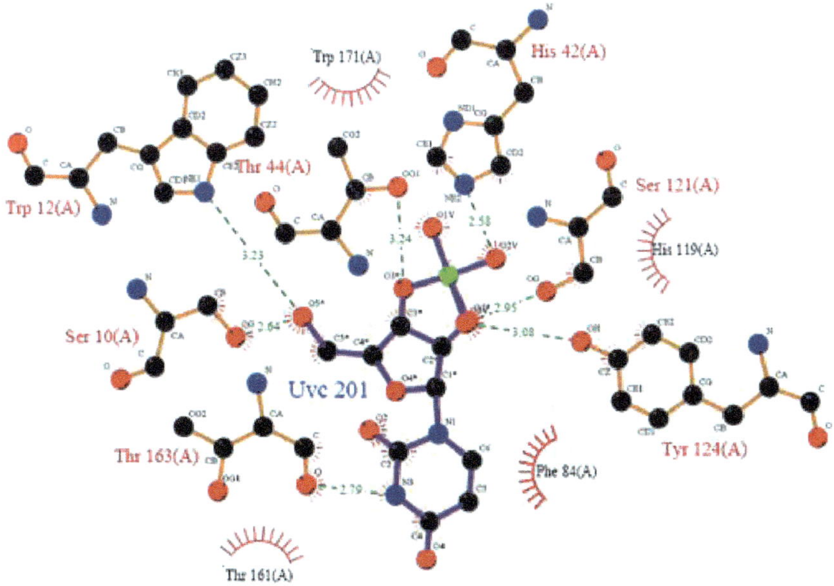

Color Plate 4. Protein–ligand interactions using Ligplot showing the residues involved in binding to the ligand uridine-2′,3′-vanadate. (Fig. 5, Chapter 28; *see* discussion on p. 484).

21

PepSeeker: Mining Information from Proteomic Data

Jennifer A. Siepen, Julian N. Selley, and Simon J. Hubbard

Summary

Driven by advances in mass spectrometry and analytical chemistry, coupled with the expanding number of completely sequenced genomes, proteomics is becoming a widely exploited technology for characterizing the proteins found in living systems. As proteomics becomes increasingly more high-throughput there is a parallel need for storage of the large quantities of data generated, to support data exchange and allow further analyses. The capture and storage of such data, along with subsequent release and dissemination, not only aid in sharing of the data throughout the proteomics community but also provide scientific insights into the observations between different laboratories, instruments, and software. Growing numbers of resources offer a range of approaches for the capture, storage, and dissemination of proteomic experimental data reflecting the fact that proteomics has now come of age in the postgenomic era and is delivering large, complex datasets that are rich in information. This chapter demonstrates how one such resource, PepSeeker, can be used to mine useful information from proteomic data, which can then be exploited for peptide identification algorithms via a better understanding of how peptides fragment inside mass spectrometers.

Key Words: Mass spectrometry; ion fragmentation; peptide identification; proteomic databases.

1. Introduction

Proteomics is self-evidently the technique of choice for scientists wishing to study the proteins present in cells and tissues. Although the level of mRNA transcripts can be monitored via the expanding microarray-based techniques currently available, proteins are the functional molecules in the cell and are usually the focus of target discovery and drug design. Although there are a

From: *Methods in Molecular Biology, vol. 484: Functional Proteomics: Methods and Protocols*
Edited by: J. D. Thompson et al., DOI: 10.1007/978-1-59745-398-1, © Humana Press, Totowa, NJ

growing group of protein array technologies becoming available, proteins do not share the simple base-pairing rules of nucleotides that have enabled recombinant technologies in DNA/RNA systems and protein arrays are more complex. Instead, a large component of the proteomics field relies on mass spectrometry (MS) as an analytical technique. Indeed, MS and tandem mass spectrometry (MS/MS) have proved invaluable in the identification of peptides and proteins in biological samples.

2. Methods

Mass spectrometers are used to measure the mass-to-charge (*m/z*) ratio of proteins and, more usually, peptides (from enzymatic cleavage and/or chemical digestion) and/or peptide fragment ions to produce a characteristic mass spectra. A theoretical spectrum is shown in **Fig. 1**, showing how a peptide can be fragmented into constituent ion series; in most instances, this is typically a list of *b* and *y* ions, subtended at the peptide amino- and carboxy-terminus, respectively. The mass spectrum is essentially a list of *m/z* values and corresponding peak intensities, which can be compared to theoretical spectra from a database of known sequences to find the sequence that best matches the experimental spectrum, using a variety of popular database search tools ***(1–5)***. The ability of these tools to identify peptides and proteins relies upon an understanding of how molecules are first ionized, activated, and detected, and second, in tandem MS, the chemistry of the gas phase: which bonds are broken and the factors that may affect this. The main goal of PepSeeker is to provide a framework for mass spectrometrists and proteome scientists to investigate these phenomena, and to analyze the patterns observed in real peptide spectra that have produced high-quality peptide identifications. This section will provide a very brief introduction to the different stages of an MS experiment through to the protein identification stage, which are relevant to the data and queries users can perform in PepSeeker. This context is essential in order to appreciate the data contained in the repository, and to construct sensible queries and mine the database for patterns and information. For more details on the general techniques involved in acquiring peptide identifications and mass spectrometry data, consult other chapters in this volume.

2.1. Sample Preparation, Ionization, and Mass Analysis

Extracted proteins may be analyzed directly or separated via liquid chromatography or gel electrophoresis, either 1 or 2D gels; prior to hydrolysis protein samples are typically digested with a proteolytic enzyme to generate constituent peptides prior to MS analysis. Usually this enzyme is trypsin,

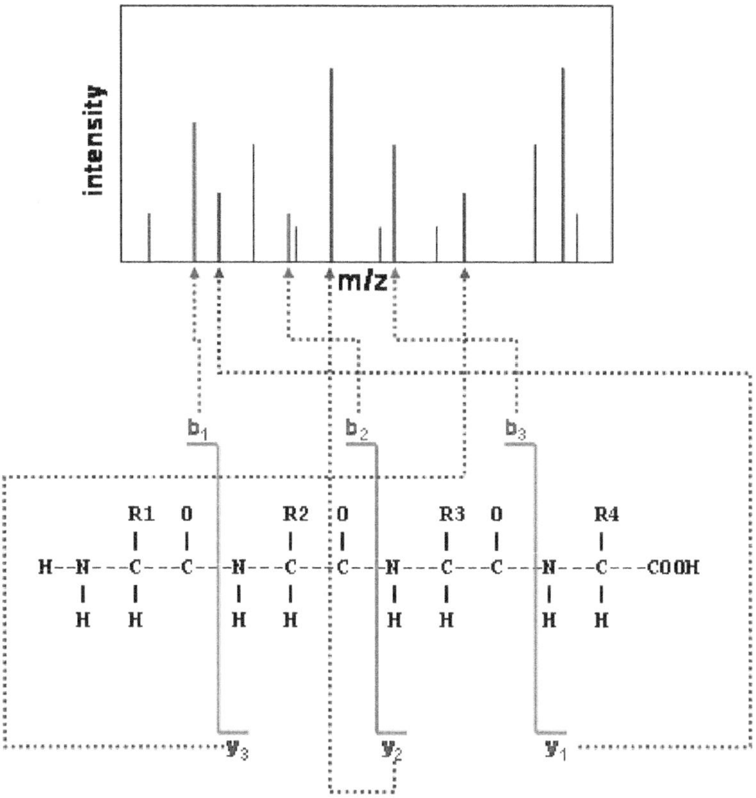

Fig. 1. A theoretical mass spectrum showing how fragmentation at peptide bonds leads to ion *b* and *y* series, which are characteristic of a given amino acid sequence. "R" represent the characteristic amino acid side chains.

which will normally cleave at every peptide bond C-terminal to arginine and lysine amino acids, except where either of these residues is followed by a proline residue. In practice, complete digestion by the protease is not always achieved and "missed cleavages" are also observed, where some peptide bonds susceptible to proteolysis are not cleaved. These tryptic peptides can then be separated by, for example, capillary electrophoresis *(6,7)* or MuDPIT (multidimensional protein identification technology) *(8)*, prior to ionization and analysis in the mass spectrometer. The two most widely used ionization techniques in proteomics are electrospray ionization (ESI), often following a chromatographic method directly coupled to MS, and matrix-assisted laser desorption/ionization (MALDI). Typically ESI induces a range of charge states, whereas only singly charged ions are observed in MALDI.

2.2. MS/MS Fragmentation

The development of tandem MS and potentially MS^n has provided a powerful identification strategy that has become the method of choice for most proteomics laboratories. Specific peptide ions are selected following the first round of MS and then fragmented further by methods such as gas-phase activation or electron capture disassociation (ECD) and the *m/z* of the fragment ions measured. Fragmentation of a peptide is believed to occur through charge-directed pathways *(9)*. In the absence of solvent in the gas phase the carbonyl oxygen of the backbone can effectively act as a solvent, facilitating the transfer of mobile protons to cleavage sites throughout the peptide. Cleavage can occur at different bonds along the peptide backbone leading to different types of ion, which are summarized in **Fig. 1**. Typically cleavage occurs at the amide bond, producing *b* ions if the amino-terminal retains the charge or *y* ions if the carboxy-terminal fragment retains the charge. Where the peptide is multiply charged (2+ or higher), cleavage can occur leading to complementary ion pairs; for example, a doubly charged ion fragment can produce a *b*/*y* ion pair, although both ion types are not always detected in equal abundance due to instrument variability or their stability against further fragmentation. The different ion types, *y* and *b*, can also have neutral losses; these include the loss of NH_3 and H_2O groups, both of which cause a shift in the peak on the resulting mass spectrum and need to be considered in the identification *(10)*.

2.3. Spectra Interpretation

The types of ions that are observed in a spectrum are very much dependent upon the instrumentation used *(10)*, the peptide sequence *(10)*, and many other factors associated with the experiment. Although these processes are, in some part, quite well understood, much is still unknown concerning the mechanisms through which certain amino acid combinations lead to suppressed or promoted fragmentation at given peptide bonds. However, an understanding of the fragmentation pathways promoted or induced in the gas phase can lead to improvements in the peptide identification algorithms. There are a number of different scoring systems available to match experimental spectra to theoretical spectra. Some examples include Mascot from Matrix Science *(2)*, Sequest from Thermo Finnigan *(3)*, X!Tandem *(1)*, Phenyx *(5)*, and OMMSA *(4)*, among others. These scoring systems predominantly ignore the actual intensity of the ions observed in daughter ion spectra, and rely largely on just the *m/z* values in order to compare experimental spectra to theoretical ones derived from a database of candidate protein sequences. The scoring systems usually provide some tool-specific score, as well as some likelihood that each match was achieved by chance (e.g., an expectation value), both of which are used as a measure of the quality of the identification. At the time of development, no single score or consistent probability value was available from all the search tools. PepSeeker captures minimally the tool-dependent score (usually the Mascot Ion Score) along with some likelihood measure such as an expectation value that the peptide identification was not a chance one. In some cases, a further probabilistic p-value derived from the PeptideProphet tool *(11)* is also available as a measure of quality.

The identification process is further complicated by the presence of posttranslational modifications (PTMs), which may or not be present on the peptide. An exhaustive search of all possible PTMs is far too computationally expensive; as a result, search engines usually allow the user to search for a small number of these in each given search. Again, these are captured from search engine output by PepSeeker.

2.4. Proteomics Databases

The growth in proteomic technologies has led to the development of a number of repositories *(12–17)*, with a parallel drive to develop standard reporting formats for exchange and data capture needs *(18)*. Data sharing between different laboratories offers the potential for the discovery of valuable insights into the underlying chemistry and also the reduction of repetition between experiments. The growing numbers of repositories essentially capture the same information, although differing in their primary focus and each supporting different formats. Data standards for mass spectrometric data and molecular interactions have matured in proteomics *(18,19)*, but the identifications standard is still currently a work in progress. Until this is resolved, each repository offers the user something different, providing a wealth of information on related experiments performed in laboratories throughout the world. The principal proteomic databases contain a combination of the original spectra, in a variety of different formats that include mzData (the standard format from the Proteomics standard intiative [PSI] *(18,19)*), mzXML *(20)* (from the Institute of Systems Biology in Seattle), and other formats including nonstandard XML and MySQL. Some databases also contain the protein and peptide identifications from individual experiments in a variety of instrument/search tool-specific formats. All of the databases enable searching of the data at varying levels of detail, from simple searches relating to only specific experimental details to complex searches at the peptide level. Some data repositories offer even more complex searching, for example, PepSeeker *(16)* is the only repository to enable complex queries of the fragment ions produced in the mass spectrometer and identified by the search engine. This chapter will focus on PepSeeker *(16)* as an example of why these data resources can provide a useful tool in developing the field of proteomics.

3. The PepSeeker Database

3.1. Motivation and Focus

Given the interest in investigating the peptide fragmentation patterns observed in the gas phase, PepSeeker focuses on peptide identifications and associated fragment ion information, as well as basic details on the putative protein, experimental spectra, and search parameters. The PepSeeker database schema is shown in **Fig. 2**.

The current implementation of PepSeeker has been developed using a MySQL platform with a schema designed to capture identification data obtained primarily from a local Mascot-based proteomics pipeline. The schema includes information concerning the search parameters, the original spectra, protein and peptide identifications, and the fragment ion details. A second database, PepSeekerGOLD, has also been developed alongside PepSeeker. This database contains only high-quality identifications, whereby only top-ranking peptides with an expectation score of better than 0.05 are considered. This database is considerably smaller and as a result much quicker to query. Recently an improved interface to the PepSeeker and PepSeekerGOLD databases has been developed using BioMart *(21)* to enable enhanced search capabilities.

3.2. BioMart

BioMart *(21)* has been developed jointly by the European Bioinformatics Institute (EBI) and the Cold Spring Harbor Laboratory. It is a query-oriented data management system that enables a range of advanced query interfaces and administration tools. It can be downloaded from http://www.ebi.ac.uk/biomart.

BioMart consists of three tiers; the first is a set of one or more relational databases. Each of the databases contains one or more marts that, in turn, can contain a number of individual datasets. For PepSeeker there are two databases, one for PepSeeker and a second for PepSeekerGOLD. Each of these contains several marts that were built using the martBuilder tool, including, for example, a peptide mart. This mart contains all of the information directly connected to each of the peptide identifications, including the protein identified, posttranslational modifications, and precursor ions. Each mart has an associated dataset that defines what is seen on the interface, including the optional search parameters and outputs to be included in the results. Individual marts can also be connected. More complex queries can be implemented by adding extra columns to the underlying mart. For example, the addition of a single column to the mart was done to enable searches for unique peptide sequences in PepSeeker—a specific query expected to be popular with users.

The second tier of BioMart consists of the application programming interface (API), which in the case of PepSeeker was Perl based. Finally, the third tier consists of the query interface and has different instances including a stand alone GUI tool, a web services tool, and a web browser interface. The latter was implemented for the PepSeeker databases and is shown in **Fig. 3**.

3.3. The Query Interface

The query interface, shown in **Fig. 3**, has been designed with users in mind to allow complex searching of the data. **Fig. 3** demonstrates how the PepSeeker database can be used to build complex queries and the different ways in which the results can be presented. This supports the idea that the PepSeeker repository provides a means to explore the ion fragmentation patterns in mass spectrometry at the amino acid level over many thousands of different spectra. A comprehensive explanation of all the possible queries and features is beyond the scope of this chapter, but **Fig. 3** shows a stepped walk-through of a query, which is essentially self-explanatory and demonstrates many of PepSeeker's features.

3.4. PepSeeker Applications

The basis of MS identification methods involves the correlation or comparison of experimental spectra with theoretical spectra of proteolytic peptides derived from sequenced proteins, evaluating the similarity between fragment ions produced in the experimental and theoretical spectra. The interpretation of MS/MS spectra continues to improve as advances are made in the understanding of peptide chemistry.

As discussed earlier in this chapter, cleavage of the peptide backbone occurs typically at the amide bond, producing *b* ions if the amino-terminal fragment retains the charge, or *y* ions if the carboxy-terminal fragment retains the charge ***(10)***. Other types of ions are also observed and these include *a* ions, corresponding to the loss of CO from a *b* ion. Which ion types are observed in an MS^n experiment varies depending on a number of factors including the peptide, the activation step, the instrument's observation time frame, and/or the instrument discrimination factors ***(10)***. An advantage of the PepSeeker database ***(16)*** is that the observed peptide fragmentation patterns are retained in addition to the peptide and instrument information. The resource therefore makes it possible over a large data set to fully investigate peptide fragmentation patterns in relation to the peptide sequence, the instrument, and other phenomena that affect the fragmentation. An example is discussed below.

3.4.1. The Proline Effect

The proline effect describes the abundance of intense fragment ions formed by preferential fragmentation of a peptide N-terminal to a proline residue ***(22,23)***.

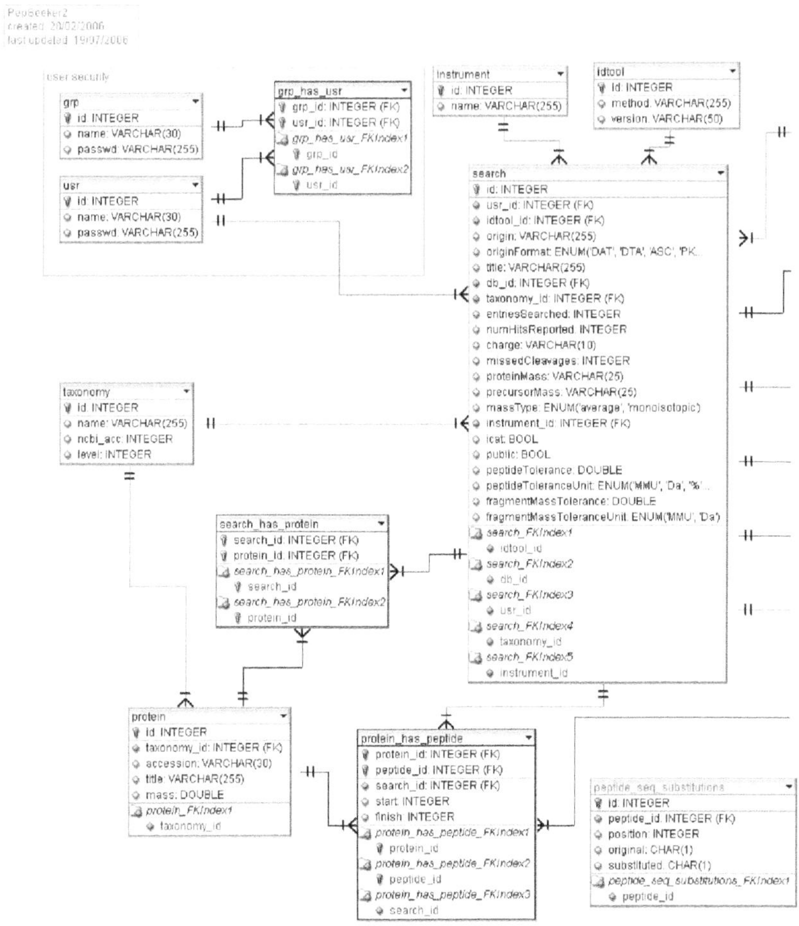

Fig. 2. The PepSeeker database schema, showing the tables of the database and the relationships between them.

Protonated peptides containing proline are known to exhibit distinct fragmentation patterns upon collision-induced dissociation (CID) ***(24)***, which seem to be due to a combination of factors including the effect on the ion structure and the high proton affinity of the proline residue ***(24)***.

Breci and colleagues ***(22)*** investigated fragmentation patterns N-terminal to proline. They found that cleavage at the Xxx-Pro bond formed more readily than at other locations in the peptides. They had a database of 316 peptides

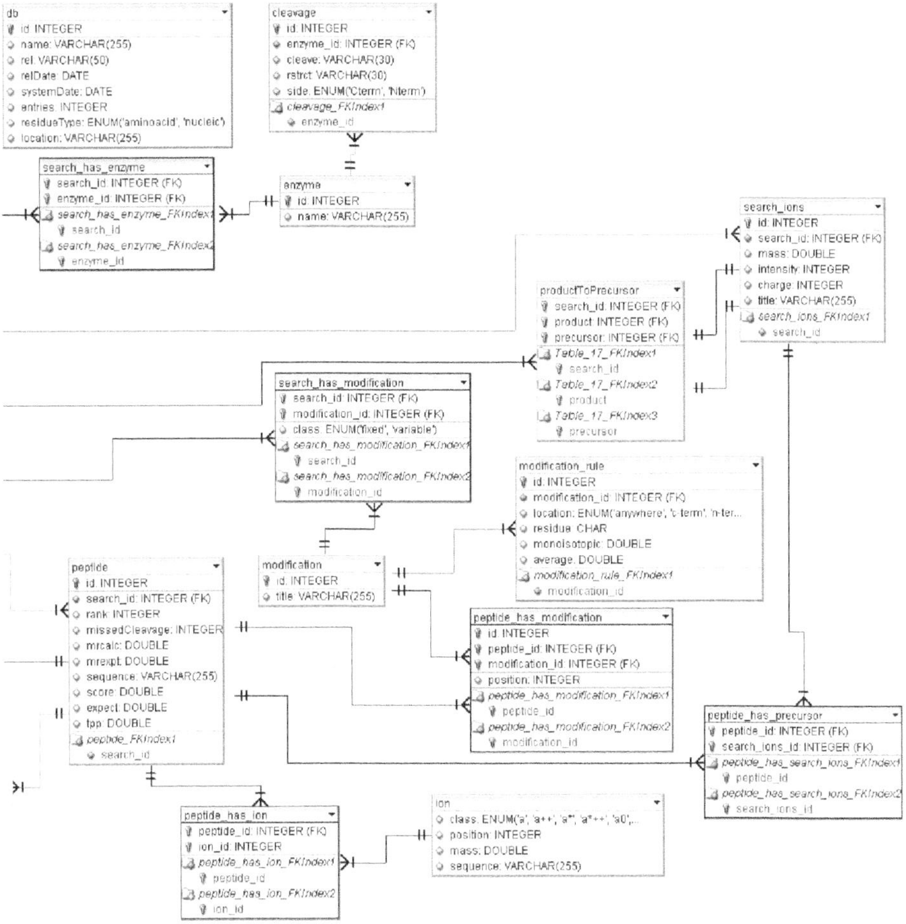

Fig. 2. (Continued)

investigated for Pro-Xxx cleavage and 5126 peptides to investigate Xxx-Pro fragmentation. They found that 36.3% of the total *a*, *b*, and *y* ion intensity was due to cleavage at the Xxx-Pro bonds in proline-containing peptides. They investigated in detail the amino acids surrounding the fragmentation and saw some interesting patterns. Although currently it is challenging to match individual ion intensities to each peptide identification in PepSeekerGOLD, a study similar to that described above ***(22)*** can be performed to investigate

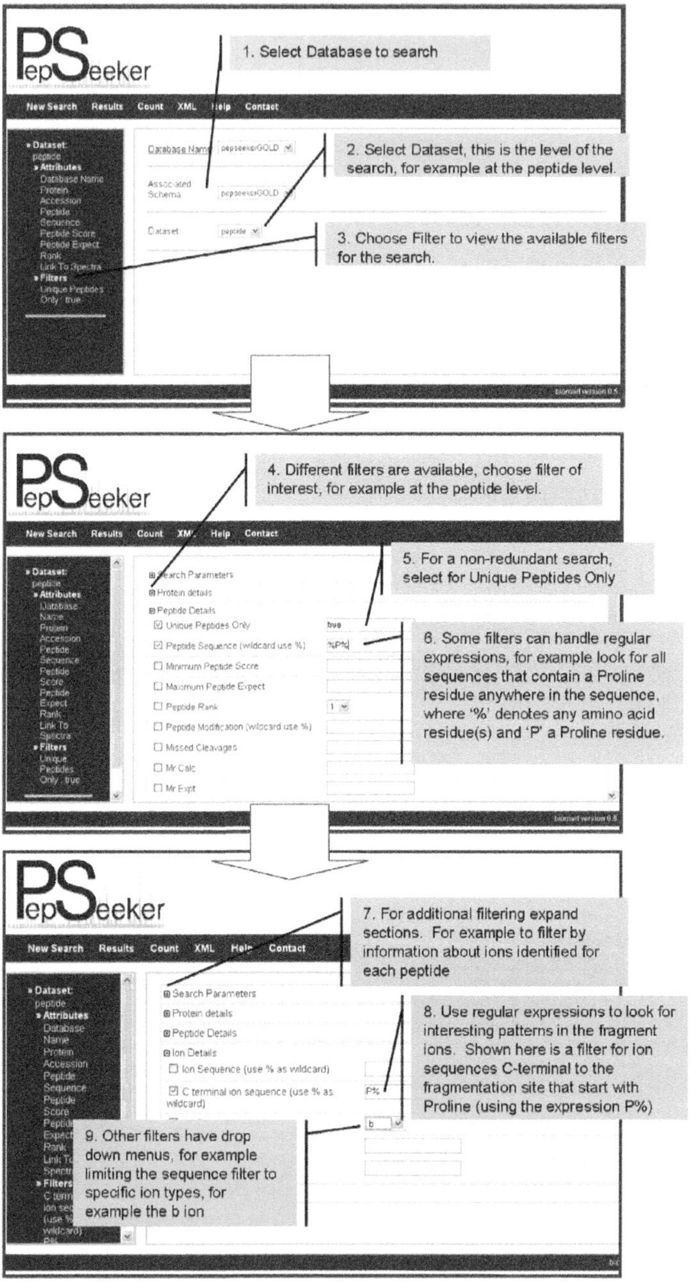

Fig. 3. Screen shots of PepSeeker, demonstrating how the BioMart interface can be used to implement complex queries through specific filters and the different ways in which the results can be viewed.

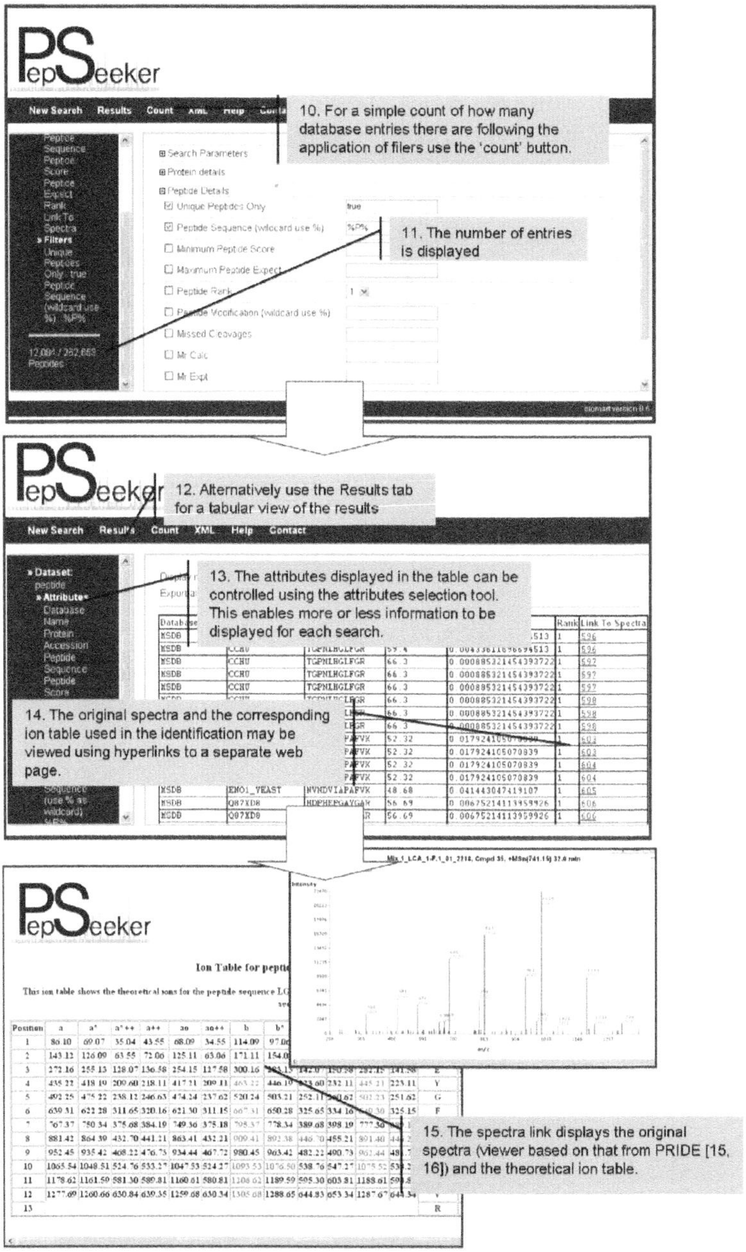

Fig. 3. (Continued).

patterns based on the number of fragment ions observed in PepSeeker, although in this case over a much large dataset. The clear advantage of PepSeeker for such a study is the number of high confidence peptide identifications. There are over 11,000 proline-containing nonredundant peptides in PepSeeker with better than 95% confidence as estimated from the associated Mascot expectation values. A similar study using the PepSeeker interface (shown in **Fig. 3**) reveals that in PepSeekerGOLD a little over 12% of proline-containing peptides show fragmentation at Xxx-proline, with the next largest fragmentation occurring at Xxx-leucine in 7% of these peptides. A preliminary look at the amino acid residues surrounding the cleavage site suggests that leucine–proline, alanine–proline, and valine–proline are the three most abundant fragmentation patterns in PepSeekerGOLD at Xxx-Pro and methionine–proline, cysteine–proline, and tryptophan-proline are the three least common patterns. These findings are similar to those of Breci and colleagues *(22)* in which valine–proline had the highest relative bond cleavage ratio, whereas cysteine–proline and methionine–proline had the lowest.

4. Notes

The intention of this chapter is to show the reader how the PepSeeker database can be mined to gain information on the fragmentation patterns observed in peptides in the gas phase, as part of wider proteomics projects. The "proline effect" presented here provides a good example of this. The use of the BioMart interface built on top of the current schema supports simple queries that can return large and complex datasets readily, and in user-definable formats. PepSeeker itself also provides a simple spectral viewer that allows the user to browse the peptide identification in more detail, examining the relative peak heights of the fragment ions of interest. We hope this will be of use to mass spectrometrists who wish to validate their own data, and will be of general interest to the proteomics community. The PepSeeker database can be found at http://www.ispider.manchester.ac.uk/pepseeker.

Acknowledgments

The authors would like to thank the BioMart team at the EBI for helpful advice from their mailing list. This work has been supported by several BBSRC grants to the authors, ISPIDER (BBSB17204) to J.A.S. and S.J.H., EGM17685 to S.J.H., and BBD0069961 to J.N.S.

References

1. Craig, R. and Beavis, R. C. (2004) TANDEM: matching proteins with tandem mass spectra. *Bioinformatics* **20,** 1466–1467.
2. Perkins, D. N., Pappin, D. J. C., Creasy, D. M., and Cottrell, J. S. (1999) Probability-based protein identification by searching sequence databases using mass spectrometry data. *Electrophoresis* **20,** 3551–3567.
3. Eng, J. K., Mccormack, A. L., and Yates, J. R. (1994) An approach to correlate tandem mass-spectral data of peptides with amino-acid-sequences in a protein database. *J. Am. Soc. Mass Spectrom.* **5,** 976–989.
4. Geer, L. Y., Markey, S. P., Kowalak, J. A., Wagner, L., Xu, M., Maynard, D. M., Yang, X. Y., Shi, W. Y., and Bryant, S. H. (2004) Open mass spectrometry search algorithm. *J. Proteome Res.* **3,** 958–964.
5. Colinge, J., Masselot, A., Cusin, I., Mahe, E., Niknejad, A., Argoud-Puy, G., Reffas, S., Bederr, N., Gleizes, A., Rey, P. A., and Bougueleret, L. (2004) High-performance peptide identification by tandem mass spectrometry allows reliable automatic data processing in proteomics. *Proteomics* **4,** 1977–1984.
6. Guo, T., Lee, C. S., Wang, W. J., DeVoe, D. L., and Balgley, B. M. (2006) Capillary separations enabling tissue proteomics-based biomarker discovery. *Electrophoresis* **27,** 3523–3532.
7. Huang, Y. F., Huang, C. C., Hu, C. C., and Chang, H. T. (2006) Capillary electrophoresis-based separation techniques for the analysis of proteins. *Electrophoresis* **27,** 3503–3522.
8. Kislinger, T., Gramolini, A. O., MacLennan, D. H., and Emili, A. (2005) Multidimensional protein identification technology (MudPIT): technical overview of a profiling method optimized for the comprehensive proteomic investigation of normal and diseased heart tissue. *J. Am. Soc. Mass Spectrom.* **16,** 1207–1220.
9. Yates, J. R. (1998) Database searching using mass spectrometry data. *Electrophoresis* **19,** 893–900.
10. Wysocki, V. H., Resing, K. A., Zhang, Q. F., and Cheng, G. L. (2005) Mass spectrometry of peptides and proteins. *Methods* **35,** 211–222.
11. Keller, A., Nesvizhskii, A. I., Kolker, E., and Aebersold, R. (2002) Empirical statistical model to estimate the accuracy of peptide identifications made by MS/MS and database search. *Anal. Chem.* **74,** 5383–5392.
12. Craig, R., Cortens, J. P., and Beavis, R. C. (2004) Open source system for analyzing, validating, and storing protein identification data. *J. Proteome Res.* **3,** 1234–1242.
13. Desiere, F., Deutsch, E. W., King, N. L., Nesvizhskii, A. I., Mallick, P., Eng, J., Chen, S., Eddes, J., Loevenich, S. N., and Aebersold, R. (2006) The PeptideAtlas project. *Nucleic Acids Res.* **34,** D655–D658.
14. Jones, P., Côté, R. G., Martens, L., Quinn, A. F., Taylor, C. F., Derache, W., Hermjakob, H., and Apweiler, R. (2006) PRIDE: a public repository of protein and peptide identifications for the proteomics community. *Nucleic Acids Res.* **34,** D659–D663.

15. Martens, L., Hermjakob, H., Jones, P., Adamski, M., Taylor, C., States, D., Gevaert, K., Vandekerckhove, J., and Apweiler, R. (2005) PRIDE: the proteomics identifications database. *Proteomics* **5,** 3537–3545.
16. McLaughlin, T., Siepen, J. A., Selley, J., Lynch, J. A., Lau, K. W., Yin, H. J., Gaskell, S. J., and Hubbard, S. J. (2006) PepSeeker: a database of proteome peptide identifications for investigating fragmentation patterns. *Nucleic Acids Res.* **34,** D649–D654.
17. Prince, J. T., Carlson, M. W., Wang, R., Lu, P., and Marcotte, E. M. (2004) The need for a public proteomics repository. *Nat. Biotechnol.* **22,** 471–472.
18. Taylor, C. F., Hermjakob, H., Julian, R. K., Garavelli, J. S., Aebersold, R., and Apweiler, R. (2006) The work of the Human Proteome Organisation's Proteomics Standards Initiative (HUPO PSI). *OMICS* **10,** 145–151.
19. Hermjakob, H., Montecchi-Palazzi, L., Bader, G., Wojcik, R., Salwinski, L., Ceol, A., Moore, S., Orchard, S., Sarkans, U., von Mering, C., Roechert, B., Poux, S., Jung, E., Mersch, H., Kersey, P., Lappe, M., Li, Y. X., Zeng, R., Rana, D., Nikolski, M., Husi, H., Brun, C., Shanker, K., Grant, S. G. N., Sander, C., Bork, P., Zhu, W. M., Pandey, A., Brazma, A., Jacq, B., Vidal, M., Sherman, D., Legrain, P., Cesareni, G., Xenarios, L., Eisenberg, D., Steipe, B., Hogue, C., and Apweiler, R. (2004) The HUPOPSI's Molecular Interaction format—a community standard for the representation of protein interaction data. *Nat. Biotechnol.* **22,** 177–183.
20. Pedrioli, P. G. A., Eng, J. K., Hubley, R., Vogelzang, M., Deutsch, E. W., Raught, B., Pratt, B., Nilsson, E., Angeletti, R. H., Apweiler, R., Cheung, K., Costello, C. E., Hermjakob, H., Huang, S., Julian, R. K., Kapp, E., McComb, M. E., Oliver, S. G., Omenn, G., Paton, N. W., Simpson, R., Smith, R., Taylor, C. F., Zhu, W. M., and Aebersold, R. (2004) A common open representation of mass spectrometry data and its application to proteomics research. *Nat. Biotechnol.* **22,** 1459–1466.
21. Durinck, S., Moreau, Y., Kasprzyk, A., Davis, S., De Moor, B., Brazma, A., and Huber, W. (2005) BioMart and Bioconductor: a powerful link between biological databases and microarray data analysis. *Bioinformatics* **21,** 3439–3440.
22. Breci, L. A,, Tabb, D. L., Yates, J. R., and Wysocki, V. H. (2003) Cleavage hboxN-terminal to proline: analysis of a database of peptide tandem mass spectra. *Anal. Chem.* **75,** 1963–1971.
23. Schaaff, T. G., Cargile, B. J., Stephenson, J. L., and McLuckey, S. A. (2000) Ion trap collisional activation of the (M+2H)(2+)-(M+17H)(17+) ions of human hemoglobin beta-chain. *Anal. Chem.* **72,** 899–907.
24. Vaisar, T. and Urban, J. (1996) Probing the proline effect in CID of protonated peptides. *J. Mass Spectrom.* **31,** 1185–1187.

22

Toward High-Throughput and Reliable Peptide Identification via MS/MS Spectra

Jian Liu

Summary

One fundamental problem in proteomics study is to identify proteins and determine their expression levels in cells. Coupled with advanced liquid chromatography, tandem mass spectrometry has become the standard tool for peptide sequencing. In the past decade, many different algorithms and software packages have been developed to support high-throughput proteomics studies. This chapter reviews and compares the computational methods and software for the interpretation of tandem mass spectra. We also present techniques to assess the reliability of peptide identification. Finally, future directions and new research paradigms in tandem mass spectrometry are discussed.

Key Words: Tandem mass spectrometry; peptide sequencing; computational proteomics; algorithms; software programs; bioinformatics.

1. Introduction

The completion of multiple genome projects has fueled great interest in proteomics research. Even armed with vital genetic information, however, improving the existing methods and developing new ones are still essential to characterize the proteins expressed in cells during different times, at different levels, and in different forms. The variations of cellular activities are often reflected in changes at protein expression levels. In particular, the expression of proteins is not always consistent with the corresponding mRNA expression. Protein identification is therefore a cornerstone for disease diagnoses and drug design. Facilitated by high-performance liquid chromatography (HPLC), mass spectrometry is currently the predominant approach to identify proteins in a

From: *Methods in Molecular Biology, vol. 484: Functional Proteomics: Methods and Protocols*
Edited by: J. D. Thompson et al., DOI: 10.1007/978-1-59745-398-1, © Humana Press, Totowa, NJ

cell *(1,2)*. In particular, this technology is capable of detecting posttranslational modifications (PTMs) to proteins, which cannot be acquired directly from genomic studies.

The building blocks of proteins are 20 different amino acids. The primary structure of a protein is a chain of amino acids connected by peptide bonds. In other words, a peptide is a subsequence of a protein. Generally speaking, there are two analytical techniques to identify proteins through mass spectrometry. The first one is peptide mass fingerprinting (PMF). During PMF, the unknown protein of interest is digested into peptides by a protease, such as trypsin. A unique signature of the protein is made up of a group of peptides resulting from the digestion. A mass spectrometer is used to measure the masses of these peptides, and the generated mass spectrum is then *in silico* compared to the protein sequences in a protein database. This approach is based on the assumption that all the detected peptides come from a single protein. Thus, if the proteins cannot be separated from the sample mixtures, the matching process can be seriously misled. The second technique is tandem mass spectrometry (MS/MS). As its name implies, the peptides undergo a second mass analyzer to determine the composition of its amino acids. Since the sequence is determined at the amino acid level, it is more reliable than PMF, especially when PTMs must be taken into account. Thus, MS/MS has become the standard tool to identify peptides.

Typically, peptide sequencing through MS/MS involves multiple steps in shotgun proteomics. First, protein mixtures are digested by proteases, and the

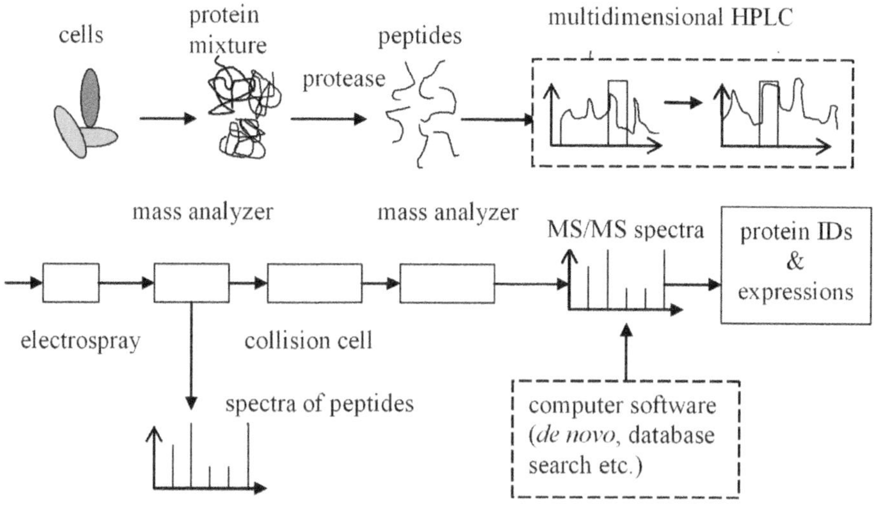

Fig. 1. Flow chart of shotgun proteomics.

resulting peptides are separated by liquid chromatography. Those peptides of interest are then selected by mass spectrometers. These peptides are further fragmented during collision-induced dissociation, producing ions of various types after they are broken at different positions. If some charges remain on an ion, its mass/charge ratio and intensity are detected as a peak. Finally, tandem mass spectra are produced by recording the peak list of various ions; computer programs are then invoked to reconstruct the peptide sequences from the mass spectra. On the basis of the successfully interpreted spectra, the protein contents in the sample mixture can eventually be identified. **Figure 1** illustrates such a multistep procedure of shotgun proteomics.

2. Challenges in MS/MS Spectra Analysis

Upon the generation of MS/MS spectra, the peptide identification problem is reduced to sequence the peptides from spectra. Unfortunately, for various reasons it is not always easy to interpret the MS/MS spectra. First, the fragmentation of the peptides is determined by their physicochemical characteristics as well as many other factors. Consequently, a peptide may be broken more than once, resulting in ions of internal fragmentations. Each ion can be multiply charged; thus multiple peaks in a spectrum may correspond to the same ions. Second, some ions may be missing in the experimental spectra, while noise peaks spoil the peak series. The intensity of the same ion can vary drastically for different runs. Third, ions can also lose certain chemical compounds, such as an ammonium or water group, while other minor types of ions (i.e., a- and c-ions) appear at different rates. Besides these peptide- and instrument-dependent factors, PTMs often occur to the proteins, leading to shifts of many peaks along the *m/z* axis.

Taken together, the experimental MS/MS spectra usually display very limited resemblance to their corresponding theoretical spectra. While the instruments possess the high-throughput capacity to produce massive MS/MS spectra, software tools can be the bottleneck in the pipeline of proteomics study. Traditionally, *de novo* and database searching are the two most widely used approaches to sequence peptides via MS/MS spectra. In the following sections, we review these methods and popular software packages.

3. *De Novo* Peptide Sequencing

This approach attempts to reconstruct the peptide sequence solely from a given experimental MS/MS spectrum. Theoretically, *de novo* sequencing needs to consider all possible linear combinations of amino acids, which is computationally intractable. To make the goal practical, software programs in this class first carefully tune the objective function under specific assumptions and restrictions, and then incorporate an efficient algorithm to search the optimal peptides. Typically, a graph model is derived from the spectrum. In such a graph, each vertex denotes a peak related to a possible ion. An edge is added to connect to a pair of vertices if the mass difference between the peaks is approximately equal to that of an amino acid. Each vertex or edge is assigned a weight, which is usually correlated to the corresponding ion intensity. Therefore, the problem is transformed to the search for the optimal path traversing the spectrum graph *(3)*. Various programs have been developed to implement a specific *de novo* algorithm within such a framework. Although mostly based on the spectral graphs explicitly or implicitly, they vary in objective functions and treatment of peak selections throughout the graph. Such subtle differences lead to significant discrepancies concerning their performances.

Among them, PEAKS *(4)* is one of the most successful *de novo* tools. It exploits the fact that the complementary *b/y* ions are the most abundant and develops a unique sandwich algorithm to scan a given spectrum. By simultaneously exploring both prefix and suffix of the peptide sequence from two ends of the peak list, it significantly boosts the sensitivity by avoiding false paths. In addition, PEAKS employs advanced data structures and algorithms to improve the speed and prune the search space. Although it internally has a dynamic programming algorithm to compute the matching score for a peptide, PEAKS is capable of analyzing a spectrum in less than 1 s on a modern desktop computer.

Probabilistic models are also used in *de novo* sequencing. PepNovo *(5)* incorporates the information of supporting ions into the Bayesian framework to distinguish observed matches from random matches of ions. Continuing in the same direction, a more complicated algorithm *(6)* has also been introduced to establish a hidden Markov model (HMM) to accurately estimate the likelihood of producing the experimental spectrum from a given peptide. In this study, the hidden states represent the amino acids in protein sequences, while the observable outputs indicate the ion peaks. The HMM has the advantage of tolerating some missing peaks of ions in the spectrum as they are not always observed. The parameters of the probabilistic networks are obtained from machine learning over annotated spectra. Therefore their performances in practice are also subject to the training data.

4. Database Searching

Despite its fast speed, the *de novo* approach has some inevitable limitations. First, it requires high-quality spectra with almost complete *b/y* ion ladders. Since similar amino acid sequences may share close or even identical masses, it is unlikely to determine the whole sequence of peptides correctly when the *b*- and *y*-ions series are incomplete. In practice, the spectra produced by low end mass spectrometers are hard to interpret by *de novo* methods. Second, the predicted peptides may not really exist, even though their theoretical spectra demonstrate a very strong similarity to experimental ones.

Database searching provides an alternative to interpret the tandem mass spectra. This approach explores a protein sequence database to find the peptides whose theoretical spectra best match experimental ones. With the improved quality and coverage of protein databases, it has become the prevailing method to analyze MS/MS spectra. For a given spectrum, a set of candidate peptides can be found from the protein database whose masses are within the mass error tolerance to the precursor ion mass. Given a large protein sequence database, the candidate set could contain hundreds of thousands of tryptic peptides. Therefore, a high-resolution scoring function plays a key role in identifying the correct peptide from such a large candidate set.

In the past decade, a range of database search programs has been developed to analyze tandem spectra. In general, this type of software first cleans the spectrum by removing putative noisy peaks, and then evaluates the degree of similarity between experimental and theoretical spectra. Among them, Mascot *(7)* and Sequest *(8)* are the earliest and most used in academia and industry. Their central idea is to use statistical or probabilistic measures to assess the pairwise spectral similarity. Mascot considers the matches between peptide fragments and peaks in the experimental spectrum as random events. Therefore, for each candidate peptide, the probability that it matches the spectrum can be computed. Such a probability is extremely small for true positives as most peaks are matched. Whereas Sequest computes the cross-correlation between the experimental and theoretical peak lists, their pairwise similarity for true positive peptides is anticipated to be very high.

Since peptide fragmentation is also an instrument-specific process, machine learning is a natural choice to optimize the scoring function as in *de novo* methods. PRIMA *(9)* is such a database search tool to construct a linear scoring function based on machine learning techniques. It selects statistically significant features of ion matching and then formulates the problem of peptide identification as a classification task. Finally, it uses a linear programming to determine the coefficients in the scoring function. Another similar algorithm, PepReap *(10)* adopts support vector machines (SVMs) as an implicit scoring function to classify peptides. To improve the sensitivity of the SVM scorer, a heuristic assessment is conducted as a preprocessing step to remove the majority of candidate peptides by roughly evaluating the degrees of their matches to the MS/MS spectrum.

5. Advanced Methods for MS/MS Interpretation

Database searching and *de novo* sequencing have been in use for more than a decade. As described above, both of them have their own merits and drawbacks. Recently, researchers have made tremendous efforts to explore new solutions to boost the speed and correctness of peptide identifications. Some new approaches and exciting breakthroughs have been reported.

5.1. Combination of de Novo and Database Methods

The *de novo* approach provides a fast but potentially vulnerable method to sequence peptides. If the spectra contain incomplete *b* and *y* ion ladders, it may return false peptides. However, in such cases, the predicted peptides often contain correct subsequences, which are also known as tags, with a length of a few amino acids. If such tags are highly reliable and detectable, the valuable information can be used to improve the database search. Different software programs such as PepNovo *(11)* have been developed to generate peptide tags of high confidence. In general, tags are determined through searching significant peaks with intervals equal to masses of specific amino acids. Therefore, the tags are also characterized by their locations on the *m/z* axis. Other facts, such as variant of ions and complementary peaks, are also leveraged to enhance the reliability of tags. Because generating peptide tags is independent of any custom protein databases, this step can be accomplished very fast. Once the tags are derived from MS/MS spectra, the peptides that do not contain any predicted tags can be eliminated directly from the database search. Such a filtering step can substantially reduce the time for spectral alignment during matching spectra against the database and reduce the possibility of false positives. The software InsPecT *(12)* used such tags to speed up the blind search of PTMs. Although theoretically the computational complexity is prohibitively expensive for variable modifications, it is reported that InsPecT is two orders of magnitude faster than the traditional SEQUST tool *(12)*. Such a breakthrough makes it feasible to support high-throughput proteomics studies with desktop computers, which previously required high-performance computer clusters.

Nevertheless, sequence tags may still contain possible errors, especially when PTMs complicate the tag generation. There are two ways to enhance the correctness of predicted tags. Software programs such as PepNovo usually produce a list of short tags in a conservative manner to ensure the true positives are labeled correctly at least once. Such software also allows users to specify the length of the tags. The other strategy is to develop ad hoc programs such as SPIDER *(13)* to tolerate the errors in tags. When the *de novo* sequences are mapped to protein sequences, homology mutations and substitutions are permitted to match the subsequences.

5.2. Direct Comparison of Experimental Spectra

The success of traditional methods, either *de novo* or database searching, relies on the models of chemical and physical rules governing peptide fragmentation. Due to the complexity of fragmentation models, complicated algorithms have been used in the database search methods to recognize the spectra. While they indeed can improve the sensitivity, the algorithms also require intensive computation. This problem becomes more serious for unrestrictive search of PTMs, which exponentially increases the search space, leading potentially to considerable damage to the accuracy of peptide identification.

To deal with these challenges, peptide identification by direct comparison of experimental spectra has drawn much attention recently. This type of approach has the advantage of directly taking into account instrument-dependent or peptide-specific contributing factors in spectra generation. Consequently, it is not necessary to explicitly build a complicated kinetic model to characterize the peptide fragmentation. Therefore, direct comparison of experimental spectra provides an appealing alternative to support high-throughput peptide sequencing due to its simplicity and speed. In principle, these methods vectorize the spectra and employ some statistical measures, such as correlation coefficient or inner product, as the scores of pairwise spectral similarity.

Tools of this category allow comparison of the protein/peptide contents of different sample mixtures without actual identification of peptides. The other advantage of this approach is that it can be used to cluster the spectra of the same peptides. Duplicate spectra are ubiquitous in large-scale proteomic data as many proteins may share the same peptides. Furthermore, the same peptides may be fragmented multiple times or repeated in different runs. In practice, 20–50% of interpretable spectra could be duplicates. Therefore, it would also reduce the search time substantially by recognizing the duplicates. NoDupe ***(14)*** is such a software package used to detect duplicate peptide spectra.

It is noteworthy that the spectra of the same peptide may share low similarity, although some ion fragmentation patterns are reproduced. It is thus desirable to collapse a cluster of spectra to a strong representative spectrum. Some tools, such as Pep-Miner ***(15)*** and MS2grouper ***(16)*****,** have been designed to cluster spectra by their similarity and derive a representative spectrum. The tools attempt to find the most significant peaks that are common to the spectra of the same peptide. To achieve this goal, dedicated algorithms are designed to filter noise and align the peaks of high intensities. Another recent study even demonstrates that an effective representative can be constructed by ensemble averaging the spectra in the cluster ***(17)***. Although this method is straightforward, its performance steadily improves for larger clusters as the noisy peaks are downplayed asymptotically after averaging. Moreover, this study shows that some spectra that initially failed *de novo* or database search programs can be identified correctly by using their average representatives.

Different from *de novo* or database search, pairwise similarity is based on the entire peak list (some software tools may filter noisy peaks) instead of a small subset of most significant peaks. Indeed, the statistical measures cannot ensure satisfactory sensitivity and specificity when the number of candidate peptides is large as they are affected by peaks of noise and minor ions. However, because practical database searches are usually limited to a specific taxonomy, the number of candidate peptides is reasonable small. Under such circumstances, direct comparison of spectra is a fast means to identify peptides. Another obvious concern is whether representative spectra are instrument neutral. The studies of X! Hunter ***(18)*** and BiblioSpec ***(19)*** confirm the robustness of this method as spectra produced by different instruments are practically comparable, although they perform best when the spectra are collected from the same type of mass spectrometers.

In summary, **Table 1** provides a list of recent and widely used software packages and their availabilities for peptide sequencing via tandem mass spectrometry. The research community and bioinformatics industry constantly upgrade or release new software tools; updates on MS/MS search engines can be

Table 1
Popular Software Programs for Peptide Identification via MS/MS Spectra

Category	Software	URL	Availability
De novo	PEKAS	http://www.bioinfor.com/peaksonline	Commercial, free online
	PepNovo	http://peptide.ucsd.edu/pepnovo.py	Open source, free online
Database search	Sequest	http://fields.scripps.edu/sequest	Commercial
	Mascot	http://www.matrixscience.com/	Commercial, free online
	X! Tandem	http://www.thegpm.org/tandem	Open source
Tag-based hybrid system	PepNovo	http://peptide.ucsd.edu/pepnovo.py	Open source, free online
	InspecT	http://peptide.ucsd.edu/inspect.py	Open source, free online
	SPIDER	http://bif.csd.uwo.ca/spider	Free online
Spectral comparison	X! Hunter	http://www.thegpm.org/HUNTER	Free online
	BiblioSpec	http://proteome.gs.washington.edu/bibliospec/documentation/	Free online

found at http://www.proteomecommons.org/. Although each of these software programs has its own advantages with regards to accuracy and speed, none of them is perfect. Given the same dataset, it is conceivable that each program may fail to recognize a subset of spectra. Therefore, some proteomics research laboratories run multiple search engines in parallel when the computational resources are available, and then compile the consensus results from outputs of different programs. It has been reported *(20)* that such a meta-search strategy is capable of significantly improving the accuracy and coverage of peptide identification in practice.

In addition to the approaches described above, some other software programs have also been developed to facilitate peptide identification from other perspectives, such as determining the quality of spectra, the charge states, and purifying the raw spectra. Recent studies show that appropriately configuring these tools can enhance both the accuracy and speed of MS/MS analysis considerably.

6. Reliability Assessment of Peptide Identification

Given a spectrum, the software mentioned above generally returns a list of peptides, each associated with a matching score. The algorithms do not always return true positives. Therefore, it is necessary to develop methods to assess the reliability of peptide identifications. For the database search approach, one commonly used method is to also search the same spectra against the inversed protein sequence database *(21)*. In such a methodology, each protein sequence in the original database is reversed. The procedure of reversing guarantees that the new database maintains some vital characteristics of the protein sequences, such as the number of candidate peptides and the homology among the protein sequences. Searching against this spurious protein database provides a score distribution for the false positives. By further employing the Bayesian analysis, the reliability of peptide identifications can be determined. In other words, given a score, we can estimate the probability of the identified peptides being a true positive. Some other methods *(22,23)* further improve this approach by deriving a new synthetic score; they also consider other factors, such as charge states and spectral quality, to assess the reliability of peptide identification.

A more sophisticated strategy is presented in a recent study *(24)*, which is also based on the search against the inversed protein sequence database. It assumes that if a search algorithm cannot return true positives, it has an equal chance to return a false positive from regular or inverted protein sequences. Some other factors, such as the lengths of peptides and differences between scores of the top and second ranked peptides, are also taken into account. The multidimensional space is then partitioned into a set of smaller rectangles. For each of the rectangular regions, the ratio of false positives from the reversed peptides is calculated, and then an accurate estimate of reliability can be derived based on the above assumption.

7. Summary and Future Directions

With the continuous advances in both hardware and software, tandem mass spectrometry has become the mainstay for high-throughput proteomics study. It is computationally challenging to analyze the gigantic spectra data produced from the instruments worldwide. A wide range of fast and effective computer programs has been designed to identify peptides via MS/MS spectra. These software tools have steadily improved and now are capable of processing enormous spectra data in a timely manner. This chapter provides an up-to-date review of several of the most recognized algorithms and methods from computational perspectives.

The ultimate objective of tandem mass spectrometry is to determine the underlying protein complex and estimate its abundance. The reliable identification of peptides provides a solid basis for this goal. Some heuristic models are presented to identify the proteins of maximal likelihood based on simplified mathematical principles ***(22,25)***. However, after protein cleavage not all peptides have an equal likelihood of being detected by current MS-based techniques. Given a protein, only the *proteotypic* peptides are reproducible from a particular proteomic pipeline, whereas other peptides are very difficult to find. Several pioneering quantitative proteomics approaches explored the possibility of solving the problem in the framework of systems biology ***(26–28)***. It is anticipated that integrating data from genomic, proteomic, and other sources will eventually determine the contents of protein mixtures in a biologically meaningful manner. This will greatly help us to reveal the functionality and interactions of various proteins under normal physiological conditions as well as in diseased states.

References

1. Kinter, M. and Sherman, N. E. (2000) *Protein Sequencing and Identification Using Tandem Mass Spectrometry*. Wiley-Interscience, New York.
2. Snyder, A.P. (2002) *Interpreting Protein Mass Spectra: A Comprehensive Resource*. Oxford University Press, New York.
3. Chen, T., Kao, M. T., Tepel, M., Rush, J., and Church, G. M. (2001) A dynamic programming approach to *de novo* peptide sequencing via tandem mass spectrometry. *J. Comput. Biol.* **8(3),** 325–337.
4. Ma, B., Zhang, K., Hendrie, C., Liang, C., Li, M., Doherty-kirby, A., and Lajoie, G. (2003) PEAKS: powerful software for peptide *de novo* sequencing by tandem mass spectrometry. *Rapid Commun. Mass Spectrom.* **17(20)**, 2337–2342.

5. Frank, A. and Pevzner, P. (2005) Pepnovo: *de novo* peptide sequencing via probabilistic network modeling. *Anal. Chem.* **77(4),** 964–973.
6. Fischer, B., Roth, V., Roos, F., Grossmann, J., Baginsky, S., Widmayer, P., Gruissem, W., and Buhmann, J. M. (2005) NovoHMM: a hidden Markov model for *de novo* peptide sequencing. *Anal. Chem.* **77(22),** 7265–7273.
7. Perkins, D. N., Pappin, D. J., Creasy, D. M., and Cottrell, J. S. (1999) Probability-based protein identification by search sequence databases using mass spectrometry data. *Electrophoresis* **20(18),** 2551–3567.
8. Eng, J. K., McCormack, A. L., and Yates, J. R. (1994) An approach to correlate tandem mass spectral data of peptides with amino acid sequences in the protein database. *J. Am. Soc. Mass Spectrom.* **5(11),** 976–989.
9. Liu, J., Ma, B., and Li, M. (2006) PRIMA: peptide robust identification from MS/MS spectra. *J. Bioinform. Comp. Biol.* **4(1),** 125–138.
10. Wang, H., Fu, Y., Sun, R., He, S., Zeng, R., and Gao, W. (2006) An SVM Scorer for more sensitive and reliable peptide identification via tandem mass spectrometry. *Proc. Pacific Symp. Biocomput.* 304–213.
11. Frank, A., Tanner, S., Bafna, V., and Pevzner, P. (2005) Peptide sequence tags for fast database search in mass spectrometry. *J. Proteome Res.* **4(4),** 1287–1295.
12. Tsur, D., Tanner, S., Zandi, E., Bafna, V., and Pevzner, P. (2005) Identification of post-translational modifications by blind search of mass spectra. *Nat. Biotechnol.* **23(15),** 1562–1567.
13. Han, Y., Ma, B., and Zhang, K. (2005) SPIDER: software for protein identification from sequence tags with *de novo* sequencing error. *J. Bioinform. Comp. Biol.* **3(3),** 697–716.
14. Tabb, D. L., MacCoss, M. J., Wu, C. C., Anderson, S. D., and Yates, J. R. (2003) Similarity among tandem mass spectra from proteomic experiments: detection, significance and utility. *Anal. Chem.* **75(10),** 2470–2477.
15. Beer, I., Barnea, E., Ziv, T., and Admon, A. (2004) Improving large-scale proteomics by clustering of mass spectrometry data. *Proteomics* **4(4),** 950–960.
16. Tabb, D. L., Thompson, M. R., Khalsa-Moyers, G., VerBermoes, N. C., and McDonald, W. H. (2005) MS2Grouper: group assessment and synthetic replacement of duplicate proteomic tandem mass spectra. *J. Am. Soc. Mass Spectrom.* **16(8),** 1250–1261.
17. Liu, J., Bell, A. W., Bergeron, J. J. M., Yanofsky, C. M., Carrillo, B., Beaudrie C. E. H., and Kearney, R. E. (2007) Methods for peptide identification by spectral comparison. *Proteome Sci.* **5(3).**
18. Carig, R., Corteins, J. C., and Beavis, R. C. (2006) Using annotated peptide mass spectrum libraries for peptide identification. *J. Proteome Res.* **5(8),** 1843–1849.
19. Frewen, B. E., Merrihew, G. E., Wu, C. C., Noble, W. S., and MacCoss, M. J. (2006) Analysis of peptide MS/MS spectra from large-scale proteomics experiments using spectrum libraries. *Anal. Chem.* **78(16),** 5678–5684.
20. Resing, K. A., Meyer-Ardent, K., Mendoza, A. M., Aveline-Wolf, L. D., *et al.* (2004) Improving reproducibility and sensitivity in identifying human proteins by shotgun proteomics. *Anal. Chem.* **76(13),** 3556–3568.

21. Keller, A., Nesvizhskii, A. I., Kolker, E., and Aebersold, R. (2002) Empirical statistical model to estimate the accuracy of peptide identifications made by MS/MS and database search. *Anal. Chem.* **74(20),** 5383–5392.
22. Razumovskaya, J., Olman, V., Xu, D., Uberbacher, E., Verbermoes, N., and Xu, Y. (2004) A computational method for assessing peptide identification reliability in tandem mass spectrometry analysis with SEQUEUST. *Proteomics* **4(4),** 961–969.
23. Li, F., Sun, W., Gao, Y., and Wang, J. (2004) RScore: a peptide randomicity score for evaluating tandem mass spectra. *Rapid Commun. Mass Spectrom.* **18(14),** 1655–1659.
24. Kislinger, T., Rahman, K., Radulovic, D., Cox, B., Rossant, J., and Emili, A. (2003) PRISM: A generic large-scale proteomics investigation strategy for mammals. *Mol. Cell. Proteomics* **2(2),** 96–106.
25. Sadygov, R. G., Liu, H., and Yates J. R. (2004) Statistical models for protein validation using tandem mass spectral data and protein amino acid sequence databases. *Anal. Chem.* **76(6),** 1664–1671.
26. Chu, W., Ghahramani, Z. Krause, R., and Wild, D. L. (2006) Identifying protein complexes in high-throughput protein interaction screens using an infinite latent feature model. *Proc. Pacific Symp. Biocomput.* 214–242.
27. Ho, Y., Gruhler, A., Heilbut, A., *et al.* (2002) Systematic identification of protein complexes in Saccharomyces cerevisiae by mass spectrometry. *Nature* **415(6868),** 180–183.
28. Lu, P., Vogel, C., Wang, R., Yao, X., and Macotte, E. M. (2007) Absolute protein expression profiling estimates the relative contribution of transcriptional and translational regulation. *Nat. Biotechnol.* **25(1)**, 117–124.

23

MassSorter: Peptide Mass Fingerprinting Data Analysis

Ingvar Eidhammer, Harald Barsnes, and Svein-Ole Mikalsen

Summary

MassSorter is a software tool that sorts, systemizes, and analyzes data from peptide mass fingerprinting (PMF) experiments on proteins with known amino acid sequences. Several experiments can be simultaneously analyzed for sequence coverage and posttranslational modifications occurring during sample handling, induced chemical modifications, and unexpected cleavages. Experimental *m/z* values are compared with *m/z* values from an *in silico* digestion, taking modifications into account. Filters can be defined by users for marking autolytic protease peaks and other contaminating peaks. MassSorter functions as a database of all the detected peptides. It includes tools for visualization of the results, such as sequence coverage, accuracy plots, statistics, and 3D models.

Key Words: Peptide mass fingerprinting; MassSorter; analyzing MS data; comparing MS experiments.

1. Introduction

Though there is an enormous increase in large-scale proteomics, it is still necessary to perform small-scale experiments concentrating on one or a small number of proteins. This is of particular interest when the aim is to characterize posttranslational modifications in a protein. Tools for analyzing data from such experiments are needed. A number of programs can be used for small-scale protein identification, e.g., MS-Fit *(1)*, Mascot *(2)*, Profound *(3)*, Aldente *(4)*, Phenyx *(5)*, and GPMAW *(6)*. For some of them the search parameters include modifications believed to be present in the proteins analyzed, achieving a partial characterization of the protein in question. Programs directed more toward further characterization of identified proteins are FindMod *(7)*

From: *Methods in Molecular Biology, vol. 484: Functional Proteomics: Methods and Protocols*
Edited by: J. D. Thompson et al., DOI: 10.1007/978-1-59745-398-1, © Humana Press, Totowa, NJ

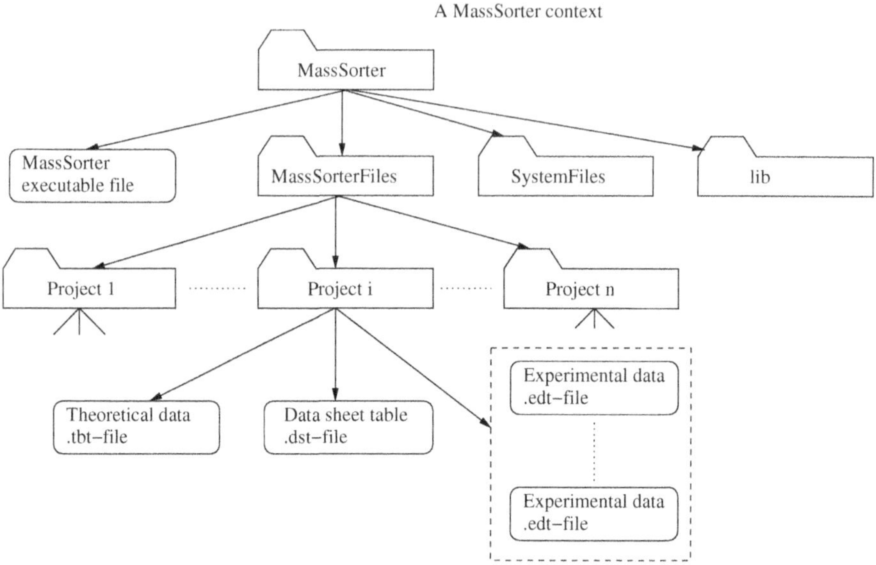

Fig. 1. Overview of the MassSorter File System.

and FindPept *(8)*. However, only Phenyx includes an administrative unit for collecting and analyzing data from several experiments, and is mostly directed toward large-scale identification.

MassSorter *(9)* is especially developed for analyzing and comparing the results of several experiments on known proteins, "known" meaning that the sequence is available. It consists of a set of analytical tools integrated around an administrative unit that functions as a database (**Fig. 1**). Experimental and theoretical data are compared in a table (spreadsheet), and all the analytical tools have a uniform and user-friendly style, making the transformation of data between the different tools easy. The known protein can be analyzed for sequence coverage and different forms of modifications.

2. Materials

The goal of MassSorter is to maximize the number of reasonable matches between experimental and theoretical *m/z* values, taking into account different types of modifications, missed cleavages, and potentially unexpected cleavages. This means that as many of the experimental *m/z* values as possible should be explained. The results of the analyses are collected in a table, and presented in an easily understandable form. MassSorter is platform independent, and the graphic user interface (windows, menus, etc.) is created in a standard way, making it easy to use.

For simplicity we sometimes refer to *m/z* values as masses. As we here are handling peptide mass fingerprinting (PMF) data with charge +1, the mass corresponds to $(m + H^+)$.

2.1. The Conceptual View of MassSorter

MassSorter performs the analyses in *Projects*. One Project consists of Project Data, Theoretical Data, Experimental Data, and a Data Sheet Table showing the connection between the theoretical and experimental data. A Project is usually concentrated on one protein (but not necessarily). There is one set (file) of theoretical data, but typically several sets (files) of experimental data.

2.1.1. Project Data

The Project Data includes the following:

1. Project name that identifies a Project.
2. Project description.
3. Accuracy acceptance level for matches between experimental and theoretical masses.

2.1.2. Theoretical Data

The Theoretical Data includes the following:

1. The sequence of the protein in the project.
2. The protease used for *in silico* digestion.
3. The modifications to take into account in the *in silico* digestion.
4. The maximum number of missed cleavages (per peptide) in the *in silico* digestion.
5. A list of theoretical peptides from the digestion, each element in the list containing the following:
    a. The mass.
    b. Start and end position of the peptide.
    c. Modifications applied.
    d. Number of missed cleavages.
    e. Sequence of the peptide.

2.1.3. Experimental Data

The Experimental Data contains data from one or several experiments. The data for one experiment include the following:

1. The name of the experiment.
2. The protein name.
3. Date for the experiment.
4. Comments (optional).
5. Expected and possible modifications.
6. A list of data for the peaks of the experimental spectrum, each peak element containing
   a. The mass.
   b. Intensity (optional).
   c. Comments (optional).

As mentioned, the data in one project usually belong to one protein.

2.1.4. Data Sheet Table

The data sheet table (DST) contains the result from comparing the experimental data and the theoretical data. It shows the matches between the experimental masses and the masses of the theoretical peptides. This is explained in more detail in the Methods section.

2.2. The Tools

The main function of MassSorter is to compare the experimental and theoretical data in the data sheet table and give a reasonable presentation of the result. For these operations MassSorter is constructed as a set of tools, of which the most important are briefly mentioned below, and explained in more detail in the Methods section.

1. ProteinDigester is the tool used for *in silico* digestion.
2. Filter is used for specifying masses that may come from contaminants and other noise sources.
3. SequenceSuggester can be used when an experimental mass does not match a theoretical peptide mass or a filter mass. The reason may be unexpected cleavages, and it is therefore possible to compare the unidentified mass with the theoretical mass of all subsequences of the protein sequence, searching for a match.
4. MassFinder is a tool that given an amino acid sequence and a list of modifications can calculate the (theoretical) mass of the sequence.
5. UniModSearch is used for investigating whether unmatched masses may correspond to modifications not considered in the first round of analysis. The modifications are defined in a local version of the UniMod modification database ***(10)***. This tool is not available from the Tools menu; it can be obtained only by right clicking an experimental mass in a DST (*see* **Subheading 3.4.1**).
6. Report is an alternative presentation of the results for the comparison of the theoretical and experimental data. All the matched and unmatched masses are

grouped and counted, and the sequence coverage is calculated and visualized both per experiment and combined for all the experiments included in the project.

7. ProteinViewer presents a three-dimensional (3D) model of the protein structure (if known), indicating the detected parts of the sequence. The 3D structure files of many proteins are found in the Protein Data Bank (PDB) structure database ***(11)***.
8. Statistics presents four types of statistics for the comparisons in a Project.

2.3. Installing MassSorter and the MassSorter File System

MassSorter is freely available for academic users at www.bioinfo.no/software/massSorter, where a detailed procedure for downloading and installing is also found.

To increase the benefit of MassSorter it is necessary to have an understanding of how it works and the (sub)folders and files it uses. For the description we assume that MassSorter is installed in a folder called "MassSorter." "MassSorter" with its subfolders and files defines the MassSorter *context.*

In addition to the executable MassSorter file, the folder "MassSorter" contains the system subfolders:

1. SystemFiles that contain the system parameters (modifications, filters, etc.).
2. lib that contains library functions.
3. MassSorterFiles, which for each Project contains a subfolder with the Project name. A Project subfolder contains the theoretical data in a .tbt file, the experimental data in .edt files, and the data sheet table in a .dst file.

A Project inside the MassSorter context is easily available from other Projects in the same context. It is, for example, possible, when importing theoretical or experimental data into a Project, to use data from other Projects.

3. Methods

Here we explain how to use MassSorter for defining Projects, performing comparisons with different parameters, and presenting results and statistics. There is space for only the main procedures; details and more specific possibilities are described in the tutorial at MassSorter's home page and in the help pages in MassSorter.

3.1. Creating a New Project

The first time you start MassSorter a "Welcome" window appears above the main window, in which you select the "New Project" button (you can also select it from the "File" menu of the main window if you choose to close the welcome window). A wizard, consisting of four steps, will guide you through the import of the necessary data.

3.1.1. Step 1: Project Details

Provide a Project name and description. Only the name is mandatory, but inserting a description is highly recommended for later use.

3.1.2. Step 2: Theoretical Data

You now have two choices: you can either create a new theoretical data file from scratch using MassSorter's own tool ProteinDigester or you can select one from the list of the existing data files (.tbt) that are presented in the window. Note that those files are theoretical data from other Projects. In the latter case you select the one you want by clicking the circular button to the right, and then clicking on the "Next" button. The theoretical data file (.tbt) is stored into the new project folder regardless of whether this is a new file created for the purpose or it is picked up from another folder. If you want to create new data, you click on the "ProteinDigester" button, and a new window appears. Now you can either fill in the sequence (by typing or copy and paste) or import from a (text) file by selecting "Import Sequence" from the "File" menu. Then select the parameters for the digestion, the considered modifications, etc. (If you want to see more information about the modifications right click on the given abbreviation.) Then you click on the "Digest Protein" button. To preview the contents of the file, right click on the given row and select "Preview Theoretical Data File" from the popup menu.

3.1.3. Step 3: Experimental Data

Again you have two choices: either import new experimental data files or select one or more from the list of already available data files. The files can be sorted according to the contents of any of the columns by clicking on the column title. If you are going to import new experimental data then click on the "Import Experimental Data" button. You have three choices for importing: Delimited Text File, XML File, and Cut and Paste. Delimited Text Files are text files in which the text is ordered in columns separated by some delimiter, for example, space or ",". XML files are more structured text files containing so-called tags explaining the content of each line of text. In either case you must make sure that the parameters, column number or tag names, are correct. Cut and Paste simply means copying the data from a spreadsheet or a text file. When you have collected the data for an experiment, the last import window appears. Insert the correct protein name, make sure that the correct enzyme is selected (the enzyme should normally be the same for all experiments in a project), and insert any comments if wanted. Choose the modifications that are expected in the experiment and click on "Import". Repeat the procedure to import additional experimental data.

In the current window you now have a list of available experimental data files (you can sort them as explained above). To see the contents of the files, right click on the desired row and select "Preview Experimental Data File" from the popup menu. Make sure that the wanted experiments are selected, and click on "Next."

3.1.3.1. Manual Editing of the Data

During the import procedure described above, it is possible to manually edit the data, e.g., to delete peaks that are recognized as noise or to add a peak that the spectrum analysis program has not recognized. To do this you must preview the experimental data you want to edit (by right clicking and selecting "Preview Experimental Data File"). In the preview window you can now delete a peak by selecting the row; go to the "Edit" menu and select "Delete Row." For adding a peak you select a row and choose "Insert Row After" or "Insert Row Before" from the "Edit" menu, and then the data (*m/z* value and optionally the intensity) can be manually filled in.

3.1.4. Step 4: Create the Data Sheet Table

This final step has two purposes: to obtain an overview of the data you have selected to be included in the Project and to choose the accuracy limit (ppm or Da) to be used for the comparison of theoretical and experimental masses. Click on "Finish" and the data sheet table for the Project is created.

3.2. The Data Sheet Table

The main view of a Project is the DST containing all the comparisons of the experimental and theoretical peptides. The logic behind the performed comparisons is now described. Each experimental peptide's *m/z* value is first compared to the theoretical *m/z* values. If a match is found within the given accuracy limit, the program checks to see if the given theoretical peptide is modified. If it is, the modifications also have to be in the list of possible/realistic modifications for the given MS experiment. If the modifications are in this list, or the theoretical peptide is not modified, the two peptides are considered "equal" and positioned on the same row in the table. If an experimental *m/z* value does not match any of the theoretical *m/z* values it is compared to the *m/z* values from the other MS experiments if any, and placed on the same row if they are within the selected accuracy limit.

The DST can also color code the experimental values according to the detected intensities by selecting "Intensity Grading" on the "View" menu. The experimental values are then divided into three groups and each group is given a specified color. Default colors are different shadings of green where the most intense peaks have the darkest shading. The peak with a normalized intensity of 100 is colored blue. The colors used can be altered by selecting "Edit color" on the "View" menu, and the limits for each of the shadings can be edited in the same window.

When comparing the *m/z* values, it is possible to obtain more than one match against the theoretical *m/z* values (within the accuracy limit) for a given experimental *m/z* value. The best match (smallest absolute difference) is automatically selected as a "primary match" and the others are labeled "secondary matches." If the match automatically selected as primary is for some reason wrong, you can manually select one of the others. First make the secondary matches visible by deselecting "Hide secondary matches" from the "View" menu. The secondary matches are colored dark green. Choose one of the secondary matches, i.e., one of the dark green cells, and right click on the corresponding third column of the secondary match. A window appears in which you can choose the match you

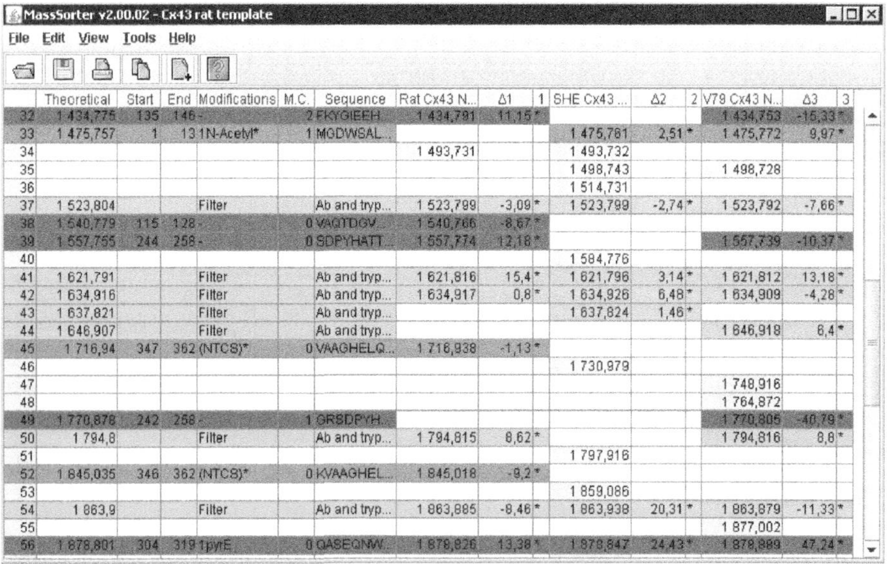

	Theoretical	Start	End	Modifications	M.C.	Sequence	Rat Cx43 N...	Δ1	1	SHE Cx43 ...	Δ2	2	V79 Cx43 N...	Δ3	3
32	1 434,775	135	146	-	2	FKYGIEEH...	1 434,791	11,15	*				1 434,753	-15,33	*
33	1 475,757	1	13	1N-Acetyl*	1	MGDWSAL...				1 475,761	2,51	*	1 475,772	9,97	*
34							1 493,731			1 493,732					
35										1 498,743			1 498,728		
36										1 514,731					
37	1 523,804			Filter		Ab and tryp...	1 523,799	-3,09	*	1 523,799	-2,74	*	1 523,792	-7,66	*
38	1 540,779	115	128	-	0	VAQTDGV...	1 540,766	-8,67	*						
39	1 557,755	244	258	-	0	SDPYHATT...	1 557,774	12,18	*				1 557,739	-10,37	*
40										1 584,776					
41	1 621,791			Filter		Ab and tryp...	1 621,816	15,4	*	1 621,796	3,14	*	1 621,812	13,18	*
42	1 634,916			Filter		Ab and tryp...	1 634,917	0,8	*	1 634,926	6,48	*	1 634,909	-4,28	*
43	1 637,821			Filter		Ab and tryp...				1 637,824	1,46	*			
44	1 646,907			Filter		Ab and tryp...							1 646,918	6,4	*
45	1 716,94	347	362	(NTCS)*	0	VAAGHELQ...	1 716,938	-1,13	*						
46										1 730,979					
47													1 748,916		
48													1 764,872		
49	1 770,878	242	258	-	1	GRSDPYH...							1 770,805	-40,79	*
50	1 794,8			Filter		Ab and tryp...	1 794,815	8,62	*				1 794,816	8,8	*
51										1 797,916					
52	1 845,035	346	362	(NTCS)*	0	KVAAGHEL...	1 845,018	-9,2	*						
53										1 859,086					
54	1 863,9			Filter		Ab and tryp...	1 863,885	-8,46	*	1 863,938	20,31	*	1 863,879	-11,33	*
55													1 877,002		
56	1 878,801	304	319	1pyrE	0	QASEQNW...	1 878,826	13,38	*	1 878,847	24,43	*	1 878,889	47,24	*

Fig. 2. A fraction of the DST comparing Cx43 from rat, Syrian hamster, and Chinese hamster. For simplicity, only one sample is shown for each species. The rows 33, 45, 46, 47, 52, 53, and 55 are specifically mentioned in the text. Rows 32 and 38 are examples of unmodified matches. Row 56 is an example of a modified match and row 37 corresponds to a filter peak. Row 35 is an example of two experimental masses that are unmatched, but identical to each other within the chosen accuracy (in this case 50 ppm). (*See* Color Plate 2)

want as a primary match or remove the matches all together for this particular peptide. Note that removing all the matches is irreversible. An example of a DST is given in **Fig. 2** (*see* Color Plate 2).

3.2.1. Filtering of Data

In MS experiments there is a possibility that the samples may contain proteins other than the one you are studying, for example, keratin or parts of the enzyme used for digestion. To avoid disturbances due to these nonrelevant peptides you can add a filter that will mark such *m/z* values in gray and remove them from further consideration. To perform filtering, select "Filter(s)" from the "Edit" menu in the main window. Now you can either select from the list of available filters or you can create a new one. To create a new filter, click on the "New Filter" button. A new window then appears where you insert a name, a description for the filter, and a list of masses (optionally with comments). After saving the filter will appear in the list of available filters. From this list you select the filter(s) you want to use for the given Project and click on "Update." The filters are then applied on the data. Filters are removed by deselecting them in the list.

3.3. Updating the Data of a Project

It is possible to update both the theoretical and experimental data. This can be done if you want to look for modifications in an experiment and those modifications were not included in the theoretical digestion or in the list of possible modifications for the experiment. The updating is performed from the DST. The theoretical data file can be changed by right clicking on the header of the column in the DST labeled "Theoretical" and selecting "View Theoretical Data" from the popup menu. The theoretical data are displayed and the contents can be altered. If you want to completely change the data, select "Re-Digest" from the "Tools" menu. The data for an experiment can be altered in the same way by right clicking on the column in the DST labeled with the experiment name.

Adding or removing experiments can be done by selecting "Experimental Data" from the "Edit" menu. Then you select or deselect the experiments you want to add or remove. You can also change the order of the experimental data files in the DST by changing the sorting criteria by clicking on the column name or by moving specific experiments up or down in the list.

3.4. Increasing the Number of Matches

Two ways of increasing the number of matches are included in MassSorter.

3.4.1. Considering More Modifications: UniModSearch

One way of increasing the number of matches would be to include many modifications in the theoretical digestion and make them all possible in all the MS experiments. This would probably make the digestion and comparison significantly slower and would also create many incorrect matches, simply by chance; much work must be done to find the correct ones. A better approach is therefore to include in the *in silico* digestion only the modifications that are expected and test for others later. MassSorter includes a local version of the database UniMod ***(10)*** that contains data on a number of different modifications. To search this database for modifications, right click on one of the yellow (unmatched) masses and select "Modification search." The UniModSearch window then appears. Select the relevant settings and click on "Search," and a list of possible modifications that may explain the unmatched *m/z* value is shown. The list is created as follows: All the theoretical *m/z* values between "Search mass + lower limit" and "Search mass + upper limit" are compared to the (unmatched) search mass and the difference is calculated. This difference is compared to the list of mass changes from all the modifications in the UniMod database. If the difference between the "theoretical *m/z* value" + "the mass change of a modification" and the experimental *m/z* value is within the chosen accuracy limit, we have a possible match. If you click on "Insert into DST" the selected modification is inserted into the DST, and the row is colored blue. A match inserted in this way can be removed by right clicking on the given mass and selecting "Remove Match."

3.4.2. Unexpected Cleavage Sites: SequenceSuggester

Another way of increasing the number of identified *m/z* values is to check for "nontheoretical" cleavage sites. When MassSorter digests an amino acid sequence it cleaves only at the theoretically correct sites of the enzyme selected, e.g., trypsin cleaves after R and K, unless followed by P. When digesting in experiments the enzyme sometimes cleaves at other sites as well, or a peptide may be sensitive to chemical cleavage. These two cases, combined or alone, may result in peptides that have one or two terminals that do not match any theoretically digested peptides. To search for these kinds of peptides right click on one of the yellow masses and select "Suggest Sequence(s)." A window similar to the ProteinDigester appears. Choose the relevant parameters and click on "Suggest Sequences." A list of the possible peptides from the given protein sequence, with nontheoretical cleavage sites, appears. If you click on a row in the table, the selected peptide will be marked blue in the frame in the upper right. The red parts of this sequence are the already covered parts. After selecting a row, the match can be inserted into the DST by selecting "Insert Selected Mass into DST" from the "File" menu. The row will be marked NTCS in the modifications column (see row 45 in **Fig. 2**). These matches can be removed by right clicking on the given mass and selecting "Remove Match."

SequenceSuggester is also useful if the protein has an unexpected truncation N- or C-terminally due to a posttranslational maturation of the protein. An example is included in the tutorial at MassSorter's homepage.

3.5. Report and Statistics

The presentations are divided into reports and statistics.

3.5.1. Reports

By use of the Report tool the information in the DST can be presented in a different way. The information is compressed into an html file where (for each experiment and all experiments combined) the matches are divided into different categories: matches with unmodified theoretical peptides, matches with modified theoretical peptides, matches with filter(s), and so on. Additional information is also shown, such as % match (of all the *m/z* values in the given experiment, how many match theoretical values within the given accuracy limit) and sequence coverage. The sequence coverage is also shown in a model of the sequence. The red parts are the covered parts. Underscored residues are residues that may be modified. By right clicking on a covered residue, information about the peptides containing the selected residue is shown. Modification details can be accessed in the same way.

The Report contains a model of the amino acid sequence of the protein. If a PDB file of the protein in question is available, a 3D model can also be shown by clicking on the "View as 3D model" link in the report. A file chooser appears where you select a PDB file from which the 3D model is created. The structural information from the PDB file is then coupled with the coverage data from the Report and a 3D model is created. The 3D model uses the same color-coding scheme as in the Report, but can also be extended to coloring modifications, residues, and/or amino acids.

3.5.2. Statistics

MassSorter includes four types of statistics:

1. Peptide Statistics shows for each Project and experiment the distributions of hydropathy, sequence coverage, average peptide length, average mass, cleavage site frequencies, and amino acid frequencies. It can be used to investigate the impact the different peptide properties have for a peptide to be detected in the mass spectrometer. This has also been previously investigated *(12)*.
2. Accuracy Statistics shows the accuracy with which the matches are found. This can, for example, be used to discover calibration error.

3. Accuracy Plot shows a plot of the accuracy of the matches. Systematic errors in calibration are easily visualized.
4. Fractional Masses shows a plot of the fractional masses. It can be used to indicate whether unmatched masses may be due to nonpeptide ions *(13)*. It may also be used to deduce some peptide properties if the accuracy is high enough (15–20 ppm or better).

3.6. Changing the System Parameters

The system contains many parameters that can be changed. The system parameters are different mass values, peptide terminals, amino acid property values, and available enzyme properties. The cleavage rules of the enzymes can be changed and new enzymes can be added. The standard procedure for changing system parameters is selecting "Options" from the "Tools" menu, but most of them can also be changed from the windows in which they are used. New definitions of modifications can also be added.

3.7. Examples

We will illustrate some of the features in MassSorter by using experimental data. The resulting DST is shown in **Fig. 2**. The integral membrane protein connexin43 (Cx43) was purified by immunoprecipitation from four sources and three species: Syrian hamster embryo (SHE) cells, Chinese hamster V79 cells, Wistar rat embryo cells (here called R5), and HeLa cells transfected with a construct encoding rat Cx43. HeLa cells do not express the endogenous human Cx43. The samples were run on 1D sodium dodecyl sulfate polyacrylamide gel electrophoresis (SDS–PAGE) together with samples that contained only the antibody used for immunoprecipitation. The samples that we here call antibody correspond to gel pieces excised from the antibody lanes at exactly the same position at which Cx43 migrates in the neighboring lane. In this context, our aim is to show that we have been able to purify the correct protein from the four sources (this includes indicating which of the detected peptides are identical or different in the three species, thus showing that PMF is able to distinguish between the conserved protein Cx43 from three closely related species), and further to do a partial characterization of Cx43. The peak lists have been pruned to avoid an excessive discussion of the results.

As we expected the antibody to give some background in the analysis of Cx43, we would like to subtract this background before a more detailed analysis is performed on Cx43. All peak lists were collected in text files and then pasted into MassSorter at the appropriate places. The antibody background will also contain peaks from trypsin, the protease used in these experiments.

1. Defining the background peaks: First, a new project was established for the antibody samples. These samples had been trypsinized in parallel with the Cx43 samples. In this case, we used antibody samples from four experiments. Trypsin was chosen as the theoretical cleavage file, because autolytic trypsin peaks are present in the spectra, and have been used for internal calibration. The four antibody peaklists were imported from a text file by the "copy and paste" function described in **Subheading 3.1.3.** The DST was then created, making it simple to detect peaks found in more than one sample. In this case, we decided that the peaks had to be present in two or more of the samples to be included into a new consensus peak list. Some of the peaks are due to autolytic trypsin peaks. As exact *m/z* values are available for these peaks ***(1)***, the experimental values were replaced by the theoretical values. This peak list functioned as our filter. Note that this approach also can be used for checking the reproducibility of PMF experiments even for unknown proteins.
2. Initial comparison between the theoretical rat Cx43 sequence and experimental rat samples: A tryptic digest of rat Cx43 (NP_036699) was chosen as the basis for the initial comparison with the two samples containing rat Cx43. Another project was established for these samples. MassSorter suggested that 12 peptides are common to the two samples within 50 ppm of the theoretical *m/z* values. Four potential Cx43 peptides are found in either one or the other sample.
3. Application of a filter: However, the majority of experimental masses did not fit the Cx43 sequence. We therefore added the filter defined in *step 1* as described in **Subheading 3.2**. The majority of previously unmatched masses found their hits with the filter. In fact, one of the peptides from the HeLa samples found a better hit with the filter, slightly decreasing the sequence coverage.
4. Partial characterization of unmatched masses: We concentrated on the four pairs of unmatched masses found in both samples. First, the possibility of unexpected cleavages was investigated. The appropriate cell was selected and right-clicked as described in **Subheading 3.4.2**. In most cases, several peptides may fit within the selected accuracy (here 50 ppm), especially if many modifications are allowed during the analysis. The user must decide whether one or none of the peptides could be a realistic possibility, and we recommend a very strict judgment, e.g., restricting the acceptance to previously published unexpected cleavages for the protease used. In our case, two of the four pairs of unmatched peptides fitted two overlapping peptides, 347–362 (*m/z* 1716.94) and 346–362 (*m/z* 1845.02), having a correct cleavage at the N-terminus, but a cleavage between R and P at the C-terminus. The two remaining pairs were analyzed by "Modification search," but no realistic alternatives were suggested.
5. A brief comparison with closely related species: We first added two peak lists from SHE cell samples to the DST created above. Eight peptides coincided with those detected in one or the other of the rat samples. In addition, four peptides not detected in rat Cx43 were found in SHE cells. We then added two peak lists from Chinese hamster Cx43. Eleven peptides coinciding with one or the other rat Cx43 sample. A peptide at 1475.76 in the hamster samples could potentially be

the acetylated N-terminus of Cx43. At present, we have no further support for this suggestion. Overall, there is good reproducibility of the detected peaks between closely related species.

Some peptides clearly showed species-specific distribution in that they are reproducibly found in one species but not in another species. We will here mention only one example, but it consists of three overlapping peptides in each species. In rat Cx43, the peptide 347-VAAGHELQPLAIVDQRPSSR-366 is found at 2144.16. This peptide overlaps peptides 347–362 and 346–362, indicated by SequenceSuggester, as described in *point 4* above. Peptides of *m/z* values 2158.18 and 2176.13 were found in Syrian hamster and Chinese hamster Cx43, respectively. These peptides are usually among the more intense peaks in the Cx43 spectra from the different species. The mass differences are 14.02 Da (corresponding to amino acid changes N→Q, D→E, or V→L/I) and 31.97 Da (A→C or V→M) between rat and the two hamster species. N is not present in this rat peptide, but D, V, and A are. The changes D→E, V→L/I, and V→M would require only one nucleotide difference in the affected codon. Interestingly, we found a peptide at 1730.96 in SHE cells and 1748.93 in V79 cells. These peptides would fit with the unexpected cleavage at 362-RP-363 in the rat samples, having a 14 and 32 Da higher mass than the rat peptide. Similarly, we found peptides 1845.02 (rat), 1859.09 (Syrian hamster), and 1877.00 (Chinese hamster) have the same mass difference. **Figure 2** shows a part of the DST. Subsequent cDNA sequencing showed that the amino acid sequence is 347-IAAG… in Syrian hamster and 347-MAAG… in Chinese hamster ***(14)***. In principle, amino acid-changing short nucleotide polymorphisms are basically similar to this example.

References

1. ProteinProspector, http://prospector.ucsf.edu/
2. Perkins, D. N., Pappin, D. J. C., Creasy, D. M., and Cottrell, J.S. (1999) Probability-based protein identification by searching sequence databases using mass spectrometry data. *Electrophoresis* **20,** 3551–3567.
3. Zhang, W. and Chait, B. T. (2000) ProFound—an expert system for protein identification using mass spectrometric peptide mapping information. *Anal. Chem.* **72,** 2482–2489.
4. Tuloup, M., Hemandez, C., Coro, I., Hoogland, C., Binz, P-A., and Appel, R. D. (2003) Aldente and BioGraph: an improved peptide mass fingerprinting protein identification environment. In *Understanding Biological Systems through Proteomics*. Swiss Proteomics Society, pp. 174–176.
5. Phenyx, http://www.phenyx-ms.com/.
6. Peri, S., Steen, H., and Pandey, A. (2001) GPMAW—a software tool for analyzing proteins and peptides. *Trends Biochem. Sci.* **11,** 687–689.

7. FindMod, http://au.expasy.org/tools/findmod/.
8. Gattiker, A., Bienvenut, W. V., Bairoch, A., and Gasteiger, E. (2002) FindPept, a tool to identify unmatched masses in mass fingerprinting protein identification. *Proteomics* **2,** 1435–1444.
9. Barsnes, H., Mikalsen S-O., and Eidhammer, I. (2006) MassSorter: a tool for administrating and analyzing data from mass spectrometry experiments on proteins with known amino acid sequences. *BMC Bioinform.* **7,** 42–50.
10. UniMod, http://unimod.org/fields.html.
11. RCSB PDB: http://www.rcsb.org./pdb/home/home.do.
12. Schmidt, F., Schmid, M., Jungblut P. R., Mattow, J., Facius, A., and Pleissner, K. P. (2003) Iterative data analysis is the key for exhaustive analysis of peptide mass fingerprints from proteins separated by two-dimensional electrophoresis. *J. Am. Soc. Mass* **14,** 943–956.
13. Wool, A. and Smilansky, Z. (2002) Precalibration of matrix-assisted laser desorption/ionization-time of flight spectra for peptide mass fingerprinting. *Proteomics* **2,** 1365–1373.
14. Cruciani, V., Heintz, K-M., Husøy, T., Hovig, E., Warren, D. J., and Mikalsen, S-O. (2004) The detection of hamster connexins: a comparison of expression profiles with wild-type mouse and the cancer-prone *Min* mouse. *Cell Commun. Adhes.* **11,** 155–171.

24

Database Similarity Searches

Frédéric Plewniak

Summary

With genome sequencing projects producing huge amounts of sequence data, database sequence similarity search has become a central tool in bioinformatics to identify potentially homologous sequences. It is thus widely used as an initial step for sequence characterization and annotation, phylogeny, genomics, transcriptomics, and proteomics studies. Database similarity search is based upon sequence alignment methods also used in pairwise sequence comparison. Sequence alignment can be global (whole sequence alignment) or local (partial sequence alignment) and there are algorithms to find the optimal alignment given particular comparison criteria. However, as database searches require the comparison of the query sequence with every single sequence in the database, heuristic algorithms have been designed to reduce the time required to build an alignment that has a reasonable chance to be the best one. Such algorithms have been implemented as fast and efficient programs (Blast, FastA) available in different types to address different kinds of problems. After searching the appropriate database, similarity search programs produce a list of similar sequences and local alignments. These results should be carefully examined before coming to any conclusion, as many traps await the similarity seeker: paralogues, multidomain proteins, pseudogenes, etc. This chapter presents points that should always be kept in mind when performing database similarity searches for various goals. It ends with a practical example of sequence characterization from a single protein database search using Blast.

Key Words: Similarity; homology; database; search; sequence alignment; sequence comparison.

1. Introduction

When reading this chapter you might expect to find some methods for performing database sequence similarity searches. There is, however, a large number of different web sites providing similarity search services and it would

From: *Methods in Molecular Biology, vol. 484: Functional Proteomics: Methods and Protocols*
Edited by: J. D. Thompson et al., DOI: 10.1007/978-1-59745-398-1, © Humana Press, Totowa, NJ

be impossible to provide exhaustive instructions for using them all. Furthermore, as the sites are modified and improved over time, this chapter might soon be obsolete. Finally, you may even have access to private, local database similarity search services for which I could definitely not provide any instruction.

Therefore, this chapter will not provide any technical recipe for performing database similarity searches. My goal is rather to present a methodology for similarity searching and interpretation of results, including caveats and rules of thumbs, that could help you to obtain the best out of your searches. The ability to extract information from similarity search results is of major importance: why would you want to perform a similarity search if you cannot obtain any information from it?

2. Similarity versus Homology

Before we search for similar sequences, we must understand what similarity is. First, we should always keep in mind that similarity is not a synonym for homology. Similarity can be defined as a measure of the degree to which two sequences look alike. Similarity is therefore quantitative and is represented by a score or a percentage.

On the other hand, homology is an evolutionary relationship between sequences: two sequences are said to be homologous if they share a common ancestor. Thus, sequences are either homologous or they are not: homology cannot be measured and there is no such thing as a percentage of homology.

So why then do we sometimes refer to distant or closely related homologues as if the homology relationship between sequences could be quantified? What is actually quantified in this case is not homology. Homologues are homologous sequences, no less, no more; but depending on the time that separates them from their common ancestor, they may be more or less similar to each other. When two homologues separate from each other during speciation, their sequences are identical or very close to it. However, in the course of evolution, mutations accumulate over time independently in both homologues, and homologous sequences gradually diverge. Therefore, although similarity is a good indicator of homology, two sequences may be homologous and still be more or less similar to each other. Thus, distant homologues and closely related homologues are short-cut terms designating homologues whose sequences are very dissimilar or very similar, respectively.

3. Defining a Sequence Similarity Measure

Similarity is defined above as a measure of how much two sequences look like each other. Therefore, to assess similarity between two sequences, we first need to be able to compare them and then to evaluate the result of this comparison. But what does this mean for biological sequences? As we already stated above, homologous sequences gradually diverge and their similarity decreases during evolution as mutations accumulate. Thus, similarity can be estimated by the amount of potential evolutionary events that occurred since the putative homologues separated from their hypothetical common ancestor: point mutations, insertions, and deletions. And that is actually the underlying rationale for the most widely used sequence similarity measure tool: sequence alignment.

3.1. Sequence Alignment

Basically, a sequence alignment is a representation of possible evolutionary events that may have occurred since the separation of two homologues. In a sequence alignment, it is assumed that stacked residues are equivalent in terms of evolution, structural role, or function, i.e., they are thought to correspond to the same original residue in the common ancestor sequence or play the same role in the protein's function or structural stability. Residues that were probably involved in insertion or deletion events are aligned with gaps.

Sequence alignments may be global or local. In global alignments sequences are aligned over their full length. In this case, sequences are considered to be comparable from their N-terminal end to their C-terminal end. Thus a global alignment requires both sequences to be homologous. On the other hand, only the most similar parts of the sequences are aligned in local alignments. Thus, as sequences do not need to be comparable over their whole length for local alignments, these are suitable for comparing sequences of proteins having only domains or small regions in common.

3.2. Similarity Score

Similarity is quantitative and we need a numerical value computed from the sequence alignment. Many different methods have been proposed and used to address the question of an appropriate measure of similarity.

The simplest method involves counting the proportion of identical residues in aligned sequences relative to the alignment overall length, including gaps. This provides a percentage of identity that also takes into account the size of all gaps in the alignment (**Fig. 1**).

Another method computes a score for the sequence alignment by summing individual scores for stacked residues and subtracting a penalty for gaps (**Fig. 1**). Individual scores for aligning residues are provided by scoring matrices, the simplest one being the identity matrix scoring 1 for identical residues and 0 otherwise. Many other matrices have also been designed to reflect amino acid properties. These replacement scores were either computed from physical and chemical properties *(1)* or from observed frequencies of replacement of an amino

<table>
<tr><th>Scoring method</th><th>LNAWM-ESRC
  || ||
YQAWIVES--</th><th>LNAW-------FGDCGHLNY
  || | ||
YQAWIVESRTGF-DC-----</th></tr>
<tr><td>% identity/alignment length</td><td>4/10 = 40%</td><td>5/20 = 25%</td></tr>
<tr><td>% identity/longest sequence</td><td>4/9 = 44.4%</td><td>5/14 = 35.7%</td></tr>
<tr><td>% identity/shortest sequence</td><td>4/8 = 50%</td><td>5/13 = 38.5%</td></tr>
<tr><td>Identity scoring matrix
gop = 0.5, gep = 0.1</td><td>(0) + (0) + (1) + (1) + (0) + (0) + (1) + (1)
− 2 × 0.5 − 3 × 0.1 = 2.7</td><td>(0) + (0) + (1) + (1) + (1) + (0) + (1) + (1)
−3 × 0.5 − 13 × 0.1 = 2.2</td></tr>
<tr><td>Identity scoring matrix
gop = 0.5, gep = 0.5</td><td>(0) + (0) + (1) + (1) + (0) + (0) + (1) + (1)
− 2 × 0.5 − 3 × 0.5 = 1.5</td><td>(0) + (0) + (1) + (1) + (1) + (0) + (1) + (1)
− 3 × 0.5 − 13 × 0.5 = −3</td></tr>
<tr><td>BLOSUM62 scoring matrix
gop = 4, gep = 1</td><td>(− 1) + (−2) + (4) + (11) + (1) + (5) + (4) −
2 × 4 − 3 × 1 = 11</td><td>(−1) + (−2) + (4) + (11) + (6) + (6) + (9) −
3 × 4 - 13 × 1 = 8</td></tr>
<tr><td>BLOSUM62 scoring matrix
gop = 4, gep = 4</td><td>(− 1) + (−2) + (4) + (11) + (1) + (5) + (4) −
2 × 4 − 3 × 4 = 2</td><td>(−1) + (−2) + (4) + (11) + (6) + (6) + (9) −
3 × 4 − 13 × 4 = −31</td></tr>
</table>

Fig. 1. Examples of alignment scores. Considering the above alignments, their percentage identity relative to alignment length is given by the number of aligned identical residues divided by the length of the reference (alignment, longest or shortest sequence); their similarity score is given by $s(a, b) - gop \cdot n_g - gep \cdot l_g$, where $s(a, b)$ is the individual score for aligning residue a with residue b, gop is the gap opening penalty, gep is the gap extension penalty, n_g is the number of gaps, and l_g is the total length of the gaps. It is clear from the examples above that different methods yield different similarity scores and it is important to specify how a similarity score was computed when producing one. It also appears that increasing the gep strongly penalizes alignments with large gaps.

acid by another in related proteins. Although real properties would seem to provide the most rational similarity scale, statistical scores actually reflect the effect of these properties on protein evolution and mutations allowed by natural selection. Statistical matrices eventually proved to be the most efficient ones *(2)* and today, most similarity search programs use the statistical BLOSUM *(3)* or PAM *(4)* matrices built from reference alignments.

The most widely used gap penalty is the so-called affine gap penalty. It is computed as a linear function of the number of gaps and their total length. Parameters provide control over the relative importance of number and length of gaps: a larger "gap opening penalty" will favor fewer but somewhat larger gaps, whereas a larger "gap extension penalty" would give preference to small gaps (**Fig. 1**).

Most similarity search programs now provide statistics allowing the user to estimate the significance of a similarity score. Expected values computed by

Blast *(5)* or FastA *(6)* from an extreme value distribution *(7)* give the number of times one expects to find by chance an alignment achieving the same score. If such a value is much smaller than 1, it means that the searched database is not large enough to expect to obtain by chance one alignment with this score and the alignment should be considered as significant.

3.3. Alignment Algorithms

Building a sequence alignment involves not simply stacking sequences one over another. Equivalent residues need to be identified and gaps inserted at the proper place to allow this. Several algorithms have been designed to build sequence alignments suitable for sequence similarity determination.

Given a pair of sequences, a scoring matrix, and gap penalties, optimal algorithms return the alignment with the highest possible score. But keep in mind that this does not mean that the alignment produced is the most appropriate one for subsequent biological interpretation and must be taken for granted, but simply that within the defined context no other alignment can be found with a better similarity score. A global optimal alignment was designed *(8)* and is now implemented in the EMBOSS package as the "needle" command. The original algorithm was later modified to produce local alignments *(9)*. This algorithm is implemented as the "water" command in the EMBOSS package. However, as optimal alignments need to explore the whole search space in order to find the best similarity score, they are time consuming and are not suitable for database searches unless highly parallel computers are used.

This is the reason why database similarity search programs use heuristic algorithms. Such algorithms are based upon heuristics, i.e., rules, in order to reduce the time required to build an alignment having a reasonable chance to be the best one. Basically, this is achieved by filtering out regions in which one would reasonably not expect any interesting similarity and by comparing only those regions having a good chance of being equivalent. The consequence of such rules is that there is no guarantee that the best alignment will be found; however, if the rules are reasonable enough there is a good chance of obtaining an appropriate alignment in an acceptable time. The well-known programs FastA *(6)* and Blast *(5)* are implementations of such heuristic database search algorithms.

4. Searching Databases for Similar Sequences: For What?

Similarity search is clearly a central tool in bioinformatics. Its principal use is to identify known homologous sequences for genomic or structural studies, phylogeny. Information gathered from the identified homologues can also help the characterization and annotation of the query sequence.

4.1. Detection of Homologous Sequences

Genomic studies, phylogeny, and structural modeling all require the identification of homologous sequences.

In genomic studies, the presence or absence of homologues of a set or a family of proteins (complex, pathway) in different species may provide invaluable hints about the role of the sequence in a system-oriented context, when examined in light of biological knowledge. Such studies require the availability of a set of complete proteomes or genomes, whose choice depends on the biological problem of interest. If the completeness of the available proteomes is suspicious, then it might be more effective to search the corresponding genomes even if the presence of introns may hinder homologue detection. A sensitive similarity search method is also required in order to avoid missing remote homologues and drawing wrong conclusions in their absence. For instance, in Blast, the expected threshold for returning hits should be set high enough and a thorough examination of returned alignments should be performed before concluding the absence of a homologue. Subsequent multiple alignment of the detected, potentially homologous sequences may help in drawing final conclusions.

Phylogeny requires a set of representative homologous sequences covering a wide enough range of similarity. Searching a generic protein database such as Uniprot should normally be sufficient to gather the necessary sequences that can be subsequently selected according to species. Nonredundant databases such as Uniref90 may facilitate the selection of sequences. It may also be interesting sometimes to search an available complete proteome or genome in order to reduce noise and facilitate the detection of a potentially very distant homologue. Homologue detection is only a prerequisite of phylogeny studies, which involve more specialized computations that will not be covered in this chapter.

Structural modeling by homology exploits the structure/sequence relationship paradigm. Homologues are supposed to share the same structure. Therefore if the structure of a given protein is known, it should be possible to predict the structure of its homologue. Specialized programs exist for doing this once a homologue with a known structure has been detected by searching 3D structure databases such as the Protein Data Bank (PDB).

4.2. Sequence Annotation

Let us assume that we are faced with an uncharacterized protein sequence and we would like to obtain as much information about it as possible before we decide whether to undertake further biological experiments. As homologous sequences derive from a common ancestor, it is quite reasonable to think that their function has not changed much since they separated. We also already know that homologues have similar sequences so that the more similar two sequences are, the more chance they have of being homologous. Thus, the relationship of similarity between sequences defines a relationship of homology between proteins, which can in turn be used to deduce the function of the uncharacterized protein: if two sequences are sufficiently similar the corresponding proteins can be said to be homologous and have the same function. This is the well-known sequence/function relationship.

However, there are quite a few limitations to this paradigm due to the existence of paralogues and the modular organization of proteins. Paralogues can be defined as homologues originating from a duplication event and often have a different, though similar, function. On the other hand, orthologues are strict homologous equivalents in two different species and have the same function. Thus, when a similar sequence is found in another species, it is not always clear whether it is the true orthologue or a potential paralogue. The final decision usually requires more in-depth studies that extend beyond the scope of the present chapter: conserved genomic localization and short-range synteny favoring orthology, expression pattern, wet biology experiments, etc.

Most proteins are organized in domains that can be seen as elementary modules from which new proteins are built in the course of evolution. Thus, two different proteins may share one or several common domains even if they are not strictly homologous and do not have the same function. In this case, similarity may be locally very high over the common domains, but it would be wrong to assume homology and identity of function based on these similarity results. However, the common domains might somehow be considered as homologous modules having a similar function or role, such as the DNA-binding domain. Thus, even if similarity between two sequences is only partial, it is nonetheless possible to deduce some information about the protein function. This is why the NCBI Blast server searches the Conserved Domain Database before performing the actual Blast search in order to produce a map of potential domains for the query protein sequence.

In the case of highly diverged sequences (whole sequence or domains), similarity may have become extremely low. However, as selection pressure exerts most of its influence on sequence segments or residues that are important for the function (catalytic sites, binding sites), locally conserved segments or "words" can often be identified in database similarity search results. Thus, very distant homologues may be detected due to the presence of conserved words. Furthermore, as these conserved segments can be associated with protein function, their detection provides useful hints in protein sequence annotation.

However, although it is possible to obtain much information from a database similarity search (especially today as there are more and more characterized sequences in databases), it is often necessary to refine annotation through the use of a multiple alignment of detected similar sequences. Sequence annotation and multiple alignment are discussed in more detail elsewhere in this book.

4.3. Sequence Identification

Given a sequence, or a portion of it, obtained from proteomics experiments, or a cDNA library, the corresponding protein can be identified by searching an up-to-date generic protein or mRNA sequence database such as Uniprot or Refseq mRNA. Of course, in a perfect world, sequence identification would not necessitate similarity search as we are actually looking for identical sequences. However, the required sequence may not be available yet in databases, or there might be errors in the sequence to be identified. Thus, because similarity search is able to detect not only identical sequences but also very similar ones, it is able to overcome these problems.

Gene expression studies, chromosome localization, and exon mapping are also based upon sequence identification. However, the problem is reversed: the query sequence is known and the object is to identify the corresponding sequences in a database. For gene expression, an expressed sequence tags (EST) database can be searched to provide information about where and when a given gene is expressed. Note that EST databases can also be useful to identify alternatively spliced sequences. For chromosome localization and exon mapping, the complete genome should be searched. However, be aware that in eukaryotes, pseudogenes lacking introns may score better than the actual gene because introns introduce large gaps in the alignment.

5. Searching Databases for Similar Sequences: How?

5.1. Which Programs for Which Purpose?

Smith-Waterman or Needleman-Wunsch optimal algorithm implementations are very time consuming and cannot reasonably be applied to database similarity searching without the help of massively parallel machines. Database searching is thus best performed by specialized programs such as FastA *(6)* and Blast *(5)* using heuristics to detect similarities in databases. The main advantage of Blast over FastA for protein database searching is that the Blast algorithm uses a scoring matrix from the very first step, when defining elementary words and their synonyms before searching the database dictionary to detect potential similar regions. FastA, on the other hand, searches for identical words at this step.

PsiBlast *(5)* is an iterative version of the Blast algorithm that is highly sensitive and useful for remote homologue detection in protein databases. This algorithm starts with a regular blastp search and then builds a position-specific scoring matrix (PSSM) from the best hits. It then uses this PSSM to search the database again and detect more distant similarities. At each step, PsiBlast refines its PSSM from the best hits found so far and searches the database for even more distantly similar sequences until it converges and no new hit is found or it reaches a predefined number of iterations. This method is very sensitive at the expense of computation time since each iteration takes at least as long as a single Blast search.

Blast and FastA algorithms have been implemented in different types for different purposes. **Table 1** shows the available Blast and FastA programs for different goals.

Blast programs can be used on-line on the NCBI server:
  http://www.ncbi.nlm.nih.gov/BLAST/.
  A command-line version of Blast can be downloaded from ftp://ftp.ncbi.nih.gov/blast/.

FastA programs can be used on-line on the University of Virginia web server:
  http://fasta.bioch.virginia.edu/fasta_www2/fasta_list2.shtml.
  A command-line version of fasta can be downloaded from
  http://fasta.bioch.virginia.edu/fasta_www2/fasta_down.shtml.

5.2. Database Choice

5.2.1. Generic or Specialized Database

The choice between a generic or a specialized database actually depends on your goal and what is available. You should use a generic database such as Uniprot ***(10)*** or Refseq_Protein ***(11)*** if you know nothing about your protein and if the appropriate specialized database is too small or does not exist. For structure homology modeling searching, a suitable database of sequences can be extracted from the PDB of 3D structures ***(12)***. Full proteomes, if available, are databases of choice when doing phylogeny or genomics studies.

Many generic databases are somewhat redundant for technical reasons, because of research trends or simply because of the existence of large multigene families. For instance, version 11.2 of the Uniprot database contains over 18,500 gag proteins, around 16,000 of which come from the human immunodeficiency virus, including more than 14,500 fragments. Thus, there is an overrepresentation of some sequences in databases and on some occasions interesting similarity results may be lost or hidden in the vast amount of redundant information. This problem may be addressed by using nonredundant databases such as Uniref100, Uniref90, or NCBI's nrdb. Uniref100 and Uniref90 yield a database size reduction of approximately 10% and 40%, respectively ***(13)***.

Nonredundant databases are also useful if you need a representative sample of similar sequences ranging from close relatives down to distant homologues.

Table 1
Available Blast and FastA Programs for Different Goals[a]

Goals	Query	Database	Comparison	Programs
Homologue search for annotation, phylogeny, etc. of noncoding sequences (promoters)	Nucleotide	Nucleotide	Nucleotide	Blastn fasta
Homologue search for annotation, phylogeny, structural modeling, etc.	Protein	Protein	Protein	Blastp psiblast fasta
Homologue search for annotation, phylogeny Expression Alternative splicing sites Exon map Chromosome localization	Protein	Nucleotide (translated in all six phases)	Protein	tblastn tfasta
Homologue search for annotation, phylogeny of coding sequences Find open reading frames in the query	Nucleotide (translated in all six phases)	Protein	Protein	Blastx Fastx
Homologue search for annotation of coding sequences Expression Alternative splicing sites Exon map Chromosome localization	Nucleotide (translated in all six phases)	Nucleotide (translated in all six phases)	Protein	tblastx tfastx

[a]Blast and FastA algorithms have been implemented in different types adapted to different purposes.

You may also create your own database if you have access to a local version of FastA or Blast. All you have to do is extract the required sequences in fasta format using any database querying system (SRS or NCBI Entrez or any other tool available). FastA is able to search fasta formatted files and Blast comes with a command (formatdb) to build a personal Blast database from fasta sequence files.

5.2.2. Nucleotide Databases (ESTs, Genomics)

Even when you are interested only in protein studies, nucleotide databases can still be useful. Full genomes may indeed provide a valuable alternative to incomplete or unavailable proteomes for genomics studies. ESTs and high-throughput cDNA (HTC) databases may be interesting for expression and alternative splicing studies.

5.2.3. Size Matters

One thing to keep in mind is that the size of the database has some effect on Blast or FastA statistics. In small databases the expected value for a given score is smaller than the expected value for the same score obtained from searching a large database. This sounds perfectly logical because one can expect to find an alignment with a given score in a large database more often than in a small one. Expected values obtained from different searches should therefore not be compared unless the size of the search space was identical for both searches. This is possible with the -z Blast parameter that allows the user to set the size of the search space. Thus, it is possible to search a small database, and obtain statistics as if they were computed from searching a larger dataset.

Database size should not be much of a problem for relatively close sequences, but it may make a difference for distant sequences: for example, when searching a small database of 838 nuclear receptors with RXRA_HUMAN Uniprot sequence, the *Caenorhabditis elegans* nuclear hormone receptor NHR9_CAEEL sequence is identified as a similar sequence with an expected value of 10^{-4}, while it has a much less significant expected value of 0.26 when searching the 4,736,514 sequences of the whole uniprot database.

5.3. Filtering Out Low Complexity Segments

Many proteins contain low complexity segments, i.e., segments containing predominantly one or a few amino acids, or very short repeats, or even runs of one amino acid. Such segments may be artificially aligned to totally unrelated sequences with a relatively high score and a significant expected value. A database search with a sequence containing low complexity segments might therefore be cluttered with many false positives and may be very difficult to exploit and interpret. There are filtering programs such as SEG ***(14)*** that are able to mask low complexity segments in sequences to reduce the number of false positives. Blast programs propose to filter the protein query sequence with the SEG algorithm in order to mask low complexity regions before performing the database search (option -F of the blastall program). However, although it is generally a good idea to filter sequences before a similarity search, filtering algorithms may mask some functional sites such as zinc fingers. For instance, the SEG algorithm used by Blast filters out the CRLKKLKCSKEKPKCKAC segment overlapping both yeast Gal4 zinc fingers.

6. Interpretation of Similarity Search Results: A Practical Approach

Let us assume that the following sequence is unknown and we would like to characterize it by searching a protein database sequence:

```
>Unknown
MEHTEIDHWLEFSATKLSSCDSFTSTINELNHCLSLRTYLVGNSLSLADLCVWATLKGNA
AWQEQLKQKKAPVHVKRWFGFLEAQQAFQSVGTKWDVSTTKARVAPEKKQDVGKFVELPG
AEMGKVTVRFPPEASGYLHIGHAKAALLNQHYQVNFKGKLIMRFDDTNPEKEKEDFEKVI
LEDVAMLHIKPDQFTYTSDHFETIMKYAEKLIQEGKAYVDDTPAEQMKAEREQRIESKHR
KNPIEKNLQMWEEMKKGSQFGHSCCLRAKIDMSSNNGCMRDPTLYRCKIQPHPRTGNKYN
VYPTYDFACPIVDSIEGVTHALRTTEYHDRDEQFYWIIEALGIRKPYIWEYSRLNLNNTV
LSKRKLTWFVNEGLVDGWDDPRFPTVRGVLRRGMTVEGLKQFIAAQGSSRSVVNMEWDKI
WAFNKKVIDPVAPRYVALLKKEVIPVNVPEAQEEMKEVAKHPKNPEVGLKPVWYSPKVFI
EGADAETFSEGEMVTFINWGNLNITKIHKNADGKIISLDAKFNLENKDYKKTTKVTWLAE
TTHALPIPVICVTYEHLITKPVLGKDEDFKQYVNKNSKHEELMLGDPCLKDLKKGDIIQL
QRRGFFICDQPYEPVSPYSCKEAPCVLIYIPDGHTKEMPTSGSKEKTKVEATKNETSAPF
KERPTPSLNNNCTTSEDSLVLYNRVAVQGDVVRELKAKKAPKEDVDAAVKQLLSLKAEYK
EKTGQEYKPGNPPAEIGQNISSNSSASILESKSLYDEVAAQGEVVRKLKAEKSPKAKINE
AVECLLSLKAQYKEKTGKEYIPGQPPLSQSSDSSPTRNSEPAGLETPEAKVLFDKVASQG
EVVRKLKTEKAPKDQVDIAVQELLQLKAQYKSLIGVEYKPVSATGAEDKDKKKKEKENKS
EKQNKPQKQNDGQRKDPSKNQGGGLSSSGAGEGQGPKKQTRLGLEAKKEENLADWYSQVI
TKSEMIEYHDISGCYILRPWAYAIWEAIKDFFDAEIKKLGVENCYFPMFVSQSALEKEKT
HVADFAPEVAWVTRSGKTELAEPIAIRPTSETVMYPAYAKWVQSHRDLPIKLNQWCNVVR
WEFKHPQPFLRTREFLWQEGHSAFATMEEAAEEVLQILDLYAQVYEELLAIPVVKGRKTE
KEKFAGGDYTTTIEAFISASGRAIQGGTSHHLGQNFSKMFEIVFEDPKIPGEKQFAYQNS
WGLTTRTIGVMTMVHGDNMGLVLPPRVACVQVVIIPCGITNALSEEDKEALIAKCNDYRR
RLLSVNIRVRADLRDNYSPGWKFNHWELKGVPIRLEVGPRDMKSCQFVAVRRDTGEKLTV
AENEAETKLQAILEDIQVTLFTRASEDLKTHMVVANTMEDFQKILDSGKIVQIPFCGEID
CEDWIKKTTARDQDLEPGAPSMGAKSLCIPFKPLCELQPGAKCVCGKNPAKYYTLFGRSY
```

To do so, we will search the protein generic Swiss-Prot database with the blastp program and, for the sake of the demonstration, we will pretend that the above sequence was not previously described and is not already present in the database.

6.1. Description Review

A quick review of the description for sequences identified as similar by blastp shows a large majority of glutamyl-tRNA synthetase and prolyl-tRNA synthetase sequences. Of the 348 hits with an expected value of less than 10^{-3} (a typical threshold to decide whether a hit is significant) 248 sequences are glutamyl-tRNA synthetases and 31 are prolyl-tRNA synthetases. We can therefore reasonably assume that our sequence is an aminoacyl-tRNA synthetase. But which one? Glutamyl or prolyl? The statistics would tend to favor glutamyl, but they could well be due to a bias in the relative number of prolyl- and glutamyl-tRNA synthetases in the database. And after all, the first hit is prolyl.

6.2. Always Remember You're a Biologist

Could it be that our unknown sequence is a prolyl-tRNA synthetase (the first hit) and would also be very similar to glutamyl-tRNA synthetase sequences? This would indeed explain the mix of the two. But here we have a problem: as biologists, we should know that there are two classes of aminoacyl-tRNA synthetases and that proteins from one class are not homologous, or similar, to those from the other class. And glutamyl-tRNA and prolyl-tRNA synthetases are not from the same classes and therefore are not similar to each other. Thus, our sequence must contain at least two regions: one similar to a glutamyl-tRNA synthetase domain and the other one similar to a prolyl-tRNA synthetase domain. This is confirmed by looking at alignments where it is possible to spot relatively conserved residue motifs specific to both classes.

6.3. Do Not Trust the First Hit Alone

The above conclusion also teaches us that the first hit should always be considered with caution and may not always be the most pertinent one. For instance, the similarity of the first hit may be only partial, but nonetheless may score better than the similarity to any known homologue. The basic assumptions for similarity searches (smaller or no gaps are preferred to long gaps) may also favor sequences other than those for which we are actually looking. This is the case when looking for chromosome localization: pseudogenes that do not contain introns usually score better than the actual corresponding genes whose coding sequence may be interrupted by large introns.

In our case, trusting the first hit alone would have led us to wrongly conclude our unknown sequence is a prolyl-tRNA synthetase without any further consideration of its glutamyl-tRNA synthetase part.

6.4. Keep an Eye on Sequence Length: Hit Versus Query

Now, let us have a look at sequence length. Our query sequence is 1440 amino acids long. The length of most of the sequences identified by blastp ranges from 500 to 800 residues. This confirms that similarity is partial. This also tells us that we could easily fit two similar sequences, one prolyl-tRNA synthetase and one glutamyl-tRNA synthetase, in our sequence.

6.5. Hit Position Is Informative

If we further examine the alignments returned by blastp we can easily notice that all glutamyl-tRNA synthetases are aligned over a large portion of their sequence with the N-terminal end of our sequence, roughly from position 120 to position 620. Prolyl-tRNA synthetases, however, are aligned with the C-terminal end of our query, approximately from residue 940 to residue 1440. Both types of alignments are around 500 amino acids long, enough to fit a full tRNA synthetase, potential extensions not included. We could therefore hypothesize that our sequence is a bifunctional tRNA synthetase: a glutamyl-prolyl-tRNA synthetase.

But this leaves us with an uncharacterized region of about 320 residues between the two characterized domains. Searching the alignments returned by blastp, we can find the hits shown in **Fig. 2**. The first hit is described as a fragment of a bifunctional glutamyl-prolyl-tRNA synthetase, but with 49 residues only. I would use this information with great caution. The second hit shows a partial similarity with a tryptophanyl-tRNA synthetase. A closer look at the alignment positions shows clearly that our uncharacterized region is actually the repetition of three modules between residues 670 and 880, separated by roughly 20 residues. We cannot say anything more from these results alone, but these modules are actually WHEP-RS repeats, as could be deduced from searching a domain or family database such as Pfam *(15)* with this region. WHEP-RS are repeats found in bifunctional tryptophanyl and histidinyl-tRNA synthetases.

To conclude this quick study, we can say that our formerly unknown protein is a bifunctional Glu-Pro-tRNA synthetase containing one N-terminal glutamyl-tRNA synthetase domain from 120 to 620 and one C-terminal prolyl-tRNA synthetase domain from 940 to 1440, separated by three WHEP RS repeats.

6.6. Beware of Fragmentary Information and Errors in Databases

Protein sequences derived from the sequence of functionally cloned cDNA are usually of high quality, although some may represent fragments of full-size proteins. With the genome sequencing projects, many protein sequences in databases are now predicted from genomic sequences by computer programs. It has become evident that these programs may produce inaccurate or invalid data. For instance, translation start sites in prokaryotes or exon boundary determination in eukaryotes have been reported to be unsatisfactory.

```
>SYEP_CRIGR (Q7SIA2) Bifunctional aminoacyl-tRNA synthetase [Includes:
          Glutamyl-tRNA synthetase (EC 6.1.1.17); Prolyl-tRNA synthetase (EC 6.1.1.15)
          (Fragment)
         Length = 49

 Score = 87.0 bits (214), Expect = 4e-16
 Identities = 40/47 (85%), Positives = 46/47 (97%)

Query:755 YDEVAAQGEVVRKLKAEKSPKAKINEAVECLLSLKAQYKEKTGKEYI 801
          YD++AAQGEVVRKLKAEK+PKAK+ EAVECLLSLKA+YKEKTGKEY+
Sbjct:1   YDKIAAQGEVVRKLKAEKAPKAKVTEAVECLLSLKAEYKEKTGKEYV 47


 Score = 70.9 bits (172), Expect = 3e-11
 Identities = 34/49 (69%), Positives = 42/49 (85%)

Query:682 YNRVAVQGDVVRELKAKKAPKEDVDAAVKQLLSLKAEYKEKTGQEYKPG 730
          Y+++A QG+VVR+LKA+KAPK  V  AV+ LLSLKAEYKEKTG+EY PG
Sbjct:1   YDKIAAQGEVVRKLKAEKAPKAKVTEAVECLLSLKAEYKEKTGKEYVPG 49


 Score = 62.0 bits (149), Expect = 1e-08
 Identities = 31/48 (64%), Positives = 37/48 (77%)

Query:833 FDKVASQGEVVRKLKTEKAPKDQVDIAVQELLQLKAQYKSLIGVEYKP 880
          +DK+A+QGEVVRKLK EKAPK +V  AV+ LL LKA+YK   G EY P
Sbjct:1   YDKIAAQGEVVRKLKAEKAPKAKVTEAVECLLSLKAEYKEKTGKEYVP 48


>SYW_MOUSE (P32921) Tryptophanyl-tRNA synthetase (EC 6.1.1.2)
          Length = 481

 Score = 69.7 bits (169), Expect = 6e-11
 Identities = 34/63 (53%), Positives = 45/63 (71%), Gaps = 3/63 (4%)

Query:671 NCTTSEDSLVLYNRVAVQGDVVRELKAKKAPKEDVDAAVKQLLSLKAEYKEKTGQEYKPG 730
          +CT+    L L+N +A QG++VR LKA  APK+++D+AVK LLSLK  YK   G+EYK G
Sbjct:9   SCTSP---LELFNSIATQGELVRSLKAGNAPKDEIDSAVKMLLSLKMSYKAAMGEEYKAG 65

Query: 731 NPP 733
            PP
Sbjct: 66  CPP 68


 Score = 60.1 bits (144), Expect = 5e-08
 Identities = 29/59 (49%), Positives = 39/59 (66%)

Query:821 PAGLETPEAKVLFDKVASQGEVVRKLKTEKAPKDQVDIAVQELLQLKAQYKSLIGVEYK 879
          P+G        LF+ +A+QGE+VR LK   APKD++D AV+ LL LK  YK+ +G EYK
Sbjct:5   PSGESCTSPLELFNSIATQGELVRSLKAGNAPKDEIDSAVKMLLSLKMSYKAAMGEEYK 63


 Score = 52.8 bits (125), Expect = 8e-06
 Identities = 25/47 (53%), Positives = 34/47 (72%)

Query:754 LYDEVAAQGEVVRKLKAEKSPKAKINEAVECLLSLKAQYKEKTGKEY 800
          L++ +A QGE+VR LKA  +PK +I+ AV+ LLSLK  YK   G+EY
Sbjct:16  LFNSIATQGELVRSLKAGNAPKDEIDSAVKMLLSLKMSYKAAMGEEY 62
```

Fig. 2. Blast hits for repeats in the query. Blast aligned the same portion of the database sequence with different segments from the query sequence.

Furthermore, even for high-quality sequences, the annotation process may yield errors if not conducted properly. Let us consider again the first hit shown in **Fig. 2**. It is obvious that such a short fragment is not enough to conclude that the full sequence is a bifunctional tRNA synthetase as stated by its description in the database, particularly as this fragment is a repeat that can be found in tRNA synthetases other than bifunctional ones such as tryptophanyl-tRNA synthetases (**Fig. 2**). This sequence was probably annotated, perhaps automatically, by propagating the description of the first hit of a database similarity search. As the sequence was obviously much shorter than the detected bifunctional Glu-Pro-tRNA synthetase, it was identified as a fragment without noticing it was a repeat sequence and could possibly also be a tryptophanyl- or histidinyl-tRNA synthetase. This is another example, if needed, indicating that it is not always possible to trust the first hit only.

6.7. Expected Value Is Just a Statistical Indicator

Finally, one last word about expected values. Blast, FastA, and other programs provide expected values calculated from an extreme value distribution. These expected values provide an indication of the statistical significance of the returned alignments. As such, they are very useful to determine quickly if an alignment is likely to be pertinent. However, databases are not random sets of random sequences and all residues in a biological sequence are not functionally or structurally equivalent. Thus, functional motifs and signatures important to the protein function may be well conserved in an alignment while no similarity could be clearly detected between motifs. Such an alignment would probably be given a poor expected value, though it would probably be biologically pertinent. On the other hand, some paralogues may be similar enough to the query to obtain an excellent expected value, much better than remote orthologues. We have an example of this in **Fig. 3**, where we can see some prolyl-tRNA synthetase paralogues (threonyl-tRNA synthetases) reaching an expected

```
...
SYE_STAAC (Q5HIE7) Glutamyl-tRNA synthetase (EC 6.1.1.17) (Gluta...   67   4e-10
SYT_LEGPL (Q5WT82) Threonyl-tRNA synthetase (EC 6.1.1.3) (Threon...   67   5e-10
SYT_LEGPH (Q5ZS05) Threonyl-tRNA synthetase (EC 6.1.1.3) (Threon...   67   5e-10
SYE_PHOPR (Q6LTT8) Glutamyl-tRNA synthetase (EC 6.1.1.17) (Gluta...   67   5e-10
...
SYT_STRMU (Q8DT12) Threonyl-tRNA synthetase (EC 6.1.1.3) (Threon...   37   0.35
SYP_RICPR (Q9ZDE7) Prolyl-tRNA synthetase (EC 6.1.1.15) (Proline...   37   0.35
EF1G_ORYSA (Q9ZRI7) Elongation factor 1-gamma (EF-1-gamma) (eEF-...   37   0.35
...
```

Fig. 3. Some paralogous sequences may reach more significant expected values than orthologues. Here, the prolyl-tRNA synthetase paralogues (threonyl-tRNA synthetases) have an expected value of 5×10^{-10}, equal to those of some true glutamyl-tRNA synthetase orthologues. On the other hand, some true prolyl-tRNA synthetases scored a mere 0.35 and are lost among remote paralogues (threonyl-tRNA synthetase) and false positives (elongation factor 1-gamma).

value of 5×10^{-10}, equivalent to the expected value of true glutamyl-tRNA synthetase orthologues. This expected value is much more significant than the 0.35 obtained by some prolyl-tRNA synthetases lost between remote paralogues and false positives.

7. Conclusion

We have seen that there is much more to database similarity searching than simply identifying one potential homologue and that much information can be extracted from the results. For this, a careful interpretation of these results in light of biological knowledge is most important to avoid errors and wrong conclusions that could hinder further studies. Sequence database similarity searching thus plays a key role in bioinformatics as the first step in sequence annotation methods, phylogeny, and structural and genomics studies that will be carried with more specialized programs and methods.

References

1. Rao, J. K. M. (1987) New scoring matrix for amino acid residue exchange based on residue characteristic physical parameters. *Int. J. Peptide Protein Res.* **29**, 276–281.
2. Henikoff, S. and Henikoff, J. G. (1993) Performance evaluation of amino acid substitution matrices. *Proteins: Structure Function Genet.* **17**, 49–61.
3. Henikoff, S. and Henikoff, J. G. (1992) Amino acid substitution matrices from protein blocks. *Proc. Natl. Acad. Sci. USA* **89**, 10915–10919.
4. Dayhoff, M. O., Schwartz, R. M., and Orcutt, B. C. (1978) A model of evolutionary change in proteins. *Atlas Protein Sequence Struct.* **5**, 345–352.
5. Altschul, S. F., Madden, T. L., Schaeffer, A. A., Zhang, J., Zhang, Z., Miller, W., and Lipman, D. J. (1997) Gapped BLAST and PSI-BLAST: a new generation of protein database search programs. *Nucleic Acids Res.* **25**, 3389–3402.
6. Pearson, W. R. and Lipman, D. J. (1988) Improved tools for biological sequence comparison. *Proc. Natl. Acad. Sci. USA* **85**, 2444–2448.
7. Gumbel, E. J. (1958) *Statistics of Extremes*. Columbia University Press, New York.
8. Needleman, S. B and Wunsch, C. D. (1970) A general method applicable to the search for similarities in the amino acid sequences of two proteins. *J. Mol. Biol.* **48**, 443–453.
9. Smith, T. F. and Waterman, M. S. (1981) Identification of common molecular subsequences. *J. Mol. Biol.* **147**, 195–197.
10. The UniProt Consortium. (2007) The Universal Protein Resource (UniProt). *Nucleic Acids Res.* **35**, D193–D197.
11. Pruitt, K. D., Tatusova, T., and Maglott, D. R. (2007) NCBI reference sequences (RefSeq): a curated non-redundant sequence database of genomes, transcripts and proteins. *Nucleic Acids Res.* **35**, D61–D65.

12. Berman, H. M., Battistuz, T., Bhat, T. N., Bluhm, W. F., Bourne, P. E., Burkhardt, K., Feng, Z., Gilliland, G. L., Iype, L., Jain, S., Fagan, P., Marvin, J., Padilla, D., Ravichandran, V., Schneide, B., Thanki, N., Weissig, H., Westbrook, J. D., and Zardecki, C. (2002) The Protein Data Bank. *Acta Crrystallogr. D Biol. Crystallogr.* **58**, 899–907.
13. Suzek, B. E., Huang, H., McGarvey, P., Mazumder, R., and Wu, C. H. (2007) UniRef: comprehensive and non-redundant UniProt reference clusters. *Bioinformatics* **23,** 282–288.
14. Wootton, J. C. and Federhen, S. (1993) Statistics of local complexity in amino acid sequences and sequence databases. *Comput. Chem.* **17**, 149–163.
15. Bateman, A., Coin, L., Durbin, R., Finn, R. D., Hollich, V., Griffiths-Jones, S., Khanna, A., Marshall, M., Moxon, S., Sonnhammer, E. L., Studholme, D. J., Yeats, C., and Eddy, S. R. (2004) The Pfam protein families database. *Nucleic Acids Res.* **32**, D138—D141.

25

Protein Multiple Sequence Alignment

Chuong B. Do and Kazutaka Katoh

Summary

Protein sequence alignment is the task of identifying evolutionarily or structurally related positions in a collection of amino acid sequences. Although the protein alignment problem has been studied for several decades, many recent studies have demonstrated considerable progress in improving the accuracy or scalability of multiple and pairwise alignment tools, or in expanding the scope of tasks handled by an alignment program. In this chapter, we review state-of-the-art protein sequence alignment and provide practical advice for users of alignment tools.

Key Words: Multiple sequence alignment; review; proteins; software.

1. Introduction

Sequence alignment is a standard technique in bioinformatics for visualizing the relationships between residues in a collection of evolutionarily or structurally related proteins (*see* **Note 1**). Given the amino acid sequences of a set of proteins to be compared, an alignment displays the residues for each protein on a single line, with gaps ("–") inserted such that "equivalent" residues appear in the same column. The precise meaning of equivalence is generally context dependent: for the phylogeneticist, equivalent residues have common evolutionary ancestry; for the structural biologist, equivalent residues correspond to analogous positions belonging to homologous folds in a set of proteins; for the molecular biologist, equivalent residues play similar functional roles in their corresponding proteins. In each case, an alignment provides a bird's eye view of the underlying evolutionary, structural, or functional constraints characterizing a protein family in a concise, visually intuitive format.

From: *Methods in Molecular Biology, vol. 484: Functional Proteomics: Methods and Protocols*
Edited by: J. D. Thompson et al., DOI: 10.1007/978-1-59745-398-1, © Humana Press, Totowa, NJ

In this chapter, we review state-of-the-art techniques for protein alignment. The literature is vast, and hence our presentation of topics is necessarily selective (*see* **Note 2**). Here, we address the problem of alignment construction: we survey the range of practical techniques for computing multiple sequence alignments, with a focus on practical methods that have demonstrated good performance on real-world benchmarks. We discuss current software tools for protein alignment and provide advice for practitioners looking to get the most out of their multiple sequence alignments.

2. Algorithms

Most modern programs for constructing multiple sequence alignments (MSAs) consist of two components: an *objective function* for assessing the quality of a candidate alignment of a set of input sequences, and an *optimization procedure* for identifying the highest scoring alignment with respect to the chosen objective function *(1)*. In this section, we describe common themes in the architecture of modern MSA programs (*see* **Fig. 1**).

2.1. The Sum-of-Pairs Scoring Model

In the problem of pairwise sequence alignment, the score of a candidate alignment is typically defined as a summation of *substitution scores*, for matched

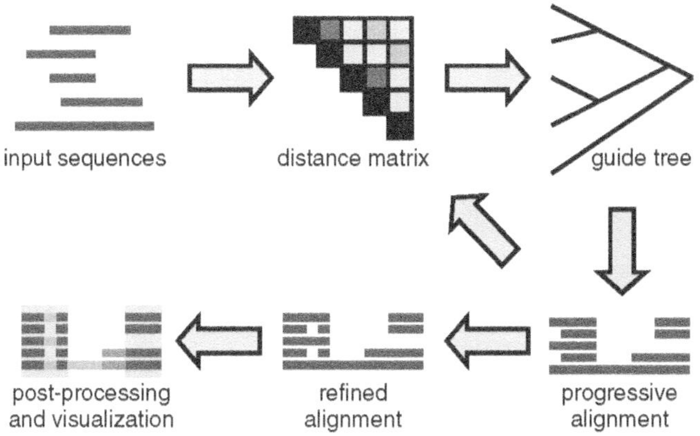

Fig. 1. Diagram of the basic steps in a prototypical modern multiple sequence alignment program: computation of matrix of distances between all pairs of input sequences; estimation of phylogenetic guide tree based on distance matrix; progressive alignment according to guide tree; guide tree reestimation and realignment; iterative refinement; and postprocessing and visualization.

pairs of characters in the sequences being aligned, and *gap penalties*, for consecutive substrings of gapped characters. Given a fixed set of scoring parameters, efficient dynamic programming algorithms (*see* **Note 3**) for computing the optimal alignment of two sequences in quadratic time and linear space have been known since the early 1980s *(2–5)*.

In the case of multiple sequence alignment for N sequences, the multiple alignment score is usually defined to be the summed scores of all $N(N-1)/2$ pairwise projections of the original candidate MSA to each pair of input sequences. This is known as the standard *sum-of-pairs* (SP) scoring model *(6)*. While other alternatives exist, such as consensus *(7)*, entropy *(8)*, or circular sum *(9)* scoring, most alignment methods rely on the SP objective and its variants. Unlike the pairwise case, multiple sequence alignment under the SP scoring model is NP-complete *(10–13)*; direct dynamic programming methods for multiple alignment require time and space exponential in N.

Some strategies for dealing with the exponential cost of multiple alignment involve pruning the space of candidate multiple alignments. The "MSA" program *(14,15)*, for instance, uses the Carrillo–Lipman bounds *(16)* in order to determine constraints on an optimal multiple alignment based on the projections of the alignment to all pairs of input sequences; similarly, the DCA program *(17–21)* employs a divide-and-conquer approach that uses pairwise projected alignments to identify suitable "cut" points for partitioning a large multiple alignment into smaller subproblems. In practice, however, these methods are impractical for more than a few sequences. Consequently, most current techniques for SP-based multiple alignment work by either applying heuristics to solve the original NP-complete optimization problem approximately, or replacing the SP objective entirely with another objective whose optimization is tractable.

2.2. Global Optimization Techniques

In general, finding a mathematically optimal multiple alignment of a set of sequences can be formulated as a complex optimization problem: given a set of candidate MSAs, identify the alignment with the highest score. Global optimization techniques, developed in applied mathematics and operations research, provide a generic toolbox for tackling complex optimization problems. Over the past several decades, application of these methods to the MSA problem has become routine.

Among these methods, *genetic algorithms* *(22)*—which maintain a population of candidate alignments that are stochastically combined and mutated through a directed evolutionary process—have been particularly popular *(23–28)*. In this technique, the SP objective (or an approximation thereof) provides a measure of fitness for individual alignments within the population. Typical mutation operations involve local insertion, deletion, or shuffling of gaps; designing these operations in a manner that allows fast traversal of the space of candidate alignments while remaining efficient to compute is the main challenge in the development of effective genetic algorithm approaches for MSA. Sequence alignment programs based on genetic algorithms include SAGA *(29)*, MAGA *(30–33)*, and PHGA *(34)*.

In *simulated annealing* *(35)*, a candidate alignment is also iteratively modified via local perturbations in a stochastic manner, which tends toward alignments with high SP scores *(36–38)*. Unlike genetic algorithms, simulated annealing approaches do not maintain a population of candidate solutions; rather, modifications made to candidate solutions may either improve or decrease the objective function, and the probability of applying a particular modification to a candidate alignment is dependent both on the resulting change in SP score and on a scaling constant known as the temperature. In theory, when using appropriately chosen temperature schedules, simulated annealing provably converges to optimal MSAs. The number of iterations required to reach an optimal alignment with appreciable probability, however, can often be exponentially large. The MSASA *(37)* program for simulated annealing-based alignment overcomes this barrier by using multiple alignments obtained via progressive alignment (described later) as a starting point.

Search-based strategies form a third class of global optimization techniques that have been applied to multiple alignment. In these methods, multiple alignment is typically formulated as a shortest path problem, where the initial state is the empty alignment (containing no columns), goal states are the set of all possible alignments of the given sequences, intermediate states represent candidate partial alignments of sequence prefixes, and state transition costs represent the change in score resulting from the addition of a column to an existing partial alignment. Despite the large state space, search techniques such as A* and branch-and-bound use heuristics to prune the set of searched alignments *(39,40)*. The MSA *(14,15)* and DCA/OMA *(17,19–21,41–43)* programs are two examples of methods based on this strategy.

2.3. Progressive Alignment

While global optimization techniques are powerful in their general applicability, they are less commonly used in modern MSA programs due to their computational expense (*see* **Note 4**). In this section, we examine a heuristic, known as *progressive alignment*, that solves the intractable problem of MSA approximately via a sequence of tractable subproblems. Unlike the techniques discussed in the last section, which find good multiple alignments directly, progressive alignment works indirectly, relying on variants of known algorithms for pairwise alignment.

In the popular *progressive alignment* strategy *(44–46)*, the sequences to be aligned are each assigned to separate leaves in a rooted binary tree (known as an alignment guide tree, *see* **Section 2.4.1**). Next, the internal nodes of the tree are visited in a bottom-up order, and each visited node is associated with an MSA of the sequences in its corresponding subtree. At the end of the traversal, the MSA associated with the root node is returned. By restricting MSAs at each internal node to preserve the aligned columns in the MSAs associated with their children nodes, the overall procedure reduces to a sequence of pairwise alignment computations: here, each pairwise alignment operates on a pair of alignments rather than a pair of sequences.

Under the most common gap scoring schemes, aligning a pair of alignments to optimize the SP score exactly is theoretically NP-hard *(47)*. Here, the complication arises from the fact that a gap opening character for some sequence in an MSA may not necessarily be present in every projected pairwise alignment involving that sequence. In practice, aligning alignments can be accomplished via procedures that optimize upper or lower bounds on the SP score *(48)*, which use a "quasinatural gap" approximation to the full SP score *(49)*, or which approximate each set of input alignments as a *profile*—a matrix of character frequencies at each position in the alignment *(50,51)*. Progressive alignment is the foundation of several alignment programs including DFALIGN *(44)*, MULTAL *(45,46)*, MAP *(52)*, PCMA *(53)*, PIMA *(54)*, PRIME *(55)*, PRRP *(56)*, MULTALIN *(57)*, CLUSTALW *(58–60)*, MAFFT *(50,61)*, MUSCLE *(51, 62)*, T-Coffee *(63,64)*, KAlign *(65)*, POA *(66–68)*, PROBCONS *(69)*, and MUMMALS/PROMALS *(70,71)*.

Profile–profile alignment techniques are routinely used in classification tasks such as remote homology detection and fold recognition *(72–75)*. In this literature, a considerable amount of effort has been placed in identifying profile–profile scoring functions that discriminate well between weakly homologous sequences and nonhomologous sequences *(76–81)*. While one might expect that a profile–profile scoring function that works well for classification should give accurate multiple sequence alignments, empirical tests have revealed only minor differences in alignment quality resulting from various profile–profile scoring schemes *(62,82–84)*.

2.4. Extensions to Progressive Alignment

The efficiency and simplicity of progressive algorithms for sequence alignment account for their widespread use in modern sequence alignment tools. Given a guide tree over N sequences, MSA construction requires $N - 1$ pairwise merge steps, hence rendering the cost of alignment effectively linear in the number of sequences (*see* **Note 5**). Nonetheless, progressive alignment strategies may also suffer from inaccuracies in the constructed guide trees or the accumulation of errors from the early pairwise alignment stages. In this section, we describe a number of heuristics used in modern MSA programs to overcome the shortcomings of vanilla progressive alignment.

2.4.1. Guide Tree Construction

In most progressive alignment programs, the guide tree used to determine the merging order for sequence groups is taken to be the phylogenetic tree relating the input sequences. Distance matrix methods for tree construction, such as the UPGMA ***(85,86)*** or neighbor-joining ***(87,88)*** algorithms, work by first estimating the evolutionary time between each pair of sequences. Then, a greedy procedure is used to construct a tree whose edge lengths correspond to evolutionary distances between points of divergence in the evolutionary history of the input sequences.

Problems with alignment guide trees generally result from either errors in the computed distance matrices or violated assumptions associated with the used tree reconstruction technique. The former case is especially common as many modern multiple alignment programs (e.g., MUSCLE, MAFFT, and MUMMALS/PROMALS) use fast approximate distance measures, such as k-mer counting, to form distance matrices for progressive alignment ***(50,58,89,90)***. Replacing these measures with more sensitive distance-estimation methods based on full pairwise alignment can be effective but slow ***(60)***. Recently, the Wu–Manber algorithm for fast inexact string matching ***(91)***, as employed in the KAlign program, has been shown to be significantly more sensitive than simple k-mer approaches for especially distant sequences ***(65)***.

Alternatively, *guide tree reestimation* can be effective for obtaining more accurate distance measures; given an approximate multiple alignment generated from the progressive alignment algorithm, it is generally possible to compute evolutionary trees of higher quality than the original guide trees formed using simple distance measures ***(50,56)***. In practice, alignment programs that use guide tree reestimation (e.g., MAFFT, MUSCLE, PRIME, PRRP, and MUMMALS/PROMALS) compute new distance matrices using an MSA obtained by progressive alignment. This revised distance matrix is then used to construct a new guide tree, which is in turn used in a second round of progressive alignment. The procedure may be iterated as many times as desired (or until convergence).

2.4.2. Modified Objective Functions

Even with perfect guide trees, errors can still occur in the pairwise merge steps of the progressive alignment. Errors made at early stages of the progressive alignment are particularly detrimental as they provide a distorted view of sequence homology that increases the chances of incorrect pairwise alignments at all higher levels of the tree. *Consistency-based* objective functions focus on improved scoring of matches in early alignments by incorporating information from outgroup sequences during each pairwise merge step ***(92–95)***. In particular, when performing a pairwise alignment of two sequences *x* and *y*, knowing that the *k*th residue of an outgroup sequence *z* aligns well with the *i*th residue of *x* and the *j*th residue of *y* provides strong evidence that the *i*th position of *x* and *j*th position of *y* should align with each other—i.e., pairwise alignments induced by a multiple alignment should be consistent (*see* **Fig. 2A**). Based on this transitivity condition, consistency-based objective functions typically modify the score for matching positions in an alignment of two groups during pairwise alignment by considering the relationship of each group to sequences not involved in the pairwise merge. Consistency-based scoring is used in the T-Coffee, DIALIGN, PROBCONS, PCMA, MUMMALS, PROMALS, and Align-m ***(96,97)*** alignment algorithms.

A number of modern programs (e.g., CLUSTALW, MUSCLE, and MAFFT) also use *position-specific gap penalties* to bias alignment algorithms toward placing gaps where previous gaps were opened during each pairwise merge step. Here, the rationale is that gap opening events that occur simultaneously in a group of sequences likely represent a single evolutionary event and hence should not be overpenalized. In addition, for globular protein sequences, hydrophobic residues are abundant in core regions where sequence indels are likely to affect proper folding, whereas hydrophilic residues are abundant on the protein surface, where extra loops are more likely to be tolerated (*see* **Fig. 2B**). CLUSTALW and MUSCLE attempt to make use of this signal by heuristically increasing gap penalties in hydrophobic regions and decreasing them in hydrophilic regions, though in practice the impact of hydropathy-based scoring on these methods is small. Recently, however, the CONTRAlign program ***(98)*** has demonstrated that rigorous statistical estimation of hydropathy-based gap penalty modifications can result in improvements in alignment accuracy of several percent for distant sequences; similar results have also been observed for detection of homology via profile alignments ***(99)***.

Sequence weighting is another common modification of the traditional SP multiple alignment objective applicable when the representation of sequence subgroups in a multiple alignment is highly skewed (*see* **Fig. 2C**). For

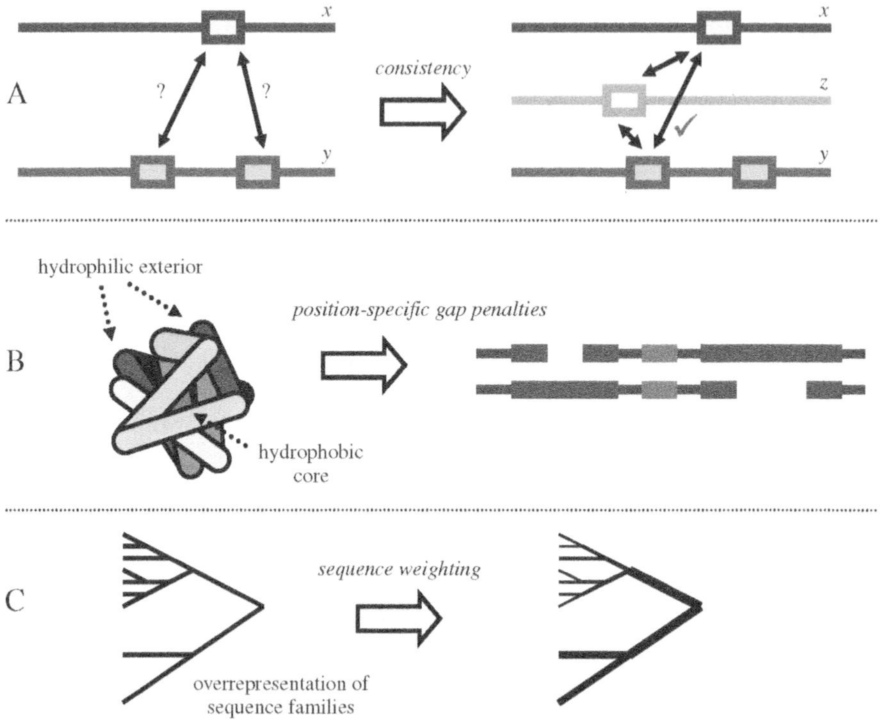

Fig. 2. Modified objective functions for sum-of-pairs alignment. (**A**) To aid in the alignment of two sequences x and y, consistency-based aligners use alignments of x and y to a third sequence z. (**B**) Gaps occur more frequently in the hydrophilic exterior than the hydrophobic core of globular proteins; position-specific gap penalties are higher in regions with hydrophobic residues and lower in regions with hydrophilic residues. (**C**) Sequence weighting corrects for sequence family overrepresentation.

example, in a multiple alignment of K sequences, if a large number of copies of a single sequence are added to the input, then an unweighted SP optimizer will emphasize the alignments of the redundant sequence to the other $K - 1$ sequences, thus effectively generating a biologically incorrect star alignment. While numerous schemes for computing sequence weights exist ***(92,100–108)***, the best choice of weights for alignment programs is unclear. In practice, the exact choice of weighting technique is generally a second-order effect; most reasonable sequence weighting techniques can greatly improve the accuracy of alignments in situations of sequence overrepresentation.

2.4.3. Postprocessing

In many cases, no amount of preprocessing is sufficient to prevent errors during progressive alignment. Postprocessing procedures, generally known as *iterative refinement* techniques, deal with progressive alignment errors by making changes to an existing alignment obtained from progressive alignment. For instance, iterative realignment techniques work by repeatedly dividing an alignment into two groups of aligned sequences, and realigning the groups ***(56, 109–111)***. In practice, iterative realignment can greatly improve the quality of an existing multiple alignment while requiring little extra programming effort. Alignment programs that make use of iterative realignment procedures include ITERALIGN ***(112)***, TULLA ***(113)***, AMPS/AMULT ***(114,115)***, MULTAN ***(116)***, OMA ***(42)***, PRRP, PROBCONS, MUSCLE, and MAFFT.

Other refinement techniques focus on correcting local errors in alignments by pattern matching or stochastic optimization, and bear strong similarity to the global optimization strategies introduced earlier ***(110,117–119)***. While global optimization techniques are generally considered less efficient than heuristic strategies such as progressive alignment in constructing multiple alignments, they can, nonetheless, be extremely effective given a good initial starting point (i.e., an existing multiple alignment).

2.5. Local Alignment

Most protein sequence alignment tools make the implicit assumption of *global homology*—the assumption that the sequences being aligned are generally related over their entire length. In many practical situations, however, two proteins may simply share a few common domains interspersed with regions of little to no homology. In these scenarios, variants of dynamic programming can be used for pairwise alignment ***(3)***. A space-efficient formulation of the dynamic programming algorithm, in particular, forms the basis of the SIM and LALIGN pairwise local alignment programs ***(120,121)***.

When speed is essential, indexing-based techniques can also be used for local alignment. These methods work by identifying segments of fixed length (known as *seeds* or *k-mers*) that are shared between two sequences; seeds meeting a certain threshold score are either chained or extended to form local alignments. This strategy is employed by the BLASTP ***(122,123)*** and LFASTA ***(124–126)*** programs.

For the problem of multiple local alignment, the DIALIGN ***(127–130)*** and DIALIGN-T ***(131)*** programs work by identifying homologous ungapped segments using a unique probabilistic segment scoring system that does not explicitly penalize for indels. Segments are then selected for inclusion in the multiple alignment via a greedy procedure that requires conserved segments to be present in the same order in each sequence. Related procedure for finding conserved "boxes" or for identifying high-confidence matches are used in the MATCH-BOX ***(132,133)*** and AMAP ***(134)*** programs.

In some proteins, however, conserved domains may appear multiple times in a single sequence (known as repeats) or may appear in a different order in different sequences (known as rearrangements). Repeated domains can generally be identified via local alignment of a sequence to itself ***(135)***; programs that specialize in the identification and alignment of protein repeats include Mocca ***(136)***, RADAR ***(137)***, REPRO ***(138)***, and TRUST ***(139)***. A more recent program called RAlign ***(140)*** performs global alignments while taking into account repeat structure.

Constructing multiple local alignments with both repeats and rearrangements is an extremely difficult problem that is usually done manually. Motif finders, such as GIBBS ***(141,142)***, MOTIF ***(143,144)***, MEME ***(145)***, and CONSENSUS ***(141)***, in principle can detect local ungapped homologies between several protein sequences. In practice, however, these methods are usually slow and can find only short, well-conserved gap-free segments of fixed length. Existing domain finding programs, such as DOMAINER ***(146)*** and MACAW ***(147)***, have similar restrictions, and the latter also requires significant manual intervention. Recently, a number of programs have addressed the challenges of representing multiple local alignments of protein sequences using partial-order ***(66)*** and A-Bruijn ***(148)*** graphs; some recent attempts to completely automate multiple local alignment construction include the ABA ***(149)*** and ProDA ***(150)*** alignment tools.

2.6. Probabilistic Models

While most alignment techniques rely abstractly on a scoring scheme that uses substitution scores and gap penalties, they do not develop an explicit model of the evolutionary process. In this section, we consider the class of probabilistic methods for aligner construction that has garnered much recent interest. Probabilistic techniques for multiple alignment generally come in three main varieties: complex evolutionary models of insertion, deletion, and mutation in multiple sequences; fixed dimensionality profile models for representing specific protein families; and hybrid methods that combine probabilistic models with traditional ad hoc alignment techniques.

Of the three approaches, evolutionary models for statistical alignment provide the most explicit representation of change in biological sequences as a stochastic process ***(151,152)***. Research in statistical alignment typically derive from the classic Thorne–Kishino–Felsenstein (TKF) pairwise alignment model ***(153)*** in which amino acid substitutions follow a time-reversible Markov process and single-gap creation and deletion are treated as birth/death processes over imaginary "links" separating letters in a sequence. Subsequent work on statistical alignment has focused on modeling multiresidue, overlapping indels ***(154–159)***, extending the TKF model to multiple alignment ***(160–167)***, and the even more complex task of coestimating alignment and sequence phylogeny ***(164,168–172)***. Unlike traditional score-based alignment approaches, statistical alignment methods provide a natural framework for estimating the parameters underlying stochastic evolutionary processes ***(173)***. However, the resulting models are often quite complex. While dynamic programming is sometimes possible, these models often require sampling-based inference procedures ***(174)*** that share many of the disadvantages of simulated annealing approaches discussed earlier. The accuracy of TKF-based techniques in alignment construction is unclear as few methods based on this approach have been comparatively benchmarked against standard programs; one exception is the Handel ***(162,163)*** program for statistical multiple alignment, which achieves substantially lower accuracy (i.e., 13% fewer correctly aligned residue pairs) than CLUSTALW, the prototypical score-based modern sequence aligner.

A second class of probabilistic modeling techniques is the profile hidden Markov model (profile HMM), a sophisticated variant of the character frequency profile matrices that takes into account position-specific indel probabilities ***(8,175–179)***. To construct a profile HMM given a set of unaligned sequences, a length is chosen for the initial profile, as well as initial emission probabilities for each position in the profile and transition probabilities for indel creation and extension after each position. Next, the model is optimized according to a likelihood criterion using an expectation–maximization (EM)-based Baum–Welch procedure ***(8)***, simulated annealing ***(38)***, deterministic annealing ***(180)***, or approximate gradient descent ***(181,182)***. Finally, all sequences are aligned to the profile using the Viterbi algorithm ***(183)*** for finding the most likely correspondence between each individual sequence and the profile, and the correspondences of each sequence to the profile are accumulated to form the multiple alignment. Profile HMMs and their variants ***(184)*** form the basis of many remote homology detection techniques ***(185–187)*** and have been used to characterize protein sequence families ***(188)***. Empirically, profile HMMs ***(177,189)*** have great appeal in practice as they provide a principled probabilistic framework, and, when properly tuned ***(190,191)***, achieve good empirical performance close to that of CLUSTALW ***(192,193)***.

Finally, hybrid techniques combine the rigor of probabilistic model parameter estimation with standard heuristics for multiple alignment. The ProAlign ***(194)***, COACH ***(81)***, and SATCHMO ***(195,196)*** progressive alignment tools, for instance, all achieve CLUSTALW accuracy; the recent PRANK aligner ***(197)*** has revealed the benefits of scoring insertions and deletions differently for the purposes of indel distribution estimation. A separate promising direction has been the development of the maximum expected accuracy (MEA) algorithm for pairwise alignment based on posterior match probabilities ***(198)***, which was generalized to consistency-based multiple alignment in the PROBCONS algorithm ***(69)***. Other programs based on the public domain PROBCONS source code include AMAP ***(199)***, which optimizes an objective function that rewards for correctly placed gaps, and ProbAlign ***(200)***, which uses a physics-inspired modification of the posterior probability calculations in PROBCONS. Finally, the MUMMALS program ***(70)***, which extends the PROBCONS approach to allow for more sophisticated HMM structures, has achieved the highest reported accuracies to date of all modern stand-alone multiple alignment programs.

2.7. Computation-Intensive Methods

In recent years, a new category of computation-intensive methods has risen in importance. Typically, these methods are designed not for high-throughput scenarios but rather for situations in which accuracy is paramount and abundant computing resources are available. Such scenarios arise in protein structure prediction, where alignment quality is the bottleneck in fold prediction accuracy, and the need for high-speed alignment is less important.

Ensemble methods (often known as meta-prediction methods in the protein structure prediction community) consider the predictions of a number of separate individual methods in order to form an aggregate prediction. M-Coffee ***(201)*** places input alignments into an alignment library and then assembles a multiple alignment using the T-Coffee progressive algorithm for solving the maximum-weight trace problem ***(202–204)***. A similar program called meta_align is also available as part of the MUMMALS package ***(70)***. In both cases, the resulting alignments generated by the ensemble predictor are more accurate than those made by any individual prediction technique.

Finally, *database-aided methods* add external information to help the aligner resolve ambiguities in alignment decisions. For instance, adding homologous sequences found in a large sequence database when the number of input sequences is small has been shown to be effective for methods such as MAFFT, PRALINE ***(205,206)***, and DbClustal ***(207)***. Alternatively, adding extra experimental or predicted information regarding the structural properties of the sequences being aligned can also improve accuracy. For example, the NdPASA ***(208)***, HHAlign ***(75)***, and PrISM.1 ***(209)*** pairwise aligners and the PSI-PRALINE ***(205)*** and SPEM ***(210)*** multiple aligners all make use of known or predicted secondary structure; similarly, the 3D-Coffee ***(211,212)*** multiple aligner incorporates structural alignments when they are available. In general, the specific program used for performing the alignment tends to be less important than the data incorporated by each alignment approach. Given this, the best database-aided method to use in any given alignment situation should generally be based on the data available.

3. Other Considerations

In studies of multiple sequence alignment, the algorithms used can be important, but they are not the only consideration that must be made. In this section, we provide a brief overview of aligner performance assessment and recent developments in parameter estimation.

3.1. Benchmarking

Techniques for assessing aligner performance typically have one of four goals: (1) demonstrating the effectiveness of a particular heuristic strategy for SP objective optimization; showing that a particular software package achieves good accuracy relative to "gold standard" reference alignments of either (2) real or (3) simulated proteins; or (4) quantifying alignment accuracy on real data in a reference-independent manner. For comparing software packages relying on different objective functions, the first validation scheme is not applicable. In this subsection, we focus on the latter three methods of aligner validation.

In real protein sequences, the true alignment of a set of sequences based on structural considerations is not necessarily the same as the true alignment based on evolutionary or functional considerations. In practice, structural alignments are relatively easy to obtain for proteins of known structure, and hence, are the *de facto* standard in most real-world benchmarks of alignment tools. Popular databases of hand-curated structural alignments include BAliBASE version 2 ***(213,214)*** and HOMSTRAD ***(215)***. Because of the difficulty and lack of reproducibility of hand curation, a number of modern alignment databases rely on automated structural alignment protocols, including SABmark ***(216)***, PREFAB ***(51)***, OxBench ***(217)***, and to a large extent, BAliBASE version 3 ***(218)***. Because the correct protein structural alignment can sometimes also be ambiguous, most alignment databases annotate select portions of their provided alignments as "core blocks"—regions for which structural alignments are known to be reliable—and measures of accuracy such as the Q score [defined as the proportion of pairwise matches in a reference alignment predicted by the aligner; other measures of accuracy also exist ***(219)***] are computed with respect to only core blocks.

The difficulties of ambiguity in structural alignments can be avoided when benchmarking with simulated evolution programs, such as SIMPROT ***(220,221)*** or Rose ***(222)***. In simulation studies, the true "evolutionary" relationships between positions in a set of a sequences are completely known. Besides allowing for the construction of large testing sets, simulation-based validation also has the advantage of enabling detailed studies of aligner performance in specific settings; for example, the IRMBase database ***(131)***, created using the Rose simulator, was built to evaluate the ability of local alignment methods to identify short implanted conserved motifs within nonhomologous sequences. Despite these advantages, simulation studies are highly prone to parameter overfitting. Furthermore, the performance of a method on simulated proteins may not be representative of its performance on real proteins, especially if the simulator fails to properly model all of the biological features used by the aligner. For instance, a method that accounts for gap enrichment in hydrophilic regions of proteins will perform relatively worse on simulations that do not account for hydropathy properties of protein sequences than on real proteins for which hydropathy plays an important role.

Finally, it is possible to avoid dealing with ambiguities in reference alignments using techniques that directly assess the quality of an alignment in terms of the resulting structural superposition. For a pair of proteins, the coordinate root-mean-square-distance (coordinate RMSD) between positions identified as "equivalent" according to an alignment (after the two protein structures have been appropriately rotated and translated) is a common measure for evaluating structural alignment quality. Several RMSD variants exist ***(223)***, including variants that account for protein length ***(224)***, that examine pairwise distances between residues in a protein ***(225)***, or that rely on alternate representations of protein backbones ***(226)***. Another recently proposed metric is the APDB measure ***(227)***, an approximation of the Q score that judges the "correctness" of aligned residue pairs based on the degree to which nearby aligned residues have similar local geometry in the sequences being aligned.

3.2. Parameter Estimation

For traditional score-based sequence alignment procedures, estimation of substitution matrices and gap penalties are usually treated separately (*see* **Note 6**). Briefly, substitution matrices are generally estimated from databases of alignments known to be reliable. Statistical estimation procedures for constructing log-odds substitution matrices vary in their details, but most methods nonetheless tend to generate sets of matrices approximately parameterized by some notion of evolutionary distance for which that matrix is optimal. Popular matrices include the BLOSUM ***(228)***, PAM ***(229,230)***, JTT ***(89)***, MV ***(231)***, and WAG ***(232)*** matrices; matrices derived from structural alignments for use with low-identity sequences also exist ***(233)***. For gap parameters, an empirical trial-and-error approach *(234)* is common as the number of parameters to be estimated is low.

Probabilistic models have the advantage that the maximum likelihood principle provides a natural mechanism for estimating gap parameters when example alignments are available *(235)*; when only unaligned sequences are available, unsupervised estimation of gap parameters can still be effective *(69)*. Alternatively, Bayesian methods *(236,237)* automatically combine the results obtained when using multiple varying parameter sets and thus avoid the need for deciding on fixed parameter sets.

Recently, the problem of parameter estimation has been the subject of renewed attention, stemming from the influence of the convex optimization and machine learning communities. Kececioglu and Kim *(238)* described a simple cutting-plane algorithm for *inverse alignment*—the problem of identifying a parameter set for which an aligner aligns each sequence in a training set correctly. Their algorithm is fast in practice, though the biological accuracy of the resulting alignments on unseen test data is unclear. Do *et al.* *(98)* developed a machine learning-based method based on pair conditional random fields (pair-CRFs) called CONTRAlign, which achieves significantly better generalization performance than existing methods for pairwise alignment of distant sequences. Most recently, Yu *et al.* *(239)* described a fast approach for training protein threading models based on support vector machines *(240)*, which shares many of the generalization advantages of CONTRAlign.

4. Advice for Practitioners

Given the multitude of choices, it can be difficult for a user of multiple alignment software to understand the situations in which a particular alignment tool is or is not appropriate. When aligning a small number (<20) of globally homologous sequences with high percent identity (>40%), most modern alignment programs will have no difficulty in returning a correct multiple sequence alignment, and no special consideration is needed. When all of these conditions do not hold, however, choosing the appropriate tools and configuration, while keeping in mind the tradeoff between accuracy and computational cost, can be difficult. In this section, we provide a list of currently popular alignment software (*see* **Table 1**) and give advice on tool selection (*see* **Fig. 3**) and effective use of alignments.

4.1. The Extreme Cases

Extreme cases for sequence alignment programs involve scenarios typically not encountered in most alignment benchmarking studies. The spectrum of

Table 1
MSA Programs

Tool	URL
CLUSTALW	http://www.clustal.org/
DIALIGN	http://bibiserv.techfak.uni-bielefeld.de/dialign/
MAFFT	http://align.bmr.kyushu-u.ac.jp/mafft/software/
MUMMALS	http://prodata.swmed.edu/mummals/
MUSCLE	http://www.drive5.com/muscle/
PRALINE	http://zeus.cs.vu.nl/programs/pralinewww/
PRIME	http://prime.cbrc.jp/
ProbAlign	http://probalign.njit.edu/standalone.html
PROBCONS	http://probcons.stanford.edu/
ProDA	http://proda.stanford.edu/
PROMALS	http://prodata.swmed.edu/promals/
SPEM	http://sparks.informatics.iupui.edu/
T-Coffee, M-Coffee, 3D-Coffee	http://www.tcoffee.org/

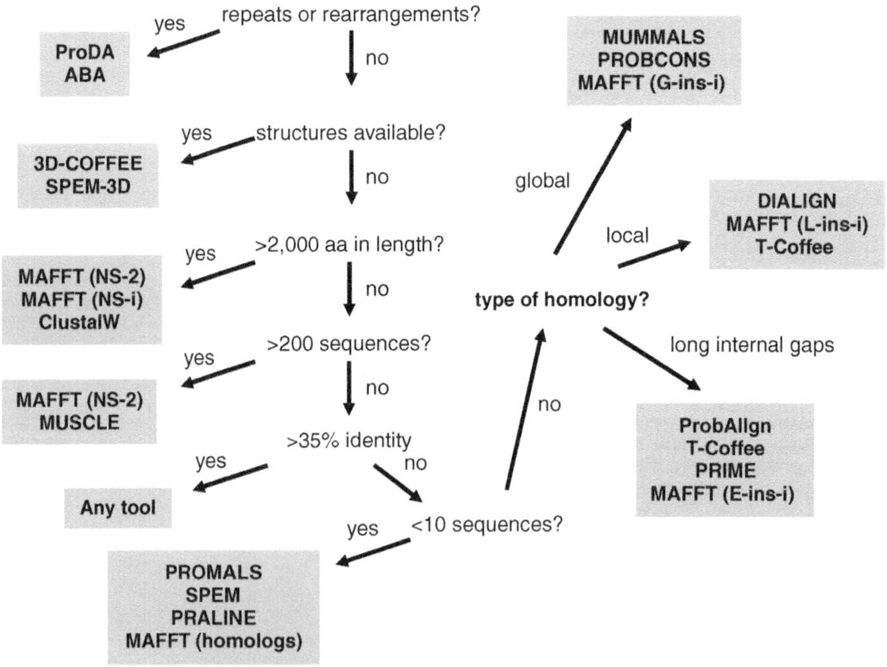

Fig. 3. Decision tree for selecting an appropriate MSA tool.

applicable tools in extreme alignment cases is generally small. We distinguish three particular situations: (1) repeated or rearranged protein domains, (2) high-throughput alignment of large numbers (>200) of input sequences, and (3) extremely long sequences (>2000 amino acids).

Currently, few programs adequately deal with alignments involving proteins with repeated or rearranged domains. While some repeat finding programs can be used for identifying repeats in protein alignments, these programs do not present a complete view of the homology in a collection of protein sequences. To date, the only programs that attempt to address this issue are ABA ***(149)*** and ProDA ***(150)***, of which we recommend the latter based on its significant advantage in accuracy on real data. While these methods are far more effective than traditional global alignment methods on sequences with repeats and rearrangements, they obtain lower accuracy on sequences where no rearrangements or repeats occur.

In high-throughput alignment scenarios, program speed can be a major bottleneck. In particular, when the number of sequences is between 200 and 1000, $O(N^2)$ distance matrix calculation (where N is the number of sequences) is generally the time-limiting factor, so progressive alignment methods with fast distance calculation, such as MAFFT (FFT-NS-2), MUSCLE (progressive), or KAlign, are recommended. For extremely large numbers of sequences (>10,000), even these fast distance calculation methods can be slow. In these cases, the PartTree ***(241)*** option in MAFFT, which relies on approximate guide tree construction in $O(N \log N)$ time based on a restricted portion of the distance matrix, is currently the only realistic option. In practice, MAFFT (PartTree), which uses approximate tree construction, achieves Q scores on average 2–3% lower than MAFFT (FFT-NS-2), which uses a full UPGMA guide tree.

For extremely long sequences (>2000 amino acids), space complexity is the main consideration in choosing an aligner. In particular, most recent multiple alignment programs tend to use dynamic programming algorithms with $O(L^2)$ memory usage (where L is the average sequence length), which is fine for most scenarios considered in benchmarking studies. For longer sequences, more efficient linear space algorithms ***(5)***, as implemented in CLUSTALW, MAFFT (FFT-NS-2), and MAFFT (FFT-NS-i), are available.

4.2. Sequences with Low Similarity

For sequences with less than 35% identity, benchmark studies under various conditions ***(221,225,242)*** have consistently identified T-Coffee, PROBCONS, and MAFFT (L-ins-i) as being the most accurate stand-alone programs currently available. More recently developed programs based on the PROBCONS framework, including MUMMALS, ProbAlign, and AMAP, have been reported to obtain even higher accuracies. In general, however, stand-alone programs tend to perform poorly for low-identity sequences. Here, we outline two main strategies for obtaining quality alignments from the point of view of an end user: careful identification of alignment scenarios and incorporation of external information to improve alignment quality.

In general, low-identity alignments may be characterized as (1) global homology over the entire length of the protein (N-terminus to C-terminus), (2) local homology surrounded by nonhomologous flanking regions, or (3) short patches of homology interrupted by long internal gaps (*see* **Fig. 4**). Case 1 is the simplest of the three situations for which the best alignment accuracy can be expected; in these situations, MUMMALS and PROBCONS are typically the most accurate. However, when large N-terminal or C-terminal extensions exist in one or more sequences (i.e., case 2), these global methods tend to perform less well than techniques that make use of local alignment; in particular, DIALIGN, T-Coffee, and MAFFT (L-ins-i) are recommended; additionally, ProbAlign is reported to work well for these situations. Finally, the third case (case 3) occurs for highly divergent sequences in which sequence similarity remains only around functionally important residues but the order of conserved regions is identical in all sequences. Here, MAFFT (E-ins-i), T-Coffee, PRIME, and DIALIGN are recommended; these methods typically make use of more sophisticated gap penalties, such as the generalized affine gap cost ***(243,244)*** in the case of MAFFT (E-ins-i), or piecewise linear gap costs in the case of PRIME.

In general, we recommend using methods tailored for case 3 when aligning full-length proteins. Once an initial alignment is obtained, then trimming the

A

```
XXXXXXXXXXX-XXXXXXXXXXXXXXXX
XX-XXXXXXXXXXXXXXXX-XXXXXXXX
XXXXX----XXXXXXXXX---XXXXX--
-XXXX-XXXXXXXXXXX----XXXXXXX
XXXXXXXXXXXXXXXXX----XXXXXXX
```

B

```
ooooooooooooooooooooooooooooooXXXXXXXXXXXXX-XXXXXXXXXXXXXXXXXX------------------
------------------------------XX-XXXXXXXXXXXXXXXXXXX-XXXXXXXXXooooooooooo-------
--------------ooooooooooooooooXXXXX-----XXXXXXXXXXX---XXXXXXXXooooooooooo-------
-----oooooooooooooooooooooooooXXXXX-XXXXXXXXXXXXXX----XXXXXXXXoooooooooooooooooo
------------------------------XXXXXXXXXXXXXXXXXXXX----XXXXXXXX------------------
```

C

```
oooooooooXXX------XXXX-----------------------------------XXXXXXXXX-XXXXXXXXXXXXXXXXooooooooooooo
---------XXXXXXXXXXXXXooo--------------------------------XXXXXXXXXXXXXXXXXX-XXXXXXX-------------
-----ooooXXXXXX---XXXXooooooooo--------------------------XXXXX---XXXXXXXXXXXXXXXXXXooooooooooooo
---------XXXXX----XXXXoooooooooooooooooooooooooooooooooooXXXXX-XXXXXXXXXXXX---XXXXX-------------
---------XXXXX----XXXX-----------------------------------XXXXX---XXXXXXXXXX---XXXXXooooo--------
```

Fig. 4. Types of alignment homology. "-" represents a gap, "X" represents an aligned amino acid residue, and "o" is an unalignable residue. (**A**) Global homology. (**B**) Local homology. (**C**) Long internal gaps.

alignment to include only the relevant homologous parts can be done manually, and then a method designed for case 1 can be applied to give the best possible accuracy. For even more accuracy, ensemble approaches, such as the M-Coffee mode of T-Coffee or the meta_align program in MUMMALS, merge numerous independently calculated multiple sequence alignments into a single combined alignment. Clearly, ensemble aligners will not perform well if the input individual multiple alignments are poor, but in general can give modest improvements in accuracy over their component aligners.

Usually, however, the best way to improve alignment accuracy is not by more sophisticated algorithms or more careful program tuning, but rather by incorporation of external information when present. For example, the structural similarity of homologous proteins is generally conserved even after sequence similarity becomes nondetectable over the course of evolution. Therefore, sequence alignment tools that make use of structural information, such as 3D-Coffee and SPEM-3D, can achieve significantly better accuracies than tools relying solely on sequence data. Additionally, when speed is not critical and the number of input sequences is small (<10), database-aided methods can achieve better accuracy by recruiting additional homologs from a sequence database. This sort of analysis is supported by DbClustal, MAFFT-homologs, PRALINE, SPEM, and PROMALS. By enhancing site-specific evolutionary constraints, homologs can improve accuracy to a level comparable to the benefits of adding structural information.

4.3. Postprocessing and Visualization

Once an alignment has been generated, visualization tools allow manual identification of regions with reliably predicted homology; many of these tools also allow for interactive alignment editing. For alignments of sequences with low similarity, postprocessing is extremely important as most regions in a low-identity alignment will not be reliably alignable. Typically, high confidence aligned regions can be identified by looking for groups of residues with strongly conserved physicochemical properties (e.g., hydropathy, polarity, and volume), using alternative alignment objective functions for identifying reliable columns, using posterior confidences generated by alignment programs such as PROBCONS, using the consensus of several alignment methods, or even better, cross-referencing aligned positions with amino acid residues in three-dimensional protein structures. Tools for integrating structural and functional information with sequence data for alignments, such as MACSIMS *(245)*, can also be helpful for analyzing multiple alignments. Other freely available alignment visualization and editing programs are listed in **Table 2**.

Table 2
Alignment Visualization Tools

Tool	URL
Jalview	http://www.jalview.org/
SeaView	http://pbil.univ-lyon1.fr/software/seaview.html
CINEMA	http://www.bioinf.manchester.ac.uk/dbbrowser/CINEMA2.1/
Kalignvu	http://msa.cgb.ki.se/
GeneDoc	http://www.nrbsc.org/gfx/genedoc/
STRAP	http://www.charite.de/bioinf/strap/
ClustalX	http://www.clustal.org/
BoxShade	http://www.ch.embnet.org/software/BOX_form.html
ALTAVIST	http://bibiserv.techfak.uni-bielefeld.de/altavist/

5. Conclusions

Despite its long history, research in sequence alignment continues to flourish. Each year, dozens of articles describing new methods for protein alignments are published. Although many of these approaches rely on the same basic principles, the details of the implementations can have dramatic effects on the performance, both in terms of accuracy and speed. A primary reason for this continued interest in protein sequence alignment is the centrality of comparative sequence analysis in modern computational biology: accurate alignments form the basis of many bioinformatics studies, and advances in alignment methodology can confer sweeping benefits in a wide variety of application domains.

In recent years, trends in the alignment field have included the development of efficient tools suited for high-throughput processing on a single PC (e.g., MUSCLE, MAFFT, POA, KAlign), the application of machine learning techniques for parameter estimation and sequence modeling (e.g., PROBCONS, CONTRAlign, MUMMALS), and the exploitation of publicly available sequence databases to improve accuracy of low-identity alignments (e.g., PRALINE, MAFFT, PROMALS). Furthermore, recent attempts to build alignment algorithms for dealing with proteins containing repeats and rearrangements (e.g., ABA, PRODA) push the boundaries of the types of scenarios considered by aligner developers. Finally, a number of groups have recognized the growing importance of integrating multiple alignments with other forms of data for presentation to biologists [e.g., MAO ***(246)***, MACSIMS]. While it is impossible to predict all the advances in sequence alignment research to come, their implications for practitioners is clear: the next generation of protein alignment tools will be faster, more accurate, and easier to use.

6. Notes

1. In this chapter, we focus on the problem of sequence alignment, which we distinguish from the related topic of homology search, in which we would like to identify homologs of a "query" sequence among a collection of "database" sequences. Unlike sequence alignment tools, homology search tools, such as BLASTP ***(122)*** or PSI-BLAST ***(123)***, rely extensively on approximate string matching techniques but do not focus on providing accurate residue-level alignments of the returned sequences.
2. We refer the reader to a number of other recent reviews on protein sequence alignment techniques ***(1,247–251)*** and their applications ***(252)***.
3. *Dynamic programming* (DP) refers to a class of algorithms that decomposes the solution for a complex optimization problem into overlapping solutions for smaller subproblems ***(253)***. By exploiting these overlaps, DP algorithms search an exponentially large space (e.g., the space of all possible alignments) by solving a small polynomial number of subproblems.
4. The SAGA algorithm, for example, was found to be 100–1000× slower than CLUSTALW in a number of typical multiple alignments ***(29)***.
5. As previously pointed out ***(62)***, although the progressive alignment procedure may be linear in the number of sequences N, typical algorithms for tree construction require $O(N^3)$ time. For large numbers of sequences (e.g., 10,000), this is intractable. An approximate $O(N^2)$ UPGMA tree construction algorithm that produces reasonable trees in practice has been described; alternatively, exact worst-case quadratic time algorithms for UPGMA ***(254)*** and neighbor-joining ***(255)*** tree construction exist. For situations with very large N, the recent PartTree algorithm ***(241)*** computes approximate trees in $O(N \log N)$ time.
6. *Parametric alignment* ***(256–258)*** is an attempt to abandon the need for parameter estimation altogether by computing optimal sequence alignments for all possible parameter sets. However, the resulting algorithms are often computationally expensive, and for most biologists, the generated alignment sets are of limited benefit when alignment quality is difficult to judge manually.

Acknowledgments

We thank Karen Ann Lee for help in preparing the manuscript. C.B.D was funded by an NDSEG fellowship.

References

1. Notredame, C. (2002) Recent progress in multiple sequence alignment: a survey. *Pharmacogenomics* **3,** 131–144.
2. Needleman, S. B. and Wunsch, C. D. (1970) A general method applicable to the search for similarities in the amino acid sequence of two proteins. *J. Mol. Biol.* **48,** 443–453.

3. Smith, T. F. and Waterman, M. S. (1981) Identification of common molecular subsequences. *J. Mol. Biol.* **147,** 195–197.
4. Gotoh, O. (1982) An improved algorithm for matching biological sequences. *J. Mol. Biol.* **162,** 705–708.
5. Myers, E. W. and Miller, W. (1988) Optimal alignments in linear space. *Comput. Appl. Biosci.* **4,** 11–17.
6. Murata, M., Richardson, J. S., and Sussman, J. L. (1985) Simultaneous comparison of three protein sequences. *Proc. Natl. Acad. Sci. USA* **82,** 3073–3077.
7. Waterman, M. S. and Jones, R. (1990) Consensus methods for DNA and protein sequence alignment. *Methods Enzymol.* **183,** 221–237.
8. Durbin, R., Eddy, S. R., Krogh, A., and Mitchison, G. (1999) *Biological Sequence Analysis: Probabilistic Models of Proteins and Nucleic Acids.* Cambridge University Press, Cambridge.
9. Gonnet, G. H., Korostensky, C., and Benner, S. (2000) Evaluation measures of multiple sequence alignments. *J. Comput. Biol.* **7,** 261–276.
10. Wang, L. and Jiang, T. (1994) On the complexity of multiple sequence alignment. *J. Comput. Biol.* **1,** 337–348.
11. Bonizzoni, P. and Della Vedova, G. (2001) The complexity of multiple sequence alignment with SP-score that is a metric. *Theor. Comput. Sci.* **259,** 63–79.
12. Just, W. (2001) Computational complexity of multiple sequence alignment with SP-score. *J. Comput. Biol.* **8,** 615–623.
13. Elias, I. (2006) Settling the intractability of multiple alignment. *J. Comput. Biol.* **13,** 1323–1339.
14. Lipman, D. J., Altschul, S. F., and Kececioglu, J. D. (1989) A tool for multiple sequence alignment. *Proc. Natl. Acad. Sci. USA* **86,** 4412–4415.
15. Gupta, S. K., Kececioglu, J. D., and Schaffer, A. A. (1995) Improving the practical space and time efficiency of the shortest-paths approach to sum-of-pairs multiple sequence alignment. *J. Comput. Biol.* **2,** 459–472.
16. Carrillo, H. and Lipman, D. (1988) The multiple sequence alignment problem in biology. *SIAM J. Appl. Math.* **48,** 1073–1082.
17. Dress, A., Fullen, G., and Perrey, S. (1995) A divide and conquer approach to multiple alignment. *Proc. Int. Conf. Intell. Syst. Mol. Biol.* **3,** 107–113.
18. Stoye, J., Perrey, S. W., and Dress, A. W. M. (1997) Improving the divide-and-conquer approach to sum-of-pairs multiple sequence alignment. *Appl. Math. Lett.* **10,** 67–73.
19. Stoye, J., Moulton, V., and Dress, A. W. (1997) DCA: an efficient implementation of the divide-and-conquer approach to simultaneous multiple sequence alignment. *Comput. Appl. Biosci.* **13,** 625–626.
20. Stoye, J. (1998) Multiple sequence alignment with the divide-and-conquer method. *Gene* **211,** GC45–56.
21. Reinert, K., Stoye, J., and Will, T. (2000) An iterative method for faster sum-of-pairs multiple sequence alignment. *Bioinformatics* **16,** 808–814.
22. Holland, J. H. (1975) *Adaptation in Natural and Artificial Systems.* University of Michigan Press, Ann Arbor.

23. Zhang, C. and Wong, A. K. (1997) A genetic algorithm for multiple molecular sequence alignment. *Comput. Appl. Biosci.* **13,** 565–581.
24. Anbarasu, L. A., Narayanasamy, P., and Sundararajan, V. (1998) Multiple sequence alignment using parallel genetic algorithms. SEAL.
25. Chellapilla, K. and Fogel, G. B. (1999) Multiple sequence alignment using evolutionary programming. Congress on Evolutionary Computation.
26. Gonzalez, R. R., Izquierdo, C. M., and Seijas, J. (1999) Multiple protein sequence comparison by genetic algorithms. SPIE-98.
27. Cai, L., Juedes, D., and Liakhovitch, E. (2000) Evolutionary computation techniques for multiple sequence alignment. Congress on Evolutionary Computation.
28. Zhang, G.-Z. and Huang, D.-S. (2004) Aligning multiple protein sequence by an improved genetic algorithm. IEEE International Joint Conference on Neural Networks.
29. Notredame, C. and Higgins, D. G. (1996) SAGA: sequence alignment by genetic algorithm. *Nucleic Acids Res.* **24,** 1515–1524.
30. Isokawa, M., Takahashi, K., and Shimizu, T. (1996) Multiple sequence alignment using a genetic algorithm. *Genome Inform.* **7,** 176–177.
31. Harada, Y., Wayama, M., and Shimizu, T. (1997) An inspection of the multiple alignment methods with use of genetic algorithm. *Genome Inform.* **8,** 272–273.
32. Hanada, K., Yokoyama, T., and Shimizu, T. (2000) Multiple sequence alignment by genetic algorithm. *Genome Inform.* **11,** 317–318.
33. Yokoyama, T., Watanabe, T., Taneda, A., and Shimizu, T. (2001) A web server for multiple sequence alignment using genetic algorithm. *Genome Inform.* **12,** 382–383.
34. Nguyen, H. D., Yoshihara, I., Yamamori, K., and Yasunaga, M. (2002) A parallel hybrid genetic algorithm for multiple protein sequence alignment. *Evol. Comput.* **1,** 309–314.
35. Kirkpatrick, S., Gelatt, J., C. D., and Vecchi, M. P. (1983) Optimization by simulated annealing. *Science* 220, 671–680.
36. Ishikawa, M., Toya, T., Hoshida, M., Nitta, K., Ogiwara, A., and Kanehisa, M. (1993) Multiple sequence alignment by parallel simulated annealing. *Comput. Appl. Biosci.* **9,** 267–273.
37. Kim, J., Pramanik, S., and Chung, M. J. (1994) Multiple sequence alignment using simulated annealing. *Comput. Appl. Biosci.* **10,** 419–426.
38. Eddy, S. R. (1995) Multiple alignment using hidden Markov models. *Proc. Int. Conf. Intell. Syst. Mol. Biol.* **3,** 114–120.
39. Ikeda, T. and Imai, H. (1999) Enhanced A* algorithms for multiple alignments: optimal alignments for several sequences and k-opt approximate alignments for large cases. *Theor. Comput. Sci.* **210,** 341–374.
40. Horton, P. (2001) Tsukuba BB: a branch and bound algorithm for local multiple alignment of DNA and protein sequences. *J. Comput. Biol.* **8,** 283–303.
41. Reinert, K., Lenhof, H.-P., Mutzel, P., Mehlhorn, K., and Kececioglu, J. D. (1997) A branch-and-cut algorithm for multiple sequence alignment. RECOMB.

42. Reinert, K., Stoye, J., and Will, T. (1999) Combining divide-and-conquer, the A*-algorithm and successive realignment approaches to speed up multiple sequence alignment. German Conference on Bioinformatics.
43. Lermen, M. and Reinert, K. (2000) The practical use of the A* algorithm for exact multiple sequence alignment. *J. Comput. Biol.* **7,** 655–671.
44. Feng, D. F. and Doolittle, R. F. (1987) Progressive sequence alignment as a prerequisite to correct phylogenetic trees. *J. Mol. Evol.* **25,** 351–360.
45. Taylor, W. R. (1987) Multiple sequence alignment by a pairwise algorithm. *Comput. Appl. Biosci.* **3,** 81–87.
46. Taylor, W. R. (1988) A flexible method to align large numbers of biological sequences. *J. Mol. Evol.* **28,** 161–169.
47. Kececioglu, J. and Starrett, D. (2004) Aligning alignments exactly. RECOMB.
48. Kececioglu, J. and Zhang, W. (1998) Aligning alignments. CPM.
49. Altschul, S. F. (1989) Gap costs for multiple sequence alignment. *J. Theor. Biol.* **138,** 297–309.
50. Katoh, K., Misawa, K., Kuma, K., and Miyata, T. (2002) MAFFT: a novel method for rapid multiple sequence alignment based on fast Fourier transform. *Nucleic Acids Res.* **30,** 3059–3066.
51. Edgar, R. C. (2004) MUSCLE: multiple sequence alignment with high accuracy and high throughput. *Nucleic Acids Res.* **32,** 1792–1797.
52. Huang, X. (1994) On global sequence alignment. *Comput. Appl. Biosci.* **10,** 227–235.
53. Pei, J., Sadreyev, R., and Grishin, N. V. (2003) PCMA: fast and accurate multiple sequence alignment based on profile consistency. *Bioinformatics* **19,** 427–428.
54. Smith, R. F. and Smith, T. F. (1992) Pattern-induced multi-sequence alignment (PIMA) algorithm employing secondary structure-dependent gap penalties for use in comparative protein modelling. *Protein Eng.* 5, 35–41.
55. Yamada, S., Gotoh, O., and Yamana, H. (2006) Improvement in accuracy of multiple sequence alignment using novel group-to-group sequence alignment algorithm with piecewise linear gap cost. *BMC Bioinform.* 7, 524.
56. Gotoh, O. (1996) Significant improvement in accuracy of multiple protein sequence alignments by iterative refinement as assessed by reference to structural alignments. *J. Mol. Biol.* **264,** 823–838.
57. Corpet, F. (1988) Multiple sequence alignment with hierarchical clustering. *Nucleic Acids Res.* **16,** 10881–10890.
58. Higgins, D. G. and Sharp, P. M. (1988) CLUSTAL: a package for performing multiple sequence alignment on a microcomputer. *Gene* **73,** 237–244.
59. Higgins, D. G. and Sharp, P. M. (1989) Fast and sensitive multiple sequence alignments on a microcomputer. *Comput. Appl. Biosci.* **5,** 151–153.
60. Thompson, J. D., Higgins, D. G., and Gibson, T. J. (1994) CLUSTAL W: improving the sensitivity of progressive multiple sequence alignment through sequence weighting, position-specific gap penalties and weight matrix choice. *Nucleic Acids Res.* **22,** 4673–4680.

61. Katoh, K., Kuma, K., Toh, H., and Miyata, T. (2005) MAFFT version 5: improvement in accuracy of multiple sequence alignment. *Nucleic Acids Res.* **33,** 511–518.
62. Edgar, R. C. (2004) MUSCLE: a multiple sequence alignment method with reduced time and space complexity. *BMC Bioinform.* **5,** 113.
63. Notredame, C., Holm, L., and Higgins, D. G. (1998) COFFEE: an objective function for multiple sequence alignments. *Bioinformatics* **14,** 407–422.
64. Notredame, C., Higgins, D. G., and Heringa, J. (2000) T-Coffee: A novel method for fast and accurate multiple sequence alignment. *J. Mol. Biol.* **302,** 205–217.
65. Lassmann, T. and Sonnhammer, E. L. (2005) Kalign–an accurate and fast multiple sequence alignment algorithm. *BMC Bioinform.* **6,** 298.
66. Lee, C., Grasso, C., and Sharlow, M. F. (2002) Multiple sequence alignment using partial order graphs. *Bioinformatics* **18,** 452–464.
67. Lee, C. (2003) Generating consensus sequences from partial order multiple sequence alignment graphs. *Bioinformatics* **19,** 999–1008.
68. Grasso, C. and Lee, C. (2004) Combining partial order alignment and progressive multiple sequence alignment increases alignment speed and scalability to very large alignment problems. *Bioinformatics* **20,** 1546–1556.
69. Do, C. B., Mahabhashyam, M. S., Brudno, M., and Batzoglou, S. (2005) ProbCons: Probabilistic consistency-based multiple sequence alignment. *Genome Res.* **15,** 330–340.
70. Pei, J. and Grishin, N. V. (2006) MUMMALS: multiple sequence alignment improved by using hidden Markov models with local structural information. *Nucleic Acids Res.* **34,** 4364–4374.
71. Pei, J. and Grishin, N. V. (2007) PROMALS: towards accurate multiple sequence alignments of distantly related proteins. *Bioinformatics* **23,** 802–808.
72. Gribskov, M., McLachlan, A. D., and Eisenberg, D. (1987) Profile analysis: detection of distantly related proteins. *Proc. Natl. Acad. Sci. US A* **84,** 4355–4358.
73. von Ohsen, N., Sommer, I., and Zimmer, R. (2003) Profile-profile alignment: a powerful tool for protein structure prediction. *Pac. Symp. Biocomput.* 252–263.
74. von Ohsen, N., Sommer, I., Zimmer, R., and Lengauer, T. (2004) Arby: automatic protein structure prediction using profile-profile alignment and confidence measures. *Bioinformatics* **20,** 2228–2235.
75. Soding, J. (2005) Protein homology detection by HMM-HMM comparison. *Bioinformatics* **21,** 951–960.
76. von Ohsen, N. and Zimmer, R. (2001) Improving profile-profile alignments via log-average scoring. WABI.
77. Yona, G. and Levitt, M. (2002) Within the twilight zone: a sensitive profile-profile comparison tool based on information theory. *J. Mol. Biol.* **315,** 1257–1275.
78. Heger, A. and Holm, L. (2003) Exhaustive enumeration of protein domain families. *J. Mol. Biol.* **328,** 749–767.
79. Mittelman, D., Sadreyev, R., and Grishin, N. (2003) Probabilistic scoring measures for profile-profile comparison yield more accurate short seed alignments. *Bioinformatics* **19,** 1531–1539.

80. Sadreyev, R. and Grishin, N. (2003) COMPASS: a tool for comparison of multiple protein alignments with assessment of statistical significance. *J. Mol. Biol.* **326,** 317–336.
81. Edgar, R. C. and Sjolander, K. (2004) COACH: profile-profile alignment of protein families using hidden Markov models. *Bioinformatics* **20,** 1309–1318.
82. Rychlewski, L., Jaroszewski, L., Li, W., and Godzik, A. (2000) Comparison of sequence profiles. Strategies for structural predictions using sequence information. *Protein Sci.* **9,** 232–241.
83. Edgar, R. C. and Sjolander, K. (2004) A comparison of scoring functions for protein sequence profile alignment. *Bioinformatics* **20,** 1301–1308.
84. Ohlson, T., Wallner, B., and Elofsson, A. (2004) Profile-profile methods provide improved fold-recognition: a study of different profile–profile alignment methods. *Proteins* **57,** 188–197.
85. Sokal, R. R. and Michener, C. D. (1958) A statistical method for evaluating systematic relationships. *Univ. Kans. Sci. Bull.* **28,** 1409–1438.
86. Sneath, P. H. and Sokal, R. R. (1962) Numerical taxonomy. *Nature* **193,** 855–860.
87. Saitou, N. and Nei, M. (1987) The neighbor-joining method: a new method for reconstructing phylogenetic trees. *Mol. Biol. Evol.* **4,** 406–425.
88. Studier, J. A. and Keppler, K. J. (1988) A note on the neighbor-joining algorithm of Saitou and Nei. *Mol. Biol. Evol.* **5,** 729–731.
89. Jones, D. T., Taylor, W. R., and Thornton, J. M. (1992) The rapid generation of mutation data matrices from protein sequences. *Comput. Appl. Biosci.* **8,** 275–282.
90. Edgar, R. C. (2004) Local homology recognition and distance measures in linear time using compressed amino acid alphabets. *Nucleic Acids Res.* **32,** 380–385.
91. Wu, S. and Manber, U. (1992) Fast text searching allowing errors. *Commun. ACM* **35,** 83–91.
92. Vingron, M. and Argos, P. (1989) A fast and sensitive multiple sequence alignment algorithm. *Comput. Appl. Biosci.* **5,** 115–121.
93. Vingron, M. and Argos, P. (1990) Determination of reliable regions in protein sequence alignments. *Protein Eng.* **3,** 565–569.
94. Vingron, M. and Argos, P. (1991) Motif recognition and alignment for many sequences by comparison of dot-matrices. *J. Mol. Biol.* **218,** 33–43.
95. Gotoh, O. (1990) Consistency of optimal sequence alignments. *Bull. Math. Biol.* **52,** 509–525.
96. Van Walle, I., Lasters, I., and Wyns, L. (2003) Consistency matrices: quantified structure alignments for sets of related proteins. *Proteins* **51,** 1–9.
97. Van Walle, I., Lasters, I., and Wyns, L. (2004) Align-m–a new algorithm for multiple alignment of highly divergent sequences. *Bioinformatics* **20,** 1428–1435.
98. Do, C. B., Gross, S. S., and Batzoglou, S. (2006) CONTRAlign: discriminative training for protein sequence alignment. RECOMB.
99. Lolkema, J. S. and Slotboom, D. J. (1998) Hydropathy profile alignment: a tool to search for structural homologues of membrane proteins. *FEMS Microbiol. Rev.* **22,** 305–322.

100. Altschul, S. F., Carroll, R. J., and Lipman, D. J. (1989) Weights for data related by a tree. *J. Mol. Biol.* **207,** 647–653.
101. Vingron, M. and Sibbald, P. R. (1993) Weighting in sequence space: a comparison of methods in terms of generalized sequences. *Proc. Natl. Acad. Sci. USA* **90,** 8777–8781.
102. Sibbald, P. R. and Argos, P. (1990) Weighting aligned protein or nucleic acid sequences to correct for unequal representation. *J. Mol. Biol.* **216,** 813–818.
103. Henikoff, S. and Henikoff, J. G. (1994) Position-based sequence weights. *J. Mol. Biol.* **243,** 574–578.
104. Eddy, S. R., Mitchison, G., and Durbin, R. (1995) Maximum discrimination hidden Markov models of sequence consensus. *J. Comput. Biol.* **2,** 9–23.
105. Gotoh, O. (1995) A weighting system and algorithm for aligning many phylogenetically related sequences. *Comput. Appl. Biosci.* **11,** 543–551.
106. Krogh, A. and Mitchison, G. (1995) Maximum entropy weighting of aligned sequences of proteins or DNA. *Proc. Int. Conf. Intell. Syst. Mol. Biol.* **3,** 215–221.
107. Karchin, R. and Hughey, R. (1998) Weighting hidden Markov models for maximum discrimination. *Bioinformatics* **14,** 772–782.
108. May, A. C. (2001) Optimal classification of protein sequences and selection of representative sets from multiple alignments: application to homologous families and lessons for structural genomics. *Protein Eng.* **14,** 209–217.
109. Hirosawa, M., Totoki, Y., Hoshida, M., and Ishikawa, M. (1995) Comprehensive study on iterative algorithms of multiple sequence alignment. *Comput. Appl. Biosci.* **11,** 13–18.
110. Wang, Y. and Li, K. B. (2004) An adaptive and iterative algorithm for refining multiple sequence alignment. *Comput. Biol. Chem.* **28,** 141–148.
111. Wallace, I. M., O'Sullivan, O., and Higgins, D. G. (2005) Evaluation of iterative alignment algorithms for multiple alignment. *Bioinformatics* **21,** 1408–1414.
112. Brocchieri, L. and Karlin, S. (1998) A symmetric-iterated multiple alignment of protein sequences. *J. Mol. Biol.* **276,** 249–264.
113. Subbiah, S. and Harrison, S. C. (1989) A method for multiple sequence alignment with gaps. *J. Mol. Biol.* **209,** 539–548.
114. Barton, G. J. and Sternberg, M. J. (1987) A strategy for the rapid multiple alignment of protein sequences. Confidence levels from tertiary structure comparisons. *J. Mol. Biol.* **198,** 327–337.
115. Barton, G. J. and Sternberg, M. J. (1987) Evaluation and improvements in the automatic alignment of protein sequences. *Protein Eng.* **1,** 89–94.
116. Bains, W. (1986) MULTAN: a program to align multiple DNA sequences. *Nucleic Acids Res.* 14, 159–177.
117. Thompson, J. D., Thierry, J. C., and Poch, O. (2003) RASCAL: rapid scanning and correction of multiple sequence alignments. *Bioinformatics* **19,** 1155–1161.
118. Chakrabarti, S., Lanczycki, C. J., Panchenko, A. R., Przytycka, T. M., Thiessen, P. A., and Bryant, S. H. (2006) State of the art: refinement of multiple sequence alignments. *BMC Bioinform.* **7,** 499.

119. Chakrabarti, S., Lanczycki, C. J., Panchenko, A. R., Przytycka, T. M., Thiessen, P. A., and Bryant, S. H. (2006) Refining multiple sequence alignments with conserved core regions. *Nucleic Acids Res.* **34,** 2598–2606.
120. Huang, X. Q., Hardison, R. C., and Miller, W. (1990) A space-efficient algorithm for local similarities. *Comput. Appl. Biosci.* **6,** 373–381.
121. Huang, X. and Miller, W. (1991) A time-efficient, linear-space local similarity algorithm. *Adv. Appl. Math.* **12,** 337–357.
122. Altschul, S. F., Gish, W., Miller, W., Myers, E. W., and Lipman, D. J. (1990) Basic local alignment search tool. *J. Mol. Biol.* **215,** 403–410.
123. Altschul, S. F., Madden, T. L., Schaffer, A. A., Zhang, J., Zhang, Z., Miller, W., et al. (1997) Gapped BLAST and PSI-BLAST: a new generation of protein database search programs. *Nucleic Acids Res.* **25,** 3389–3402.
124. Pearson, W. R. (1998) Empirical statistical estimates for sequence similarity searches. *J. Mol. Biol.* **276,** 71–84.
125. Pearson, W. R. (1990) Rapid and sensitive sequence comparison with FASTP and FASTA. *Methods Enzymol.* **183,** 63–98.
126. Pearson, W. R. (2000) Flexible sequence similarity searching with the FASTA3 program package. *Methods Mol. Biol.* **132,** 185–219.
127. Morgenstern, B., Dress, A., and Werner, T. (1996) Multiple DNA and protein sequence alignment based on segment-to-segment comparison. *Proc. Natl. Acad. Sci. USA* **93,** 12098–12103.
128. Morgenstern, B., Frech, K., Dress, A., and Werner, T. (1998) DIALIGN: finding local similarities by multiple sequence alignment. *Bioinformatics* **14,** 290–294.
129. Morgenstern, B. (1999) DIALIGN 2: improvement of the segment-to-segment approach to multiple sequence alignment. *Bioinformatics* **15,** 211–218.
130. Morgenstern, B. (2004) DIALIGN: multiple DNA and protein sequence alignment at BiBiServ. *Nucleic Acids Res.* **32,** W33–36.
131. Subramanian, A. R., Weyer-Menkhoff, J., Kaufmann, M., and Morgenstern, B. (2005) DIALIGN-T: an improved algorithm for segment-based multiple sequence alignment. *BMC Bioinform.* **6,** 66.
132. Depiereux, E. and Feytmans, E. (1992) MATCH-BOX: a fundamentally new algorithm for the simultaneous alignment of several protein sequences. *Comput. Appl. Biosci.* **8,** 501–509.
133. Depiereux, E., Baudoux, G., Briffeuil, P., Reginster, I., De Bolle, X., Vinals, C., et al. (1997) Match-Box_server: a multiple sequence alignment tool placing emphasis on reliability. *Comput. Appl. Biosci.* **13,** 249–256.
134. Schwartz, A. S. and Pachter, L. (2007) Multiple alignment by sequence annealing. *Bioinformatics* **23,** e24–29.
135. Pellegrini, M., Marcotte, E. M., and Yeates, T. O. (1999) A fast algorithm for genome-wide analysis of proteins with repeated sequences. *Proteins* **35,** 440–446.
136. Notredame, C. (2001) Mocca: semi-automatic method for domain hunting. *Bioinformatics* **17,** 373–374.
137. Heger, A. and Holm, L. (2000) Rapid automatic detection and alignment of repeats in protein sequences. *Proteins* **41,** 224–237.

138. Heringa, J. and Argos, P. (1993) A method to recognize distant repeats in protein sequences. *Proteins* **17,** 391–341.
139. Szklarczyk, R. and Heringa, J. (2004) Tracking repeats using significance and transitivity. *Bioinformatics* **20(Suppl 1)**, I311–I317.
140. Sammeth, M. and Heringa, J. (2006) Global multiple-sequence alignment with repeats. *Proteins* **64,** 263–274.
141. Lawrence, C. E., Altschul, S. F., Boguski, M. S., Liu, J. S., Neuwald, A. F., and Wootton, J. C. (1993) Detecting subtle sequence signals: a Gibbs sampling strategy for multiple alignment. *Science* **262,** 208–214.
142. Neuwald, A. F., Liu, J. S., and Lawrence, C. E. (1995) Gibbs motif sampling: detection of bacterial outer membrane protein repeats. *Protein Sci.* **4,** 1618–1632.
143. Henikoff, S., Henikoff, J. G., Alford, W. J., and Pietrokovski, S. (1995) Automated construction and graphical presentation of protein blocks from unaligned sequences. *Gene* **163,** GC17–26.
144. Smith, H. O., Annau, T. M., and Chandrasegaran, S. (1990) Finding sequence motifs in groups of functionally related proteins. *Proc. Natl. Acad. Sci. USA* **87,** 826–830.
145. Bailey, T. L. and Elkan, C. (1994) Fitting a mixture model by expectation maximization to discover motifs in biopolymers. *Proc. Int. Conf. Intell. Syst. Mol. Biol.* **2,** 28–36.
146. Sonnhammer, E. L. and Kahn, D. (1994) Modular arrangement of proteins as inferred from analysis of homology. *Protein Sci.* **3,** 482–492.
147. Schuler, G. D., Altschul, S. F., and Lipman, D. J. (1991) A workbench for multiple alignment construction and analysis. *Proteins* **9,** 180–190.
148. Pevzner, P. A., Tang, H., and Tesler, G. (2004) De novo repeat classification and fragment assembly. *Genome Res.* **14,** 1786–1796.
149. Raphael, B., Zhi, D., Tang, H., and Pevzner, P. (2004) A novel method for multiple alignment of sequences with repeated and shuffled elements. *Genome Res.* **14,** 2336–2346.
150. Phuong, T. M., Do, C. B., Edgar, R. C., and Batzoglou, S. (2006) Multiple alignment of protein sequences with repeats and rearrangements. *Nucleic Acids Res.* **34,** 5932–5942.
151. Bishop, M. J. and Thompson, E. A. (1986) Maximum likelihood alignment of DNA sequences. *J. Mol. Biol.* **190,** 159–165.
152. Hein, J., Wiuf, C., Knudsen, B., Moller, M. B., and Wibling, G. (2000) Statistical alignment: computational properties, homology testing and goodness-of-fit. *J. Mol. Biol.* **302,** 265–279.
153. Thorne, J. L., Kishino, H., and Felsenstein, J. (1991) An evolutionary model for maximum likelihood alignment of DNA sequences. *J. Mol. Evol.* **33,** 114–124.
154. Thorne, J. L., Kishino, H., and Felsenstein, J. (1992) Inching toward reality: an improved likelihood model of sequence evolution. *J. Mol. Evol.* **34,** 3–16.
155. Miklos, I. and Toroczkai, Z. (2001) An improved model for statistical alignment. WABI.

156. Miklos, I. (2003) Algorithm for statistical alignment of sequences derived from a Poisson sequence length distribution. *Disc. Appl. Math.* **127,** 79–84.
157. Miklos, I., Lunter, G. A., and Holmes, I. (2004) A "Long Indel" model for evolutionary sequence alignment. *Mol. Biol. Evol.* **21,** 529–540.
158. Knudsen, B. and Miyamoto, M. M. (2003) Sequence alignments and pair hidden Markov models using evolutionary history. *J. Mol. Biol.* **333,** 453–460.
159. Metzler, D. (2003) Statistical alignment based on fragment insertion and deletion models. *Bioinformatics* **19,** 490–499.
160. Hein, J. (2001) A generalisation of the Thorne-Kishino-Felsenstein model of statistical alignment to k sequences related by a binary tree. PSB.
161. Hein, J., Jensen, J. L., and Pedersen, C. N. (2003) Recursions for statistical multiple alignment. *Proc. Natl. Acad. Sci. USA* **100,** 14960–14965.
162. Holmes, I. and Bruno, W. J. (2001) Evolutionary HMMs: a Bayesian approach to multiple alignment. *Bioinformatics* **17,** 803–820.
163. Holmes, I. (2003) Using guide trees to construct multiple-sequence evolutionary HMMs. *Bioinformatics* **19(Suppl 1),** i147–157.
164. Steel, M. and Hein, J. (2001) Applying the Thorne-Kishino-Felsenstein model to sequence evolution on a star-shaped tree. *Appl. Math. Lett.* **14,** 679–684.
165. Miklos, I. (2002) An improved algorithm for statistical alignment of sequences related by a star tree. *Bull. Math. Biol.* **64,** 771–779.
166. Lunter, G. A., Miklos, I., Song, Y. S., and Hein, J. (2003) An efficient algorithm for statistical multiple alignment on arbitrary phylogenetic trees. *J. Comput. Biol.* **10,** 869–889.
167. Jensen, J. L. and Hein, J. (2005) Gibbs sampler for statistical multiple alignment. *Stat. Sin.* **15,** 889–907.
168. Hein, J. (1990) Unified approach to alignment and phylogenies. *Methods Enzymol.* **183,** 626–645.
169. Vingron, M. and von Haeseler, A. (1997) Towards integration of multiple alignment and phylogenetic tree construction. *J. Comput. Biol.* **4,** 23–34.
170. Fleissner, R., Metzler, D., and von Haeseler, A. (2005) Simultaneous statistical multiple alignment and phylogeny reconstruction. *Syst. Biol.* **54,** 548–561.
171. Lunter, G., Miklos, I., Drummond, A., Jensen, J. L., and Hein, J. (2005) Bayesian coestimation of phylogeny and sequence alignment. *BMC Bioinform.* **6,** 83.
172. Redelings, B. D. and Suchard, M. A. (2005) Joint Bayesian estimation of alignment and phylogeny. *Syst. Biol.* **54,** 401–418.
173. Metzler, D., Fleissner, R., Wakolbinger, A., and von Haeseler, A. (2001) Assessing variability by joint sampling of alignments and mutation rates. *J. Mol. Evol.* **53,** 660–669.
174. Allison, L. and Wallace, C. S. (1994) The posterior probability distribution of alignments and its application to parameter estimation of evolutionary trees and to optimization of multiple alignments. *J. Mol. Evol.* **39,** 418–430.
175. Krogh, A., Brown, M., Mian, I. S., Sjolander, K., and Haussler, D. (1994) Hidden Markov models in computational biology. Applications to protein modeling. *J. Mol. Biol.* **235,** 1501–1531.

176. Krogh, A. (1998) An introduction to hidden Markov models for biological sequences. In *Computational Methods in Molecular Biology* (Salzberg, S., Searls, D., Kasif, S., eds.). Elsevier Science, St. Louis, MO, pp. 45–63.
177. Hughey, R. and Krogh, A. (1996) Hidden Markov models for sequence analysis: extension and analysis of the basic method. *Comput. Appl. Biosci.* **12,** 95–107.
178. Eddy, S. R. (1996) Hidden Markov models. *Curr. Opin. Struct. Biol.* **6,** 361–365.
179. Eddy, S. R. (1998) Profile hidden Markov models. *Bioinformatics* **14,** 755–763.
180. Mamitsuka, H. (2005) Finding the biologically optimal alignment of multiple sequences. *Artif. Intell. Med.* **35,** 9–18.
181. Baldi, P. and Chauvin, Y. (1994) Smooth on-line learning algorithms for hidden Markov models. *Neural Comput.* **6,** 307–318.
182. Baldi, P., Chauvin, Y., Hunkapiller, T., and McClure, M. A. (1994) Hidden Markov models of biological primary sequence information. *Proc. Natl. Acad. Sci. USA* **91,** 1059–1063.
183. Viterbi, A. J. (1967) Error bounds for convolutional codes and an asymptotically optimum decoding algorithm. *IEEE Trans. Inform. Theory* **It13,** 260.
184. Grundy, W. N., Bailey, T. L., Elkan, C. P., and Baker, M. E. (1997) Meta-MEME: motif-based hidden Markov models of protein families. *Comput. Appl. Biosci.* **13,** 397–406.
185. Bucher, P., Karplus, K., Moeri, N., and Hofmann, K. (1996) A flexible motif search technique based on generalized profiles. *Comput. Chem.* **20,** 3–23.
186. Karplus, K., Barrett, C., and Hughey, R. (1998) Hidden Markov models for detecting remote protein homologies. *Bioinformatics* **14,** 846–856.
187. Park, J., Karplus, K., Barrett, C., Hughey, R., Haussler, D., Hubbard, T., *et al.* (1998) Sequence comparisons using multiple sequences detect three times as many remote homologues as pairwise methods. *J. Mol. Biol.* **284,** 1201–1210.
188. Sonnhammer, E. L., Eddy, S. R., Birney, E., Bateman, A., and Durbin, R. (1998) Pfam: multiple sequence alignments and HMM-profiles of protein domains. *Nucleic Acids Res.* **26,** 320–322.
189. Eddy, S. R. HMMER: a profile hidden Markov modeling package, available from http://hmmer.janelia.org/.
190. Sjolander, K., Karplus, K., Brown, M., Hughey, R., Krogh, A., Mian, I. S., *et al.* (1996) Dirichlet mixtures: a method for improved detection of weak but significant protein sequence homology. *Comput. Appl. Biosci.* **12,** 327–345.
191. Barrett, C., Hughey, R., and Karplus, K. (1997) Scoring hidden Markov models. *Comput. Appl. Biosci.* **13,** 191–199.
192. McClure, M. A., Smith, C., and Elton, P. (1996) Parameterization studies for the SAM and HMMER methods of hidden Markov model generation. *Proc. Int. Conf. Intell. Syst. Mol. Biol.* **4,** 155–164.
193. Karplus, K. and Hu, B. (2001) Evaluation of protein multiple alignments by SAM-T99 using the BAliBASE multiple alignment test set. *Bioinformatics* **17,** 713–720.
194. Loytynoja, A. and Milinkovitch, M. C. (2003) A hidden Markov model for progressive multiple alignment. *Bioinformatics* **19,** 1505–1513.

195. Edgar, R. C. and Sjolander, K. (2003) Simultaneous sequence alignment and tree construction using hidden Markov models. *Pac. Symp. Biocomput.* 180–191.
196. Edgar, R. C. and Sjolander, K. (2003) SATCHMO: sequence alignment and tree construction using hidden Markov models. *Bioinformatics* **19,** 1404–1411.
197. Loytynoja, A. and Goldman, N. (2005) An algorithm for progressive multiple alignment of sequences with insertions. *Proc. Natl. Acad. Sci. USA* **102,** 10557–10562.
198. Holmes, I. and Durbin, R. (1998) Dynamic programming alignment accuracy. *J. Comput. Biol.* **5,** 493–504.
199. Schwartz, A. S., Myers, E., and Pachter, L. (2006) Alignment metric accuracy. *arXiv 2006:q-bio.QM/0510052.*
200. Roshan, U. and Livesay, D. R. (2006) Probalign: multiple sequence alignment using partition function posterior probabilities. *Bioinformatics* **22,** 2715–2721.
201. Wallace, I. M., O'Sullivan, O., Higgins, D. G., and Notredame, C. (2006) M-Coffee: combining multiple sequence alignment methods with T-Coffee. *Nucleic Acids Res.* **34,** 1692–1699.
202. Kececioglu, J. D. (1993) The maximum weight trace problem in multiple sequence alignment. CPM.
203. Kececioglu, J. D., Lenhof, H.-P., Mehlhorn, K., Mutzel, P., Reinert, K., and Vingron, M. (2000) A polyhedral approach to sequence alignment problems. *Disc. Appl. Math.* **104,** 143–186.
204. Koller, G. and Raidl, G. R. (2004) An evolutionary algorithm for the maximum weight trace formulation of the multiple sequence alignment problem. In *LNCS*, 3242, pp. 302–311.
205. Simossis, V. A. and Heringa, J. (2005) PRALINE: a multiple sequence alignment toolbox that integrates homology-extended and secondary structure information. *Nucleic Acids Res.* **33,** W289–294.
206. Simossis, V. A., Kleinjung, J., and Heringa, J. (2005) Homology-extended sequence alignment. *Nucleic Acids Res.* **33,** 816–824.
207. Thompson, J. D., Plewniak, F., Thierry, J., and Poch, O. (2000) DbClustal: rapid and reliable global multiple alignments of protein sequences detected by database searches. *Nucleic Acids Res.* **28,** 2919–2926.
208. Wang, J. and Feng, J. A. (2005) NdPASA: a novel pairwise protein sequence alignment algorithm that incorporates neighbor-dependent amino acid propensities. *Proteins* **58,** 628–637.
209. Yang, A. S. (2002) Structure-dependent sequence alignment for remotely related proteins. *Bioinformatics* **18,** 1658–1665.
210. Zhou, H. and Zhou, Y. (2005) SPEM: improving multiple sequence alignment with sequence profiles and predicted secondary structures. *Bioinformatics* **21,** 3615–3621.
211. O'Sullivan, O., Suhre, K., Abergel, C., Higgins, D. G., and Notredame, C. (2004) 3DCoffee: combining protein sequences and structures within multiple sequence alignments. *J. Mol. Biol.* **340,** 385–395.

212. Armougom, F., Moretti, S., Poirot, O., Audic, S., Dumas, P., Schaeli, B., *et al.* (2006) Expresso: automatic incorporation of structural information in multiple sequence alignments using 3D-Coffee. *Nucleic Acids Res.* **34,** W604–608.
213. Thompson, J. D., Plewniak, F., and Poch, O. (1999) BAliBASE: a benchmark alignment database for the evaluation of multiple alignment programs. *Bioinformatics* **15,** 87–88.
214. Thompson, J. D., Plewniak, F., and Poch, O. (1999) A comprehensive comparison of multiple sequence alignment programs. *Nucleic Acids Res.* **27,** 2682–2690.
215. Mizuguchi, K., Deane, C. M., Blundell, T. L., and Overington, J. P. (1998) HOMSTRAD: a database of protein structure alignments for homologous families. *Protein Sci.* **7,** 2469–2471.
216. Van Walle, I., Lasters, I., and Wyns, L. (2005) SABmark–a benchmark for sequence alignment that covers the entire known fold space. *Bioinformatics* **21,** 1267–1268.
217. Raghava, G. P., Searle, S. M., Audley, P. C., Barber, J. D., and Barton, G. J. (2003) OXBench: a benchmark for evaluation of protein multiple sequence alignment accuracy. *BMC Bioinform.* **4,** 47.
218. Thompson, J. D., Koehl, P., Ripp, R., and Poch, O. (2005) BAliBASE 3.0: latest developments of the multiple sequence alignment benchmark. *Proteins* **61,** 127–136.
219. Sauder, J. M., Arthur, J. W., and Dunbrack, R. L., Jr. (2000) Large-scale comparison of protein sequence alignment algorithms with structure alignments. *Proteins* **40,** 6–22.
220. Pang, A., Smith, A. D., Nuin, P. A., and Tillier, E. R. (2005) SIMPROT: using an empirically determined indel distribution in simulations of protein evolution. *BMC Bioinform.* **6,** 236.
221. Nuin, P. A., Wang, Z., and Tillier, E. R. (2006) The accuracy of several multiple sequence alignment programs for proteins. *BMC Bioinform.* **7,** 471.
222. Stoye, J., Evers, D., and Meyer, F. (1998) Rose: generating sequence families. *Bioinformatics* **14,** 157–163.
223. Eidhammer, I., Jonassen, I., and Taylor, W. R. (2000) Structure comparison and structure patterns. *J. Comput. Biol.* **7,** 685–716.
224. Carugo, O. and Pongor, S. (2001) A normalized root-mean-square distance for comparing protein three-dimensional structures. *Protein Sci.* **10,** 1470–1473.
225. Armougom, F., Moretti, S., Keduas, V., and Notredame, C. (2006) The iRMSD: a local measure of sequence alignment accuracy using structural information. *Bioinformatics* **22,** e35–39.
226. Chew, L. P., Huttenlocher, D., Kedem, K., and Kleinberg, J. (1999) Fast detection of common geometric substructure in proteins. *J. Comput. Biol.* **6,** 313–325.
227. O'Sullivan, O., Zehnder, M., Higgins, D., Bucher, P., Grosdidier, A., and Notredame, C. (2003) APDB: a novel measure for benchmarking sequence alignment methods without reference alignments. *Bioinformatics* **19(Suppl 1),** i215–221.

228. Henikoff, S. and Henikoff, J. G. (1992) Amino acid substitution matrices from protein blocks. *Proc. Natl. Acad. Sci. USA* **89,** 10915–10919.
229. Dayhoff, M. O., Eck, R. V., and Park, C. M. (1972) A model of evolutionary change in proteins. In *Atlas of Protein Sequence and Structure* (Dayhoff, M. O., ed.). National Biomedical Research Foundation, Washington, DC, pp. 89–99.
230. Dayhoff, M. O., Schwartz, R. M., and Orcutt, B. C. (1978) A model of evolutionary change in proteins. In *Atlas of Protein Sequence and Structure* (Dayhoff, M. O., ed.). National Biomedical Research Foundation, Washington, DC, pp. 345–352.
231. Muller, T. and Vingron, M. (2000) Modeling amino acid replacement. *J. Comput. Biol.* **7,** 761–776.
232. Whelan, S. and Goldman, N. (2001) A general empirical model of protein evolution derived from multiple protein families using a maximum-likelihood approach. *Mol. Biol. Evol.* **18,** 691–699.
233. Prlic, A., Domingues, F. S., and Sippl, M. J. (2000) Structure-derived substitution matrices for alignment of distantly related sequences. *Protein Eng.* **13,** 545–550.
234. Reese, J. T. and Pearson, W. R. (2002) Empirical determination of effective gap penalties for sequence comparison. *Bioinformatics* **18,** 1500–1507.
235. Arribas-Gil, A., Gassiat, E., and Matias, C. (2006) Parameter estimation in pair-hidden Markov models. *Scand. J. Stat.* **33,** 651–671.
236. Liu, J. S., Neuwald, A. F., and Lawrence, C. E. (1995) Bayesian models for multiple local sequence alignment and Gibbs sampling strategies. *J. Am. Stat. Assoc.* **90,** 1156–1170.
237. Zhu, J., Liu, J. S., and Lawrence, C. E. (1998) Bayesian adaptive sequence alignment algorithms. *Bioinformatics* **14,** 25–39.
238. Kececioglu, J. and Kim, E. (2007) Simple and fast inverse alignment. RECOMB.
239. Yu, C.-N., Joachims, T., Elber, R., and Pillardy, J. (2007) Support vector training of protein alignment models. RECOMB.
240. Tsochantaridis, I., Joachims, T., Hofmann, T., and Altun, Y. (2005) Large margin methods for structured and interdependent output variables. *J. Mach. Learn. Res.* **6,** 1453–1484.
241. Katoh, K. and Toh, H. (2007) PartTree: an algorithm to build an approximate tree from a large number of unaligned sequences. *Bioinformatics* **23,** 372–374.
242. Ahola, V., Aittokallio, T., Vihinen, M., and Uusipaikka, E. (2006) A statistical score for assessing the quality of multiple sequence alignments. *BMC Bioinform.* **7,** 484.
243. Altschul, S. F. (1998) Generalized affine gap costs for protein sequence alignment. *Proteins* **32,** 88–96.
244. Zachariah, M. A., Crooks, G. E., Holbrook, S. R., and Brenner, S. E. (2005) A generalized affine gap model significantly improves protein sequence alignment accuracy. *Proteins* **58,** 329–338.
245. Thompson, J. D., Muller, A., Waterhouse, A., Procter, J., Barton, G. J., Plewniak, F., et al. (2006) MACSIMS: multiple alignment of complete sequences information management system. *BMC Bioinform.* **7,** 318.

246. Thompson, J. D., Holbrook, S. R., Katoh, K., Koehl, P., Moras, D., Westhof, E., et al. (2005) MAO: a multiple alignment ontology for nucleic acid and protein sequences. *Nucleic Acids Res.* **33,** 4164–4171.
247. Gotoh, O. (1999) Multiple sequence alignment: algorithms and applications. *Adv. Biophys.* **36,** 159–206.
248. Phillips, A., Janies, D., and Wheeler, W. (2000) Multiple sequence alignment in phylogenetic analysis. *Mol. Phylogenet. Evol.* **16,** 317–330.
249. Lambert, C., Campenhout, J. M. V., DeBolle, X., and Depiereux, E. (2003) Review of common sequence alignment methods: clues to enhance reliability. *Curr. Genom.* **4,** 131–146.
250. Wallace, I. M., Blackshields, G., and Higgins, D. G. (2005) Multiple sequence alignments. *Curr. Opin. Struct. Biol.* **15,** 261–266.
251. Edgar, R. C. and Batzoglou, S. (2006) Multiple sequence alignment. *Curr. Opin. Struct. Biol.* **16,** 368–373.
252. Morrison, D. A. (2006) Multiple sequence alignment for phylogenetic purposes. *Aust. Syst. Bot.* **19,** 479–539.
253. Cormen, T. H., Leiserson, C. E., Rivest, R. L., and Stein, C. (2001) *Introduction to Algorithms*. MIT Press, Cambridge, MA.
254. Eppstein, D. (2000) Fast hierarchical clustering and other applications of dynamic closest pairs. *J. Exp. Algorithmics* **5**, 1–23.
255. Elias, I. and Lagergren, J. (2005) Fast neighbor joining. ICALP.
256. Waterman, M. S., Eggert, M., and Lander, E. (1992) Parametric sequence comparisons. *Proc. Natl. Acad. Sci. USA* **89,** 6090–6093.
257. Waterman, M. S. (1994) Parametric and ensemble sequence alignment algorithms. *Bull. Math. Biol.* **56,** 743–767.
258. Gusfield, D., Balasubramanian, K., and Naor, D. (1994) Parametric optimization of sequence alignment. *Algorithmica* **12,** 312–326.

26

Discovering Biomedical Knowledge from the Literature

Jasmin Šarić, Henriette Engelken, and Uwe Reyle

Summary

Biomedical knowledge is to a very large extent represented only in textual form. To make this knowledge accessible to humans and/or further automatic processing, text mining applications have been developed. At the end of this chapter we present an overview of the most important open access applications and their functionality. The main part of the paper is devoted to the major problems with which all such applications have to deal. The first problem is terminology processing, i.e., recognizing biomedical terms and identifying their meanings, at least to a certain degree. The second problem is to bring together information units that are distributed over more than one sentence. The task of coreference resolution consists of identifying the entities to which the text refers in different sentences and in different ways. The third problem we discuss is that of information extraction, in particular, extraction of relational information. The representation of the domain knowledge is an indispensable component of any text mining application. We discuss different types and depths of ontological modeling and how this knowledge helps to accomplish the tasks described above. An overview of ontological resources is given at the end of the chapter.

Key Words: Natural language processing; text mining; information extraction; named entity recognition; terminology processing; ontologies; taxonomies; ambiguity; coreference.

1. Introduction

The rapid development of the life sciences domain has led to a tremendous increase in scientific and patent literature. According to Scopus (*see* **Note 1**) a researcher starting a project on *epidermal growth factor receptor* in 1980 had to inspect 10 articles. By 1985, there were 321 articles to read, and in 2006 an overwhelming number of 17,782 references had to be taken into account.

From: *Methods in Molecular Biology, vol. 484: Functional Proteomics: Methods and Protocols*
Edited by: J. D. Thompson et al., DOI: 10.1007/978-1-59745-398-1, © Humana Press, Totowa, NJ

In spite of the development of databases for all areas in the life sciences, most information in this domain is available only in unstructured form, i.e., in comment lines of databases, abstracts, or full research papers (*see* **Note 2**). Information in natural language form has thus become a greatly underutilized source of information for biologists. Automated information seeking and information extraction have become necessary for researchers in biology and biochemistry to keep up with what has been published. Applying conventional extraction technology has, however, not been successful. The complex terminology in the area requires the development of specialized lexica, parsers, and ontologies. Only then can information extraction technology developed by computational linguistics over the past 20 years be applied to this particular domain to automatically map information from text sources into unambiguously structured and automatically processable representations, such as knowledge bases.

To begin, some related challenges need to be detangled. *Text Mining* (TM) is the task of extracting knowledge from unstructured Text. If the information is stored not only as text but also as tables, graphics, etc., the task is called *Literature Mining*. TM is achieved by the following subtasks: *Information Retrieval* (IR) (also referred to as *Text Retrieval* [TR], if the documents consist primarily of text) retrieves documents relevant to a given user's query from a collection of documents (e.g., PubMed). The returned documents need further processing (by the user or by an IE system) to extract the requested information. *Information Extraction* (IE) extracts structured information (of a predefined, domain-specific kind) from a set of unstructured, natural language texts in order to populate a database with facts. *Data Mining* (DM) retrieves implicit, previously unknown information by finding correlations among pieces of structured data (in databases).

2. Processing Terminology

Named entity recognition consists of three tasks: (1) the identification of words and phrases (so-called *terms*) that specify names of persons, organizations, dates, locations, etc.; (2) the classification of terms by corresponding predefined concepts; and (3) the task of establishing links between terms and referent identifiers in data sources. This task is called *term mapping*. In the biomedical context named entities usually involve genes, proteins, cells, tissues, organisms, compounds, diseases, analysis methods, and the like.

As an example, consider the following sentence (PMID: 16919868): *In recent years, new therapeutic targets such as EGFR (epidermal growth factor receptor), COX-2, and KIT have emerged as new potential therapeutic targets.* In the first step, the terms *EGFR*, *epidermal growth factor receptor*, *COX-2*, and *KIT* should be recognized as terms. Then they should be classified as referring to protein and gene names, respectively. The third step consists of identifying the unique database accession numbers of these entities, such as *P00403* for human *COX-2* in SwissProt.

The major problem in named entity recognition is the lack of standardization. Morphosyntactic variants, use of abbreviations and acronyms, as well as variations in spelling usually generate a whole class of different names (often called *synonyms*) for one and the same entity. For *COX-2* this class contains at least *cytochrome c oxidase polypeptide II, cytochrome c oxidase 2, COX subunit II,* and *COII* (*see* **Note 3**). Synonymous gene and protein names also occur in different orthographic variants such as *COX2*, *Cox2*, and *COX2p*.

There are three main types of approaches for the identification and classification of terms. The first approach is rule-based, trying to characterize the set of terms of a certain class by a kind of shallow morphosyntactic grammar. There are various first gene name recognition approaches using such manually crafted rules ***(1–4)***. They take into account typical features of gene names such as the *ase* suffix or specific combinations of letters and numbers. As an additional hint the occurrence of words such as *protein* or *receptor* in the nearby context is used in the rules. The second type of approach tries to derive characteristic features of term classes automatically by machine learning techniques ***(3–9)***. These methods require manually annotated text collections as a starting point for a training procedure ***(10,11)***. And finally, there are methods based on comprehensive and well-curated lists of term names and their synonyms. (For gene names *see* **refs. *1,6,12–15***.)

The challenge of any type of approach is the problem of ambiguity resolution. Most of the words and phrases of natural language are ambiguous and the language user must identify their intended meaning by taking into account the context in which they are used. In principle, this resolution process requires a deeper syntactic and semantic analysis of the context. To decide, for example, whether a particular occurrence of the common, frequent, and seemingly unambiguous word "cell" is used to refer to the biological organism or is used as a synonym of the gene called *carboxyl ester lipase-like* depends on the context. But there are also statistical approaches to deal with the ambiguity resolution problem. (For advances in automatic gene name disambiguation *see* **refs. *14,16,17***.)

Once terms are identified and classified there remains the task of mapping them to referent identifiers in data sources. If the term is contained in the synonym list of a particular database entry this task is rather trivial. But if it is not, matters become much more complex and require a lifting of the mapping procedure from the level of the strings to the level of the properties that the text on the one hand and the database on the other assign to them.

The reasoning tasks that are involved in such a more complicated mapping procedure require knowledge about the domain of application. This knowledge must be provided by ontologies. Ontologies also provide the concepts to be used for the classification of terms.

3. Ontologies

The use of ontologies is indispensable for any TM application, in particular, for the tasks of named entity recognition, information extraction, and retrieval. Ontologies are also needed in the area of data integration. The incompatibilities among data formats, structure, and models (flat files, relational databases, etc.) have become a major obstacle in biological research. To overcome this problem of integration of data, e.g., by creating data warehouses or distributed federated databases, it is necessary to know and describe exactly which data entries in one data source relate to the data entries in another source and to know how they are related.

Attempts in this direction have led to the development of explicitly enumerated lists of terms (so-called *controlled vocabularies*) to annotate genes and gene products (e.g., Gene Ontology), of knowledge-based schemata and data exchange formats for data integration (e.g., BioCyc, BioPax), and of formal representations of biological and biochemical processes (e.g., HyBrow). In any of these applications ontologies serve as a means for establishing a conceptually concise basis for describing terms of languages (natural or formal) or concepts as the building blocks of categorization and reasoning.

Ontology is the study of things that exist or may exist in some domain. An ontology contains terms for the types of things in the domain. These terms represent the predicates, word senses, or concept and relational types of a language used for the purpose of talking about the domain. An informal ontology may be specified by a catalog of types that are either undefined or defined only by statements in some natural language. Examples of informal ontologies are controlled vocabularies and *thesauri*, i.e., networked collections of controlled vocabulary terms each of which is given a definition in natural language terms. A formal ontology is specified by a collection of names for concept and relational types together with explicit definitions of their meanings in terms of logical theories.

The gene ontology (GO) is an example of a controlled vocabulary. Its primary focus is to provide a practically useful framework for keeping track of the biological annotations that are applied to gene products, and not to build a logically rigorous formalization of the domain (Too little attention was paid to the formal definition of terms such as *function*, *part*, *component*, *substance*, *action*, *domain*, or *complex*, which were employed in the construction of GOs.) or on achieving reasoning efficiency in software implementations.

Formal ontologies are divided into two main classes: foundational and domain. *Foundational* (or *upper*) *ontologies* are axiomatic accounts of high-level domain-independent categories about the real world. They are based on formal theories of part–whole structure (mereology and mereotopology), temporal and spatial relations, identity, event and process structure, distinctions between abstract and physical, modality, etc. (*see* **Note 4**). They constitute toolboxes of reusable information modeling primitives for building application ontologies in specific domains. As such, they enhance semantic interoperability between agents by specifying descriptively adequate shared conceptualizations. *Domain ontologies* model specific domains or parts of the world. They attempt to provide an explicit formal definition of the senses of the terms, or concepts, they are supposed to model. Structures that organize their concepts only into hierarchies, in which each term is in one or more parent–child relationships (e.g., whole–part, genus–species, type–instance) to other terms, are called *taxonomies*. Some selected taxonomies are presented in **Section 6**.

As already mentioned above, one application area of ontologies is that of data integration. *Data integration* is concerned with unifying data that share some common semantics but originate from unrelated, heterogeneous sources. Heterogeneity may result from the use of different languages and data representations and/or the use of concepts with different or imprecise interpretations. Ontologies may serve as a stable conceptual interface to these databases that is independent of the particular database schemata. The knowledge represented by such an ontology must be sufficiently comprehensive to support translation of all the relevant information sources into its common frame of reference. It also supports consistent management and recognition of inconsistent data.

Another application area of ontologies is natural language processing, including Text Mining. In NL-processing systems ontologies provide an interface from ontological types to linguistic concepts and meanings of linguistic terms and phrases in general. The demands on this interface vary with the type of application: at the lowest level of detail (as required, e.g., for information retrieval tasks) it may consist of mapping NL terms to (a set of) key words only, whereas at a deeper level of analysis (ranging from Information Extraction and Question Answering to Machine Translation and Text Understanding) NL terms express complex relationships between ontological types. Typically these relationships are, in addition, dependent on the syntactic, semantic, and pragmatic context in which the NL terms occur.

The general need for the incorporation of ontological knowledge in NL processing is *ambiguity control*, i.e., the reduction or even elimination of ambiguities in linguistic expressions. This included all kinds of ambiguities ranging from lexical and structural to semantic and pragmatic ones. At the lexical level word sense selection and lexical disambiguation may be achieved by exploiting selectional restrictions, i.e., ontological constraints that have to be fulfilled by arguments of verbs (e.g., the subject of *express* is required to be a gene [in the biochemical domain] and its object a protein) or relational nouns (*active side of* requires a protein as argument). Also the interpretation of noun compounds is dependent on domain knowledge. A sophisticated analysis of compound terms should result in the following inferences: *anthranilate synthase component*/is part of *anthranilate synthase*, and the term *mitogen-activated protein kinase kinase* denotes a kinase that acts on a protein kinase. From a morphosyntactic perspective the ambiguities of noun compounds have much in common with ambiguities on a structural level. Here the most prominent and ubiquitous problems are those provided by PP-attachment and coordination. Coordination ambiguities arise systematically if the coordinating conjunction occurs between compound nouns. Compare the *heterodimer of the* α *and* β *subunit* with the *precursor of STAT4 and protein kinase*. The first phrase can refer to a complex consisting of only two subunits, called α and β. It cannot mean something like "heterodimer of α subunit and heterodimer of β subunit," nor can it talk about two substances referred to by the phrases "heterodimer of α" and "β subunit." In contrast, it is exactly this latter structure that fits with the second compound, which denotes the two peptides *precursor of STAT4* and *protein kinase*. Semantic ambiguities arise very often at the lexical level (the term *cell*, for example, may denote either the biological cell or the gene *TBP-associated factor 1* of *Drosophila melanogaster*—and note that certainly not all of the additional meanings of *cell*, i.e., *cadre*, *cubicle*, *prison cell*, *electric cell*, and *cell phone*, can be ignored in the context of life sciences). Most of the other semantic ambiguities relevant for rediscovering information from the literature are related to scope. The interpretation of a phrase such as *Two-thirds of children reported a dental impact "often" or "everyday"* depends, among other things, on the solution to the question "Who suffered from the dental impacts reported." Although it could be anybody else that was mentioned in the text preceding this sentence, it is quite reasonable to assume that the children reported their own impacts. In addition, in this case it is not clear whether it was the dental impact that occurred often or each day, or whether the reports were made often or every day; in either case there is the additional question of whether each of the two-thirds of children reported their own dental impact, or if all reported the dental impact of one and only one specific child. Questions of this type are very difficult to answer; fortunately they belong to the area of Text Understanding and not so much to Information Extraction and Question Answering tasks; at present there are no means to do this computationally. Pragmatic ambiguities occur whenever intersentential relationships are triggered, most commonly by anaphoric expressions. We will discuss this issue in the following section.

Using ontological knowledge in NL-processing systems presupposes its availability in explicit form processible by computers. Formalisms for knowledge representation and its deductive manipulation are in place. But what about the knowledge acquisition problem? Extensive research in the 1980s and early 1990s has shown that it is possible to extract knowledge from texts with a more or less regular structure, such as machine-readable dictionaries. Today's online dictionaries, such as WordNet or WikiPedia, facilitate this task because they already provide systems of categories that can be used as a conceptual basis for building a semantic network. The dictionaries differ, however, in the degree to which the relationships between categories are semantically typed and well defined. If an ontology is to be learned from free and completely unstructured text sources the first step is to learn the concepts themselves. This step may be divided into three subtasks: acquisition of the relevant terminology, identification of synonyms and linguistic variants, and concept formation proper (**Table 1**) *(18)*.

Once the concept hierarchy has been constructed the next tasks in ontology learning consist of acquiring relevant domain-specific relationships and organizing them in a hierarchical order together with the relationships given by some domain-independent upper-level ontology. Now, the general axiom schemata have to be populated and finally we may try to acquire more specific axioms, a task that for the time being has to be done manually.

Within the biomedical domain knowledge acquisition is more complicated due to the restricted availability of machine-readable dictionaries, thesauri, and texts. Some selected ontologies are presented in **Section 6**.

Table 1
Layers of Ontology Learning *(18)*

$\forall x(\mathrm{BODY}(x) \rightarrow \exists y(\mathrm{CAPUT}(y) \bigwedge \ldots)$	Aquisition of domain-specific Axioms
disjoint(THORAX, VENTER)	Instantiation of upper-level Axiom Schemata
internus $\sqsubset_r$ part_of	Relation Hierarchy
superior(THORAX, VENTER)	Relation Aquisition
VENTER $\sqsubset_c$ TRUNCUS	Concept Hierarchy
VENTER	Concepts
abdomen, belly, venter	Synonyms
skin, abdomen, belly, disease, cancer,...	Term Aquisition

4. Coreferring Expressions in Text

Noun phrases, such as names, demonstratives, pronouns, and definite noun phrases, are means provided by natural language to refer to objects in the real world. The term *coreference* is used to indicate the fact that two of these textual elements refer to the same object. A typical example of coreference as established by an anaphoric expression is *Gpi1p is of further significance: it is required for growth of S. cerevisiae ....* In the preferred reading of this passage, the pronoun *it* is a sort of "abbreviated mention" of the individual "Gpi1p," which is denoted by the expression *Gpi1p*. Following standard terminology we will say that *Gpi1p* is the *antecedent* of the anaphor *it*.

Antecedents need not always be overt and must often be inferred from context. So-called *bridging* references are expressions that denote objects related only to the denotation of their antecedent by (shared) generic knowledge as in *Each hair on your body grows out of a tiny tube in the skin.* We are able to interpret the description *the skin* as coreferring with *the skin of your body* because we know that the skin is a part of the body, and *body* was mentioned at the beginning of the sentence. Beneath the part–whole relationship between a definite noun phrase and its antecedent the bridging reference may also refer to the object filling a role in an event, whether implicitly or explicitly introduced. [The resolution of bridging references ***(19)*** is based on a semantic classification of some 100 biochemical verbs ***(20)***.]

Both pronouns and bridging references are frequently used in the biomedical literature. [One hundred distinct anaphors of both types in a set of 60–70 Medline abstracts have been found ***(21)***.] This makes dealing with them crucial for any text- processing system. *Coreference Resolution* is the task of resolving these anaphoric noun phrases to the entities to which they refer and thus establishing a link between them and their antecedent(s). Much work has been done in the past in this area. Linguistically motivated approaches are based on syntax, focus, and what is called Centering Theory. Machine learning techniques include unsupervised methods such as clustering and supervised methods such as decision trees.

Because pronouns do not have any semantic content on their own, morphological and syntax-based constraints on their resolution are prominent. Morphological constraints involve congruence of number and gender between the pronoun and its antecedent. (Note, however, that this is a default that may be overridden by the logical numerus of the antecedent [*the quartet—they*], by "abstraction" [*each hair—they*], or by "summation" [*Ribose is related to deoxyribose. They ...*].) Syntax-based constraints rule out certain sentence-internal coreference relationships (*Its structure is similar to β-sitosterol.* Its $\neq$ β-sitosterol). In spite of these morphosyntactic constraints in most cases more than one possible antecedent remains and syntactic, semantic, and pragmatic preference rules are applied in order to identify the intended one. The rules comprise the notions of thematicity, salience, and grammatical role of the antecedent (subject is preferred over object, what is the text/passage "talking about," etc.). They make a difference as to whether the antecedent occurs in a main or subordinate sentence, or whether it occurs at a greater distance from the anaphoric expression or not. Finally, the rules also take ontological knowledge into account (*If the baby doesn't like the milk then boil it.*).

The theory most influential in tracking the saliency of entities, when processing a text, is Centering Theory ***(22,23)***. [Centering is algorithmically used ***(24)***.] This approach takes into account the fact that entities that are mentioned are more or less at the center of attention and thus serve to link one utterance to another. Each utterance is assigned a set of forward-looking centers ranked by its grammatical role and a unique backward-looking center. Anaphors are then resolved by connecting the backward-looking center of an utterance to the highest ranked forward-looking center of the previous utterance. An alternative to Centering is presented in **ref.** ***(25)***. This algorithm considers possible antecedents of pronouns globally, depending on whether they are new or old to the hearer. Hearer-old discourse entities are preferred for anaphor resolution, instead of choosing a local antecedent such as in the Centering Theory. Strube's algorithm resolves pronouns with an F-measure (*see* **Note 5**) of 75.1 and 61.6 on the MUC-6 and MUC-7 coreference resolution corpora (*see* **Note 6**), respectively.

Statistical approaches train their models on annotated text corpora in order to compute probabilities of candidate antecedents for a given anaphor. The model of Ge *et al.* ***(26)***, for example, takes into account the distance, the number of mentions, and syntactic and semantic factors. The candidate with the highest probability is then selected. On a test corpus this model achieved a success rate of 82.9% for pronoun resolution.

Coreference resolution may also be treated as a clustering task ***(27)***. This approach automatically groups the noun phrases, i.e., possible anaphors and antecedents, of a document into equivalence classes. As a basis, each noun phrase is represented by a set of 11 features (position, gender, semantic class, etc.), which enables the distance between pairs of noun phrases to be computed. If this distance is less than a certain clustering radius threshold, their two classes are merged (unless they contain an incompatible element) and the noun phrases are thus considered coreferent. On MUC-6 this clustering approach achieved an F-measure of 53.6%.

A different technique is based on decision trees, regarding coreference resolution as a binary classification task ***(28,29)***. In contrast to clustering, this employs supervised learning methods. The classifier is trained on some pairs of noun phrases, which had been manually marked as either positive or negative instances of coreference. After training, the classifier is used to select the antecedent for each anaphor in a text, e.g., by choosing the nearest preceding noun phrase whose likelihood of being coreferent with the anaphor is above some value ["closest-first" *(28)* or apply "best-first" *(29)*]. Being evaluated on MUC-6 and MUC-7 the system of Ng and Cardie reached F-measures of 70.4 and 63.4, respectively, outperforming Soon's system, which achieved F-measures of 62.6 and 60.4. [It was shown *(30)* that the performance of a system like Soon's could be improved by applying additional features for matching noun phrases and modifying the selection of the training instances. On a set of Medline documents they attained an F-measure of 68.9.]

While the latter approach considers only a single candidate antecedent at a time, a twin-candidate learning model has been proposed *(31)*. The advantage is that each candidate noun phrase is compared to any other, ensuring that the best one is selected. The candidate that wins most of these pairwise competitions is regarded as the antecedent. Evaluation resulted in F-measures of 71.3 on MUC-6 and 60.2 on MUC-7, not showing a significant improvement compared with the single-candidate model of Ng and Cardie *(29)*. After all, performance of the machine learning and the linguistic approaches to coreference resolution shows that there is room for improvement. This could be achieved by integrating semantic and ontological knowledge into the systems. The resolution algorithm *(21)*, for example, relies on the UMLS (*see* **Note 7**) typing system, enabling the system to recognize type relationships between noun phrases. This particularly supports the resolution of sortal anaphors (bridging references) (*Glutamine regulates gastrointestinal cell growth. The amino acid ...*). By increasing the salience measure of an antecedent candidate if it is of the same UMLS type as the anaphor, the system achieved an F-measure of 73.8 on MUC-7. (A similar approach of integrating knowledge about the semantic relatedness of possibly coreferring noun phrases is given in **ref.** *(32)*, using WordNet [*see* **Note 8**].)

5. Extracting Relational Information

Basically three approaches can be identified for the extraction of facts (e.g., "regulates[A,B]"): cooccurrence, rule-based, and machine learning approaches. The simplest approach to extract candidates for relationships between entities is to say that two entities stand in some relationship to each other, if they cooccur in the same sentence or paragraph. As any two entities might be mentioned together in the same sentence without being related semantically, most systems use additional features such as frequency-based weighting or scoring to rank the extracted relationships *(14,25,46–58)*. The more often candidate pairs of cooccurring entities are found the more likely it is that they are indeed related. Cooccurrence-based methods are, however, not able to determine the relationship that holds for a particular candidate pair ***(49,52)***. As a consequence, this approach provides a high recall but poorer precision. To increase precision, cooccurrence results are often combined with further (e.g., experimental) evidence.

Substantial efforts in biomedical relationship extraction have been devoted to manually crafting rules to identify and extract relationships by natural language processing (NLP) techniques. The rules may include syntactic as well as semantic regularities ***(33–35)***. They operate on a preprocessed input, i.e., the text is first tokenized (i.e., identification of word and sentence boundaries) and then part of speech (i.e., categories such as noun or verb) and semantic or ontological tags are assigned to the words in a sentence. The syntactic analysis operates on the part-of-speech tags and identifies phrases, i.e., meaningful subsequences of the sentence (such as N[oun]P[hrases], P[repositional]P[hrases], or V[erb]P[hrases]), and recognizes their grammatical role (subject, object, etc.). Syntactic analyses may differ with respect to their depth. Shallow analyses group only minimal phrases (so-called *chunks*), i.e., do not involve any recursion. They may be carried out very fast and yield an output for any sentence. Deep analyses lack this robustness, may involve a great amount of processing time, and typically produce many ambiguous representations that then must be reduced by statistical means. Syntactic and semantic analyses are often interwoven in that the semantic tags assigned to single words are projected to semantic tags of the phrase to which they belong. If, for example, the two terms *Yck1p* and *Yck2p* are tagged by PROT, meaning that they denote proteins, then the chunk *Yck1p and Yck2p* may be tagged PROT as well (thereby not making any assumption about the actual meaning of the phrase, which may be taken to denote either two proteins or a protein assembly; the tag states only that both parts of the conjunction are proteins). The crucial step for the extraction of the factual information operates on the semantically labeled chunks. This step uses manually crafted rules to normalize the combination of semantic labels and syntactic structures. Rule-based approaches have proven to be successful, particularly because they reach good precision rates.

The complete processing architecture of an NLP-based system is shown in **Table 2**. The sentence analyzed is *Yck1p and Yck2p participate in the phosphorylation of the Ste3p CTD*. The factual information that is extracted on the basis of the syntactic–semantic processing indicates that (1) there is a chemical reaction ID-0 that causes the phosphorylation of Ste3p CTD, and (2) ID-0 has Yck1p and Yck2p as participants (**Table 3**).

A combination of linguistic and statistical methods has been presented ***(36)***. The system was integrated in the IBM Unstructured Information Management Architecture (UIMA) ***(37)***. It automatically identifies biomedical

Table 2
Relevant Levels of an NLP Processing System[a]

Token	POS	STAG	Chunks		
				B-SENT	
Yck1p	NN	PROT			
and	CC				
Yck2p	NN	PROT			NP
participate	VVP	INV_{verb}			VC
In	IN				
the	DT				
phosphorylation	NN	$PHORYL_{nom}$	NP		
of	IN	OF			
the	DT				
Ste3p	NN	PROT			
CTD	NN	PROT-part	NP	NP	PP
				E-SENT	

[a]The words (i.e., Token) get annotated with part-of-speech tags (i.e., POS), semantic labels (i.e., STAG), and syntactic labels (i.e., Chunks). This linguistic annotation is used for the extraction of factual information in the next step.

entities such as "genes," "proteins," "compounds," or "drugs" and facts related to them. An unsupervised approach has been presented to automatically discover English expression patterns for the extraction of protein–protein interactions ***(38,39)***. It should be noted that statistical approaches require manually curated training data that have to be tailored to the users needs, which is a very time-consuming task. Although some annotated corpora already exist (e.g., the GENIA corpus ***(10)***, the Yapex corpus [*see* **Note 9**]), for most questions relevant for biology no corresponding annotated data

Table 3
Filled Templates

Interaction: ID-1	
Type:	Phosphorylation
Causer:	ID-0
Patient:	Ste3p CTD
Interaction: ID-0	
Type:	Reaction
Participants:	{Yck1p, Yck2p}
Caused-Reaction:	ID-1

yet exist for textual resources. Although rule-based approaches are usually considered labor intensive and difficult to adapt to entirely new domains, these systems are transparent and semantic criteria can be enforced more easily.

6. Resources

6.1. Open Access Text Mining Applications

1. iHOP [Information Hyperlinked over Proteins *(40)*]. iHOP makes it possible to search for abstract sentences that mention a specific gene (including synonyms), and to create an interaction network of comentioned genes. It can be accessed through http://www.ihop-net.org/.
2. EBIMed *(41)*. EBIMed is a web application that combines information retrieval and extraction from Medline. EBIMed finds Medline abstracts in the same way PubMed does. Then it goes a step beyond and analyzes them to offer a complete overview on associations between UniProt protein/gene names, GO annotations, Drugs, and Species. The results are shown in a table that displays all the associations and links to the sentences that support them and to the original abstracts. It can be accessed through http://www.ebi.ac.uk/Rebholz-srv/ebimed/.
3. MedLEE: Medical Language Extraction and Encoding System. MedLEE extracts, structures, and encodes clinical information in textual patient reports so that the data can be used by subsequent automated processes. A demonstration can be accessed through http://lucid.cpmc.columbia.edu/medlee/.
4. MEDIE: Semantic Retrieval Engine for MEDLINE. MEDIE is an intelligent search engine used to retrieve biomedical correlations from MEDLINE. You can find abstracts/sentences in MEDLINE by asking for arguments of specific relationships, for example, "What activates p53" and "What causes colon cancer." The application is accessible through http://www-tsujii.is.s.u-tokyo.ac.jp/medie/.
5. Ali Baba: Graphical Summarization of Interactions *(42)*. Ali Baba is an interactive tool for graphic summarization of data on protein–protein interactions, gene–disease associations, and subcellular locations of proteins. Ali Baba shows the relationships as a graphic network. The online version of the tool can be accessed through http://alibaba.informatik.hu-berlin.de/.
6. Ontogene: High-precision robust syntactic parsing for protein interactions *(43)*. Ontogene focuses on the extraction of semantic relationships such as *bind*, *activate*, or *block* between genes and proteins from the scientific literature. The application can be accessed through http://www.ontogene.org/.
7. Chemsearch, IBM, Almaden *(44)*. The IBM Chemical Search Engine alpha site enables the user to search U.S. patents and applications using molecular similarity. As of May 2007 the index contained over 4 million unique chemical structures that occur more than 70 million times. These compounds were extracted from the U.S. Patent corpus years 1976–2005. Chemsearch can be accessed through https://chemsearch.almaden.ibm.com/chemsearch/SearchServlet.

6.2. Biomedical Ontology Resources

1. Gene Ontology (GO) was developed by the Gene Ontology Consortium at the European Bioinformatics Institute (EBI) *(45)*. Gene ontology addresses the need for consistent organism description of gene products across databases. GO consists of three structured controlled vocabularies on *molecular functions*, *biological processes*, and *cellular components*. The Gene Ontology can be browsed and downloaded at http://www.geneontology.org/.
2. OBO (Open Biological Ontologies). The Open Biomedical Ontologies (OBO) Foundry is a collaborative experiment to produce well-structured vocabularies for shared use across different biological and medical domains. OBO offers a collection of ontologies for the biological and biomedical domain. The ontologies repository can be accessed via http://obo.sourceforge.net/.
3. MedDRA (Medical Dictionary for Regulatory Activities) is an international medical terminology developed under the auspices of the International Conference on Harmonization of Technical Requirements for Registration of Pharmaceuticals for Human Use (ICH). It is accessible through http://www.meddramsso.com.
4. FMA [Foundational Model of Anatomy *(46)*] is an ontology that explicitly distinguishes physical entities that have or do not have mass, and treats anatomical spaces, surfaces, lines, and points as universals or classes. It has more than 75,000 multiply located anatomical entities (universals) that exist in the idealized (canonical) instances that they subsume. In addition to the taxonomy component of the FMA, it gives an account of the structural and developmental relationships that exist between anatomical entities. It is accessible through http://sig.biostr.washington.edu/projects/fm/index.html.
5. MGED (Microarray Gene Expression Data ontology group). The MGED Society is an international organization of biologists, computer scientists, and data analysts that aims to facilitate the sharing of microarray data generated by functional genomics and proteomics experiments. It can be accessed via http://www.mged.org.
6. Protégé Ontologies Library (Stanford Medical Informatics, Stanford University School of Medicine) offers a series of ontologies that were developed either at Stanford or by the Protégé user community. It is available through http://protege.cim3.net/cgi-bin/wiki.pl?ProtegeOntologiesLibrary.
7. Systematized Nomenclature of Medicine (SNOMED) was developed by the College of American Pathologists. SNOMED is a systematically organized computer processable collection of medical terminology. It covers most areas of clinical information such as diseases, findings, procedures, microorganisms, and pharmaceuticals. It can be accessed through http://www.snomed.org/.
8. Unified Medical Language System (UMLS) offered by the National Library of Medicine *(47)* is a controlled compendium of several vocabularies and ontologies providing a mapping structure between them. It contains in Version 2005AA more than 1 million biomedical concepts with more than 5 million concept names. It can accessed through http://www.nlm.nih.gov/research/umls/.

9. Biomedical Ontology is an initiative by the National Center for Biomedical Ontology. It is a consortium of leading biologists, clinicians, informaticians, and ontologists. They are developing innovative technology and methods that allow scientists to create, disseminate, and manage biomedical information and knowledge in machine-processable form. Ontologies and further information are accessible through http://www.bioontology.org/.
10. The Ontoselect Ontology Library monitors the web to provide an access point for ontologies on any possible topic or domain. It is automatically updated, organized in a meaningful way, and offers support for ontology search and selection. It contains several biomedical ontologies. Ontoselect was developed by the Competence Center Semantic Web at DFKI (German Research Center for Artificial Intelligence). The repository is available through http://olp.dfki.de/ontoselect/.
11. Cooperative Ontologies Programme (Co-ode). The Co-ode and HyOntUse projects aim to provide support for communities interested in OWL (Web Ontology Language) by developing materials, enhancing tools, and looking at some of the theoretical problems to help form usable solutions before big problems arise. Co-ode offers software downloads, tutorial materials, and a calendar of events at http://www.co-ode.org/.

6.3. Ontology Editors

1. OILEd (Ontology Inference Layer) is an ontology editor allowing the user to build ontologies using DAML+OIL. DAML+OIL is a semantic markup language for Web resources. It can be downloaded from http://oiled.man.ac.uk/.
2. Protégé is a free, open source ontology editor and knowledge-based framework supported by the National Library of Medicine. Protégé supports two main ways of modeling ontologies via the Protégé-Frames and Protégé-OWL editors. It is based on Java, is extensive, and provides a plug-and-play environment that makes it a flexible base for rapid prototyping and application development. It can be downloaded from http://protege.stanford.edu/.
3. SWOOP is a tool for creating, editing, and debugging OWL ontologies. It was produced by the MIND laboratory at the University of Maryland, College Park, but is now an open source project with contributers from all over. It can be downloaded from http://code.google.com/p/swoop/.
4. OntoEdit, Ontoprise GmbH. It can be downloaded from http://www.ontoprise.de.
5. Chimera, Stanford Ontology Builder. It can be downloaded from http://ksl.stanford.edu/software/chimaera/.
6. KAON OI-Modeller. It can be downloaded from http://kaon.semanticweb.org/.

7. Notes

1. Scopus is one of the largest abstract and citation database of research literature. It is provided by Elsevier. http://www.scopus.com.

2. Prabhakar Raghavan, Head of Yahoo! Research, reported in the keynote talk at the *7th International Workshop on the Web and Databases* ***(48)***, indicates that 80% of information in companies is unstructured, in the form of spoken or written language, images, etc.
3. A good collection of synonymous gene and protein names is offered through http://prothesaurus.bio.ifi.lmu.de/. Details have previously been described ***(15)***.
4. In the context of the Semantic Web effort, several so-called *foundational* or *upper level* ontologies [e.g., Descriptive Ontology of Congnitive and Linguistic Engineering ***(49)***, Basic Formal Ontology (http://www.ifomis.uni-saarland.de/bfo)] have emerged.
5. Common performance measures: **F-measure**: the harmonic mean of recall and precision (2 × *precision* × *recall*/[*precision* + *recall*]). **Recall** (sensitivity): the proportion of correct answers (or relevant documents) returned by the system out of all correct answers (or relevant documents) available in the data set (retrieved relevant documents/relevant documents). **Precision** (specificity): the proportion of answers (or documents) that is correct (or relevant) out of all the answers (or documents) returned by the system (retrieved relevant documents/retrieved documents).
6. The Message Understanding Conference (MUC) provides an annotated training corpus and a standard for the evaluation of information extraction systems, including the task of coreference resolution, since the sixth conference (MUC-6, 1995). See http://www.itl.nist.gov/iaui/894.02/related_projects/muc/index.html or http://www.cs.nyu.edu/cs/faculty/grishman/muc6.html.
7. *See* **Section 6**: Resources.
8. *See* http://wordnet.princeton.edu/.
9. The Yapex corpus is available through http://www.sics.se/humle/projects/prothalt/.

Acknowledgment

H. E. is funded by the Klaus Tschira Foundation.

References

1. Bhalotia, G., Nakov, P. I., Schwartz, A. S., and Hearst, M. A. (2003) BioText team report for the TREC 2003 genomics track. *Proc. TREC 2003*, Vol. 12.
2. Fukuda, K., Tamura, A., Tsunoda, T., and Takagi, T. (1998) Toward information extraction: identifying protein names from biological papers. In *Pacific Symposium of Biocomputation*, Hawaii, Vol. 3, pp. 707–718. World Scientific, Singapore.
3. Tanabe, L. and Wilbur, W. J. (2002) Tagging gene and protein names in biomedical text. *Bioinformatics* **18**, 1124–1132.
4. Franzen, K., Eriksson, G., Olsson, F., Asker, L., Liden, P., and Coster, J. (2002) Protein names and how to find them. *Int. J. Med. Inform.* **67**, 49–61.

5. Collier, N., Nobata, C., and Tsujii, J. (2000) Extracting the names of genes and gene products with a hidden Markov model. *Int. Conf. Comput. Linguistics* **18,** 201–207.
6. Chang, J. T., Schütze, H., and Altman, R. B. (2004) GAPSCORE: finding gene and protein names one word at a time. *Bioinformatics* **20**, 216–225.
7. McDonald, R. and Pereira, F. (2005) Identifying gene and protein mentions in text using conditional random fields. *BMC Bioinformatics* **6**, S6.
8. Settles, B. (2005) ABNER: an open source tool for automatically tagging genes, proteins, and other entity names in text. *Bioinformatics* **21**, 3191–3192.
9. Zhou, G., Shen, D., Zhang, J., Su, J., and Tan, S. (2005) Recognition of protein/gene names from text using an ensemble of classifiers. *BMC Bioinform.* **6**, S7.
10. Kim, J.-D., Ohta, T., Tateisi, Y., and Tsujii, J. (2003) GENIA corpus—a semantically annotated corpus for bio-textmining. *Bioinformatics* **19(Suppl. 1)**, i180–i182.
11. Gaizauskas, R. J., Demetriou, G., Artymiuk, P. J., and Willett, P. (2003) Protein structures and information extraction from biological texts: the PASTA system. *Bioinformatics* **19**, 135–143.
12. Krauthammer, M., Rzhetsky, A., Morozov, P., *et al.* (2000) Using blast for identifying gene and protein names in journal articles. *Gene* **259**, 245–252.
13. Fundel, K., Güttler, D., Zimmer, R., and Apostolakis, J. (2005) A simple approach for protein name identification: prospects and limits. *BMC Bioinform.* **6**, S15.
14. Hanisch, D., Fundel, K., Mevissen, H. T., Zimmer, R., and Fluck, J. (2005) ProMiner: rule-based protein and gene entity recognition. *BMC Bioinform.* **6**, S14.
15. Fundel, K. and Zimmer, R. (2006) Gene and protein nomenclature in public databases. *BMC Bioinform.* **7**, 372.
16. Gaudan, S., Kirsch, H., and Rebholz-Schuhmann, D. (2005) Resolving abbreviations to their senses in medline. *Bioinformatics* **21**, 3658–3664.
17. Schijvenaars, B. J. A., Mons, B., Weeber, M., Schuemie, M. J., van Mulligen, E. M., Wain, H. W., and Kors, J. A. (2005) Thesaurus-based disambiguation of gene symbols. *BMC Bioinform.* **6**, 149.
18. Cimiano, P. (2006) *Ontology Learning and Population from Text: Algorithms, Evaluation and Applications*. Springer, New York.
19. Cimiano, P., Reyle, U., and Saric, J. (2004) Ontology driven discourse analysis for information extraction. *Data Knowledge Eng. J.* **55(1),** 59–83.
20. Cimiano, P. (2002) On the resolution of bridging references within information extraction systems. Master's Thesis.
21. Castaño, J., Zhang, J., and Pustejovsky, J. (2002) Anaphora resolution in biomedical literature. International Symposium on Reference Resolution.
22. Grosz, B., Joshi, A. K., and Weinstein, S. (1983) Providing a unified account of definite noun phrases in discourse. *Proceedings of the 21st Annual Meeting of the Association for Computational Linguistics*, pp. 44–50.
23. Grosz, B., Joshi, A. K., and Weinstein, S. (1995) Centering: a framework for modeling the local coherence of discourse. *Comput. Linguistics* **2(21)**, 203–225.
24. Brennan, S. E., Friedman, M. W., and Pollard, C. J. (1987) A centering approach to pronouns. *Proceedings of the 25th Annual Meeting of the Association for Computational Linguistics*, pp. 155–162.

25. Strube, M. (1998) Never look back: an alternative to centering. *Proceedings of the 17th International Conference on Computational Linguistics*, pp. 1251–1257.
26. Ge, N., Hale, J., and Charniak, E. (1998) A statistical approach to anaphora resolution. *Proceedings of the 6th ACL Workshop on Very Large Corpora*, pp. 161–170.
27. Cardie, C. and Wagstaff, K. (1999) Noun phrase coreference as clustering. *Proceedings of the Joint Conference on Empirical Methods in Natural Language Processing and Very Large Corpora*, pp. 82–89.
28. Soon, W. M., Ng, H. T., and Lim, D. C. Y. (2001) A machine learning approach to coreference resolution of noun phrases. *Comput. Linguistics* **27(4)**, 521–544.
29. Ng, V. and Cardie, C. (2002) Improving machine learning approaches to coreference resolution. *Proceedings of the 40th Annual Meeting of the Association for Computational Linguistics*, pp. 104–111.
30. Yang, X., Zhou, G., Su, J., and Tan, C. L. (2004) Improving noun phrase coreference resolution by matching strings. *Proceedings of the 1st International Joint Conference of Natural Language Processing, Lecture Notes in Computer Science*, Vol. 3248, pp. 22–38.
31. Yang, X., Zhou, G., Su, J., and Tan, C. L. (2003) Coreference resolution using competition learning approach. *Proceedings of the 41st Annual Meeting of the Association for Computational Linguistics*, pp. 176–183.
32. Harabagiu, S. M., Bunescu, R. C., and Maiorano, S. J. (2001) Text and knowledge mining for coreference resolution. *Proceedings of the 2nd Conference of the North American Chapter of the Association for Computational Linguistics*, pp. 55–62.
33. Blaschke, C., Andrade, M., Ouzounis, C., and Valencia, A. (1999) Automatic extraction of biological information from scientific text: protein-protein interactions. *Proceedings of the Fifth International Conference on Intelligent Systems for Molecular Biology (ISMB99)*, pp. 60–67.
34. Koike, A. and Takagi, T. (2004) PRIME: automatically extracted PRotein Interactions and Molecular Information database. *In Silico Biol.* **5**, 0004.
35. Šarić, J., Jensen, L. J., Ouzounova, R., Rojas, I., and Bork, P. (2005) Extraction of regulatory gene/protein networks from Medline. Doi:10.1093/ bioinformatics/bti597.
36. Mack, R., *et al.* (2004) Text analytics for life science using the unstructured information management architecture. *IBM Syst. J.* **43**, 490–515.
37. Ferrucci, D. and Lally, A. (2004) Uima: an architectural approach to unstructured information processing in the corporate research environment. *Nat. Lang. Eng.* **10**, 327–348.
38. Huang, M., Zhu, X., Hao, Y., Payan, D. G., Qu, K., and Li, M. (2004) Discovering patterns to extract protein-protein interactions from full texts. *Bioinformatics* **20**, 3604–3612.
39. Hao, Y., Zhu, X., Huang, M., and Li, M. (2005) Discovering patterns to extract protein-protein interactions from the literature: Part ii. *Bioinformatics* **21**, 3294–3300.

40. Hoffmann, R. and Valencia, A. (2004) A gene network for navigating the literature. *Nat. Genet.* **36**, 664.
41. Rebholz-Schuhmann, D., Kirsch, H., Arregui, M., Gaudan, S., Riethoven, M., and Stoehr, P. (2007) Ebimed—text crunching to gather facts for proteins from medline. *Bioinformatics* **23**, 237–244.
42. Plake, C., Schiemann, T., Pankalla, M., Hakenberg, J., and Leser, U. (2006) Alibaba: Pubmed as a graph. *Bioinformatics* **22**, 2444–2445.
43. Rinaldi, F., Schneider, G., Kaljurand, K., Hess, M., and Romacker, M. (2006) An environment for relation mining over richly annotated corpora: the case of GENIA. *BMC Bioinform.* **7**, S3.
44. Rhodes, J., Boyer, S., Kreulen, Y., J. Chen, and Ordonez, P. (2007) Mining patents using molecular similarity search. *12th Pacific Symposium on Biocomputing*, Hawaii, Vol. 12, pp. 304–315. World Scientific, Singapore.
45. Ashburner, M., *et al.* (2000) Gene ontology: tool for the unification of biology. *Nat. Genet.* **25**, 125–29.
46. Rosse, C. and Mejino, J. L. V. (2003) A reference ontology for biomedical informatics: the foundational model of anatomy. *J. Biomed. Inform.* **36**, 478–500.
47. U.S. Department of Health and Human Services, N.L.O.M., NIH (2002) Unified medical language system. URL: http://www.nlm.nih.gov/research/umls/.
48. Raghavan, P. (2004) Text centric structure extraction and exploitation (abstract only). *WebDB '04: Proceedings of the 7th International Workshop on the Web and Databases*, New York.
49. Gangemi, A., Guarino, N., Masolo, C., Oltramari, A., and Schneider, L. (2002) Sweetening ontologies with dolce. *Proceedings of the 13th International Conference on Knowledge Engineering and Knowledge Management,* pp. 166–181.

27

Protein Subcellular Localization Prediction Using Artificial Intelligence Technology

Rajesh Nair and Burkhard Rost

Summary

Proteins perform many important tasks in living organisms, such as catalysis of biochemical reactions, transport of nutrients, and recognition and transmission of signals. The plethora of aspects of the role of any particular protein is referred to as its "function." One aspect of protein function that has been the target of intensive research by computational biologists is its subcellular localization. Proteins must be localized in the same subcellular compartment to cooperate toward a common physiological function. Aberrant subcellular localization of proteins can result in several diseases, including kidney stones, cancer, and Alzheimer's disease. To date, sequence homology remains the most widely used method for inferring the function of a protein. However, the application of advanced artificial intelligence (AI)-based techniques in recent years has resulted in significant improvements in our ability to predict the subcellular localization of a protein. The prediction accuracy has risen steadily over the years, in large part due to the application of AI-based methods such as hidden Markov models (HMMs), neural networks (NNs), and support vector machines (SVMs), although the availability of larger experimental datasets has also played a role. Automatic methods that mine textual information from the biological literature and molecular biology databases have considerably sped up the process of annotation for proteins for which some information regarding function is available in the literature. State-of-the-art methods based on NNs and HMMs can predict the presence of N-terminal-sorting signals extremely accurately. *Ab initio* methods that predict subcellular localization for any protein sequence using only the native amino acid sequence and features predicted from the native sequence have shown the most remarkable improvements. The prediction accuracy of these methods has increased by over 30% in the past decade. The accuracy of these methods is now on par with high-throughput methods for predicting localization, and they are beginning to play an important role in directing experimental research. In this chapter, we review some of the most important methods for the prediction of subcellular localization.

From: *Methods in Molecular Biology, vol. 484: Functional Proteomics: Methods and Protocols*
Edited by: J. D. Thompson et al., DOI: 10.1007/978-1-59745-398-1, © Humana Press, Totowa, NJ

Key Words: Protein subcellular localization prediction; sorting signals; neural networks; support vector machines; hidden Markov models; amino acid composition; text analysis.

1. Introduction

1.1. Proteins Are the Machinery of Life

Proteins are the workhorses that are responsible for transforming the genetic information for life, stored in the nucleic acids (DNA), into physical reality. A protein molecule consists of a long unbranched chain of 20 amino acid residues; each amino acid is linked to its neighbor through a covalent peptide bond. The most distinguishing characteristic of proteins is that they have well-defined three-dimensional (3D) structures. A stretched-out polypeptide chain has no biological activity. Protein function arises from the "conformation" of the protein, which is the 3D arrangement, or shape, of the molecules in the protein. Proteins are the most structurally complex and functionally sophisticated macromolecules known and they perform a wide array of tasks in organisms, such as the catalysis of biochemical reactions, transport of nutrients, and recognition and transmission of signals. The coordinated effort of many different types of proteins is required to realize the genetic program that is encoded in DNA. All the multiple aspects of the role of any particular protein is referred to as its "function."

1.2. Predicting Protein Function: A Major Challenge for Modern Biology

To date, the genome (DNA) sequences of over 400 organisms, including the draft sequence of the human genome *(1)*, has been completed. The number of entirely sequenced genomes has been growing exponentially for many years, and this growth is expected to continue for at least the next several years. With the availability of genome sequences of entire organisms, we are, for the first time, in a position to understand the expression, function, and regulation of the entire set of proteins encoded by an organism. This information will be invaluable for understanding how complex biological processes occur at a molecular level, how they differ in various cell types, and how they are altered in disease states. Identifying protein function is a large step toward understanding diseases and identifying novel drug targets *(2)*. However, experimentally determining protein function continues to be a laborious and painfully slow task requiring enormous resources. For example, more than a decade after its discovery, we still do not know the precise and entire functional role of the prion protein *(3)*. To compound this problem, the rate at which expert annotators add experimental information into more or less controlled vocabularies of databases snails along at an even slower pace. This has left a huge and rapidly widening gap between the amount of protein sequence information deposited in databases and the experimental characterization of the corresponding proteins *(4)*. Bioinformatics plays a central role in bridging this sequence–function gap through the development of tools for faster and more effective prediction of protein function *(5)*.

1.3. "Protein Function" Has Myriad Meanings

During the past decade, advanced artificial intelligence (AI) techniques have proved remarkably successful in addressing numerous problems in molecular biology. However, excluding some bright spots, AI-based tools for predicting protein function have in general lagged in performance. One of the major problems hindering the development of methods for predicting protein function is that proteins are multifunctional and can perform different functions in different cellular contexts. Proteins can perform molecular functions such as catalyzing metabolic reactions and transmitting signals to other proteins or to DNA. At the same time they can also be responsible for performing physiological functions as a set of cooperating proteins, such as the regulation of gene expression, metabolic pathways, and signaling cascades *(6)*. The function of a protein can be associated with many mutually overlapping levels: chemical, biochemical, cellular, organism mediated, developmental, and physiological *(7)*. These levels are related in complex ways; for example, a protein kinase can be related to different cellular functions (such as cell cycle) and to a chemical function (transferase) plus a complex control mechanism by interaction with other proteins. The same kinase may also be the culprit that leads to misfunction, or disease. The variety of functional roles of a protein often results in confusing database annotations, which makes it difficult to develop tools for predicting protein function *(8)*. What we need for reliable automatic predictions are computer-readable hierarchical descriptions of function *(9)*. However, defining an ontology for protein function has proved to be an extremely difficult task.

1.4. Subcellular Localization Is an Important Aspect of Protein Function

Living organisms can be divided into two types: prokaryotes and eukaryotes. The defining feature of the prokaryotic cell is its absence of a nucleus and any other membrane-bound organelles. During the course of evolution, the cells of higher organisms, namely the eukaryotes, became progressively divided into more elaborate subcompartments. The major constituents of eukaryotic cells are extracellular space, cytoplasm, nucleus, mitochondria, Golgi apparatus, endoplasmic reticulum (ER), peroxisome, vacuoles, cytoskeleton, nucleoplasm, nucleolus, nuclear matrix, and ribosomes *(10)*. Each membrane-bound subcompartment, or organelle, performs some specialized cellular functions. For example, mitochondria are the powerhouses of the cell while the nucleus houses its genetic material. Proteins must be localized in the same subcellular compartment to cooperate toward a common physiological function. Knowledge of the subcellular localization of a protein can significantly improve target identification during the drug discovery process. For example, secreted proteins and plasma membrane proteins are easily accessible by drug molecules due to their localization in the extracellular space or on the cell surface. Aberrant subcellular localization of proteins has been observed in the cells of several diseases such as cancer and Alzheimer's disease. The consequences of mislocalization and mistargeting are manifested in a number of human genetic diseases, including cystic fibrosis *(11)*, Wilson's disease *(12)*, and juvenile pulmonary emphysema *(13)*.

1.5. Subcellular Localization Prediction Is an Ideal Testing Ground for Function Prediction Methods

The most comprehensive effort at developing a controlled vocabulary for describing protein function originates from the Gene Ontology (GO) consortium *(9)*. Subcellular localization is one of the three main classes used to organize protein function within the GO hierarchical classification scheme, the other two classes being molecular function and biological process. Predicting subcellular localization has become one of the main testing grounds for the development of function prediction methods for three major reasons: (1) subcellular classes are well defined in contrast to other aspects of protein function, (2) ample experimental data on localization are available from both traditional biochemical as well as large-scale experiments, and (3) although some proteins can localize in multiple compartments, the majority of proteins localize to a single compartment for the largest part of their lifetime. To date, high-throughput localization experiments cannot be performed for mammalian or other higher eukaryotic proteomes *(14)*. As a result, predicting subcellular localization has become one of the central challenges in bioinformatics *(15,16)*.

1.6. Protein Trafficking Proceeds via Sorting Signals

A basic knowledge of the protein-sorting mechanism is essential for understanding the different methods for predicting subcellular localization. Most eukaryotic proteins are encoded in the nuclear genome and synthesized in the cytosol, from which they need to be sorted to their final destinations. The sorting system is reasonably well understood for some organelles *(15,17)*. The system has two main branches *(18)*. On one branch, proteins are synthesized on cytoplasmic ribosomes, and from there can go to the nucleus, mitochondria, or peroxisomes. The second branch leads from the ER ribosomes to the Golgi apparatus, and from there to lysosomes, or secretory vesicles, and on to the extracellular space. At each branch point a "decision" must be made for each protein: either retain the protein in the current compartment or transport it to the next. These "decisions" are made by membrane transport complexes, which respond to targeting signals on the proteins themselves. In most cases, these targeting signals are short stretches of amino acid residues. The best understood branch point is the second one leading to secretion. Many proteins destined for this branch have an N-terminal sorting signal, referred to as the signal peptide, which is cleaved off proteolytically either during or after protein translocation through the membrane. Proteins lacking this signal are retained in the cytoplasm. The discovery of signal peptides was the first major breakthrough in understanding protein targeting and Günter Blobel was awarded the Nobel Prize in 1999 for this discovery. The targeting signals used at the other branch points are not always that clear for two reasons. First, the signals are encoded in the 3D structure of the protein, and hence are not always contiguous in amino acid sequence. Second, even where the signals are contiguous in sequence, not all targeting signals have been documented. In the absence of a clear understanding of the principles governing protein translocation, computational methods for predicting subcellular localization have pursued a number of conceptually distinct approaches.

1.7. Advanced AI Techniques Pave the Way for Predicting Protein Subcellular Localization

Sequence similarity is perhaps the most frequently used method to annotate function for unknown proteins and accounts for the majority of annotations about protein function in public databases *(19)*. The method works by first identifying a database protein of experimentally known function with significant sequence similarity to the query protein, U, and then transferring the experimental annotations of function from the homologue to the unknown query U. Homology transfer remains one of the most accurate methods available for inferring function, although it suffers from many problems, including surprisingly high error rates at even 60–80% sequence identity *(20)*. However, homology transfer alone cannot bridge the sequence–function gap. Annotation transfer by sequence homology is applicable in only a limited number of cases. For example, in completely sequenced eukaryotic proteomes subcellular localization of fewer than 25% of proteins can be inferred using homology (*Caenorhabditis elegans* in **Fig. 1**). Recent progress in predicting

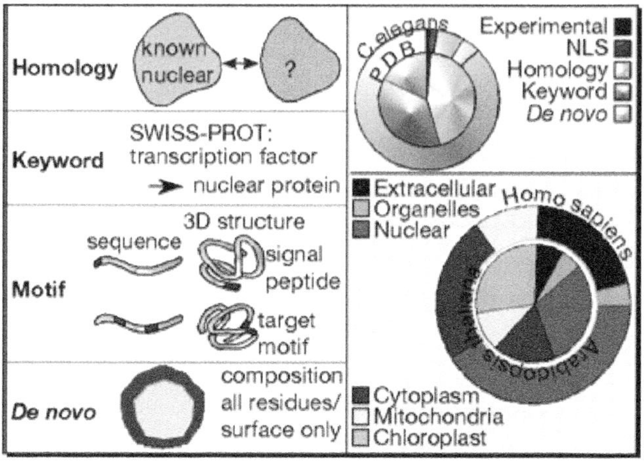

Fig. 1. Different methods for predicting localization. Left panel: four methods. (1) Homology transfer: if we know that protein A is nuclear and we find protein B very similar in sequence to A, we can usually infer that B is also nuclear. (2) Key word-based transfer: if we know that A is a transcription factor, we can automatically infer that A is a nuclear protein. (3) Motif-based prediction: many proteins are shuttled between different compartments by carrier proteins that recognize short sequence motifs. Some motifs are sequence consecutive (e.g., signal peptide) and others are conformational, i.e., discernible only from the folded structure (e.g., lysosomal retention signals). (4) *De novo* methods exploit the correlation between sequence features and localization. Right panel: results. The majority of all annotations of proteins with known structure (PDB, top right) resulted from homology transfer or lexical analysis (inner circle of top pie chart). When applying the same methods to the entire proteome of *C. elegans*, this picture changed completely (outer circle, top right): about 87% of all proteins could be handled only by *de novo* methods. Note that the fraction of actual experimental results (black) is almost invisible for both pies. The two circles on the lower right (outside, human; inside, weed) illustrate the differential breakdown into compartments. Most surprising may be the high fraction of nuclear proteins. Note that prediction methods tend to underestimate the fraction of proteins in organelles.

protein function has been the result of applying advances in AI, namely machine learning (ML) methods such as support vector machines (SVMs), neural networks (NNs), Bayesian networks, Decision Trees, and hidden Markov models (HMMs). However, many issues, some of which are peculiar to biological data, must be considered before off-the-shelf ML techniques can be applied to biological problems. The major problem is the "noisiness" of biological data. At the biological level, noise arises because proteins are multitasking; the observation of one function for a protein does not rule out other functional roles. At the experimental level, noise arises due to incorrect observations. At the curation level, noise enters databases due to misinterpretation of experimental observations by curators. A second issue is sequence similarity of training and test data sets, since even low levels of sequence similarity can imply homology, i.e., a pair of proteins are evolutionarily related and hence likely to be functionally related. This makes it difficult to assess accurately the performance of AI-based methods on blind samples that share differing degrees of sequence similarity with training samples. This is an important point since prediction using AI-based methods is most needed when function is to be inferred for a protein that shares no homology to known proteins. These issues necessitate significant retooling of off-the-shelf AI methods before application to a biological problem. To be successful, the method has to be rigorously tested and prediction accuracy carefully assessed.

2. *In Silico* Approaches to Predicting Subcellular Localization

2.1. No Straightforward Strategy for Predicting Localization

AI-based methods for predicting the subcellular localization of proteins have primarily explored three avenues: (1) predicting the sorting signals that the cell uses as "address labels," (2) mining the functional information deposited in databases and scientific literature, and (3) using the observation that the subcellular localization depends in subtle ways on the global properties of the amino acid sequence. Recently there has been a surge of interest in the third category since this strategy shows the most promise for genome-wide annotations. Additionally, there are meta-methods that combine the outputs from a number of primary methods in an optimal way to enhance accuracy and coverage. Since protein trafficking relies on the presence of sorting signals, we ideally would like to predict the signals responsible for targeting. However, our current knowledge of sorting signals is far from perfect and recent cell biological studies seem to indicate that the protein-sorting mechanism is far more complex than previously thought. This makes it extremely difficult to accurately identify sorting signals ***(21)***. In spite of their limited applicability, methods that predict sorting signals provide the most useful predictions since by pinpointing the "targeting signal" they shed light on the molecular mechanisms of protein translocation. In practical terms, they enable experimentalists to devise mutation experiments that can identify the amino acid residues responsible for protein sorting. Traditionally, expert human annotators have been responsible for interpreting experimental data in the scientific literature and annotating protein function in public databases ***(22,23)***. However, recent advances in data-mining techniques have made it possible to deploy automatic methods to complement the role of "expert annotators" and extract functional information directly from biological databases, MEDLINE abstracts, and even full scientific papers. Many recent advances in predicting subcellular localization have been the result of using global sequence properties. These *ab initio* methods do not rely on "targeting signals" or any other feature directly associated with protein sorting. Since they utilize only features derived or predicted from the primary sequence such as the amino acid composition or predicted secondary structure, they have the advantage of being applicable to all protein sequences. Methods that can accurately predict subcellular localization from the amino acid sequence alone are invaluable in interpreting the wealth of data generated by large-scale sequencing projects. Furthermore, predictions of localization can assist high-throughput techniques to determine localization from cDNAs *(24)*. However, prediction accuracy for *ab initio* methods still lags behind other approaches. This has led to the development of combination methods, commonly referred to as Meta predictors, that assign localization to a protein based on the most accurate method available for that protein. Combination methods often employ ML techniques to integrate predictions from the individual methods.

2.2. Specialized Prediction Systems Required for the Different Kingdoms

During the course of evolution, the amino acid sequences of proteins, which are evolutionarily related and perform the same function in different organisms, tend to diverge due to accumulated mutations in the genes. This tendency of protein sequences to diverge makes it hard to accurately identify a protein with annotated function that is evolutionarily related to an unknown query protein, especially when the organisms they belong to are only distantly related. The strength of AI methods for function prediction lies in their exceptional ability to detect and exploit remote similarities in various features of protein sequences. In general, the accuracy of machine learners increases with the size of the training data. Combining experimental localization data for proteins from all organisms would result in the largest available training dataset. Ideally, such a combined predictor would provide the most accurate predictions for proteins from all organisms. However, in reality, protein sequences from distantly related organisms have diverged to such an extent that the inclusion of extremely diverged sequences in a combined predictor results in significantly reduced prediction accuracies. Hence, prediction methods employ specialized predictors for the most diverged kingdoms. In choosing the number of specialized predictors, the goal is to maximize both the size of the available training data and the prediction accuracy. Since evolution is the driving force behind sequence and functional divergence, the phylogenetic classification of organisms is most often used to define kingdoms for training specialized predictors. The majority of subcellular localization prediction methods employ specialized predictors for gram-positive and gram-negative bacteria, plants, and animals. Fungi are usually included in the animals group but some methods employ specialized predictors for them. Most prediction methods do not explicitly consider archaeal proteins since they usually inhabit uncommon and extreme environments and very few experimental data are available for them.

We first review the PSORT system, which was the first publicly available method for predicting subcellular localization. PSORT was essentially a classical system, relying on an extensive knowledge base of experimentally determined "targeting signals," but in its newer incarnations it has evolved into a combination method. The targeting signals were manually extracted through an extensive literature search. The new ML-based methods routinely use PSORT predictions as a benchmark.

2.3. PSORT: Expert System for Predicting Localization

The PSORT system *(25,26)* predicts the localization of proteins from gram-negative bacteria, gram-positive bacteria, yeasts, animals, and plants. The PSORT knowledge base consists of an extensive collection of sequence motifs and other features (**Table 1**) known to be involved in protein targeting. The original version was designed as an expert system with localization being assigned based on a set of "if-then" rules, with each rule testing the protein for the presence or absence of various features associated with localization. Though PSORT pioneered the field of predicting protein subcellular localization, the original version suffered from several shortcomings, in particular, our limited knowledge of sorting signals. The majority of sorting signals published in the literature suffer from either lack of generalizability or low specificity. The lack of generalizability results in low coverage levels while lack of specificity leads to low prediction accuracy. Some of these problems in PSORTII were overcome by assigning propensities for a protein to belong to the different localization classes given the presence of a feature, rather than using simple "yes/no" decisions based on the presence/absence of features. The hard-coded "if-then" rules were replaced by the k-nearest-neighbor algorithm, which interprets the set of propensity values obtained for the different features and estimates the likelihood of the protein being sorted to each candidate site. Finally, the algorithm displays some of the most probable localization sites. The PSORTII program achieved an overall prediction accuracy of 57% for distinguishing 11 subcellular classes on a testing dataset of a few hundred proteins available at the time of the development of the method. Recent improvements to the PSORT algorithm have mainly been the result of incorporating advanced ML techniques. Several extensions to PSORTII have also been proposed: iPSORT *(27)* for

Table 1
Features Detected by PSORTII

Feature	Criteria
N-terminal signal peptide	Modified McGeoch's method and the cleavage-site consensus
Mitochondrial-targeting signal	Amino acid composition of the N-terminal 20 residues and some weak cleavage site consensus
Nuclear-localization signals	Combined score for various empirical rules
ER lumen-retention signal	The KDEL-like motif at the C-terminus
ER membrane-retention signal	Motifs: XXRR-like (N-terminal) or KKXX-like (C-terminal)
Peroxisomal-targeting signal	PTS1 motif at the C-terminus and the PTS2 motif
Vacuolar-targeting signal	[TIK]LP[NKI] motif
Golgi transport signal	The YQRL motif (preferentially at the cytoplasmic tail)
Tyrosine-containing motif	Number of tyrosine residues in the cytoplasmic tail
Dileucine motif	At the cytoplasmic tail
Membrane span(s)/topology	Maximum hydrophobicity and the number of predicted spans; charge difference across the most N-terminal transmembrane segment
RNA-binding motif	RNP-1 motif
Actinin-type actin-binding motifs	From PROSITE
DNA-binding motifs	63 motifs from PROSITE
Ribosomal-protein motifs	71 motifs from PROSITE
Prokaryotic DNA-binding motifs	33 motifs from PROSITE
N-Myristoylation motif	At the N-terminus
Amino acid composition	Neural network score that discriminates between cytoplasmic and nuclear proteins
Coiled-coil structure length	Number of residues in the predicted coiled-coil state
Length	Length of sequence

extensive feature detection of N-terminal sorting signals, PSORT-B *(28)* for predicting localization of gram-negative bacteria, and WolfPSORT *(29)*, which is a meta predictor that incorporates many different predictions.

In the following sections, we review the different AI-based approaches for predicting subcellular localization and their strengths and their weaknesses, and describe the state-of-the-art methods currently available for predicting localization.

3. Predicting Sequence Motifs Involved in Protein Targeting

3.1. Sorting Signal-Based Predictions Possible for Some Cellular Classes

A number of methods try to predict localization by identifying local sequence motifs, such as signal peptides *(26,30)* or nuclear localization signals (NLS) *(15,31)* that are responsible for protein targeting. The prediction of N-terminal sorting signals has a long history originating from the early work on secretory signal peptides of Gunnar von Heijne *(32)*. In eukaryotes, N-terminal signals are responsible for the transport of secretory proteins from the cytosol to the ER and for sorting proteins to the mitochondria *(33)* and to chloroplasts *(34)*. In bacteria, N-terminal signals are responsible for translocating proteins across the cytoplasmic membrane. A distinguishing feature of N-terminal transit peptides is that the peptide itself is cleaved off once the protein has been translocated. Early methods for predicting signal peptides were essentially based on consensus signals, using linear discriminant functions with weight matrices. Modern ML techniques can predict whether a protein contains an N-terminal targeting peptide or not by automatically extracting correlations from the sequence data without any prior knowledge of targeting signals. One tradeoff in using these techniques is the impossibility of gaining any idea about the protein-sorting mechanism by looking at the output from these predictors. The introduction of ML techniques such as NNs and HMMs *(35–37)* has resulted in spectacular improvements in prediction accuracy. These methods learn to discriminate automatically from the data, using only a set of experimentally verified examples as input. It is now possible to reliably predict secretory signal peptides (SPs) using ML techniques *(38)*. Mitochondrial targeting peptides (mTPs) and chloroplast targeting peptides (cTPs) can be predicted with a somewhat lower accuracy, due to their longer lengths and greater variability in the signal sequence *(39)*. A particular problem in applying methods detecting N-terminal signals to entire genomes is that start codons are predicted with less than 70% accuracy by genome projects *(40)*. For additional details, the reader can consult a number of excellent reviews on N-terminal-sorting signal prediction *(39,41)*. Sorting signals also mediate the import of proteins into the nucleus. A protein is imported into the nucleus if it contains an NLS, which is a short stretch of amino acids. Extensive experimental research on nucleocytoplasmic transport *(42)* indicates that NLS can occur anywhere in the amino acid sequence and in general have an abundance of positively charged residues *(43)*. Since the entire protein sequence has to be searched for NLS, the application of ML techniques has proven difficult. The first effort at predicting NLS was by Cokol *et al.* *(31)*, who successfully applied "*in silico* mutagenesis" to discover new NLS. Recently, Brameier *et al.* *(44)* utilized regular expression matching and multiple program classifiers induced by genetic algorithms to predict NLS with some success. Overall, localization can be inferred for about 30% of the proteins in six eukaryotic proteomes ***(45)*** using known and predicted signaling motifs.

In the following sections, we review SignalP 3.0 and PredictNLS, which are the most accurate tools for predicting signal peptides and NLS.

3.2. SignalP 3.0: Predicting N-Terminal Signal Peptides

SingalP ***(46,47)***, which is currently in its third version, consists of two different predictors based on NNs and HMM algorithms and is among the most widely used research tools in Bioinfomatics. SignalP consists of specialized predictors for three organism groups, namely eukaryotes, gram-negative bacteria, and gram-positive bacteria. Signal peptides have long been known to have a tripartite design consisting of a short positively charged amino-terminal segment (n-region), a central hydrophobic segment (c-region), and a more polar C-terminal segment that is recognized by the signal peptidase enzyme (SPase). The peptides are 20–25 residues long on average and are cleaved off by signal peptidase during the export process ***(39)***. The signal peptide prediction problem is posed to the NNs in two ways: (1) classification of individual amino acids as belonging to the signal peptide or not, and (2) recognition of the cleavage sites against the background of all other sequence positions. For both types of networks sequence data were presented using a sliding window technique; a window of residues is presented and the network is trained to predict the state of the central residue. The sliding window approach is remarkably successful at capturing sequence features correlated over long stretches of residues ***(48)***. The window is then moved along the amino acid sequence and predictions are made in turn for each successive residue. In earlier versions of SignalP, a protein was classified as containing a signal peptide if the mean prediction score for the N-terminal amino acid residues exceeded a predefined threshold. In SignalP 3.0, a somewhat more complicated improved scoring scheme has replaced this score ***(47)***. Window sizes used in SignalP 3.0 ranged from a symmetric window of 27 residues for the eukaryotic signal peptide discrimination networks to a window of 20 positions upstream and 4 positions downstream of the cleavage site for the cleavage site predictor. SignalP 3.0 can discriminate proteins containing signal peptides with an accuracy of 93% for eukaryotes, 95% for gram-negative bacteria, and 98% for gram-positive bacteria. The major improvement over the years has been in predicting the cleavage site for which gains in prediction accuracy range from 6% to 7% for all three organism classes. Improvements in discriminating the presence/absence of signal peptides have mainly been due to the incorporation of amino acid composition as an additional input feature to the NN. The NN outperforms the HMM-based classifier in discriminating signal peptides and predicting cleavage sites. However, the HMM-based classifier better distinguishes signal peptides from signal anchors, which anchor proteins to the membrane *(49)*.

3.3. PredictNLS: Predicting Nuclear Localization Signals

Fewer than 10% of the NLS responsible for the nuclear sorting of known nuclear proteins have been experimentally determined. To remedy this, PredictNLS *(31)* uses a procedure of "*in silico* mutagenesis" to discover new motifs with potential NLS function. This procedure works as follows. (1) Change or remove some residues from the experimentally characterized NLS motifs and monitor the resulting true (nuclear) and false (nonnuclear) matches. Obviously, allowing alternative residues at particular positions increases the number of nuclear proteins found. However, often this also results in an increased number of matches to nonnuclear proteins. (2) Discard any potential NLSs that are found in known nonnuclear proteins (false matches). (3) Require that potential NLS be found in at least two distinct nuclear protein families. This procedure ensures high reliability of the resulting potential NLS motifs. The 194 potential NLS discovered using this procedure could be used to explain the nuclear sorting of over 43% of known nuclear proteins. These potential NLS can be accessed through the NLSdb database *(50)*. NLSdb contains over 6000 predicted nuclear proteins from the Protein Data Bank (PDB) and Swiss-Prot databases along with their corresponding localization signals. The observation that in DNA-binding proteins the NLS motif often overlaps the DNA-binding region *(51)* is utilized to identify proteins that bind DNA. Approximately 20% of the potential NLS motifs discovered were observed to colocalize with the experimentally determined DNA-binding region of proteins *(31)*. Using this observation, over 1500 proteins were annotated as binding DNA.

4. Mining Databases for Functional Information

Molecular biology databases contain a wealth of information regarding proteins, including different aspects of their function. Protein function databases can be categorized into two types, those that contain textual information describing protein function, for example, the experimental information in PubMed abstracts and the protein records in UniProt *(52)*, and those that contain information regarding protein families, domains, and functional sites *(8)*. In the past few years there has been a proliferation in the number of protein databases due to the emergence of large-scale experimental techniques probing protein function. This has led to the development of several methods that mine this functional information with a view to annotating subcellular localization. These methods can be broadly classified into two categories based on the type of database and information they mine: (1) methods that mine databases containing textual information, and (2) methods that mine databases of functional motifs and domains. Reliable predictions obtained by mining databases can greatly complement already existing subcellular localization annotations in databases. Below we review some of the machine learning strategies that have been employed to mine databases for annotating protein function.

4.1. Automatic Lexical Analysis of Controlled Vocabularies

Functional annotations contained in protein databases are mostly written in plain text using a rich biological vocabulary that often varies in different areas of biomedical research. This makes it extremely difficult to parse protein function annotations using computer programs. In contrast, the GO database ***(9)*** and the function key word terms in the Swiss-Prot database ***(53)*** offer structured annotations that can be easily parsed by computers. Depending on the nature of the textual information contained in a database, automatic text analysis methods can be divided into two categories: (1) those extracting information directly from plain text, for example, contained in PubMed abstracts, and (2) those that infer function from controlled vocabularies in protein databases. New experimental discoveries are first published in scientific journals. Mining scientific literature to automatically retrieve information is an appealing goal and a number of groups have worked on different aspects of this problem, such as machine selection of articles of interest ***(54)***, automated extraction of information using statistical methods ***(55)***, and natural language processing techniques for extracting pathway information ***(56)***. Recently, a few methods have attempted to infer subcellular localization directly from scientific publications ***(57–59)***. However, these methods have had only limited success. At the same time, methods that infer subcellular localization from controlled vocabularies have proven to be more successful. Many of these methods infer missing pieces of information regarding cellular function using semantic analysis of the functional annotations in GO ***(60–62)*** and the keyword annotations in Swiss-Prot ***(63–65)***. Both fully automated and semiautomated methods of semantic analysis have been developed for predicting subcellular localization. The fully automatic methods extract rules from key words by using statistical learning methods such as probabilistic Bayesian models ***(66)*** and M-ary (multiple category) classifiers such as the k-nearest neighbor *(67)*. Some of the major methods in this category are LOCkey *(63)*, Proteome Analyst *(64)*, and Spearmint *(68)*. The semiautomated methods are based on building dictionaries of rules associating a certain pattern of occurrence of key words to a functional class. The major methods in this category are EUCLID *(65)*, Meta_A *(69)*, and RuleBase *(70)*. Function annotations from RuleBase and Spearmint have been integrated into UniProt *(52)*, which is the world's most comprehensive catalogue of information on proteins. Below we review the LOCkey algorithm for predicting subcellular localization.

4.2. LOCkey: Information Theory-Based Classifier

The LOCkey system *(63)* is a novel M-ary classifier that predicts the subcellular localization of a protein based on Swiss-Prot key words. The LOCkey algorithm can be divided into two steps (**Fig. 2**): (1) building data sets of trusted vectors for proteins with known localization, and (2) classifying proteins with unknown localization. First, the list of key word terms associated with all proteins whose subcellular localization is known is extracted from Swiss-Prot.

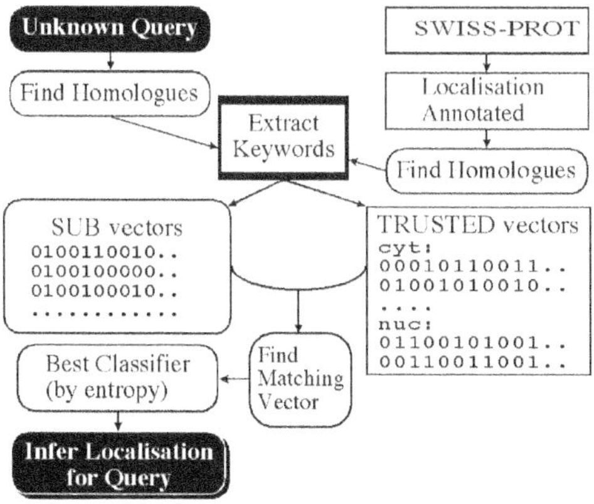

Fig. 2. The LOCkey algorithm. A sequence unique data set of localization-annotated Swiss-Prot proteins was first compiled. Key words were extracted for these proteins and merged with any key words found in homologues. The key words were represented as binary vectors in the "Trusted Vector Set." An unknown query was first annotated with key words through identification of Swiss-Prot homologues. Key words for the query were represented as binary vectors. All possible key word combinations were constructed (the SUB vectors). The best matching vector was found based on entropy criteria. This vector was used to infer localization for the query.

A data set of binary vectors is generated for each protein by representing the presence of a specific key word term in that protein's annotation by 1 and its absence by 0. Second, to infer subcellular localization of an unknown protein U all key words for U are read from Swiss-Prot. These key words are translated into a binary key word vector using the procedure described above. From this key word vector, LOCkey generates the set of all possible combinations of alternative vectors by flipping vector components of value 1 (presence of a key word) to 0 in all possible ways. For example, for a protein with three key words, there are $2^3 - 1 = 7$ possible subvectors: 111, 110, 101, 011, 100, 010, and 001. These subvectors constitute all possible key word combinations for protein U. The key word combination, i.e., subvector, that yields the best classification of U into one of 10 classes of subcellular localizations is then found. This is done by retrieving all exact matches of each of the subvectors to any of the proteins in the trusted set, i.e., by finding all proteins in the trusted set that contain all the key words present in the subvector. By construction, the proteins retrieved in this way may also contain key words not found in U. Next, LOCkey estimates the "surprise value" of a given assignment. Toward this end, the algorithm simply compiles the number of proteins belonging to each type of subcellular localization. This procedure is repeated in turn for each of the subvectors and localization is finally assigned to a protein by minimizing an entropy-based objective function. LOCkey achieved an accuracy of more than 82% in a full cross-validation test. For five entirely sequenced eukaryotic proteomes, namely yeast, worm, fly, plant (*Arabidopsis thaliana*), and human proteins, the LOCkey system automatically found about 8000 new annotations for subcellular localization.

4.3. Prediction Based on Functional Motifs, Domains, and Other Signatures

Motif and domain databases such as PROSITE, InterPro, PFAM, SMART, and others associate sequence signatures in a protein with specific types of function the protein may perform or with the family of functionally related proteins. Though not all of these motifs and domains are specific to subcellular localization, many preferentially occur in some subcellular classes. Mott *et al.* *(71)* were the first to apply a domain projection method to SMART domains associated with a protein to infer its localization. They restricted their method to the 300 SMART domains that cooccur with at least two other domains. By assigning probabilities for the individual domains to occur in the three major subcellular classes, they were able to predict the location of over 50,000 eukaryotic proteins. This idea has been incorporated by many other methods. LOCSVMPSI *(72)* predicts subcellular class by identifying the PROSITE motifs that are correlated with localization, while pTarget ***(73)*** utilizes location-specific PFAM domains. The MotifSearch algorithm in MultiLoc ***(59)*** relies on sequence motifs from PROSITE and NLSdb ***(50)*** that are specific to proteins sorted to a particular location. By modifying the LOCkey algorithm to handle sequence signatures, the LOCtree system ***(74)*** carries this one step further by inferring localization based on the combination of PROSITE motifs and PFAM domains that characterize a protein.

5. *Ab Initio* Prediction from Sequence

5.1. *Ab Initio Methods Predict Localization for All Proteins at Lower Accuracy*

The breakthrough for *ab initio* prediction came from the pioneering works of Nishikawa and co-workers ***(75,76)***. They observed that the total amino acid composition of a protein is correlated with its subcellular localization. An explanation for this observation was provided by Andrade *et al.* ***(77)*** who observed that the signal for subcellular localization was almost entirely due to the surface residues. Throughout evolution each subcellular compartment has maintained its characteristic physicochemical environment, so it is not very surprising that protein surfaces have evolved to adapt to these conditions. A wide array of methods has been developed to exploit this correlation of subcellular localization with sequence composition. The first tool to use amino acid composition was the PSORT expert system from Nakai and Kanehisa ***(78)***, which used standard statistical methods. However, it is only with the recent application of machine learning techniques that composition-based methods have started approaching the prediction accuracy of other methods. One of the earliest methods to use a machine learning approach was the NNPSL predictor ***(79)***, which used feedforward NNs trained on the amino acid composition. The network classified proteins from eukaryotic organisms into one of four possible subcellular compartments with an accuracy of 66% and prokaryotic proteins into one of three compartments with an accuracy of 81%. Reinhardt *et al.* also showed that the NN predictions were relatively insensitive to sequencing errors near the N-terminal, thereby adding weight to the importance of these predictions for whole genome annotations. Hua and Sun ***(80)*** showed that SVMs were even better at predicting localization from the amino acid composition. One reason for this is that SVMs outperform NNs at extracting correlations when the data set is relatively small and noisy ***(81)***. By training SVMs on the data set of Reinhardt and Hubbard ***(79)*** their SubLoc system was able to improve prediction accuracy by over 13%. Park and Kanehisa ***(82)*** found that adding residue pair compositions to the amino acid composition improved prediction accuracy by over 5%. Their PLOC system classifies proteins into one of nine subcellular compartments with an accuracy of over 79%. One problem with using residue pair compositions is the large number of input units it generates, over 400 including the amino acid composition. This can result in overfitting by the machine learner. To overcome this problem, Cai and co-workers ***(83,84)*** have tried to incorporate higher order correlations (residues i and [I + n], n = 2, 3, 4) by introducing the pseudo-amino acid composition. They applied a mathematical function to the amino acid pairs and then summed over the entire sequence yielding only one or two extra parameters for each separation distance. They observed some improvement in prediction accuracy. pSLIP ***(85)*** divides a protein sequence into an a priori fixed number of subsequences and uses the average physicochemical properties of these subsequences as additional inputs to the SVM while Ogul and Mumcuogu ***(86)*** use n-peptide compositions with reduced amino acid alphabets and pairwise sequence similarity scores based on a whole sequence and an N-terminal sequence. For additional details regarding some of the methods discussed here, refer to the review by Donnes and Hoglund ***(87)***.

The LOCtree system of Nair and Rost ***(74)*** significantly improves prediction accuracy by introducing several novel ideas: (1) hierarchical machine learning architecture that predicts localization by mimicking the cellular sorting machinery, (2) using predicted localization from high accuracy methods to complement experimental database annotations for training machine learners, (3) using features based on predicted secondary structure, and (4) incorporating evolutionary information through the use of profile-based composition. Below, we review the LOCtree system, which is one of the most accurate *ab initio* methods currently available for predicting localization from sequence.

5.2. LOCtree: Predicting Localization by Mimicking the Cellular Sorting Machinery

The LOCtree system was the first to implement a hierarchical ontology of localization classes, which is dictated by the biology of protein sorting, within a machine-learning framework (**Fig. 3**). This represented a major departure from preexisting prediction systems, which relied on standard parallel architectures for prediction. By construction, the system penalizes confusions of subcellular classes along the same pathway (e.g., ER instead of extracellular) less than confusions between classes from different pathways (e.g., ER instead of nuclear). Technically, the hierarchical ontology was incorporated using a decision tree with SVMs as the nodes (**Fig. 3**). Due to its hierarchical architecture, the system could exploit evolutionary similarities among the different subcellular classes, thereby significantly improving prediction accuracy over standard systems. A major advantage of the system is its ability to predict intermediate subcellular classes such as cytoplasm at much higher accuracies. The use of high-quality predicted data for training the SVMs was another source of improvement. The use of predicted data for training the machine learner could potentially improve prediction accuracy by significantly increasing the size of the available training data. However, such a result can by no means be taken for granted and many reviews strictly advise using only curated datasets for training *(88)*. However, the LOCtree system recorded a 7% gain in accuracy using highly accurate inferred localization based on Swiss-Prot key words and PROSITE/PFAM signatures for training the *ab initio* composition-based predictor. The secondary structure prediction tools available today are accurate enough that LOCtree could improve subcellular location prediction by using predicted secondary structure as an input feature to the SVM. LOCtree achieved an impressive accuracy of over 74% for discriminating among the five classes of nonplant eukaryotic proteins (Level 2 in **Fig. 3A**), but the accuracy is much higher for the intermediate localizations, for example, nuclear proteins could be discriminated from other intracellular proteins with an accuracy of over 78%.

6. Integrated Methods for Predicting Localization

6.1. Improving Accuracy Through Combinations

The different strategies for predicting localization have their own strengths and weaknesses. High accuracy methods, such as those based on sequence motifs and homology, are plagued by the problem of low coverage and can provide annotations for fewer than one-third of known sequences. In this era of whole-genome sequencing, we need solutions that can provide high-quality annotations for all proteins in an organism. Currently the best solution available is to combine high accuracy but low coverage methods with state-of-the-art high coverage methods, such as those based on composition, which have lower accuracies. This approach was pioneered by Nakai and Horton *(26)* with their PSORT system. PSORTII, as described earlier, combines a comprehensive database of sorting signals with predictions based on composition. The majority of the most recent methods belong to this category; these include WoLFPSORT *(29)*, CELLO *(89)*, pTarget *(90)*, MultiLoc *(59)*, and the LOCtree WWW server *(74)*. These methods combine predictions based on features such as sequence motifs, sequence homology to a protein of known localization, text analysis of terms related to protein function in database annotations, and, when all other methods fail, *ab initio* prediction based on compositional features of the amino acid sequence (**Table 2**).

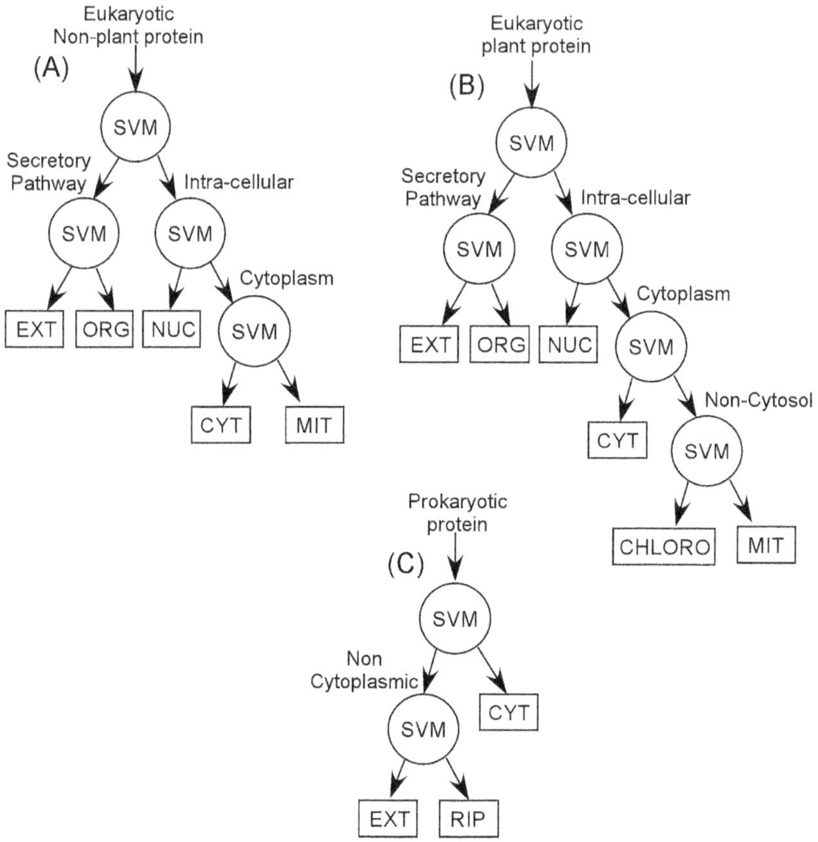

Fig. 3. Hierarchical architecture of LOCtree. LOCtree uses specialized architectures for proteins belonging to organisms of different types: (**A**) architecture for eukaryotic nonplant proteins, (**B**) architecture for plant proteins, and (**C**) architecture for prokaryotic proteins. At each branch point a support vector machine (SVM) is used to accomplish a binary classification (either protein belongs to localization class L or does not belong to L). The hierarchical architecture was designed to mimic the biological protein-sorting mechanism as closely as possible. The branches of the tree represent intermediate stages in the sorting machinery while the nodes represent the decision points in the sorting machinery. The top node SVM (denoted by Level 0) discriminates between secretory pathway proteins and other intracellular proteins (**A** and **B**) or proteins that remain in the cytoplasm from the rest (**C**). The intermediate node SVMs in the next level (Level 1) are responsible for separating extracellular proteins from proteins sorted to the organelles and nuclear proteins from cytoplasmic proteins (**A** and **B**). For the prokaryotic architecture (**C**), Level 1 is the terminal level for gram-negative bacteria and separates extracellular proteins from periplasmic proteins. In addition, Level 1 also contains the cytoplasmic leaf that is propagated without branching from Level 0. For gram-positive bacteria, Level 0 is the terminal level and separates cytoplasmic proteins from extracellular proteins (noncytoplasmic branch). The leaves

Table 2
Services for Subcellular Localization Prediction

Method	URL
Sequence homology-based localization annotations	
LOChom *(94)*	cubic.bioc.columbia.edu/db/LOChom/
Methods based on N-terminal sorting signals	
SignalP *(95)*	www.cbs.dtu.dk/services/SignalP/
TargetP *(36)*	www.cbs.dtu.dk/services/TargetP/
IPSORT *(27)*	biocaml.org/ipsort/iPSORT/
Predotar *(96)*	www.inra.fr/Internet/Produits/Predotar/
Prediction and analysis of nuclear localization signals	
PredictNLS *(31)*	cubic.bioc.columbia.edu/predictNLS/
Inferring localization using text analysis	
LOCkey *(63)*	cubic.bioc.columbia.edu/services/LOCkey/
Proteome Analyst	www.cs.ualberta.ca/~bioinfo/PA/
GeneQuiz *(65)*	jura.ebi.ac.uk:8765/ext-genequiz/
Meta_A *(69)*	mendel.imp.univie.ac.at/CELL_LOC/
Methods based on amino acid composition	
BaCelLo *(91)*	gpcr.biocomp.unibo.it/bacello/info.htm
SubLoc *(80)*	www.bioinfo.tsinghua.edu.cn/SubLoc/
PLOC *(82)*	www.genome.jp/SIT/ploc.html
ProtComp	www.softberry.ru/berry.phtml?topic=index&group =programs&subgroup=proloc
General methods	
PSORTII *(26)*	psort.nibb.ac.jp/
PSORT-B *(28)*	www.psort.org/psortb/
WolfPSORT *(29)*	wolfpsort.cbrc.jp/
LOCtree *(74)*	cubic.bioc.columbia.edu/services/LOCtree/

Fig. 3. (Continued) of the tree, represented by rectangular boxes, represent the final localization classes for which prediction is made. If a leaf has a depth smaller than the overall depth of the tree it is propagated without branching for the remainder of the tree. The terminal level (Level 2) for the eukaryotic nonplant architecture (**A**) is responsible for sorting proteins into one of five subcellular classes (mitochondria and cytosol plus the three leaves from Level 1), while Level 3 is the terminal level for the plant architecture (**C**) and separates proteins into one of six classes (mitochondria and chloroplast plus the four leaves from Level 2). The prediction accuracy of the parent nodes is higher than the child nodes, which results in significantly improved prediction accuracies for the intermediate localization states. EXT, extracellular; NUC, nucleus; CYT, cytosol; MIT, mitochondria; CHLORO, chloroplast; RIP, periplasm; ORG, organelle. Organelles are the endoplasmic reticulum, Golgi apparatus, peroxysomes, lysosomes, and vacuolar compartments.

7. Conclusions

7.1. Several Pitfalls in Assessing Quality of Annotations

To draw reliable inferences from a prediction it is essential that the accuracy of the method is properly established. To obtain accurate estimates of performance the testing procedure should mimic a blindfold prediction exercise as far as possible. One way of ensuring this is to choose training datasets such that test sequences have little or no sequence similarity to proteins in the training set, since the majority of proteins for which function is to be predicted have very little sequence similarity to the limited number of proteins with known function. However, in practice, this is often not the case and many methods test their performance on only a small sample of randomly selected proteins, resulting in overestimates of prediction accuracy. Often this results in gross overestimates of prediction accuracy. The NNPSL dataset, which has been used in the development of over a dozen subcellular localization predictors, is inherently flawed due to the high degree of sequence similarity between training and test proteins. In fact, Pierleoni *et al.* ***(91)*** recently showed that subcellular localization can be best predicted for this dataset by using a simple BLAST search, where each test protein is simply assigned the localization of the closest sequence homolog in the training set. Another problem that affects prediction accuracy is the number of redundant sequences in public databases. Adequate care must be taken during development to avoid biased predictions toward large families of redundant protein sequences by using sequence unique test sets. Otherwise the estimated accuracy is likely to be much higher than the true prediction accuracy. Benchmarking prediction methods proves to be a difficult task since different methods have been developed at different times and database annotations of function are constantly growing. In addition, there are no standard procedures for reporting prediction accuracy with some methods reporting only the overall prediction accuracy, which can be quite uninformative due to the large differences in the sizes of the datasets for the different subcellular classes. In spite of these problems, some recent reviews ***(92,93)*** have attempted to benchmark some of the publicly available prediction servers for predicting localization. One of the conclusions is that different methods perform best for different situations. Another problem, without any obvious solution, is choosing an appropriate tradeoff between sensitivity and specificity. Depending on the application, either high specificity or sensitivity might be desirable. Hence, caution should be exercised when using predictions from automatic servers, especially in cases in which little is known about the function of the protein and the sequence signals that are involved in sorting. It is sometimes instructive to compare predictions from multiple servers that use different prediction strategies. Similar predictions from the servers might indicate some propensity of the protein for the predicted localization, while conflicting predictions might call for further research.

7.2. Curated Datasets of Subcellular Localization Are Essential

Functional annotations in standard databases usually contain large numbers of incorrect annotations, which makes the development of prediction tools all the more difficult. A key element when constructing any prediction method is the quality of the data. Extracting a reliable training set from publicly available databases implies a large amount of work and carries a number of critical decisions and pitfalls. As an example, if the "subcellular location" comment in Swiss-Prot contains "endoplasmic reticulum," the protein may be dissolved in the ER lumen, embedded in the ER membrane, or even associated with the cytoplasmic face of the ER membrane; these alternatives are quite different with respect to the sorting signals involved *(88)*. The only solution is a comprehensive ontology to describe subcellular location annotations. The development of GO is a step in the right direction.

7.3. Prediction Accuracy Continues to Grow

In spite of the difficulties in correctly assessing the accuracy of prediction methods, during the past few years significant strides have been made in tackling the problem of subcellular localization prediction. One reason is the application of advanced ML techniques that can recognize subtle correlations among different kinds of sequence features. The second reason is the steady growth in the amount of functional information deposited in databases. Prediction tools are already proving useful for automatic annotations of sequence databases and for screening potentially interesting genes from genome data. In the near future it might be possible to predict the subcellular localization of almost any protein with high confidence *(94–96)*. Future improvements are likely to result through the use of integrated prediction methods that cleverly combine the output from programs that predict different functional features to provide a comprehensive prediction of subcellular localization. Integrated prediction methods better capture biological reality since events affecting the fate of proteins are interrelated. For example, it is evident that a modification enzyme will not modify its potential substrates when the membrane separates them. Moreover, combination methods can be designed to naturally fall into an ontological scheme that would help us achieve the goal of a unified framework for protein function prediction.

Acknowledgments

We thank Dr. Dariusz Przybylski, Dr. Jinfeng Liu, and Kazimierz Wrzeszczynski for helpful discussions and Christina Schlecht for proofreading the manuscript. Last, but not least, we thank all those who deposit their experimental data in public databases and those who maintain these databases and the world wide web for making so many resources easily accessible.

References

1. Venter, J. C., Adams, M. D., Myers, E. W., *et al.* (2001) The sequence of the human genome. *Science* **291(5507),** 1304–1351.
2. Brutlag, D. L. (1998) Genomics and computational molecular biology. *Curr. Opin. Microbiol.* **1(3),** 340–345.
3. Harrison, P. M., Bamborough, P., Daggett, V., Prusiner, S., and Cohen, F. E. (1997) The prion folding problem. *Curr. Opin. Struct. Biol.* **7,** 53–59.
4. Bork, P. and Koonin, E. V. (1998) Predicting functions from protein sequences—where are the bottlenecks? *Nat. Genet.* **18(4),** 313–318.
5. Luscombe, N. M., Greenbaum, D., and Gerstein, M. (2001) What is bioinformatics? A proposed definition and overview of the field. *Methods Inf. Med.* **40(4),** 346–358.
6. Bork, P., Dandekar, T., Diaz-Lazcoz, Y., Eisenhaber, F., Huynen, M., and Yuan, Y. (1998) Predicting function: from genes to genomes and back. *J. Mol. Biol.* **283(4),** 707–725.
7. Rost, B., Liu, J., Nair, R., Wrzeszczynski, K. O., and Ofran, Y. (2003) Automatic prediction of protein function. *Cell. Mol. Life Sci.* **60(12),** 2637–2650.
8. Apweiler, R., Attwood, T. K., Bairoch, A., *et al.* (2000) InterPro—an integrated documentation resource for protein families, domains and functional sites. *Bioinformatics* **16(12),** 1145–1150.
9. Ashburner, M., Ball, C. A., Blake, J. A., *et al.* (2000) Gene ontology: tool for the unification of biology. The Gene Ontology Consortium. *Nat. Genet.* **25(1),** 25–29.
10. Lodish, H., Berk, A., Baltimore, D., and Darnell, J. (2000) *Molecular Cell Biology*, 4th ed. W. H. Freeman & Co, New York.
11. Skach, W. R. (2000) Defects in processing and trafficking of the cystic fibrosis transmembrane conductance regulator. *Kidney Int.* **57(3),** 825–831.
12. Payne, A. S., Kelly, E. J., and Gitlin, J. D. (1998) Functional expression of the Wilson disease protein reveals mislocalization and impaired copper-dependent trafficking of the common H1069Q mutation. *Proc. Natl. Acad. Sci. USA* **95(18),** 10854–10859.
13. Parfrey, H., Mahadeva, R., and Lomas, D. A. (2003) Alpha(1)-antitrypsin deficiency, liver disease and emphysema. *Int. J. Biochem. Cell Biol.* **35(7),** 1009–1014.
14. Davis, T. N. (2004) Protein localization in proteomics. *Curr. Opin. Chem. Biol.* **8(1),** 49–53.
15. Nakai, K. (2000) Protein sorting signals and prediction of subcellular localization. *Adv. Protein Chem.* **54,** 277–344.

16. Schneider, G. and Fechner, U. (2004) Advances in the prediction of protein targeting signals. *Proteomics* **4(6),** 1571–1580.
17. Schatz, G. and Dobberstein, B. (1996) Common principles of protein translocation across membranes. *Science* **271(5255),** 1519–1526.
18. Darnell, J., Lodish, H., and Baltimore, D. (1990) *Molecular Cell Biology*, 2nd ed. W. H. Freeman & Co, New York.
19. Valencia, A. and Pazos, F. (2002) Computational methods for the prediction of protein interactions. *Curr. Opin. Struct. Biol.* **12(3),** 368–373.
20. Wu, C. H., Nikolskaya, A., Huang, H., *et al.* (2004) PIRSF: family classification system at the Protein Information Resource. *Nucleic Acids Res.* **32(1),** D112–114.
21. Nakai, K. (2001) Review: prediction of in vivo fates of proteins in the era of genomics and proteomics. *J. Struct. Biol.* **134(2–3),** 103–116.
22. Apweiler, R., Gateau, A., Contrino, S., *et al.* (1997) Protein sequence annotation in the genome era: the annotation concept of SWISS-PROT+TREMBL. *Proc. Int. Conf. Intell. Syst. Mol. Biol.* **5,** 33–43.
23. Bairoch, A. and Apweiler, R. (1997) The SWISS-PROT protein sequence data bank and its new supplement TrEMBL. *Nucleic Acids Res.* **25,** 31–36.
24. Simpson, J. C., Wellenreuther, R., Poustka, A., Pepperkok, R., and Wiemann, S. (2000) Systematic subcellular localization of novel proteins identified by large-scale cDNA sequencing. *EMBO Rep.* **1(3),** 287–292.
25. Nakai, K. and Kanehisa, M. (1992) A knowledge base for predicting protein localization sites in eukaryotic cells. *Genomics* **14(4),** 897–911.
26. Nakai, K. and Horton, P. (1999) PSORT: a program for detecting sorting signals in proteins and predicting their subcellular localization. *Trends Biochem. Sci.* **24(1),** 34–36.
27. Bannai, H., Tamada, Y., Maruyama, O., Nakai, K., and Miyano, S. (2002) Extensive feature detection of N-terminal protein sorting signals. *Bioinformatics* **18(2),** 298–305.
28. Gardy, J. L., Spencer, C., Wang, K., *et al.* (2003) PSORT-B: Improving protein subcellular localization prediction for Gram-negative bacteria. *Nucleic Acids Res.* **31(13),** 3613–3617.
29. Horton, P., Park, K. J., Obayashi, T., *et al.* (2007) WoLF PSORT: protein localization predictor. *Nucleic Acids Res.* **35(Web Server issue),** W585–587.
30. von Heijne, G. (1995) Protein sorting signals: simple peptides with complex functions. *EXS* **73,** 67–76.
31. Cokol, M., Nair, R., and Rost, B. (2000) Finding nuclear localization signals. *EMBO Rep.* **1(5),** 411–415.
32. von Heijne, G. (1985) Signal sequences. The limits of variation. *J. Mol. Biol.* **184,** 99–105.
33. Voos, W., Martin, H., Krimmer, T., and Pfanner, N. (1999) Mechanisms of protein translocation into mitochondria. *Biochim. Biophys. Acta* **1422(3),** 235–254.
34. Bruce, B. D. (2000) Chloroplast transit peptides: structure, function and evolution. *Trends Cell Biol.* **10(10),** 440–447.

35. Nielsen, H., Brunak, S., and von Heijne, G. (1999) Machine learning approaches for the prediction of signal peptides and other protein sorting signals. *Protein Eng.* **12,** 3–9.
36. Emanuelsson, O., Nielsen, H., Brunak, S., and von Heijne, G. (2000) Predicting subcellular localization of proteins based on their N-terminal amino acid sequence. *J. Mol. Biol.* **300(4),** 1005–1016.
37. Boden, M. and Hawkins, J. (2005) Prediction of subcellular localization using sequence-biased recurrent networks. *Bioinformatics* **21(10),** 2279–2286.
38. Kall, L., Krogh, A., and Sonnhammer, E. L. (2004) A combined transmembrane topology and signal peptide prediction method. *J. Mol. Biol.* **338(5),** 1027–1036.
39. Emanuelsson, O. and von Heijne, G. (2001) Prediction of organellar targeting signals. *Biochim. Biophys. Acta* **1541(1–2),** 114–119.
40. Gaasterland, T. and Oprea, M. (2001) Whole-genome analysis: annotations and updates. *Curr. Opin. Struct. Biol.* **11(3),** 377–381.
41. Durbin, R., Eddy, S. R., Krogh, A., and Mitchison, G. (1998) *Biological Sequence Analysis.* Cambridge University Press, Cambridge, UK.
42. Mattaj, I. W. and Englmeier, L. (1998) Nucleocytoplasmic transport: the soluble phase. *Annu. Rev. Biochem.* **67,** 265–306.
43. Jans, D. A., Xiao, C. Y., and Lam, M. H. (2000) Nuclear targeting signal recognition: a key control point in nuclear transport? *BioEssays* **22(6),** 532–544.
44. Brameier, M., Krings, A., and MacCallum, R. M. (2007) NucPred—predicting nuclear localization of proteins. *Bioinformatics* **23(9),** 1159–1160.
45. Liu, J. and Rost, B. (2002) Target space for structural genomics revisited. *Bioinformatics* **18(7),** 922–933.
46. Nielsen, H., Engelbrecht, J., Brunak, S., and von Heijne, G. (1997) Identification of prokaryotic and eukaryotic signal peptides and prediction of their cleavage sites. *Protein Eng.* **10(1),** 1–6.
47. Bendtsen, J. D., Nielsen, H., von Heijne, G., and Brunak, S. (2004) Improved prediction of signal peptides: SignalP 3.0. *J. Mol. Biol.* **340(4),** 783–795.
48. Qian, N. and Sejnowski, T. J. (1988) Predicting the secondary structure of globular proteins using neural network models. *J. Mol. Biol.* **202,** 865–884.
49. Nielsen, H. and Krogh, A. (1998) Prediction of signal peptides and signal anchors by a hidden Markov model. *Proc. Int. Conf. Intell. Syst. Mol. Biol.* **6,** 122–130.
50. Nair, R., Carter, P., and Rost, B. (2003) NLSdb: database of nuclear localization signals. *Nucleic Acids Res.* **31(1),** 397–399.
51. LaCasse, E. C. and Lefebvre, Y. A. (1995) Nuclear localization signals overlap DNA- or RNA-binding domains in nucleic acid-binding proteins. *Nucleic Acids Res.* **23(10),** 1647–1656.
52. Apweiler, R., Bairoch, A., Wu, C. H., *et al.* (2004) UniProt: the Universal Protein knowledgebase. *Nucleic Acids Res.* **32(Database issue),** D115–119.
53. Bairoch, A. and Apweiler, R. (2000) The SWISS-PROT protein sequence database and its supplement TrEMBL in 2000. *Nucleic Acids Res.* **28(1),** 45–48.
54. Iliopoulos, I., Enright, A. J., and Ouzounis, C. A. (2001) Textquest: document clustering of Medline abstracts for concept discovery in molecular biology. *Pac. Symp. Biocomput.* 384–395.

55. Stephens, M., Palakal, M., Mukhopadhyay, S., Raje, R., and Mostafa, J. (2001) Detecting gene relations from Medline abstracts. *Pac. Symp. Biocomput.* 483–495.
56. Friedman, C., Kra, P., Yu, H., Krauthammer, M., and Rzhetsky, A. (2001) GENIES: a natural-language processing system for the extraction of molecular pathways from journal articles. *Bioinformatics* **17(Suppl. 1),** S74–82.
57. Stapley, B. J., Kelley, L. A., and Sternberg, M. J. (2002) Predicting the subcellular location of proteins from text using support vector machines. *Pac. Symp. Biocomput.* 374–385.
58. Shatkay, H., Hoglund, A., Brady, S., Blum, T., Donnes, P., and Kohlbacher, O. (2007) SherLoc: high-accuracy prediction of protein subcellular localization by integrating text and protein sequence data. *Bioinformatics* **23(11),** 1410–1417.
59. Hoglund, A., Blum, T., Brady, S., *et al.* (2006) Significantly improved prediction of subcellular localization by integrating text and protein sequence data. *Pac. Symp. Biocomput.* 16–27.
60. Lu, Z. and Hunter, L. (2005) Go molecular function terms are predictive of subcellular localization. *Pac. Symp. Biocomput.* 151–161.
61. Raychaudhuri, S., Schutze, H., and Altman, R. B. (2002) Using text analysis to identify functionally coherent gene groups. *Genome Res.* **12(10),** 1582–1590.
62. Chalmel, F., Lardenois, A., Thompson, J. D., *et al.* (2005) GOAnno: GO annotation based on multiple alignment. *Bioinformatics* **21(9),** 2095–2096.
63. Nair, R. and Rost, B. (2002) Inferring sub-cellular localization through automated lexical analysis. *Bioinformatics* **18(Suppl. 1),** S78–S86.
64. Lu, Z., Szafron, D., Greiner, R., *et al.* (2004) Predicting subcellular localization of proteins using machine-learned classifiers. *Bioinformatics* **20(4),** 547–556.
65. Tamames, J., Ouzounis, C., Casari, G., Sander, C., and Valencia, A. (1998) EUCLID: automatic classification of proteins in functional classes by their database annotations. *Bioinformatics* **14(6),** 542–543.
66. Lewis, D. D. and Ringuette, M. (1994) Comparison of two learning algorithms for text categorization. Proceedings of the Third Annual Symposium on Document Analysis and Information Retrieval (SDAIR'94). Las Vegas, NV, April 11–13, 1994.
67. Dasarathy, B. V. (1991) *Nearest Neighbor (NN) Norms: NN Pattern Classification Techniques*. IEEE Computer Society Press, Los Alamitos, CA.
68. Kretschmann, E., Fleischmann, W., and Apweiler, R. (2001) Automatic rule generation for protein annotation with the C4.5 data mining algorithm applied on SWISS-PROT. *Bioinformatics* **17(10),** 920–926.
69. Eisenhaber, F. and Bork, P. (1999) Evaluation of human-readable annotation in biomolecular sequence databases with biological rule libraries. *Bioinformatics* **15(7–8),** 528–535.
70. Fleischmann, W., Moller, S., Gateau, A., and Apweiler, R. (1999) A novel method for automatic functional annotation of proteins. *Bioinformatics* **15(3),** 228–233.
71. Mott, R., Schultz, J., Bork, P., and Ponting. C. P. (2002) Predicting protein cellular localization using a domain projection method. *Genome Res.* **12(8),** 1168–1174.
72. Xie, D., Li, A., Lin, X., Wang, M., Jiang, Z., and Feng, H. (2005) Using motifs in the prediction of eukaryotic protein subcellular localization. *Conf. Proc. IEEE Eng. Med. Biol. Soc.* **3,** 2802–2804.

73. Guda, C. and Subramaniam, S. (2005) pTARGET: a new method for predicting protein subcellular localization in eukaryotes. *Bioinformatics* **21(21),** 3963–3969.
74. Nair, R. and Rost, B. (2005) Mimicking cellular sorting improves prediction of subcellular localization. *J. Mol. Biol.* **348(1),** 85–100.
75. Nishikawa, K. and Ooi, T. (1982) Correlation of the amino acid composition of a protein to its structural and biological characteristics. *J. Biochem.* **91,** 1821–1824.
76. Nakashima, H. and Nishikawa, K. (1994) Discrimination of intracellular and extracellular proteins using amino acid composition and residue-pair frequencies. *J. Mol. Biol.* **238(1),** 54–61.
77. Andrade, M. A., O'Donoghue, S. I., and Rost, B. (1998) Adaptation of protein surfaces to subcellular location. *J. Mol. Biol.* **276(2),** 517–525.
78. Nakai, K. and Kanehisa, M. (1991) Expert system for predicting protein localization sites in gram-negative bacteria. *Proteins* **11,** 95–110.
79. Reinhardt, A. and Hubbard, T. (1998) Using neural networks for prediction of the subcellular location of proteins. *Nucleic Acids Res.* **26(9),** 2230–2236.
80. Hua, S. and Sun, Z. (2001) Support vector machine approach for protein subcellular localization prediction. *Bioinformatics* **17(8),** 721–728.
81. Vapnik, V. N. (1995) *The Nature of Statistical Learning Theory.* Springer-Verlag, New York.
82. Park, K. J. and Kanehisa, M. (2003) Prediction of protein subcellular locations by support vector machines using compositions of amino acids and amino acid pairs. *Bioinformatics* **19(13),** 1656–1663.
83. Cai, Y. D., Liu, X. J., Xu, X. B., and Chou, K. C. (2002) Support vector machines for prediction of protein subcellular location by incorporating quasi-sequence-order effect. *J. Cell. Biochem.* **84(2),** 343–348.
84. Chou, K. C. and Cai, Y. D. (2003) Prediction and classification of protein subcellular location-sequence-order effect and pseudo amino acid composition. *J. Cell. Biochem.* **90(6),** 1250–1260.
85. Sarda, D., Chua, G. H., Li, K. B, and Krishnan, A. (2005) pSLIP: SVM based protein subcellular localization prediction using multiple physicochemical properties. *BMC Bioinform.* **6,** 152.
86. Ogul, H. and Mumcuogu, E. U. (2007) Subcellular localization prediction with new protein encoding schemes. *IEEE/ACM Trans. Comput. Biol. Bioinform.* **4(2),** 227–232.
87. Donnes, P. and Hoglund, A. (2004) Predicting protein subcellular localization: past, present, and future. *Genomics Proteomics Bioinform.* **2(4),** 209–215.
88. Emanuelsson, O., Brunak, S., von Heijne, G., and Nielsen, H. (2007) Locating proteins in the cell using TargetP, SignalP and related tools. *Nat. Protoc.* **2(4),** 953–971.
89. Yu, C. S., Chen, Y. C., Lu, C. H., and Hwang, J. K. (2006) Prediction of protein subcellular localization. *Proteins* **64(3),** 643–651.
90. Guda, C. (2006) pTARGET: a web server for predicting protein subcellular localization. *Nucleic Acids Res.* **34(Web Server issue),** W210–213.

91. Pierleoni, A., Martelli, P. L., Fariselli, P., and Casadio, R. (2006) BaCelLo: a balanced subcellular localization predictor. *Bioinformatics* **22(14),** e408–416.
92. Sprenger, J., Fink, J. L., and Teasdale, R. D. (2006) Evaluation and comparison of mammalian subcellular localization prediction methods. *BMC Bioinform.* **7(Suppl. 5),** S3.
93. Gardy, J. L. and Brinkman, F. S. (2006) Methods for predicting bacterial protein subcellular localization. *Nat. Rev. Microbiol.* **4(10),** 741–751.
94. Nair, R. and Rost, B. (2002) Sequence conserved for subcellular localization. *Protein Sci.* **11(12),** 2836–2847.
95. Nielsen, H., Engelbrecht, J., Brunak, S., and von Heijne, G. (1997) A neural network method for identification of prokaryotic and eukaryotic signal peptides and prediction of their cleavage sites. *Int. J. Neural Syst.* **8(5–6),** 581–599.
96. Small, I., Peeters, N., Legeai, F., and Lurin, C. (2004) Predotar: a tool for rapidly screening proteomes for N-terminal targeting sequences. *Proteomics* **4(6),** 1581–1590.

28

Protein Functional Annotation by Homology

Raja Mazumder, Sona Vasudevan, and Anastasia N. Nikolskaya

Summary

Genome sequencing projects have resulted in a rapid accumulation of predicted protein sequences. With experimentally verified information on protein function lagging far behind, computational methods are used for functional annotation of proteins. Here we describe a number of protocols for protein sequence and structure analysis that can be used to infer function of uncharacterized proteins. These protocols rely on publicly available computational resources and tools and can be utilized by anyone with an Internet access.

Key Words: Protein database; annotation; homology; protein families; protein sequence analysis; protein structure analysis; domain architecture; protein family classification; sequence alignment; database search.

1. Introduction

Genome sequencing projects have resulted in a rapid accumulation of predicted protein sequences. To fully realize the value of the data, scientists need to understand how these proteins function in making up a living cell. With experimentally verified information on protein function lagging far behind, computational methods are needed for reliable functional annotation of proteins. A general approach for functional assignment of unknown proteins is to infer protein functions based on homology to existing experimentally characterized and annotated proteins in sequence databases. Homology (common evolutionary origin) provides significant information for annotation of uncharacterized proteins. Homology can be inferred by evaluating sequence and/or structure similarity between two proteins or a group of proteins.

From: *Methods in Molecular Biology, vol. 484: Functional Proteomics: Methods and Protocols*
Edited by: J. D. Thompson et al., DOI: 10.1007/978-1-59745-398-1, © Humana Press, Totowa, NJ

When an experiment yields a sequence (or a set of sequences), a scientist tries to extract as much information as possible about the protein represented by this sequence, especially about its known or possible function. All available information has to be integrated and used in such analysis. This is especially important for poorly characterized or uncharacterized ("hypothetical") proteins where reliable data on the protein are not available in literature. The quality of this analysis is often critical for interpreting experimental results and formulating a working hypothesis, followed by designing future experiments.

Sequence similarity can be assessed on a protein-by-protein basis by using a pairwise comparison algorithm such as BLAST *(1)* to search the sequence of interest against a sequence database. Inferences about homology and hence functional predictions can be made after analyzing results of such searches.

More systematically, these questions can be addressed by defining protein families and matching proteins to curated protein families. The concept of protein family based on homology, first articulated by Margaret Dayhoff *(2)*, is widely used in protein functional prediction, in creating protein family and domain databases, and in evolutionary and other research. Protein families are usually defined and represented by multiple alignments. Importantly, position-specific scoring matrices (PSSMs) *(3)* and hidden Markov models (HMMs) *(4)* that are used for sequence searches in protein family and domain databases reflect sequence divergence within each protein family. This makes comparing a protein sequence against a protein/domain family database (a library of PSSMs or HMMs) much more sensitive than any pairwise comparisons (such as in the BLAST algorithm).

2. Computational Tools for Functional Annotation

Numerous resources and tools are available for protein sequence, structure, and function analysis. A total of 968 protein databases have been included into the NAR Online Molecular Biology Database Collection as of 2007 *(5)*. These include protein sequence, structure, domain, motif, family, function, as well as individual genomes, disease, and proteomics databases. Only a subset of major resources is used in the protocols in this chapter, but, depending on the specific task at hand, many others can be utilized.

Protein sequence databases contain individual protein records with annotations. Importantly, there are two types of major sequence databases. Databases of the first type serve as repositories [such as NCBI's GenBank *(6)*], which accept any submissions and never change, upgrade, or update annotations: only the original submitter is allowed to make any changes in the record. Hence, the annotations of old entries often stay the same over time. Databases of the second type create improved protein records; their mission is to provide quality annotations. These databases [such as UniProtKB *(7)* or NCBI's RefSeq *(8)*] undertake major annotation efforts involving expert manual curation of gene models, gene names, protein names, functions, and many other features.

Protein family and domain databases are indispensable for annotation and functional analysis of protein sequences. Some of the commonly used ones are listed in **Table 1**. Homology-based protein classifications are "natural" classifications that aim to reflect evolutionary relationships of proteins, as inferred by sequence and structure similarity. Proteins have modular structure, where domain is an evolutionary, structural, and functional unit, and domain shuffling creates multiple domain architectures. Therefore, each domain has its own evolutionary history, and domain classification makes it possible to build a hierarchy that can trace the evolution of a domain to its last common ancestor, the last point of traceable homology. However, in domain classification databases it is usually possible to annotate only a general biochemical function, and not a specific biological function (pathway or process in which this protein is involved). On the other hand, full-length protein classification systems often allow annotation of specific biological functions when some experimental data are available.

Every database comes with search tools to enable information retrieval. There are sequence search and text search options, both equally important. Probably the most indispensable sequence analysis tool is BLAST *(1)*. There is also a wide variety of sequence and structure analysis tools, geared toward various specific purposes.

Resources and tools cited in this chapter and some others generally recommended for the purposes of functional annotation and analysis are listed in **Table 1**. Note that all URLs, database references, web sites descriptions, and search results in this chapter are as of the time of writing. We apologize for any confusion that might arise from subsequent changes in database content, sequence, and/or annotation updates.

3. Methods

3.1. Overview

For the purpose of functional annotation, protein sequences can be divided into the following groups:

1. *Experimentally characterized proteins*. For these sequences, correct annotation is achieved by finding and accurate interpretation of available experimental data from the literature.
2. *Proteins characterized by similarity*. These are sequences closely related to experimentally characterized proteins. There are no hard rules as to the sequence

Table 1
Computational Tools and Resources for Protein Sequence and Structure Analysis

Resource	URL	Comment	Ref
Sequence databases			
GenBank	http://www.ncbi.nlm.nih.gov/Genbank/	Collection of all publicly available DNA sequences	*(6)*
RefSeq	http://www.ncbi.nlm.nih.gov/RefSeq/	Curated nonredundant collection of sequences representing genomes, transcripts, and proteins	*(8)*
NCBI Protein database	http://www.ncbi.nlm.nih.gov/entrez/query.fcgi?db=Protein	Protein sequences: translated from GenBank and imported from other protein databases	*(34)*
UniProtKB	http://www.pir.uniprot.org/database/knowledgebase.shtml	UniProt knowledge base: merged Swiss-Prot, TrEMBL, and PIR protein sequence databases	*(7)*
UniRef	http://www.uniprot.org/database/nref.shtml	UniProt Reference Clusters databases; UniRef combines closely related sequences into a single record to speed sequence searches	*(35)*
iProClass	http://pir.georgetown.edu/pirwww/dbinfo/iproclass.shtml	Value-added information reports for UniProtKB and unique NCBI protein sequences in UniParc	*(36)*
Protein domain databases			
Pfam	http://www.sanger.ac.uk/Software/Pfam/	An extensive collection of protein domains	*(37)*
SMART	http://smart.embl.de/	Mostly signaling domains	*(38)*
ProDom	http://prodom.prabi.fr/	Automatically generated protein domains	*(39)*
CDD	http://www.ncbi.nlm.nih.gov/Structure/cdd/cdd.shtml	Conserved domains with curated alignments that are based on available 3D structures	*(40)*
Protein family databases (full-length proteins)			
COG	http://www.ncbi.nlm.nih.gov/COG/	Clusters of orthologous groups that represent either whole proteins or individual domains	*(18)*

PIRSF	http://pir.georgetown.edu/pirsf/	Families of proteins with shared domain architecture and full-length similarity	*(31)*
Sequence motif databases			
PROSITE	http://www.expasy.org/prosite/	Protein sequence patterns and profiles that define protein domains	*(41)*
PRINTS	http://www.bioinf.man.ac.uk/ dbbrowser/PRINTS/	Protein fingerprints: groups of conserved motifs used to characterize each protein family	**42**
Integrated motif, domain, and family database			
InterPro	http://www.ebi.ac.uk/interpro/	An umbrella database integrating results from several classification databases	*(43)*
Protein structure databases			
PDB	http://www.rcsb.org/pdb	3D structures of macromolecules	*(44)*
PDBsum	http://www.ebi.ac.uk/thornton-srv/databases/pdbsum/	Pictorial representation of structures in PDB that includes protein–ligand interactions	*(28)*
MMDB	http://www.ncbi.nlm.nih.gov/Structure/MMDB/mmdb.shtml	Molecular modeling database	*(45)*
DALI	http://www.ebi.ac.uk/dali/	Comparison of protein structures in 3D	*(46)*
SCOP	http://scop.mrc-lmb.cam.ac.uk/scop/	Stuctural classification of proteins	*(47)*
Literature database			
PubMed	http://www.ncbi.nlm.nih.gov/sites/entrez?db=PubMed	Collection of biomedical literature citations the from U.S. National Library of Medicine	*(34)*
Tools			
BLAST/PSI-BLAST	http://www.ncbi.nlm.nih.gov/BLAST/Blast.cgi?PAGE=Proteins&PROGRAM=blastp&BLAST_PROGRAMS=blastp&PAGE_TYPE=BlastSearch&SHOW_DEFAULTS=on	Protein BLAST or PSI-BLAST search against NCBI databases	*(13)*
PIR tools	http://pir.georgetown.edu/pirwww/search/	Sequence analysis tools available at PIR	*(36)*

similarity and/or identity parameters, but the degree of sequence similarity between the sequence of interest and the experimentally characterized sequence should be rather high, preferably with the corresponding regions covering the majority of the length of both proteins (ideally, full-length, end-to-end similarity). For these sequences, correct annotation is achieved by identifying closely related homologs for which experimental data are available.
3. *Proteins for which function can be predicted.* In the absence of close sequence similarity to experimentally characterized proteins, an attempt should be made to predict at least a generic function by analyzing distant relationships. For these sequences, correct annotation is achieved by identifying distant homologs with experimental data using sensitive sequence and/or structure analysis methods as well as non-homology-based methods in some cases. The emphasis is on extracting maximum possible information from different sources and different methods. This is a category of sequences where important novel insights can be made via computational methods with subsequent experimental verification, such as finding new functions or filling the gaps in metabolic pathways.
4. *Proteins of unknown function* (conserved in distantly related organisms or unique to a specific organism) are sequences for which all prediction methods fail.

To annotate a protein sequence, the following basic steps should be taken: domain analysis, BLAST against sequence databases, search against protein family databases, and literature search (examples of resources used for these searches are listed in **Table 1**). If this is not sufficient, advanced sequence analysis may include PSI-BLAST search, profile searches (HMM and PSSM), pattern search (conserved motif analysis), manual multiple sequence alignments, and phylogenetic tree reconstruction. Finally, structure analysis can complement sequence analysis results and shed light on the very distantly related homologs that are not possible to detect with sequence analysis methods alone. Defining a protein family may be a useful step in this process. For many groups of proteins, good quality annotated families are available from one or more of the family and/or domain databases listed in **Table 1**. In other cases, this work has to be done *de novo*. This includes cases of novel domains not defined previously or cases in which redefining or expanding a domain superfamily is needed for the purpose of annotating distant homologs.

3.2. Sequence Analysis: Annotation of Tetraodon nigroviridis Gsteng00025548001 as a Bifunctional Chondroitin-Glucuronate C5-Epimerase/Chondroitin Sulfate Sulfotransferase

Assigning protein function based on the function of its closest experimentally characterized homolog(s) is a standard methodology utilized, with appropriate modifications, for close or remote homologs, for full-length protein similarity or individual domains. The following example of GSTENG00025548001, a sequence from *Tetraodon nigroviridis* (Green puffer) (UniProt accession no. Q4S1I2, NCBI gi|47221179 or accession no. CAG05500), demonstrates this procedure using the resources at NCBI, UniProt, and PIR.

1. Retrieve the sequence of GSTENG00025548001 from the NCBI protein database and from UniProtKB and examine the corresponding protein entries, checking if they contain any functional information. Note that the protein is currently annotated as "unnamed protein product" in the NCBI protein database (Genpept) and as "Chromosome 6 SCAF14768, whole genome shotgun sequence (Fragment)" in UniProtKB. Look at the associated publication(s) given in the entry. The reference associated with the entry Q4S1I2 ***(9)*** contains no functional information or experimental evidence about this protein, and there are no other hints as to its function in the protein entries. Therefore, the function of Q4S1I2 has to be predicted using available computational tools.
2. *Domain analysis: domain database search.* It is routine to compare a protein sequence against a number of the domain databases (**Table 1**) as a starting point in sequence and functional analysis. Then compare a protein sequence against a number of the domain databases (**Table 1**). If known conserved domains are found, their boundaries should be noted and meaningful annotations from these sources should be collected and cross-checked by examining available experimental evidence. Searching Pfam with the Q4S1I2 sequence returns no recognized domains at accepted thresholds, but in the NCBI CDD database search, the sulfotransferase domain is recognized in the C-terminal part of the protein.
3. BLAST the protein sequence of interest against major protein sequence databases (UniProtKB, PIR iProClass, and NCBI nr). At the NCBI BLAST web page (http://www.ncbi.nlm.nih.gov/BLAST/), choose "protein BLAST," which brings you to the search page http://www.ncbi.nlm.nih.gov/BLAST/Blast.cgi?PAGE=Proteins&PROGRAM=blastp&BLAST_PROGRAMS=blastp&PAGE_TYPE=BlastSearch&SHOW_DEFAULTS=on). Paste the accession number, gi, or sequence into the BLAST search page and click "BLAST." Similarly, paste the accession number or sequence into the PIR BLAST page (http:// pir.georgetown.edu/pirwww/search/blast.shtml) and UniProt BLAST page. Analyze BLAST outputs and examine protein records found therein. Note that the BLAST results are likely to be nonuniform across the three major databases as the sequences present in the databases and the search parameters may vary. Find the most relevant experimental paper(s) associated with one or more of the protein records brought up by BLAST, with the preference for experimental evidence associated with highly similar sequences. Keep in mind that the annotation should include the biochemical/biological function whenever possible.

   3.1. Select the most similar sequence with experimental data in the protein record as the first choice. This means the hit with the best e-value/score (topmost hit) out of those with good experimental data. The best hit with some information is Q8IZU8/DSEL_HUMAN, annotated as "NCAG1 protein." However, the associated paper ***(10)*** describes only the relationship of this protein to a medical condition (chromosome 18q-linked bipolar disorder) and not its biochemical or biological function. Therefore, it is necessary to find another protein with experimental data.

3.2. The best hit with biochemical data is the human dermatan-sulfate epimerase precursor, EC 5.1.3.19 (Q9UL01, DSE_HUMAN, NP_037484, gi|7019521) (synonyms: DS epimerase, chondroitin-glucuronate C5-epimerase, SART2). The associated reference ***(11)*** describes the characterization of Q9UL01 (SART2) as a chondroitin-glucuronate C5-epimerase. Given the close similarity of this protein to Q4S1I2, it would be reasonable to predict that the protein Q4S1I2 also has chondroitin-glucuronate C5-epimerase activity. Note that the similarity between Q4S1I2 and Q9UL01 is in the N-terminal part of Q4S1I2 (see the graphic display column in PIR BLAST results). Therefore, the epimerase function can be assigned to the N-terminal domain of Q4S1I2.

4. Analyze the graphic display in the PIR BLAST output. Note that the protein has the C-terminal and N-terminal parts. These show similarity to different sets of proteins (see the graphic display column in PIR BLAST results). This confirms results of the initial domain analysis (*step 2*). Importantly, in the absence of known domains defined by the domain databases, this observation would be the key to determining that the protein contains N-terminal and C-terminal domains.
5. Predict a possible function and determine the best annotation for the second (C-terminal) part of the protein. Note that many proteins that are recognized as similar to the C-terminal domain of Q4S1I2 are annotated as (predicted) sulfotransferases. Find a protein record with an associated reference providing experimental data to support the sulfotransferase annotation (in PIR BLAST results, go to page 2): the human carbohydrate sulfotransferase 3 (synonym: chondroitin 6-sulfotransferase, UniProt accession no. Q7LGC8). The associated experimental paper ***(12)*** demonstrates that human carbohydrate sulfotransferase 3 catalyzes the transfer of a sulfate group from the sulfate donor PAPS (3′-phosphoadenosine-5′-phosphosulfate) exclusively to the C6 position GalNAc nonsulfated disaccharide unit of chondroitin. Note that even if the sulfotransferase domain was not recognized by domain databases, the homology of the C-terminal domain of Q4S1I2 to the experimentally studied sulfotransferases could have been established.
6. *Conclusion.* Therefore, it may be tentatively predicted that Q4S1I2 has dual epimerase and *O*-sulfotransferaseactivities involved in dermatan sulfate biosynthesis. A tentative name for this protein may be "predicted bifunctional chondroitin-glucuronate C5-epimerase/chondroitin sulfate sulfotransferase."
7. It has to be noted that in some cases, proceeding directly to search against full-length protein family databases (such as PIRSF) with the sequence of interest can save time. Thus, searching iProClass (http://pir.georgetown.edu/pirwww/search/textsearch.shtml) or PIRSF (http://pir.georgetown.edu/pirwww/dbinfo/pirsf.shtml) with Q4S1I2 shows that this sequence has been classified as a member of PIRSF038202 (bifunctional chondroitin-glucuronate C5-epimerase/chondroitin sulfate sulfotransferase).

By clicking on the PIRSF family id, the family report page is retrieved, which shows the curated status of this family and provides integrated information from various sources.

3.3. Advanced Domain Analysis

In cases when no domain is recognized in the sequence of interest by searching domain databases, it is still possible that the sequence is distantly related to, or is a divergent version of, an established domain. Such relationships can be investigated by using position-specific iterated BLAST (PSI-BLAST) ***(13)***. PSI-BLAST performs iterative searches, and sequences found in one round of search are used to build a position-specific scoring matrix (PSSM) for the next round. More information on how to use PSI-BLAST is given at http://www.ncbi.nlm.nih.gov/Education/BLASTinfo/psi1.html. Consider the same example of GSTENG00025548001 from *Tetraodon nigroviridis* (Q4S1I2), where the N-terminal region is not assigned to any domain.

1. Copy the 1000-aa N-terminal sequence fragment of Q4S1I2 (residues 1–1000) into the PSI-BLAST search page at the NCBI web site (open http://www.ncbi.nlm.nih.gov/BLAST/ and choose "protein BLAST" and click "PSI-BLAST" algorithm option). Run PSI-BLAST. Visually inspect the degree of sequence conservation by scrolling down from the highest-scoring to the lowest-scoring proteins to confirm that all the hits are genuine homologs. For the purpose of this example, it is not necessary to run PSI-BLAST until convergence (it becomes necessary if you want to define a novel domain; this will be described in **Subheadings 3.4** and **3.5**). Also note that for this particular case, this step is not necessary for immediate annotation purposes, but in some cases it may be the best method to find homologs with experimental data (if regular BLAST did not return any such experimentally studied homologs).
2. In PSI-BLAST iteration 2, a more distantly related sequence with experimental information is found, oligoalginate lyase from *Sphingomonas* sp. (gi|9501763, BAB03319.1). A more detailed literature analysis is needed here to confirm the function of this protein: the associated publication ***(14)*** refers to an accompanying (follow-up) paper describing the characterization of *Sphingomonas* oligoalginate lyase ***(15)***. Retrieve this latter paper from PubMed and verify that the *Sphingomonas* sp. protein is experimentally characterized as an oligoalginate lyase.
3. In PSI-BLAST iteration 3, the experimentally characterized heparinase (heparin lyase) from *Pedobacter heparinus* (gi|924923, AAB18277) is retrieved ***(16)***. These experimentally characterized proteins, oligoalginate lyase and heparinase, are too distant from Q4S1I2 for the functional information to be transferred directly, especially because there are close relatives with functional information. However, a distant relationship with the heparinase II/III-like domain

(PF07940) can be established since this domain is recognized in these distant homologs.

4. Analyze the taxonomic distribution the homologs of Q4S1I2 retrieved by BLAST and PSI-BLAST. The most closely related homologs have both the N-terminal (chondroitin-glucuronate C5-epimerase) and C-terminal (tentative chondroitin sulfate sulfotransferase) domains. These include fish, human, and mouse proteins. The epimerase group contains animal proteins. The distantly related homologs of this group (found by PSI-BLAST) come from prokaryotes (bacteria). This is an additional argument that it is better not to transfer the exact function from these distantly related (prokaryotic) proteins to eukaryotic proteins (substrates may be different and the biological process may be different). Furthermore, it is interesting to note that both the single-domain and the fusion (with sulfotransferase) forms are present in fish and in mammals, thus not being very recent.

3.4. Annotation Based on Domain Analysis: A Case of Signal Transduction Proteins

3.4.1. Overview

The standard approach of assigning protein function based on the function of its closest experimentally characterized homolog is not always applicable directly. Consider the case of signal transduction components that have modular structure with various combinations of signaling domains. Due to their complex domain composition, signaling proteins are often misannotated or underannotated only as "conserved domain proteins." Because most protein annotations these days are made in an automated high-throughput fashion, it would be unrealistic to put too much trust into these annotations, especially when planning long-term experimental research. For many experimentally uncharacterized proteins, an imprecise annotation, such as "response regulator, OmpR type" (http://pir.georgetown.edu/cgi-bin/ipcSF?id=PIRSF003173) in the PIRSF protein family classification system *(17)* or "COG0745: Response regulators consisting of a CheY-like receiver domain and a winged-helix DNA-binding domain" in the COG database *(18)*, would actually be far more accurate than a more specific, but likely erroneous, functional assignment.

As a starting point in sequence analysis of a putative signal transduction protein, we recommend comparing it against several domain databases that are listed in **Table 1**. Each of these databases uses its own search tool, so the results are likely to be nonuniform, both in terms of domain recognition and in terms of domain boundaries for the same domain. A careful analysis of all meaningful annotations from these different sources, taking into account the similarity scores, the underlying experimental evidence, and the available references, is the best way to avoid costly mistakes. Here we provide an example of such analysis.

3.4.2. Identification of Lactobacillus reuteri lr0709 as a Response Regulator

Response regulators of the microbial two-component signal transduction systems typically consist of an N-terminal CheY-like receiver domain and a C-terminal output (usually DNA-binding) domain ***(19)***. By definition, almost any protein containing the receiver (REC) domain can be considered a response regulator. Exceptions include hybrid histidine kinases that contain a C-terminal REC domain and other multidomain signal transduction proteins that combine the REC domain with various sensory and/or output domains [*see*, for example, **Fig. 2** in **ref.** ***(19)***]. The relatively high sequence conservation of the REC domain makes its identification relatively straightforward. Comparing the protein in question against any of the domain databases, such as CDD, Pfam, SMART, InterPro, or ProDom, using their default parameters, is usually sufficient to find out whether this protein contains the REC domain and a recognized output domain [*see* **ref.** ***(19)*** for a listing of response regulator output domains].

Consider the following example of DNA-binding response regulator YcbB (GlnL), which controls the glutamine utilization operon ***(20)***.

1. Retrieve the sequence of *Lactobacillus reuteri* lr0709 from UniProt (Q38KD7_LACRE, accession no. Q38KD7) or the NCBI protein database (accession no. ABB02548 GI:77745295). Inspect the annotation of each of these databases. Note that this protein is uniformly annotated as "Response regulator."
2. Inspect the domain architecture of lr0709 as represented in Pfam. To do that, go to the Pfam search page at http://www.sanger.ac.uk/Software/Pfam/search.shtml and enter the UniProt name, accession number, or the protein sequence in the appropriate windows. Results will show at http://www.sanger.ac.uk/cgi-bin/Pfam/swisspfamget.pl?name=Q38KD7. Also, look at the domain representation in CDD (click the "Conserved Domains" link from the NCBI protein entry or go to http://www.ncbi.nlm.nih.gov/Structure/cdd/cdd.shtml) and then enter the NCBI accession number, gi number, or sequence. Results will also show at http://www.ncbi.nlm.nih.gov/Structure/cdd/wrpsb.cgi?INPUT_TYPE=precalc&SEQUENCE=77745295. Both CDD and Pfam recognize in the lr0709 sequence an N-terminal REC domain with convincing similarity scores. Furthermore, in the C-terminal region of the protein, the YcbB domain is recognized. Note that this domain has been recently defined by Pfam, and prior to that, the same analysis would have shown only the N-terminal REC domain.

3.4.3. Identification of the Output Domain of YcbB

In some cases, comparing a sequence of a response regulator against protein domain databases shows that the REC domain is the only one recognized in the given sequence and that it occupies only a certain part of the protein, leaving 50 or more amino acid residues not assigned to any domain. If these unassigned regions are conserved, they could well belong to as yet unrecognized protein domains. It may be possible to identify such domains by using the procedure outlined below. The following example continues the sequence analysis of the *Lactobacillus reuteri* lr0709 (YcbB/GlnL). The same steps used here to define the YcbB-like output domain (PF08664) can be performed to identify a novel domain not yet found in any database.

1. Copy the 159-aa C-terminal sequence fragment of lr0709 (residues 120–279) into the PSI-BLAST search page at the NCBI web site, http://www.ncbi.nlm.nih.gov/BLAST/ (choose "protein BLAST" and click the "PSI-BLAST" algorithm option). Run PSI-BLAST until convergence using the default parameters on the web page. The search should converge after three or four iterations, resulting in a list of ~35 database hits. Visually inspect the degree of sequence conservation by scrolling down from the highest-scoring to the lowest-scoring proteins to confirm that all the hits are genuine homologs.
2. By pressing the "Taxonomy reports" link on top of the BLAST output, generate and save the listing of database hits, sorted by their taxonomic representation. You will see that all high-scoring hits (bitscore of >133, which corresponds to the expectation value $E < 2 \times 10^{-30}$) belong to the *Firmicutes* (low G + C gram-positive bacteria).
3. To extract more information on the function and nomenclature of YcbB(GlnL), search PubMed and protein databases with the gene and/or protein names (YcbB/GlnL) and any key words that may come up in the related publications. Note that the currently used nomenclature is confusing, as YcbB (renamed GlnL) is unrelated to the *Escherichia coli* GlnL (NtrC) response regulator. In fact, the C-terminal domain of YcbB does not show an obvious relationship to any previously described HTH domains ***(19)***. Since DNA binding by YcbB has now been experimentally demonstrated ***(20)***, its C-terminal domain can be considered a new type of the DNA-binding HTH domain ***(19)***.

3.5. Sequence Analysis: Beyond High Similarity

The first step in functional annotation of a "hypothetical protein" is identification of characterized homologous proteins. As described above, potential homologous proteins can be retrieved by using BLAST. If no characterized proteins are retrieved by BLAST, the next step is to perform the more sensitive PSI-BLAST as well as sequence alignments and motif/pattern searches. The following example demonstrates the use of these methods for predicting the function and annotating the YjcG protein from *Bacillus subtilis* (O31629, O31629_BACSU).

3.5.1. Alignments (Pairwise and Multiple)

Pairwise alignment allows comparison of two sequences to identify regions of similarity and thereby determine if the two sequences are related to each other. Global sequence alignments try to align two sequences along their entire length whereas local sequence alignments try to find regions within two sequences that are most similar. Global sequence alignments perform best when two sequences of similar length share a high level of sequence similarity. Local sequence alignment performs better in identifying important biologically relevant regions when comparing proteins that have a low level of sequence similarity and different domain architectures.

When the sequence identity is high (for example >50%), good quality pairwise alignment can be done by using BLAST (Blast2Seq local alignment between two sequences: http://www. ncbi.nlm.nih.gov/blast/bl2seq/wblast2.cgi) or SSearch ***(21)*** (Smith-Waterman global alignment between two sequences: http://pir.georgetown.edu/pirwww/search/pairwise.shtml) (a multiple sequence alignment program such as CLUSTAL can also be used to do pairwise alignment).

Using pairwise alignments can sometimes lead to wrong conclusions because it might not be clear if two residues that are lined up in the alignment are really conserved or are aligned just by chance. In the pairwise alignment shown in **Fig. 1**, it seems that a, c, d, and e are conserved between the two sequences. When a multiple sequence alignment with divergent homologous sequences is performed, it is easy to identify c and d as the only two residues that are conserved. Multiple sequence alignment of distantly related proteins (protein that have diverged a long time ago) allows identification of residues that are important to structure and/or function (the argument being that if they are not important they would not be conserved).

A multiple sequence alignment of homologous sequences can be used to create a profile that is then used to identify additional homologs with low similarity to the query sequences. This method (using the PSI-BLAST algorithm in an iterative fashion) was used to identify and functionally annotate several new families of phosphoesterases in viruses, bacteria, plants, and animals ***(22)***.

1. Retrieve YjcG protein (O31629) from www.pir.georgetown.edu. In the iProClass entry for this protein (http://pir.georgetown.edu/cgi-bin/ipcEntry?id=O31629), click on "Related Sequences." Related sequences are precomputed BLAST results.
2. From the related sequences page select approximately 10 divergent sequences (you need to go to pages 2 and 3) that are of similar length (for pairwise alignment,

Pairwise alignment

```
–a c a d e
–a c _ d e
```

Multiple sequence alignment

```
–a c a d e
–a c _ d e
–d c a d f
```

Fig. 1. Protein sequence alignment showing that multiple sequence alignment helps detect functionally important residues.

select two sequences). Use the following sequences for multiple sequence alignment: Q2JKW4, Q8YXP6, Q4BWT4, Q63A50, Q2YWW8, Q6GAR2, Q5WH79, Q73BS0, O31629, and Q8ERS1.

3. Click on multiple alignment at the top right-hand corner to align the selected sequences. PIR Alignment viewer will appear in a separate window (**Fig. 2**).
4. Inspect the resulting alignment. There are two highly conserved motifs PHhTh (where h designates a hydrophobic residue) separated by approximately 100 residues. Note that additional analysis using PSI-BLAST (described below) will show that P is not always conserved.

3.5.2. Use of Sensitive Database Search (PSI-BLAST) to Retrieve Homologs with Low Similarity

Once you have an alignment, it is important to use this alignment to retrieve additional homologs. One way to do this is to use PSI-BLAST, which automatically constructs alignments at every iteration. Because of the sensitive nature

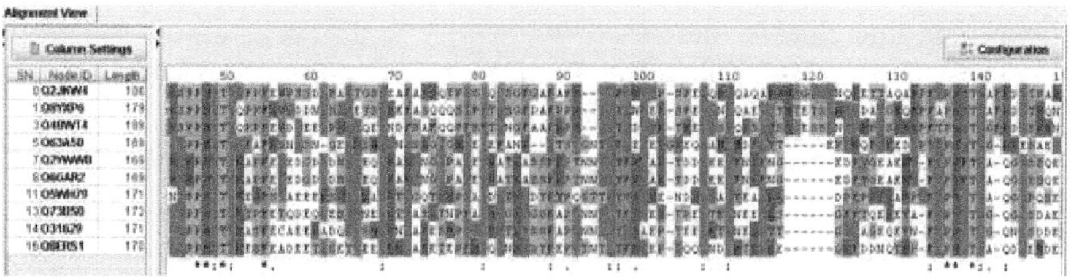

Fig. 2. Multiple sequence alignment of the proteins related to O31629_BACSU allows identification of the conserved motifs.

of PSI-BLAST, it is possible that some of the proteins retrieved are not homologous to the initial query sequence. Therefore, it is extremely important that the retrieved sequences are checked further to confirm that they are indeed homologous to the query sequence. This can be achieved by evaluating pairwise and multiple sequence alignments of the query sequence and subject sequence retrieved in the PSI-BLAST process.

The O31629_BACSU protein is used here to illustrate how PSI-BLAST can be utilized to identify homologs that have substantiated functional annotation. In the previous section it was shown that there are several bacterial proteins related to the YjcG protein (RefSeq accession NP_389067.1; UniProtKB accession O31629; PDB id 2D4G). Although many of the proteins retrieved by BLAST are annotated as "2′-5′-RNA ligase," none of them has publications associated with them providing the experimental evidence to support this annotation. Because annotation mistakes are possible in the databases, where one erroneous annotation is propagated to the entire group of proteins, such cases should be treated with care and references to experimental evidence for the annotation should be found. Note that the alignment analysis described in the previous section showed that homologs of the YjcG protein have the PHhTh motifs. This information is important in evaluating PSI-BLAST results. The following procedure demonstrates the use of PSI-BLAST to identify homologs of O31629_BACSU that have been experimentally characterized.

3.5.3. Annotation of YjcG Protein from Bacillus subtilis

1. In the PSI-BLAST search page at the NCBI web site, choose "protein BLAST" and type in NP_389067.1 (hypothetical protein BSU11850, which is the same as YjcG protein, O31629_BACSU) in the Enter Query Sequence box. Check the PSI-BLAST algorithm option and click on BLAST.
2. In the first iteration, the results show the distribution of 83 BLAST Hits on the Query Sequence. The majority of the proteins are annotated as 2'-5'-RNA ligase and several proteins are named "hypothetical proteins." As mentioned earlier, a closer examination of the retrieved proteins shows that experimental evidence is not present to substantiate this function.
3. Go carefully through the list of BLAST hits above the default threshold and make sure they have the motif: presence of two conserved Hh[ST]h motifs (h, a hydrophobic residue) separated by approximately 75–100 residues (**Fig. 3**). Although in the multiple sequence alignment (**Subheading 3.5.1**) there was a P conserved, you will find that the P is not conserved in the divergent sequences.
4. In the taxonomic distribution report (click on Taxonomy reports) you will find that the related sequences (above the bit score of 43.5; this is the lowest bit score from your BLAST threshold) are all bacteria.

```
>ref|YP_589380.1| hypothetical protein Acid345_0301 [Acidobacteria bacterium Ellin345]
 gb|ABF39306.1| hypothetical protein Acid345_0301 [Acidobacteria bacterium Ellin345]
Length=184

 Score = 46.6 bits (109),  Expect = 5e-04, Method: Composition-based stats.
 Identities = 29/122 (23%), Positives = 59/122 (48%), Gaps = 4/122 (3%)

Query  1    MKYGIVLFPSKKLQDLANSYRKRYDPSYSLIPPHLTLRASFECAEEKADQLVSHLRNIAK  60
            ++Y +V +    +       R+   P ++ +  H+T+         +AD L    R
Sbjct  7    LQYALVAYVRSPIGIFVEQLRREIHPEHAHLAAHVTVLPPRPLRGTEADALAEVQR--MA  64

Query  61   ESHP-LVLKMTKYSSFAPVNNVIYIK-AEPTEELKTLNEKLYTGVLAGEQEYNFVPHVTV  118
            E+HP   + M +  +FAP    I+I+ A     ++ L++ L TGVL  ++ + ++PH+T+
Sbjct  65   EAHPAFEIGMGEVETFAPSTPTIFIRVARAAYRMRELHDALNTGVLQYDEPWPYMPHMTI  124

Query  119  GQ  120
             +
Sbjct  125  AK  126
```

Fig. 3. BLAST alignment showing the conservation of the 2H motif in the retrieved protein.

5. Run PSI-BLAST iteration 2 by clicking on the iteration button. This will create a profile from the first BLAST alignment and search the database with the profile and retrieve additional homologs. Make sure that the conserved motifs are present in all new (NEW yellow tags in the result page) proteins retrieved in the second iteration of BLAST.
6. In the second iteration, proteins that belong to *Archaea* and have the motif identified previously based on the multiple alignment will be retrieved (*Pyrococcus horikoshii*, gi—71041774). The *Pyrococcus horikoshii* protein has a publication associated with it ***(23)*** that describes the function and the active site of the protein. These are all predicted distant homologs. Further analysis in the next section will show how additional evidence can be gathered using structural data to evaluate the relationship of the query protein to the subjects.
7. Further iterations reveal that there are more eukaryotic sequences related to this protein by virtue of having the 2H domain. You can continue with the iterations until no more new sequences are found. Root out false positives (sequences without the 2H motif) in each iteration. After four iterations you will get the following message: "Results of PSI-Blast iteration 4. No new sequences were found above the 0.005 threshold!"
8. *Conclusion*: it can be predicted that YjcG has 2′-5′-RNA ligase activity, although the physiological role of this protein is unknown since bacteria do not require this activity ***(24)***. Family classification and additional analysis (not described in this chapter) reveal that members of the YjcG family do not occur in conserved operons implicative of RNA metabolism, with the possible exception of the *Streptomyces* gene SC5G8.08, which is a gene neighbor of the tryptophanyl tRNA synthetase. Furthermore, a spatial plot of the residues, uniquely conserved in the YjcG family, does not show any extensive interaction surface associated either with the face bearing the catalytic cavity or elsewhere. This suggests that the YjcG proteins are likely to function as stand-alone proteins on as yet unknown soluble small molecules with potential 2′,3′-cyclic phosphoester linkages ***(22)***.

3.5.4. Pattern Search

Another way to identify additional homologous sequences is to do a pattern search. Based on the multiple sequence alignment, in all the proteins that are homologous to the YjcG protein (UniProtKB accession no. O31629_BACSU) the following pattern can be identified: H-[AFILMVWY]-[ST]-[AFILMVWY]-x(80,90)-H-[AFILMVWY]-[ST]-[AFILMVWY]. This pattern can be used to scan protein databases to retrieve potentially related proteins. Patterns that are not specific can result in false positives. Therefore it is important to further evaluate retrieved sequences using BLAST and/or PSI-BLAST to ensure that the motifs are indeed conserved in the retrieved sequences. For example, if the sequence has the same pattern but upon performing BLAST it is found that the motif is not conserved even among closely related sequences, it is evident that the protein is not a homolog of O31629_BACSU.

1. In the PIR Pattren Search web page (http://pir.georgetown.edu/pirwww/search/pattern.shtml), select taxon group, then select *Archaea*. This will make it possible to search for proteins with a specific pattern in all archaeal proteins.
2. Writing a pattern: use capital letters for amino acid residues and put a "-" between two amino acids (not required); use "[. . .]" for a choice of multiple amino acids in a particular position; [LIVM] means that L, I, V, or M can be in the first position; use "x" for a position that can be any amino acid; and use "(n1,n2)" for multiple or variable positions; "x (1,4)" represents "x" or "xx" or "xxx" or "xxxx."
3. Search results show that some of the retrieved proteins are indeed not homologous. For example, the protein "probable deoxyhypusine synthase" has the motif, but on performing BLAST it can be seen that the Hh[ST]h motifs are not conserved.

3.6. Using Structural Information for Functional Prediction and Annotation of Functional Sites

3.6.1. Overview

Function predictions based on sequence similarity alone work well for sequences that have high sequence identities (>50%) to a well-characterized protein. This may begin to fail for sequences that do not have any characterized homologs within this identity range. In such cases, it is often necessary to examine distant homologs that are related only at the three-dimensional structural level rather than the sequence level alone. This is not surprising since molecular evolution retains and conserves structure longer than sequence. In such cases a combined approach using structure–sequence data is crucial in accurately defining biological function and hence its annotation. A classic example that illustrates this is the diverse superfamily of 2H phosphoesterases discussed in **Subheading 3.5**. 2′,3′-Cyclic nucleotide phosphodiesterases are enzymes that catalyze at least two distinct steps in the splicing of tRNA introns in eukaryotes. The biochemistry and structure of these enzymes from various organisms have been extensively studied. They were found to share a common active site, characterized by two conserved histidines, hence the name 2H phosphoesterase superfamily. A hallmark of the 2H superfamily is extreme sequence divergence despite the conservation of the active site motifs. This presents a challenge for their identification via classical sequence analysis and calls for a combination of structure and sequence analysis methods.

This section will present an example of a structure-based position-specific systematic approach that will enable the identification of structural members of the 2H family. In addition, annotation of structural sites (active/binding sites) and the propagation of this site annotation to other members of 2H superfamily will be demonstrated using a structure-based sequence alignment. The approach that uses three-dimensional structural information can aid in function prediction for other hypothetical proteins whose functional identification fails the traditional sequence analysis.

PDB-ID 1JH7 ***(25)***, a 2′,3′-cyclic nucleotide phosphodiesterase from *Arabidopsis thaliana* that was identified as a hit below threshold by using BLAST, will be used as a starting point. Using this example, we will demonstrate how further structural information can aid in the functional annotation of some superfamily members such as O49408 and Q75II2 that are currently annotated as hypothetical.

3.6.2. Structure-Based Prediction, Functional Annotation, and Propagation of This Information to Sequence(s) of Unknown Function

1. Identification of structure neighbors of 1JH7 using the VAST algorithm ***(26)***.
   1.1. In the NCBI structure web page (http://www.ncbi.nlm.nih.gov/Structure/), enter 1JH7 into the Structure Summary box and click "go."
   1.2. Click on the pink bar labeled "Chain A" to get its structure neighbors.
   1.3. Results of the search will be displayed in a graphic form. For convenience and ease, a table is recommended. This can be obtained by using the pull-down menu options.
2. Structure-based sequence alignment using Cn3d ***(27)***. As mentioned earlier, since molecular evolution conserves three-dimensional structure, structure-based sequence alignments provide information not amenable from sequence-based methods alone. In this family of 2H, while the sequence identity is below 20%, structure–structure comparisons and alignments alone have led us to the identification of other members of this diverse family (**Fig. 4**, *see* Color Plate 3) shows the superimposition of five structures that belong to this family. Note that the sequence conservation is poor. However, the residues with the pattern HxH (highlighted in yellow), which is part of the functional site, are conserved, making it possible to use this site information.

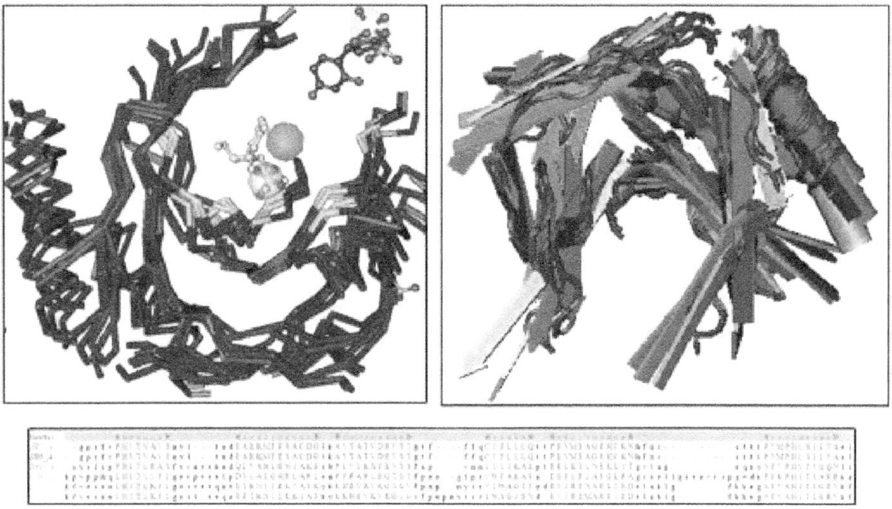

Fig. 4. Structure-guided alignment and superposition using Cn3d showing the conserved regions and conserved binding residues. (*See* Color Plate 3)

3. Identification of ligand-binding residues using LIGPLOT in PDBSum *(28)*. The structure 1JH7 is bound to its inhibitor uridine-2′,3′-vanadate. The residue level interaction of this inhibitor (identical to the substrate-binding site) with 2H can be identified as follows.
   3.1. In the PDBsum web page (http://www.ebi.ac.uk/thornton-srv/databases/pdbsum/) type 1JH7 into the PDB code box and click "Find."
   3.2. Inspect retrieved structural information. To obtain the ligand interactions, click on ligand code UVC (uridine-2′,3′-vanadate) on the left-hand side under Ligands.
   3.3. Click on the PDF file (**Fig. 5**, *see* Color Plate 4) gives the atomic-level interactions that include H-bonds (green dashed lines) and van der Waals interactions shown as half-circles.
4. Creation of site-specific HMM. The residues in 1JH7 making H-bond interactions with the inhibitor are Thr-163, Tyr-124, Ser-10, His-42, Trp-12, Thr-44, and Ser-121 as seen from **Fig. 5**. Program HMMER *(29)* is used to create HMMs from the conserved regions containing the functional site residues.
5. Propagation of annotation. The profile HMM thus built based on conserved regions makes it possible to map functionally important residues from the template structure to other members of the 2H family that do not have a solved structure. To avoid false positives, site features should be propagated automatically only if all site residues match perfectly in the conserved region by aligning both the template and target sequences to the profile HMM using HmmAlign [which is part of the HMMER package *(29)*]. Potential functional sites missing one or

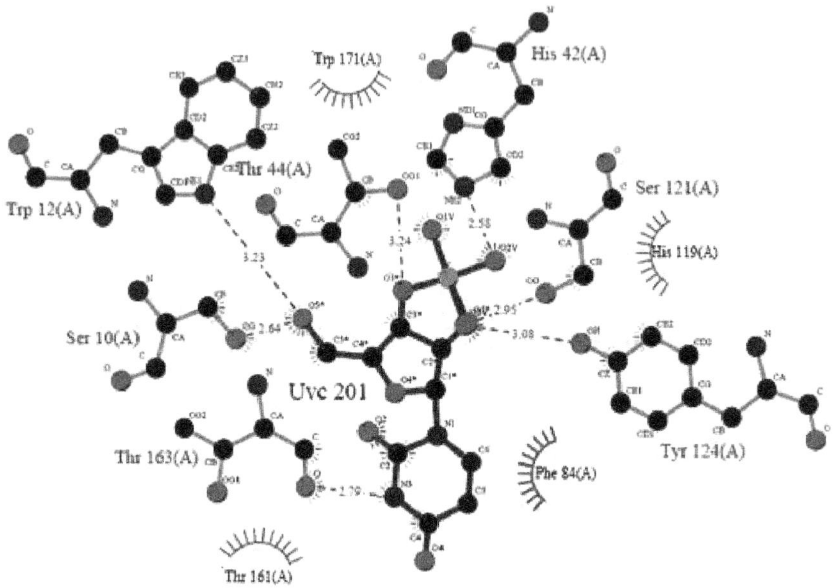

Fig. 5. Protein–ligand interactions using Ligplot showing the residues involved in binding to the ligand uridine-2′,3′-vanadate. (*See* Color Plate 4)

more residues should be annotated after expert review. In the case of 2H binding residues, annotation will be propagated only to residues His-42, Thr-44, and Ser-121 since these are the only conserved binding residues. This information can be used to identify ligand-binding residues in the family of sequences that still lacks a crystal structure. This example clearly demonstrates how a combination of sequence and structure data can be used for functional prediction and annotation.

3.7. Large-Scale Annotation

The advances in large-scale and high-throughput experimental technologies have led to the gap between available data and the ability to rapidly, accurately, and meaningfully interpret them. Sequence database resources involved in annotating protein sequences have the obvious problem of quality versus quantity, especially with respect to accurate assignment of known or predicted functions (functional annotation).

In many cases, large-scale functional annotation is based simply on BLAST best hits and is done via an automatic or semiautomatic process that carries with it many pitfalls and thus produces results that are far from perfect (*see* **Subheading 4**). Database annotation errors (often reflecting under- or overpredictions or misannotations) affect any data analysis and computational tools that rely on these annotations. To avoid annotation mistakes, human intervention (manual annotation) is needed, but it is costly and labor intensive.

Classification of proteins is widely accepted to provide valuable clues to structure, function, and evolution. Protein family classification has several advantages over traditional "genome-by-genome" or "protein-by-protein" annotation as a basic approach for large-scale annotation: (1) it improves the annotation of proteins that are difficult to characterize based on pairwise alignments since comparing a protein sequence against a family database is much more sensitive than any pairwise comparisons, (2) it assists database maintenance by promoting family-based propagation of annotation and making annotation errors apparent, (3) it provides an effective means to retrieve relevant biological information from vast amounts of data, and (4) it reflects the underlying gene families, the analysis of which is essential for comparative genomics and phylogenetics.

Employing well-curated protein families for the purpose of finding functional equivalents is a well-established approach. To be effective as a practical solution for large-scale annotation, the protein classification system should classify full-length proteins, be highly curated and annotated, provide functional predictions for uncharacterized proteins and protein families, and allow for the automatic annotation of sequences based on existing protein families. For example, the fully curated family subset of the PIRSF system is optimized for annotation propagation by being coupled with the PIR name rules and site rules for accurate and consistent transfer of annotations from the corresponding PIRSF families and subfamilies ***(30)***. PIRSF classification is used to facilitate and standardize annotations in UniProt ***(31)***.

4. Notes on Sources of Annotation Errors

A general approach for functional annotation of uncharacterized proteins is to infer protein functions based on sequence similarity to annotated proteins in sequence databases. While this is a powerful method, it may result in overprediction, underprediction, or even misannotation. Numerous genome annotation errors have been detected, many of which have been propagated throughout molecular databases. There are several sources of errors:

1. Misinterpreted experimental results (e.g., suppressors or cofactors annotated as enzymes).
2. Biologically senseless annotations arising from transfer of annotation from one major biological taxon to another without considering if function is still plausible, in cases when orthologs between the two taxons exhibit functional divergence. Examples include protein names such as "separation anxiety protein" in *Arabidopsis* and "centromere-binding protein" in *Methanococcus*.

3. Information transfer mistakes, such as substituting "abc1" for "ABC" because the latter name is found in a closely related organism without verifying that the proteins are indeed related, or truncated annotations arising from character number restrictions that lead to senseless or misleading annotations, are quite widespread. Other senseless annotations include examples such as a protein name "frameshift."
4. Low complexity sequences (coiled-coil, transmembrane, nonglobular regions) generate many spurious hits in regular BLAST searches, and therefore are prone to be misannotated on the basis of these hits.
5. Errors often occur when identification is made based on local domain similarity or similarity involving only parts of the query and target molecules. Moreover, the similarity may be to a known domain that is tangential to the main function of the protein or to a region with compositional similarity, such as transmembrane domains. Furthermore, specific biological functions can seldom be inferred solely from the generic functions of the constituent domains, and proteins with different biological functions may have a similar domain organization.
6. Special cases of enzyme evolution:
   6.1. Rapid divergence in sequence and function when minor mutations in active sites change the exact biochemical function but may fail to be detected by a simple BLAST search.
   6.2. Nonorthologous gene displacement or convergent evolution, when two groups of enzymes with the same activity have unrelated sequences and structures ***(32)***.
7. Numerous paralogous proteins within the same organism. An example is P450, a protein family greatly expanded in plants. The *Arabidopsis* genome contains up to 246 putatively functional genes for cytochrome P450. The numerous various reactions catalyzed in plants by P450 are mostly unknown, with detailed information existing for about 30 reactions in different plant species ***(33)***.
8. Errors also occur when the best hit entry is an uncharacterized or poorly annotated protein, or is itself incorrectly predicted, or simply has a different function. Aside from erroneous annotation, database entries may be underannotated, such as a "hypothetical protein" with a convincing similarity to a protein or domain of known function, and may be overidentified, such as ascribing a specific enzyme activity when a less specific one would be more appropriate.
9. Importantly, previous low-quality annotations lead to propagation of mistakes and sometimes generate families of related proteins with identical but erroneous annotations.

As a final word of caution on using database annotations, it has to be stressed that even the best annotation methods, when applied on a large-scale basis, are bound to produce some mistakes, delays in incorporating new evidence, and partial annotations. The quality of the annotation may vary from genome to genome and from database to database. Therefore, the importance of verification of functional annotations when using them as a basis for analysis, research, or making inferences can not be overestimated.

5. Conclusions

Annotating the ever-expanding protein universe is a daunting task. Can functional annotation be fully automated? The first answer is no. There are steps in this process that require an expert review and judgment: evaluating and applying new experimental evidence, considering the whole protein and its domain components, and finding distantly related characterized homologs. Most importantly, the process involves selecting the proteins to which a particular annotation can be propagated as well as the proteins that need a different, even if related, annotation. Thus, the goal should be semiautomatic annotation, where well-described cases with established functional annotations are covered by protein families. The families, before becoming "trivial cases," should undergo a rigorous process of expert curation and annotation, and thereafter new sequences that fall into these families should be annotated semiautomatically. However, the constant flow of new experimental data on previously uncharacterized or partially characterized families will always require expert analysis and annotation by a human aware of the state of the art ***(34–47)***.

Acknowledgment

This work is supported by the UniProt grant 2 U01 HG02712-04 from the National Institutes of Health.

References

1. Altschul, S. F., Gish, W., Miller, W., Myers, E. W., and Lipman, D.J. (1990) Basic local alignment search tool. *J. Mol. Biol.* **215**, 403–410.
2. Dayhoff, M. O. (1976) The origin and evolution of protein superfamilies. *Fed. Proc.* **35,** 2132–2138.
3. Gribskov, M., McLachlan, A. D., and Eisenberg, D. (1987) Profile analysis: detection of distantly related proteins. *Proc. Natl. Acad. Sci. USA* **84**, 4355–4358.
4. Eddy, S. R., Mitchison, G., and Durbin, R. (1995) Maximum discrimination hidden Markov models of sequence consensus. *J. Comput. Biol.* **2,** 9–23.
5. Galperin, M. Y. (2007) The Molecular Biology Database Collection: 2007 update. *Nucleic Acids Res.* **35,** D3–4.
6. Benson, D. A., Karsch-Mizrachi, I., Lipman, D. J., Ostell, J., and Wheeler, D. L. (2007) GenBank. *Nucleic Acids Res.* **35**, D21–25.
7. The UniProt Consortium. The Universal Protein Resource (UniProt). (2007) *Nucleic Acids Res.* **35**, D193–197.
8. Pruitt, K. D., Tatusova, T., and Maglott, D. R. (2007) NCBI reference sequences (RefSeq): a curated non-redundant sequence database of genomes, transcripts and proteins. *Nucleic Acids Res.* **35**, D61–65.

9. Jaillon, O., Aury, J. M., Brunet, F., Petit, J. L., Stange-Thomann, N., Mauceli, E., Bouneau, L., Fischer, C., Ozouf-Costaz, C., Bernot, A., *et al.* (2004) Genome duplication in the teleost fish Tetraodon nigroviridis reveals the early vertebrate proto-karyotype. *Nature* **431**, 946–957.
10. Goossens, D., Van Gestel, S., Claes, S., De Rijk, P., Souery, D., Massat, I., Van den Bossche, D., Backhovens, H., Mendlewicz, J., Van Broeckhoven, C., and Del-Favero, J. (2003) A novel CpG-associated brain-expressed candidate gene for chromosome 18q-linked bipolar disorder. *Mol. Psychiatry* **8**, 83–89.
11. Maccarana, M., Olander, B., Malmstrom, J., Tiedemann, K., Aebersold, R., Lindahl, U., Li, J. P., and Malmstrom, A. (2006) Biosynthesis of dermatan sulfate: chondroitin-glucuronate C5-epimerase is identical to SART2. *J. Biol. Chem.* **281**, 11560–11568.
12. Tsutsumi, K., Shimakawa, H., Kitagawa, H., and Sugahara, K. (1998) Functional expression and genomic structure of human chondroitin 6-sulfotransferase. *FEBS Lett.* **441**, 235–241.
13. Altschul, S. F., Madden, T. L., Schaffer, A. A., Zhang, J., Zhang, Z., Miller, W. ,and Lipman, D. J. (1997) Gapped BLAST and PSI-BLAST: a new generation of protein database search programs. *Nucleic Acids Res.* **25**, 3389–3402.
14. Momma, K., Okamoto, M., Mishima, Y., Mori, S., Hashimoto, W., and Murata, K. (2000) A novel bacterial ATP-binding cassette transporter system that allows uptake of macromolecules. *J. Bacte*riol. **182**, 3998–4004.
15. Hashimoto, W., Miyake, O., Momma, K., Kawai, S., and Murata, K. (2000) Molecular identification of oligoalginate lyase of Sphingomonas sp. strain A1 as one of the enzymes required for complete depolymerization of alginate. *J. Bacteriol.* **182**, 4572–4577.
16. Su, H., Blain, F., Musil, R. A., Zimmermann, J. J., Gu, K., and Bennett, D. C. (1996) Isolation and expression in Escherichia coli of hepB and hepC, genes coding for the glycosaminoglycan-degrading enzymes heparinase II and heparinase III, respectively, from Flavobacterium heparinum. *Appl. Environ. Microbiol.* **62**, 2723–2734.
17. Nikolskaya, A. N., Arighi, C. N., Huang, H., Barker, W. C., and Wu, C. H. (2006) PIRSF family classification system for protein functional and evolutionary analysis. *Evol. Bioinform. Online* **2**, 209–221.
18. Tatusov, R. L., Galperin, M. Y., Natale, D. A., and Koonin, E. V. (2000) The COG database: a tool for genome-scale analysis of protein functions and evolution. *Nucleic Acids Res.* **28,** 33–36.
19. Galperin, M. Y. (2006) Structural classification of bacterial response regulators: diversity of output domains and domain combinations. *J. Bacteriol.* **188**, 4169–4182.
20. Satomura, T., Shimura, D., Asai, K., Sadaie, Y., Hirooka, K., and Fujita, Y. (2005) Enhancement of glutamine utilization in Bacillus subtilis through the GlnK-GlnL two-component regulatory system. *J. Bacteriol.* **187**, 4813–4821.
21. Pearson, W. R. and Lipman D. J. (1988) Improved tools for biological sequence comparison. *Proc. Natl. Acad. Sci. USA* **85**, 2444–2448.
22. Mazumder, R., Iyer, L. M., Vasudevan, S., and Aravind, L. (2002) Detection of novel members, structure-function analysis and evolutionary classification of the 2H phosphoesterase superfamily. *Nucleic Acids Res.* **30**, 5229–5243.

23. Gao, Y. G., Yao, M., Okada, A., and Tanaka, I. (2006) The structure of Pyrococcus horikoshii 2′-5′ RNA ligase at 1.94 A resolution reveals a possible open form with a wider active-site cleft. *Acta Crystallogr. Sect. F Struct. Biol. Cryst. Commun.* **62**, 1196–1200.
24. Arn, E. A. and Abelson, J. N. (1996) The 2′-5′ RNA ligase of Escherichia coli. Purification, cloning, and genomic disruption. *J. Biol. Chem.* **271**, 31145–31153.
25. Hofmann, A., Grella, M., Botos, I., Filipowicz, W., and Wlodawer, A. (2002) Crystal structures of the semireduced and inhibitor-bound forms of cyclic nucleotide phosphodiesterase from Arabidopsis thaliana. *J. Biol. Chem.* **277**, 1419–1425.
26. Gibrat, J. F., Madej, T., and Bryant, S. H. (1996) Surprising similarities in structure comparison. *Curr. Opin. Struct. Biol.* **6**, 377–385.
27. Wang, Y., Geer, L. Y., Chappey, C., Kans, J. A., and Bryant, S. H. (2000) Cn3D: sequence and structure views for Entrez. *Trends Biochem. Sci.* **25**, 300–302.
28. Laskowski, R. A., Chistyakov, V. V., and Thornton, J. M. (2005) PDBsum more: new summaries and analyses of the known 3D structures of proteins and nucleic acids. *Nucleic Acids Res.* **33**, D266–D268.
29. Eddy S. R. (1995) Multiple alignment using hidden Markov models. *Proc. Int. Conf. Intell. Syst. Mol. Biol.* **3**, 114–120.
30. Natale, D. A., Vinayaka, C. R., and Wu, C. H. (2005) Large-scale, classification-driven, rule-based functional annotation of proteins. In *Encyclopedia of Genetics, Genomics, Proteomics and Bioinformatics. Bioinformatics Volume* (Subramaniam, S., ed.). John Wiley & Sons, Ltd, 2004.
31. Wu, C. H., Nikolskaya, A., Huang, H., Yeh, L. S., Natale, D. A., Vinayaka, C. R., Hu, Z. Z., Mazumder, R., Kumar, S., Kourtesis, P., *et al.* (2004) PIRSF: family classification system at the Protein Information Resource. *Nucleic Acids Res.* **32,** D112–D114.
32. Galperin, M. Y., Walker, D. R., and Koonin E.V. (1998) Analogous enzymes: independent inventions in enzyme evolution. *Genome Res.* **8**, 779–790.
33. Nelson, D. R., Zeldin, D. C., Hoffman, S. M., Maltais, L. J., Wain, H. M., and Nebert, D. W. (2004) Comparison of cytochrome P450 (CYP) genes from the mouse and human genomes, including nomenclature recommendations for genes, pseudogenes and alternative-splice variants. *Pharmacogenetics* **14**, 1–18.
34. Wheeler, D. L., Barrett, T., Benson, D. A., Bryant, S. H., Canese, K., Chetvernin, V., Church, D. M., DiCuccio, M., Edgar, R., Federhen, S., Geer, L. Y., Kapustin, Y., Khovayko, O., Landsman, D., Lipman, D. J ., Madden, T. L., Maglott, D. R., Ostell, J., Miller, V., Pruitt, K. D., Schuler, G. D., Sequeira, E., Sherry, S. T., Sirotkin, K., Souvorov, A., Starchenko, G., Tatusov, R. L., Tatusova, T. A., Wagner, L., and Yaschenko, E. (2007) Database resources of the National Center for Biotechnology Information. *Nucleic Acids Res.* **35,** D5–D12.
35. Suzek, B. E., Huang, H., McGarvey, P., Mazumder, R., and Wu, C. H. (2007) UniRef: comprehensive and non-redundant UniProt reference clusters. *Bioinformatics* **23**, 1282–1288.
36. Wu, C. H., Huang, H., Nikolskaya, A., Hu, Z., and Barker, W. C. (2004) The iProClass integrated database for protein functional analysis. *Comput. Biol. Chem.* **28**, 87–96.

37. Finn, R. D., Mistry, J., Schuster-Bockler, B., Griffiths-Jones, S., Hollich, V., Lassmann, T., Moxon, S., Marshall, M., Khanna, A., Durbin, R., *et al.* (2006). Pfam: clans, web tools and services. *Nucleic Acids Res.* **34,** D247–D251.
38. Letunic, I., Copley, R. R., Pils, B., Pinkert, S., Schultz, J., and Bork, P. (2006). SMART 5: domains in the context of genomes and networks. *Nucleic Acids Res.* **34,** D257–D260.
39. Servant, F., Bru, C., Carrere, S., Courcelle, E., Gouzy, J., Peyruc, D., and Kahn, D. (2002). ProDom: automated clustering of homologous domains. *Brief Bioinform.* **3,** 246–251.
40. Marchler-Bauer, A., Anderson, J. B., Derbyshire, M. K., DeWeese-Scott C., Gonzales N. R., Gwadz, M., Hao, L., He, S., Hurwitz, D. I., Jackson, J. D., Ke, Z., Krylov, D., Lanczycki, C. J., Liebert, C. A., Liu, C., Lu, F., Lu, S., Marchler, G. H., Mullokandov, M., Song, J. S., Thanki, N., Yamashita, R. A., Yin, J. J., Zhang, D., and Bryant, S. H. (2007) CDD: a conserved domain database for interactive domain family analysis. *Nucleic Acids Res.* **35,** D237–D240.
41. Hulo, N., Bairoch, A., Bulliard, V., Cerutti, L., De Castro, E., Langendijk-Genevaux, P. S., Pagni, M., and Sigrist, C. J. (2006) The PROSITE database. *Nucleic Acids Res.* **34,** D227–D230.
42. Attwood, T. K., Bradley, P., Flower, D. R., Gaulton, A., Maudling, N., Mitchell, A. L., Moulton, G., Nordle, A., Paine, K., Taylor, P., *et al.* (2003) PRINTS and its automatic supplement, prePRINTS. *Nucleic Acids Res.* **31,** 400–402.
43. Mulder, N. J., Apweiler, R., Attwood, T. K., Bairoch, A., Bateman, A., Binns, D., Bork, P., Buillard, V., Cerutti, L., Copley, R., Courcelle, E., Das, U., Daugherty, L., Dibley, M., Finn, R., Fleischmann, W., Gough, J., Haft, D., Hulo, N., Hunter, S., Kahn, D., Kanapin, A., Kejariwal, A., Labarga, A., Langendijk-Genevaux, P. S., Lonsdale, D., Lopez, R., Letunic, I., Madera, M., Maslen, J., McAnulla, C., McDowall, J., Mistry, J., Mitchell, A., Nikolskaya, A. N., Orchard, S., Orengo, C., Petryszak, R, Selengut, J. D., Sigrist, C. J. A., Thomas, P. D., Valentin, F., Wilson, D., Wu, C. H., and Yeats, C. (2007) New developments in the InterPro database. *Nucleic Acids Res.* **35**, D224–D228.
44. Berman, H. M., Westbrook, J., Feng, Z., Gilliland, G., Bhat, T.N., Weissig, H., Shindyalov, I. N., and Bourne, P.E. (2000) The Protein Data Bank. *Nucleic Acids Res.* **28,** 235–242.
45. Wang, Y., Addess, K. J., Chen, J., Geer, L. Y., He, J., He, S., Lu, S., Madej, T., Marchler-Bauer, A., Thiessen, P. A., Zhang, N., and Bryant, S. H. (2007) MMDB: annotating protein sequences with Entrez's 3D-structure database. *Nucleic Acids Res.* **35,** D298–D300.
46. Dietmann, S., Park, J., Notredame, C., Heger, A., Lappe, M., and Holm, L. (2001) A fully automatic evolutionary classification of protein folds: Dali Domain Dictionary version 3. *Nucleic Acids Res.* **29**, 55–57.
47. Andreeva, A., Howorth, D., Brenner, S. E., Hubbard, T. J., Chothia, C., and Murzin, A. G. (2004) SCOP database in 2004: refinements integrate structure and sequence family data. *Nucleic Acids Res.* **32**, D226–D229.

29

Designability and Disease

Philip Wong and Dmitrij Frishman

Summary

Structural designability is the number of ways it is possible to encode for structure. A protein's designability has been equated with the size of sequence space encoding for the protein's structure, a measure that reflects the structure's robustness to mutation. Current evidence suggests that designability is fundamental to our understanding of the evolvability and distribution of structures in nature and is a significant factor associated with human disease. Here, we describe definitions and principles underlying the concept of designability and discuss its relation to disease.

Key Words: Protein evolution; structure classification; genome analysis; disease.

1. Designability

1.1. Defining Designability

A characteristic of all life on the planet is that some level of organization exists. Living objects form structures—an ordering of components. For example, certain proteins can form well-defined three-dimensional (3D) shapes. Others exhibit disorder *(1)*, but freedom of movement remains restricted by interactions between amino acids. Many of these proteins carry out functions via nonrandom interactions with other components in the cell. Temporal and spatial ordering of proteins has also been observed throughout the cell cycle. Ordering of cellular components facilitates life by ensuring that essential reactions occur in a timely fashion.

Order in a system can be described by a set of constraints. Designability is simply the number of solutions that satisfies such constraints. Structural designability refers to the number of ways it is possible to satisfy the constraints

From: *Methods in Molecular Biology, vol. 484: Functional Proteomics: Methods and Protocols*
Edited by: J. D. Thompson et al., DOI: 10.1007/978-1-59745-398-1, © Humana Press, Totowa, NJ

defining the structure. In other words, it is the number of possible ways the structure can be created.

Creating a measure of structural designability involves defining a basic component or unit. For proteins, one natural definition of components involves amino acids, since cells literally build proteins by covalent attachment of different amino acids. A definition of designability also involves characterizing what we mean by structure. Constraints that define structures are often specified with different levels of granularity. For example, molecular structures of proteins are often described by the 3D arrangement of amino acids. The same molecule can also be described by the arrangement of coil, helix, and β-sheet regions. The former definition involves the basic component itself: the amino acid. The latter definition is less precise and different amino acid sequences may satisfy the same secondary structure constraints. Defining structure in terms of constraints that allow for different arrangements of the basic component allows for a useful definition of designability. In other words, a useful definition of designability includes the requirement that there be more than one way to specify the same structure in terms of basic components.

1.2. Fold Designability

A hierarchical classification of protein structures can be found in the manually curated SCOP (Structural Classification of Proteins) database *(2)*. In this database, domains with highly similar sequences sharing at least 30% identity are grouped into families; families sharing a relatively close common ancestor based on high structural similarity are grouped into superfamilies, and superfamilies sharing an overall structural similarity are in turn grouped into folds (**Fig. 1**). If amino acids are defined as basic components, fold designability can be defined as the number of amino acid sequences that encode a particular fold.

A highly designable fold could be formed by a large number of different amino acid sequences while a less designable fold would have fewer possible encoding sequences. Using simple models in which proteins are modeled as chains of hydrophobic and hydrophilic residues on lattices, Li *et al.* *(3)* have shown that different structures could have vastly different designabilities. Such model also suggests that proteins with more designable structures are more robust to mutation and certain external stresses *(3,4)*. This makes sense because more designable structures can be encoded by more sequences and mutations by definition create different sequences. If a fold is more likely to be maintained when an encoding sequence is mutated, then certain environmental changes that stress the structure similarly would also be less likely to alter the fold.

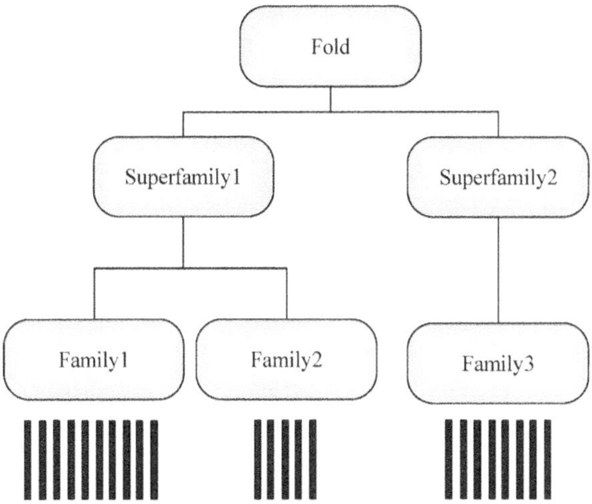

Fig. 1. SCOP hierarchy. Four levels of SCOP are shown: fold, superfamily, family, and sequence (dark rectangles). The number of sequences is equal to or greater than the number of families, which is equal to or greater than the number superfamilies, which in turn is equal to or greater than the number of folds.

It has been hypothesized that protein structures of higher designability tend to be more evolutionarily fit because such structures would allow a greater amount of sequence changes associated with a greater diversity of function ***(5)***. Evidence that more designable structures tend to be more sequence divergent and more widespread throughout proteomes has been obtained ***(6–8)*** that is consistent with the hypothesis that more designable folds should be more fit.

1.3. Estimating and Comparing Structural Designability

How can designabilities of different structures be compared? One way is to systematically perturb the structures in terms of the basic components and then observe if constraints that define the structures still hold. For example, one possibility is to systematically mutate proteins of different folds and test whether the structure is maintained after mutation. However, this task is not trivial. If 100 residues are to be systematically perturbed, there are 20^{100} combinations of sequences to test (ignoring deletions and insertions of new residues). One way to circumvent testing of all sequence combinations is to test samples of sequences for structure. However, models of how proteins behave upon mutation are only now being developed ***(9)*** and general statistical models are lacking. Another method would to be to exploit order in how structures are organized and make assumptions that simplify the space of sequences to be explored. For example, recall that SCOP is a hierarchical database in which sequences are grouped into families that share 30% identity. If the differences between the sizes of each family (the number of sequences in each family) are sufficiently small, then it is possible to compare the designability of different folds simply by comparing the number of families contained in those folds. For example, a fold containing 20 families of sequences would likely be more designable than a fold with only one family, on average. Although belief in such a hypothesis is a simplification, properties expected of more designable structures have been predicted: for example, families belonging to folds with more families tend to be more widespread in proteomes and more divergent ***(6)***.

Note that counting the number of families in a fold is an assessment of designability dependent on the diversity of families that nature has happened to evolve. Given that sequence evolution that maintains fold structure is much more frequent than evolution that creates new folds ***(10)***, the longer a particular fold exists, the greater the chance more families would have been produced with that fold. Considering known folds, older folds have significantly more families than younger ones ***(6,11)***.

It has been proposed that ancient folds are, nevertheless, more designable because they have emerged from a hot environment ***(12,13)***. Because time does affect the number of families found in folds, it is possible to either restrict comparisons between the folds of interest to those of relatively the same age or account for the age differences when estimating designability differences based on family counts.

An additional element to be considered when devising a measure of designability is the environmental conditions against which structures are tested. The environment defines additional constraints to structure exclusive to the basic components that make up the structure. For example, temperature ***(14)***, interactions with chaperones ***(15)***, proteases ***(16)***, lipids ***(17,18)***, and other protein-modifying agents ***(19)*** may influence whether a DNA sequence is eventually expressed as a functional protein. These factors are usually ignored in theoretical models concerned with only the intrinsic designability (designability measured ignoring environmental conditions) of structures, but may be important constraints to consider for practical applications. It should be emphasized that for biological systems, environmental conditions to be considered are seldom static and do fluctuate in time. An important difference between estimating designability by perturbing a structure and estimating it based on what has actually evolved in nature is that the former is carried out in artificial environmental conditions present when the structure is perturbed. Estimating designability by observing what nature has evolved (e.g., using family counts) captures the degree of success of the fold within the multitude of environments and fitness constraints experienced throughout the history of the fold; these constraints may or may not be similar to what is being measured in artificial environments.

Another example of how a set of artificial constraints may differ from what is observed in nature can be seen by examining how sequence conservation relates to structure. Structural constraints do cause amino acids to be highly conserved in distant organisms ***(20)***. However, sequence conservation in living organisms reflects fitness constraints and does not necessarily pertain to the defined level of structural constraint. For example, sequences that encode protein folds can be much more conserved than would be required by the constraints defining the fold because certain amino acids not necessary for fold formation are involved in essential reactions carried out by the protein.

1.3.1. Properties Contributing to Greater Designability

What features make one structure more designable than others? One characteristic that helps maintain structural integrity is structural modularity. Variable regions in proteins can be isolated from the rest of the protein so as not to affect overall stability when mutated or altered by the environment. For example, protein–protein binding can be mediated by binding sites that do not alter the core stability of the proteins when binding/disassociation events occur. Mutation of such sites similarly does not alter core stability ***(21)***. Certain structures are more modular than others. For example, scale-free architectures in which the majority of components are connected with few other components can be considered more modular than random networks. This architecture ensures that topological effects of random perturbations are minimized ***(22–24)*** and might explain the common appearance of this architecture in nature.

Alternatively, structural integrity can be maintained by structural dependence in which effects of perturbations are actively compensated for. The compensatory mechanisms depend on the nature of the perturbation. For positive ("gain of function") perturbations, negative feedback can help bring the system back to the desired state ***(25)***. Gate keeper residues ***(26)*** that repel nonnative contacts can be viewed as an example of residues that can provide negative feedback during folding of proteins ***(27)***. For negative ("loss of function") perturbations positive feedback ***(28)*** can ensure realization of the structure. Thus, mechanisms that promote designability can be placed into a dichotomy involving modularity and dependency.

An alternative non-mutually-exclusive classification scheme involves another dichotomy. One class of features that promotes designability is redundancy through repetition. A classic example of repetition concerns that of gene duplications. Gene duplication allows major increases in genome diversity because changes in one copy of genes can be compensated by other copies that have not changed *(29)*. Similarly, high gene expression or the occurrence of positive feedback loops can ensure robustness of the associated phenotype because failure of certain molecules to execute function can be compensated for by repeated execution by the same molecules. Compensation can also occur without repetition. Different pathways producing the same chemical reactions can compensate for each other when one is disrupted. The loss of certain intramolecular interactions that are essential for the maintenance of fold structure might be compensated for by alternative interactions through other contacts in the protein structure. Having a larger number of contacts may confer greater stability to a molecule and greater stability can confer robustness to mutation *(30,31)*. Contact information has been shown to correlate with properties of designability *(7,8,12,32,33)*. Thus, mechanisms that promote designability can be placed into another dichotomy involving redundant and alternative mechanisms. Kitano *(34)* provides a more detailed review of these mechanisms.

These different classification schemes provide different views to explain designability. For example, as previously discussed, increasing the number of contacts in a protein can provide alternative interactions to compensate for those that are lost upon mutation. Alternatively, increasing the number of contacts can be viewed as increasing the modularity of the protein if the contact density of the core increases such that the stability of the protein becomes more independent of random mutations elsewhere.

Designability can be increased by increasing the level of stability-enhancing interactions within the structure of interest. The most designable structures may be those that optimally balance modularity with structural dependency *(35)*. For crystal structures, both connectivity in terms of the number of contacts molecules make with each other and modularity in terms of the number of rigid-body degrees of freedom have been cited as reasons explaining why some crystal space groups are favored over others *(36,37)*. Because most of these proteins are exposed to water, the nature of the contacts, in particular the nonpolar interactions that shield the protein core from solvent, is likely to play a role in determining designability *(38)*.

1.3.2. Designability Estimation by Parts

For complex systems, only the robustness of the parts may be known. For example, how can the designability of proteins be estimated if only the designability of the domains is known? This is the situation that occurs when using the SCOP domain family counts as an estimate for designability. Assuming that protein domains are relatively independent, the designability of a protein can be estimated by summing or averaging the designability scores of its domains ***(6,8,39)***. This approach, however, is not appropriate when the assumption of modularity of parts does not hold. For instance, parts of proteins known as prodomains or intramolecular chaperones can assist in folding of other parts of the protein ***(40,41)***. Assuming that all parts are highly dependent upon one another, another approach would be to estimate the designability of the whole by the designability of just one part. Estimating the designability of a protein by the least designable domain has been undertaken by Wong *et al.* ***(6,39)***. Examining whether correlated mutations exists between different domains may make it possible to gain insight on interdomain dependency and perhaps to partition structures into independent parts. However, for structures in which such analysis is not feasible, it may be insightful to estimate designability using both approaches: by the designability score of all of its parts and by just one part.

2. Disease

2.1. Associating Designability with Disease

Protein function is influenced by its structure. Loss of structure at the fold level often results in significant changes to function. Such a loss may be caused by protein destabilization, which may lead to aggregation or degradation. Because mutation or environmental change can cause such destabilization, and given that a large proportion of mutations seems to affect protein structure ***(42)***, hereditary disease-related proteins were hypothesized to more often contain structures of relatively low fold designability as compared to nondisease proteins (proteins without disease annotation). Interestingly, in comparison to all human proteins, proteins associated with diseases listed in the Online Mendelian Inheritance in Man (OMIM) database (mostly hereditary diseases of high penetrance) ***(43)*** were found to have SCOP folds with fewer families ***(6,39)***. Using a database of disease properties ***(44)***, many of these diseases associated with proteins with few families were found to occur at relatively high frequencies (**Table 1**). Thus, it seems that designability as measured by SCOP family counts has a significant association with disease propensity.

Two-thirds of folds with only one family found in disease proteins are relatively young (found mostly in mouse and human) while one-third is found spread out in both prokaryotic and eukaryotic genomes ***(6)***. These latter folds are relatively ancient and the absence of many families in these folds suggests that they are relatively less designable.

On the one hand, less robust proteins would be more likely to receive disease-associated mutations. However, being less robust may also mean that diversity in terms of the structure and stability of the proteins would be greater

Table 1
Designability and Disease[a]

Protein group	(I) Mean designability of the least designable folds		(II) Mean designability across folds	
	Score	Number of proteins	Score	Number of proteins
Nondisease	13.3	9274	12.1	2543
Disease	11.6	801	10.4	218
Common disease (freq. <1:10,000)	10.2	33	7.2	15
Rare disease (freq. >1:10,000)	12.7	265	13	88

[a]ENSEMBL human proteins ***(66)*** with detectable SCOP folds were divided into disease and nondisease categories (proteins without any OMIM-based disease annotation). Disease proteins were further divided into common and rare disease categories according to Jimenez-Sanchez *et al.* ***(44)***. Mean designability scores for each of these categories are shown. Designability for each protein was measured as (I) the family count of the least designable fold and (II) the mean family count across all folds in proteins highly covered by SCOP. According to these scores, disease proteins tend to be less designable than nondisease proteins. Common disease proteins tend to be less designable than rare disease proteins ***(6)***.

in a population. Such diversity may facilitate the survival of members of the population in different environments. Certain mutations may cause disease, but if they confer a selective advantage in certain environments, subsequent expansion of the population with such mutations will associate proteins with such folds with a common disease ***(6)***.

2.2. Perturbation Frequency Affects Disease Propensity

Structural designability is not the only determinant of disease propensity. Also important is the frequency in which the structure is perturbed. Certain folds may be associated with common diseases because they are more often exposed to environmental perturbations or the DNA encoding such folds are predisposed for mutation ***(45)***. The hyperperturbation of structures of low designability may be conserved to facilitate diversity in populations.

2.3. Alternate Structures Associated with Disease

A perturbation that destroys a certain structure may not always cause total loss of the structure. The structure may be converted (perhaps from a structure of lower designability) to another stable form and it is this form that may cause disease. A noted example is that of cancer, in which perturbations result in highly robust but deleterious cells ***(46)***. Similarly, perturbation of proteins may also create stable proliferating aggregates ***(47)***. Harmless microbial communities, once genetically perturbed to become pathogens robust to different environments, are an ongoing threat ***(48,49)***. The robustness of other disease states poses a challenge for prevention and therapy. Studies on how robustness evolves may facilitate the prediction of alternative highly designable structures.

2.4. Equating Structure with Constraints

The equation of constraints to structure has certain advantages. Knowing that similar constraints exists, it is possible to predict similarities in structure. This is the most often cited explanation for convergent evolution. For example, similarities in chaperone structure have been associated with similarities in substrate properties ***(50)***. The physical constraints of visual perception have limited the structural variability of eyes ***(51)***. The magnitude of structural similarity is expected to correlate with the magnitude of constraint. Moreover, given similar constraints, the evolution of structures may share some similarity. Interestingly, proteins within the same functional modules have been found to evolve at rates more similar than those between different modules ***(52)*** in line with this idea.

Knowing that two structures are similar, it is possible to predict similarities in constraints. For example, structurally similar human proteins have been found to share analogous disease-causing positions ***(53–56)***. Interestingly, duplicates of disease proteins were found to be significantly more associated with disease than expected ***(39)***. Duplicated genome regions may be predisposed for disease via nonallelic homologous recombination ***(57)***. But because duplicated disease proteins can also share interaction partners ***(58,59)*** and functions ***(60,61)***, they may be predisposed for disease in similar ways.

2.5. Further Work

Knowing the designabilities of various parts of a system, and knowing how often these parts interact and are altered by mutation or environmental factors, it is possible to predict which parts are most likely to fail. Hence, there is a clear connection between designability and disease. For proteins, a major disadvantage of using fold family counts to predict designability is that it is an imprecise measure. It is likely that different proteins with the same folds, and hence the same family count scores, can have very different designabilities. Moreover, if the fold is relatively young, the number of families contained in that fold may be too small to reflect its designability. Although not a direct measure of evolutionary success ***(62)***, contact and stability-based measures of designability can be more precise and it would be of interest to relate these measures to disease. An alternative to these measures is that of simulation. For example, methods such as finite element analysis has gained some maturity allowing predictions such as fracture points in vertebrate systems ***(63)*** or mechanisms of optic nerve trauma ***(64)***. For proteins, *ab initio* folding of large sequence samples ***(65)*** may also become a possibility to estimate designability. The advantage of simulation methods is that they allow the user to test structures under controlled conditions not possible in reality ***(66)***. It would be interesting to see how well predicted anatomical designability and exposure to stresses correlate with a propensity for injury in such simulations.

3. *Summary*

Because life requires structure, loss of such structure can result in disease. Designability measures how robust a structure is to perturbations and can help define a structure's susceptibility to disease. Although structures from proteins to whole organisms are diverse, there are unifying concepts that help explain designability. We have outlined some of these concepts and related them to disease. We hope this review will inspire the development of methodologies to estimate designability and improve our understanding of diseases.

Acknowledgments

We thank members of BFam, the Institute of Bioinformatics and Systems Biology (MIPS), and others for inspiration and support. This work was funded by a grant from the German Federal Ministry of Education and Research (BMBF) within the BFAM framework (031U112C).

References

1. Uversky, V. N., Oldfield, C. J., and Dunker, A. K. (2005) Showing your ID: intrinsic disorder as an ID for recognition, regulation and cell signaling. *J. Mol. Recognit.* **18**, 343–384.
2. Andreeva, A., Howorth, D., Brenner, S. E., Hubbard, T. J., Chothia, C., and Murzin, A. G. (2004) SCOP database in 2004: refinements integrate structure and sequence family data. *Nucleic Acids Res.* **32**(Database issue), D226–D229.
3. Li, H., Helling, R., Tang, C., and Wingreen, N. (1996) Emergence of preferred structures in a simple model of protein folding. *Science* **273**, 666.
4. Besenmatter, W., Kast, P., and Hilvert, D. (2007) Relative tolerance of mesostable and thermostable protein homologs to extensive mutation. *Proteins* **66**, 500–506.
5. Kussell, E. The designability hypothesis and protein evolution. (2005) *Protein Pept. Lett.* **12**, 111–116.

6. Wong, P. and Frishman, D. (2006) Fold designability, distribution, and disease. *PLoS Comput. Biol.* **2**, e40.
7. Bloom, J. D., Drummond, D. A., Arnold, F. H., and Wilke, C. O. (2006) Structural determinants of the rate of protein evolution in yeast. *Mol. Biol. Evol.* **23**, 1751–1761.
8. Shakhnovich, B. E. (2006) Relative contributions of structural designability and functional diversity in molecular evolution of duplicates. *Bioinformatics* **22**, e440–e445.
9. Bloom, J. D., Arnold, F. H., and Wilke, C. O. (2007) Breaking proteins with mutations: threads and thresholds in evolution. *Mol. Syst. Biol.* **3**, 76.
10. Grishin, N. V. (2001) Fold change in evolution of protein structures. *J. Struct. Biol.* **134**, 167–185.
11. Abeln, S. and Deane, C. M. (2005) Fold usage on genomes and protein fold evolution. *Proteins* **60**, 690–700.
12. Shakhnovich, B. E., Deeds, E., Delisi, C., and Shakhnovich, E. (2005) Protein structure and evolutionary history determine sequence space topology. *Genome Res.* **15**, 385–392.
13. Zeldovich, K. B., Berezovsky, I. N., and Shakhnovich, E. I. (2006) Physical origins of protein superfamilies. *J. Mol. Biol.* **357**, 1335–1343.
14. Zeldovich, K. B., Berezovsky, I. N., and Shakhnovich, E. I. (2006) Protein and DNA sequence determinants of thermophilic adaptation. *PLoS Comput. Biol.* **3**, e5.
15. Ellis, R. J. and Minton, A. P. (2006) Protein aggregation in crowded environments. *Biol. Chem.* **387**, 485–497.
16. Groll, M., Bochtler, M., Brandstetter, H., Clausen, T., and Huber, R. (2005) Molecular machines for protein degradation. *Chembiochem.* **6**, 222–256.
17. Tourasse, N. J. and Li, W. H. (2000) Selective constraints, amino acid composition, and the rate of protein evolution. *Mol. Biol. Evol.* **17**, 656–664.
18. Taylor, M. S., Ponting, C. P., and Copley, R. R. (2004) Occurrence and consequences of coding sequence insertions and deletions in mammalian genomes. *Genome Res.* **14**, 555–566.
19. Petrescu, A. J., Wormald, M. R., and Dwek, R. A. (2006) Structural aspects of glycomes with a focus on N-glycosylation and glycoprotein folding. *Curr. Opin. Struct. Biol.* **16**, 600–607.
20. Donald, J. E., Hubner, I. A., Rotemberg, V. M., Shakhnovich, E. I., and Mirny, L. A. (2005) CoC: a database of universally conserved residues in protein folds. *Bioinformatics* **21**, 2539–2540.
21. Reichmann, D., Rahat, O., Albeck, S., Meged, R., Dym, O., and Schreiber, G. (2005) The modular architecture of protein-protein binding interfaces. *Proc. Natl. Acad. Sci. USA* **102**, 57–62.
22. Albert, R., Jeong, H., and Barabasi, A. L. (2000) Error and attack tolerance of complex networks. *Nature* **406**, 378–382.
23. Greene, L. H. and Higman, V. A. (2003) Uncovering network systems within protein structures. *J. Mol. Biol.* **334**, 781–791.

24. Deeds, E. J. and Shakhnovich, E. I. (2005) The emergence of scaling in sequence-based physical models of protein evolution. *Biophys. J.* **88**, 3905–3911.
25. Becskei, A. and Serrano, L. (2000) Engineering stability in gene networks by autoregulation. *Nature* **405**, 590–593.
26. Matysiak, S. and Clementi, C. (2006) Minimalist protein model as a diagnostic tool for misfolding and aggregation. *J. Mol. Biol.* **363**, 297–308.
27. Berezovsky, I. N., Zeldovich, K. B., and Shakhnovich, E. I. (2007) Positive and negative design and thermal adaptation of natural proteins *PLoS Comput. Biol.* doi:10.1371/journal.pcbi.0030052.eor.
28. Brandman, O., Ferrell, J. E., Jr., Li, R., and Meyer, T. (2005) Interlinked fast and slow positive feedback loops drive reliable cell decisions. *Science* **310**, 496–498.
29. Ohno, S. (1970) *Evolution by Gene Duplication*. Springer-Verlag, Heidelberg.
30. Bloom, J. D., Silberg, J. J., Wilke, C. O., Drummond, D. A., Adami, C., and Arnold, F. H. (2005) Thermodynamic prediction of protein neutrality. *Proc. Natl. Acad. Sci. USA* **102**, 606–611.
31. Bloom, J. D., Labthavikul, S. T., Otey, C. R., and Arnold, F. H. (2006) Protein stability promotes evolvability. *Proc. Natl. Acad. Sci. USA* **103**, 5869–5874.
32. England, J. L. and Shakhnovich, E. I. (2003) Structural determinant of protein designability. *Phys. Rev. Lett.* **90**, 218101.
33. Deeds, E. J. and Shakhnovich, E. I. (2007) A structure-centric view of protein evolution, design, and adaptation. *Adv. Enzymol. Relat. Areas Mol. Biol.* **75**, 133–191.
34. Kitano, H. (2004) Biological robustness. *Nat. Rev. Genet.* **5**, 826–837.
35. Hansen, T. F. (2003) Is modularity necessary for evolvability? Remarks on the relationship between pleiotropy and evolvability. *Biosystems* **69**, 83–94.
36. Wukovitz, S. W. and Yeates, T. O. (1995) Why protein crystals favour some space-groups over others. *Nat. Struct. Biol.* **2**, 1062–1067.
37. Andersson, K. M. and Hovmoller, S. (2000) The protein content in crystals and packing coefficients in different space groups. *Acta Crystallogr. D. Biol. Crystallogr.* **56**, 789–790.
38. Fernandez, A. (2004) Functionality of wrapping defects in soluble proteins: what cannot be kept dry must be conserved. *J. Mol. Biol.* **337**, 477–483.
39. Wong, P., Fritz, A., and Frishman, D. (2005) Designability, aggregation propensity and duplication of disease-associated proteins. *Protein Eng. Des. Sel.* **18**, 503–508.
40. Ignatova, Z., Wischnewski, F., Notbohm, H., and Kasche, V. (2005) Pro-sequence and Ca2+-binding: implications for folding and maturation of Ntn-hydrolase penicillin amidase from E. coli. *J. Mol. Biol.* **348**, 999–1014.
41. Yabuta, Y., Subbian, E., Oiry, C., and Shinde, U. (2003) Folding pathway mediated by an intramolecular chaperone. A functional peptide chaperone designed using sequence databases. *J. Biol. Chem.* **278**, 15246–15251.
42. Yue, P., Li, Z., and Moult, J. (2005) Loss of protein structure stability as a major causative factor in monogenic disease. *J. Mol. Biol.* **353**, 45–473.
43. Hamosh, A., Scott, A. F., Amberger, J. S., Bocchini, C. A., and McKusick, V. A. (2005) Online Mendelian Inheritance in Man (OMIM), a knowledgebase of

human genes and genetic disorders. *Nucleic Acids Res.* **33(Database issue),** D514–517.

44. Jimenez-Sanchez, G., Childs, B., and Valle, D. (2001) Human disease genes. *Nature* **409**, 853–855.
45. Rogozin, I. B., Babenko, V. N., Milanesi, L., and Pavlov, Y. I. (2003) Computational analysis of mutation spectra. *Brief Bioinform.* **4**, 210–227.
46. Kitano, H. (2004) Cancer as a robust system: implications for anticancer therapy. *Nat. Rev. Cancer* **4**, 227–235.
47. Chiti, F. and Dobson, C. M. (2006) Protein misfolding, functional amyloid, and human disease. *Annu. Rev. Biochem.* **75**, 333–366.
48. Woolhouse, M. E., Taylor, L. H., and Haydon, D. T. (2001) Population biology of multihost pathogens. *Science* **292**, 1109–1112.
49. Walther, B. A. and Ewald, P. W. (2004) Pathogen survival in the external environment and the evolution of virulence. Biol. Rev. *Camb. Philos. Soc.* **79**, 849–869.
50. Stirling, P. C., Bakhoum, S. F., Feigl, A. B., and Leroux, M. R. (2006) Convergent evolution of clamp-like binding sites in diverse chaperones. *Nat. Struct. Mol. Biol.* **13**, 865–870.
51. Fernald, R. D. (2006) Casting a genetic light on the evolution of eyes. *Science* **313**, 1914–1918.
52. Chen, Y. and Dokholyan, N. V. (2006) The coordinated evolution of yeast proteins is constrained by functional modularity. *Trends Genet.* **22**, 416–419.
53. Stevens, F. J., Pokkuluri, P. R., and Schiffer, M. (2000) Protein conformation and disease: pathological consequences of analogous mutations in homologous proteins. *Biochemistry* **39**, 15291–15296.
54. Wolff, N., Gilquin, B., Courchay, K., Callebaut, I., Worman, H. J., and Zinn-Justin S. (2001) Structural analysis of emerin, an inner nuclear membrane protein mutated in X-linked Emery-Dreifuss muscular dystrophy. *FEBS Lett.* **501**, 171–176.
55. Albrecht, M., Lengauer, T., and Schreiber, S. (2003) Disease-associated variants in PYPAF1 and NOD2 result in similar alterations of conserved sequence. *Bioinformatics* **19**, 2171–2175.
56. Myers, J. K., Beihoffer, L. A., and Sanders, C. R. (2005) Phenotology of disease-linked proteins. *Hum. Mutat.* **25**, 90–97.
57. Bailey, J. A. and Eichler, E. E. (2006) Primate segmental duplications: crucibles of evolution, diversity and disease. *Nat. Rev. Genet.* **7**, 552–564.
58. Yu, H., Luscombe, N. M., Lu, H. X., Zhu, X., Xia, Y., Han, J. D., Bertin, N., Chung, S., Vidal, M., and Gerstein, M. (2004) Annotation transfer between genomes: protein-protein interologs and protein-DNA regulogs. *Genome Res.* **14**, 1107–1118.
59. Oti, M., Snel, B., Huynen, M. A., and Brunner, H. G. (2006) Predicting disease genes using protein-protein interactions. *J. Med. Genet.* **43**, 691–698.
60. Franke, L., Bakel, H., Fokkens, L., de Jong, E. D., Egmont-Petersen, M., and Wijmenga, C. (2006) Reconstruction of a functional human gene network, with

an application for prioritizing positional candidate genes. *Am. J. Hum. Genet.* **78**, 1011–1025.

61. Oti, M. and Brunner, H. G. (2007) The modular nature of genetic diseases. *Clin. Genet.* **71**, 1–11.
62. O'Loughlin, T. L., Patrick, W. M., and Matsumura, I. (2006) Natural history as a predictor of protein evolvability. *Protein Eng. Des. Sel.* **19**, 439–442.
63. Ross, C. F. (2005) Finite element analysis in vertebrate biomechanics. *Anat. Rec. A. Discov. Mol. Cell Evol. Biol.* **283**, 253–258.
64. Cirovic, S., Bhola, R. M., Hose, D. R., Howard, I. C., Lawford, P. V., Marr, J. E., and Parsons, M. A. (2006) Computer modelling study of the mechanism of optic nerve injury in blunt trauma. *Br. J. Ophthalmol.* **90**, 778–783.
65. Yang, J. S., Chen, W. W., Skolnick, J., and Shakhnovich, E. I. (2007) All-atom ab initio folding of a diverse set of proteins. *Structure* **15**, 53–63.
66. Hubbard, T., Andrews, D., Caccamo, M., Cameron, G., Chen, Y., *et al.* (2005) Ensembl 2005. *Nucleic Acids Res.* **33(Database issue),** D447–D453.

30

Prism: Protein–Protein Interaction Prediction by Structural Matching

Ozlem Keskin, Ruth Nussinov, and Attila Gursoy

Summary

Prism (protein interactions by structural matching) is a system that employs a novel prediction algorithm for protein–protein interactions. It adopts a bottom-up approach that combines structure and sequence conservation in protein interfaces. The algorithm seeks possible binary interactions between proteins through structure similarity and evolutionary conservation of known interfaces. It is composed of a database containing protein interface structures derived from the Protein Data Bank (PDB) and predicted protein–protein interactions. It also provides related information about the proteins and an interactive protein interface viewer. In the current version, 3799 structurally nonredundant interfaces are used to predict the interactions among 6170 proteins. A substantial number of interactions are verified in two publicly available interaction databases (DIP and BIND). As the verified interactions demonstrate the suitability of our approach, unverified ones may point to undiscovered interactions. Prism can be accessed through a user-friendly website (http://prism.ccbb.ku.edu.tr) and it will be updated regularly as new protein structures become available in the PDB. Users may browse through the nonredundant dataset of representative interfaces on which the prediction algorithm depends, retrieve the list of structures similar to these interfaces, or see the results of interaction predictions for a particular protein. Another service provided is the interactive prediction. This is done by running the algorithm for the user input structures.

Key Words: Protein interactions; protein interaction prediction; protein interfaces; protein databases.

From: *Methods in Molecular Biology, vol. 484: Functional Proteomics: Methods and Protocols*
Edited by: J. D. Thompson et al., DOI: 10.1007/978-1-59745-398-1, © Humana Press, Totowa, NJ

1. Introduction

Molecular and cellular operations are largely carried out by interactions between proteins. Interactions are physical associations of protein structures through weak, noncovalent bonds. Two proteins interact through particular regions on their surfaces, called binding sites, or interfaces. Identifying protein-binding sites and knowing which proteins interact with which other proteins are crucial for a better understanding of the bases of many biological processes. Despite the ongoing effort to decipher the complex nature of protein interactions, they are still not entirely understood ***(1–5)***. Protein-binding sites have been thoroughly analyzed for the presence of certain physicochemical and geometric properties that can be used to distinguish these regions from the noninteracting surface regions. Notable differences have been found both in the chemical composition and geometric properties of these sites ***(6–10)***. Almost a decade ago, Wells and his colleagues discovered the existence of "energy hot spots," that is, residues that contribute significantly (over 2 kcal/mol) to the binding free energy ***(11)***. These residues have been identified through alanine scanning mutagenesis. Subsequently, computational methods have been developed to predict these residues. In a landmark paper, Bogan and Thorn ***(1)*** proposed that hot spots are surrounded by what they called "O-rings." These are hydrophobic regions that may serve to exclude water from the hot spot residue. Combined, binding sites have been described by amino acids that interact across the two-chain interface. However, not all amino acids contribute equally. Some contribute marginally or not at all ***(12)***. On the other hand, a few others dominate the stability of the complex. These hot spot residues were observed to correlate with structurally conserved residues ***(13,14)***. Prediction of binding sites using these specific properties can be used for improving docking algorithms. Besides the experimental methods for detecting and analyzing protein–protein interactions ***(7,15,16)***, computational approaches are becoming increasingly important venues as large amounts of data become available. The development of predictive methods is a major goal in computational biology that will lead to protein engineering and drug discovery ***(9,10,17)***. The structural classification of protein interfaces provides insight into the possible ways proteins may interact ***(18,19)***. Hence, an efficient computational technique with acceptable error rates that can be utilized to predict the binding sites and binding partners in proteins will surely be of great value ***(20–22)***.

We present Prism (protein interactions by structural matching), a system incorporating a novel protein–protein interaction algorithm ***(20,23)*** and a web server that can be used to explore protein interfaces and predict protein–protein interactions. Our algorithm principally seeks pairs of proteins that may potentially interact in a dataset of protein structures (target dataset) by comparing them with a dataset of interfaces (template dataset), which is a structurally and evolutionary representative subset of biological and crystal interactions present in the Protein Data Bank (PDB) *(24)*. If, after comparisons, two target structures are found to structurally and evolutionarily complement each other as do chains of any template interface, they qualify as a potentially interacting pair. Thus, a list of potentially interacting protein pairs is obtained as a final result. Prism consists of a web interface to the dataset of our interface dataset and target structures including a summary of the proteins to which the interface belongs (with cross-references to other biological databases where available), similarity matching results, solvent-accessible surface area calculation results on a residue level scale, and interface visualization of the protein using both static images and an interactive interface viewer implemented using a browser plug-in.

2. Materials

The rationale of our protein–protein prediction algorithm is that if any two structures contain particular regions on their surfaces that resemble the complementary partners of a known interface, they "possibly interact" through these regions. In other words, if protein *A* is known to interact with protein *B*, and *A′* shares similarity with the binding site of *A* and *B′* shares similarity with the binding site of *B*, then we predict that *A′* interacts with *B′*. This resemblance indicates the ability of these structures to structurally and evolutionarily complement each other along an interface, as chains of any template interface might do. The algorithm requires a "template" dataset, i.e., the representative dataset of "available" interfaces, and a "target" dataset, to seek every potential binary interaction between its members *(20)*.

2.1. Interface Dataset

This dataset contains a structurally nonredundant dataset of protein–protein interfaces. Interfaces consist of interacting residues between the two polypeptide chains (of a complex protein) and those residues that are in their spatial vicinity (neighboring residues), representing the scaffold of the interface. Two residues from the opposite chains were marked as interacting, if there was at least a pair of atoms, one from each residue, at a distance smaller than the sum of their van der Waals radii plus a threshold of 0.5 Å. If the C_α of a noninteracting residue lies at a distance of at most 6.0 Å from a C_α of an already assigned interface residue in the same chain, it was marked as a neighboring residue. All interfaces between two protein chains obtained from higher complexes of proteins available in the PDB were extracted *(18)* resulting in 21,684 two-chain interfaces. These interfaces were clustered structurally using a structural alignment in a sequence order-independent way *(25)*. At the end of the iterative structural clustering procedure, 3799 interface clusters were obtained. Each cluster includes a representative interface structure and members similar to the representative interface.

2.2. Template Interface Dataset

Evolutionary conservation of certain residues at protein interfaces is a strong characteristic of binding sites. Ma *et al.* ***(26)*** reported that particular residues are conserved on structurally similar interfaces. Moreover, they found that these conserved residues were highly correlated with polar residue hotspots, residues that are more important than others in defining the affinity and stability of an interaction. Therefore, the interface dataset was further filtered using a dataset of computational hotspots. Computational hotspots are the critical residues for binding on representative interfaces. The members of the 3799 interface clusters were processed by a filtering process that eliminated the redundant sequences from the clusters. A cluster was defined as nonredundant if it contained at least five nonhomologous sequences. Then, simultaneous structural alignments among the nonhomologous members of each cluster were performed ***(27)***. If a residue was conserved at a particular spot among interfaces of similar architectures with a frequency of 50% or more, it was flagged as a computational hot spot ***(13)***. As a result, we could detect the hot spots of 67 clusters out of 3799, since most of the clusters could not pass the nonhomologous filtering. The prediction algorithm serviced by Prism uses only these 67 template interfaces for similarity matching. Hence, Prism considers both shape complementarities and evolutionary conservation while searching for binding sites on the surface of a target protein.

2.3. Target Dataset

This dataset is a sequentially nonredundant subset (with a sequence identity upper limit of 50%) of all the polypeptide chains and complexes existing in the PDB. Every pair of member structures in this dataset is checked for potential interactions. The protein chains may be in the form of monomers or in the form of isolated chains from multimeric complexes. As of January 27, 2004, the target dataset contained 6170 structures ***(20)***. The generation of this dataset is a two-step process. The first is a preprocessing step that involves downloading of the set of proteins obtained by applying a sequence identity filter of 50% to all existing protein structures in the PDB. This resulted in a list containing 5427 proteins. Then, the multimeric proteins are split into constituent chains where homologous chains are counted only once; the target dataset consists of 6170 structures. Of these 1981are multimeric and 4189 are monomeric. Of the monomeric structures, 2483 are derived from complexes. All these structures are on our web as "Target Structure Dataset."

3. Methods

This section describes the algorithm to determine novel protein–protein interactions using the shape complementarities and conservation in protein interfaces. A web server that makes it possible to search the interface, the target datasets, and the predicted interactions is presented as well. The web server also makes it possible to run the algorithm on a new target protein that is not in our target database.

3.1. Protein-Protein Interaction Prediction

The prediction algorithm is based on searching pairs of proteins that share structure and conserved residue (hotspot) similarity to our known interface template data. First, we extract surfaces of target proteins and perform successive structural alignments between these surfaces and the partner chains of interfaces in the template interface dataset, in an all-against-all manner. If surfaces of two target proteins (*A* and *B*) contain regions similar to complementary partner chains of a template interface *I*, in other words, one side of the interface *I* is similar to target *A* and the other side is similar to target *B*, then we say *A* and *B* may interact through these similar regions (or through interface *I*). **Figure 1**

```
  for all protein in target dataset do
     if protein structure determined by NMR then
          protein ← take first NMR structure
       target surfaces ← extract surface of protein
    if protein is multimeric then
        monomers ← split protein into chains and eliminate homolog
        target surfaces ← target surfaces + surfaces of monomers

for all interface in template interface do
     for left and right partner in interface do
        for all surface in target surfaces
           if (size of surface) ≥ 0.7 x (size of partner) then
              alignments ← align surface with partner
              best alignment ← similarity _score(alignments)
                if similarity score(best alignment) ≥ threshold then
                     similarity list_partner ← flag protein for prediction
```

Fig. 1. Main steps of the Prism prediction algorithm.

shows the top level pseudocode and the schematic flow of our algorithm. The algorithm starts by extracting surfaces of target structures by invoking the NACCESS program *(28)*. Along with the atomic accessible surface, NACCESS calculates relative surface accessibilities (RSA) of residues. Residues whose RSAs (percent accessibility compared to the accessibility of the residue type X in an extended ALA-X-ALA tripeptide) are greater than 5% can be considered to be on the surface *(3)*.

The algorithm then determines whether particular regions on target surfaces resemble complementary partners of representative interfaces in the template dataset. Each partner (side) of an interface is then structurally aligned with the target surface by invoking MULTIPROT *(25,27)*. MULTIPROT detects common geometric cores between given protein structures in a sequence-order-independent way. This feature makes MULTIPROT a favorable selection for the task, since protein surfaces and protein–protein interfaces have sequence discontinuity. MULTIPROT returns the 10 best substructural matches resulting from every possible alignment. Each substructure corresponds to different regions on the surface, bearing different levels of structural similarity to the interface partner. Among these alignments, the algorithm seeks the most favorable alignment that maximizes our similarity scoring function. The similarity scoring function is defined as $\alpha f_{evolution} + (1 - \alpha) f_{structure}$, where $f_{evolution}$ and $f_{structure}$ are evolutionary and structural similarity scoring functions, respectively. The coefficient, α, represents the relative importance of evolutionary similarity to structural similarity. The first function reflects the number of identically matched hotspots and the second function reflects the size and quality of the alignment along the target–template alignment. We assume that hotspots bear greater importance in defining an interface than geometric complementarity. Therefore we select α as 0.6. The condition prior to alignment restricts the interface partner size to at least 0.7 times the target surface size. (The size of a structure is defined as the number of residues it contains.) This condition keeps relatively small interfaces out of computations. Such relatively small interfaces are likely to align perfectly with target surfaces and yield high similarity scores, causing biased and unselective results.

After the completion of successive structural alignments, a similarity list for each interface partner is obtained. If the similarity lists of corresponding partners of a template interface contain N and M target structures, respectively, we obtain $N \times M$ predictions for that interface. A prediction is uniquely represented by (A, B, I) triplets, where A and B are predicted targets and I is the template interface by which the interaction was predicted. The extent of favorableness of the predicted interaction (prediction score) is quantified simply by the sum of the similarity scores of the target pairs. We have run our algorithm using the template interface set and target structure set; this resulted in a total of 62,616 protein–protein interactions. The details of the algorithm and the parameters of the scoring function are available in the Prism server documentation.

3.2. Services Provided by Prism

The Prism web server provides its users with a front end to the datasets used in our prediction algorithm, an interface to the offline results of our calculations based on the most previous run of our algorithm, and also the ability to run our algorithm for a user input protein. Services provided to the user and the input types differ accordingly, so they are discussed separately.

3.2.1. *Browsing and Searching Interface Database*

In the interfaces section we make our interface dataset available to the scientific community. A total number of 21,684 interfaces are stored, divided into 3799 clusters according to their structural similarity. Users are provided with a search facility by which they can find specific interfaces in our dataset. Their inputs can be a simple search string that is searched for in the corresponding records in the title section of the PDB file of the protein to which the template interface belongs. For example, the user might be interested in interfaces that are extracted from proteins that play a role in apoptosis or the user may want to see interfaces that are extracted from enzymes only. In addition to this basic search functionality, some advanced search options can also be used, enabling the user to search for interfaces of a certain size (in terms of solvent accessible surface areas measured in $Å^2$) or interfaces that have the highest frequency for a certain type of amino acid. Once the user clicks on an interface, an output containing the following data are provided. (1) A summary of the proteins from which the interface is extracted, including cross-references to other biological databases where available. (2) Details about the interface in question, such as the names of the constituent chains, interface size (in terms of number of residues), solvent-accessible surface areas buried upon complexation, polar and nonpolar ASAs, and a listing of all interface residues with their respective interface ASA. **Figure 2** shows the web server's results on the summary of the proteins, i.e., the name of the protein, number of atoms of the protein, ASA of the interfaces, etc. (3) A visualization of the interface is also output as static images that are dynamically generated by running RasMol scripts where the interface is highlighted on the protein. The whole protein is represented with a stick representation, whereas the interface atoms are shown with spheres.

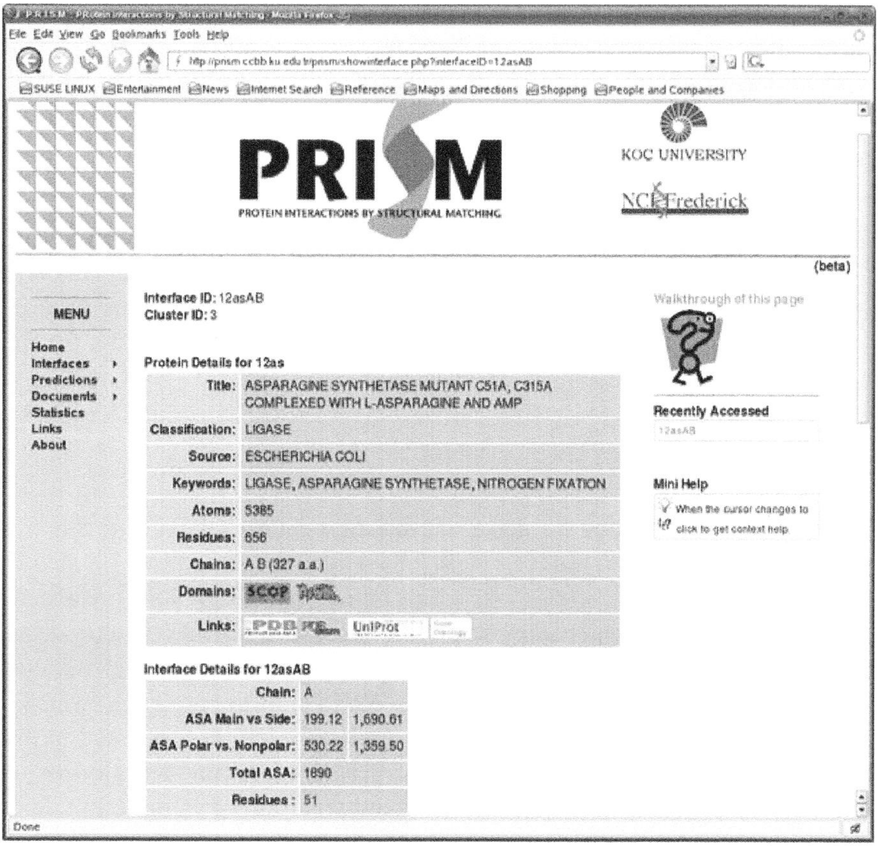

Fig. 2. The web server displaying the details of the proteins to which the interface belongs. Chain identities provide details on the two sides of the interface.

3.2.2. *Browsing and Searching Target Dataset*

In the targets section (under prediction), users are provided with a search facility with which they can find specific structures in our dataset that match a set of search criteria. The input can be a simple search string that is searched for in the corresponding records in the title section of the PDB file of the protein. In addition, using advanced search options, specific sets of target structures can be returned where, for instance, the target structure is of predefined size (size defined in number of residues) or type (monomer, complex, split chain). Once users click on a certain target protein they are provided with an output containing the following data: a summary of the target protein, a list of template interfaces for which the target structure is found to have a match, and several dynamically generated static images visualizing the target structure.

3.2.3. Searching Predicted Interactions

Under the predictions section of Prism, users can search our results in two different ways. They can directly search for the presence of similarities between a template interface and a target structure or it is possible to either input the PDB ID or the sequence of a protein [whose sequence is then aligned to the target dataset using BLAST *(29)*], which is then checked for any predicted protein–protein interactions in which the input protein participates. The target structures that match different partner chains of a template interface are then displayed to the user as a list of proteins that are candidates for an interaction. This is done by first checking to see if the input protein has a binding site similar to any one of the template interfaces as previously explained. All the target structures that are a priori found to have a binding site similar to the partner of the matched interface are listed as the predicted interacting protein.

Figure 3 shows the web server for the prediction results. The left column lists the possible binding partners for the protein with PDB code 1mr8. The second column contains links to the domain information of partners. The third column shows which template interfaces were used in the prediction phase. The last column gives the prediction score. Detailed information of the predictions is given in the respective pages. **Figure 4** displays an example output. Here one of the putative binding partners of 1mr8, 1e8a is detailed. The target here is 1mr8 and the template is 1mr8 (in the template dataset, the A chain of 1mr8 interacts with the 1mr8 B chain). The target is 1e8aA. Each row in the figure displays the residue in the template dataset that is structurally aligned with those of the target protein. The red residues (dark colored) are the computational hot spots of the template interface. It is also possible to list all proteins matching the left and right side of an interface. For example, **Fig. 5** shows all matching proteins through the interface 1mr8AB.

3.2.4. Online Prediction Calculations

The Prism web site can also be used to perform online calculations to predict binding partners of input proteins not covered by our datasets. At the moment we have implemented a preliminary service in which users can ask to see the proteins in our datasets to which their input protein interacts. Prism accepts an input protein either by its PDB code or by file upload. The online calculations build on top of our previous results. First the target dataset is replaced with the structure in question. Then the algorithm is run using the original template set and the user input structure. Upon completion of the algorithm we know which of the template interface partners are structurally similar to the surface of the structure in question. It then finds the original structures in our target set that are

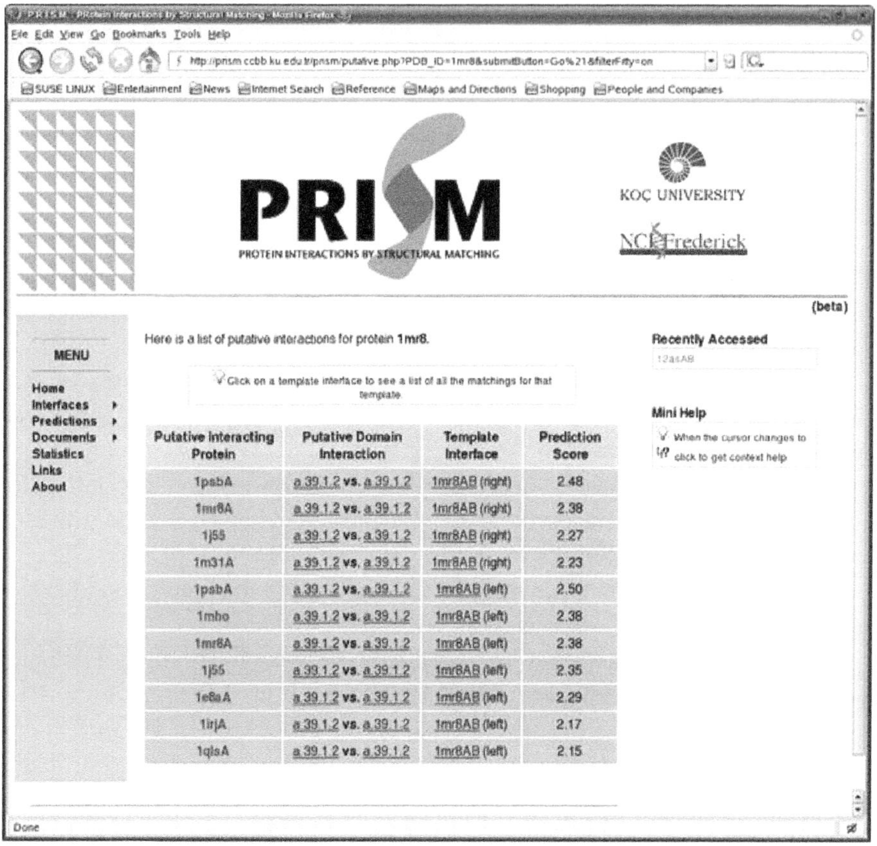

Fig. 3. List of the putative interacting proteins for protein 1mr8. The left column lists the possible binding partners. The second column contains links to the domain information of partners. The third column is the template interfaces used in the prediction. The last column is the prediction score.

similar to the partner of the template interface. These structures are then output as the proteins with which the input protein is predicted to interact.

4. Results and Discussion

Prediction results contain various interaction pairs, some of which are verified in DIP *(30)* and BIND *(31)* interaction databases as well as the PDB. Starting from 67 template interfaces we found 62,616 pairwise interactions among the 6170 target proteins. Of these interactions 31,980 are between the monomeric structures: 25,448 of them are between a monomeric protein and a complex

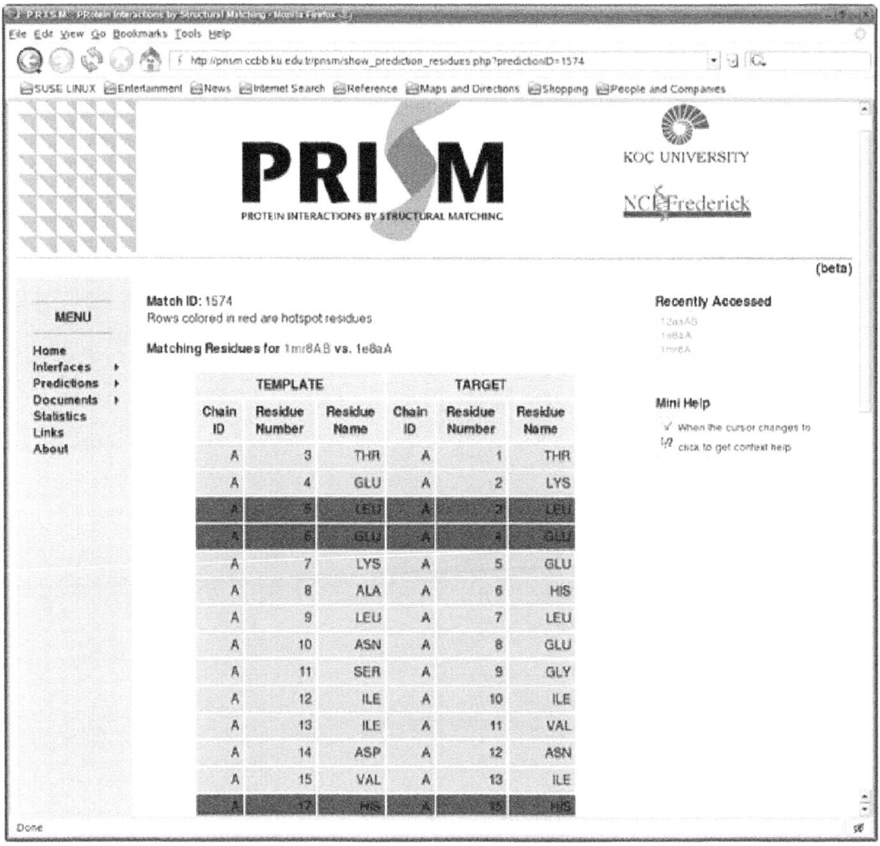

Fig. 4. The server displaying the results of the list of residues from one side of the predicted interface (target columns). The template columns are the residue listing of the template interface through which the interface was predicted. Red (dark colored) ones show the computational hot spots of the template interface.

structure and 5188 are between two complex structures. Most of these predictions are heterodimers; only 284 are homodimers (100% sequence identity between partners). This number contains predictions with partners having identical sequences within the same complex.

Table 1 displays a list of predictions with the highest scores. The first four letters in columns 1, 2, and 4 are PDB representations of proteins; the following letters are PDB chain identifiers. In columns 1 and 2, multiple chains are enclosed in curly brackets to indicate that the chains are identical and the prediction applies to all of them. In column 4, the two letters indicate the chains of the structures between which the template interface exists. Columns 5 and

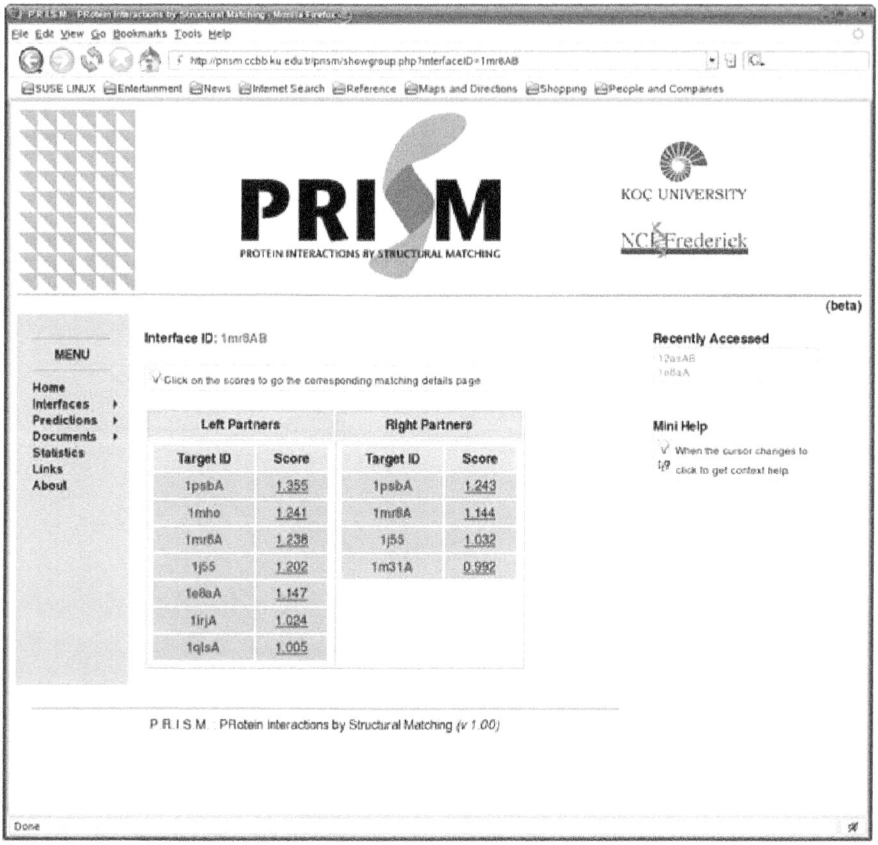

Fig. 5. Matching details of the template interface. The proteins matching the left and right side of interface 1mr8AB are listed with corresponding similarity scores.

6 are respective functions of SWISS-PROT cross-references of target partners, queried via the SWISS-PROT sequence retrieval system (SRS).

Analysis of the 62,616 predicted interactions reveals that the top five templates with the greatest number of matches contribute some 65% of the predictions (40,856 interactions). These interfaces are "fitty templates" since they scored high similarity scores and fit targets easily. Three of these come from helical proteins (1cosAC, 1aq5AC, and 1sfcBJ). They are all single domain interfaces. Furthermore, the first one (1cosAC) comes from a designed protein and is found to match most of the helical structures in the target set. Prism will normally filter these predictions from the results of search queries unless the user explicitly wants them.

Table 1
A Selected Set of Verified and Unverified Predictions[a]

Left partner	Right partner	Verified database	Template	Left function	Right function
1cov1	1h8tC		1cov13	Coxsackievirus coat protein	Echovirus 11 coat protein
1dgi	1ncqC		1cov13	Poliovirus receptor	Coat protein Vp3
1lq8{AECG}	1jjo{EF}	P	1as4AB	Plasma serine protease inhibitor	Neuroserpin
2ae2{AB}	1e7w{AB}	D, B, P	1e92AC	Tropinone reductase-II	Pteridine reductase
2sicE	1lw6I	P	2sniEI	Subtilisin BPN	Subtilisin-chymotrypsin inhibitor-2A
1mho	1psb{AB}	D,B,P	1mr8AB	S-100 protein	S-100 protein, β chain
1hj9	1jbl		1sbwAI	β-Trypsin	Cyclic trypsin inhibitor
2tnf{ABC}	1dg6	D,P	1cdaAB	TNF	TNF-related apoptosis-inducing ligand
1fxkC	1jm7B		1jm7AB	Prefoldin	Brca1-associated ring domain protein 1
1gk6{AB}	2ebo{ABC}	P	1cosAB	Vimentin	Ebola virus envelope glycoprotein
1kb9K	1n8v		1hezCE	Light chain (VI)of Fv- fragment	Chemosensory protein
1i4k1	1m5q{A..Z12}	1i4k12	Putative Snrnp Sm-like protein	Small nuclear ribonucleoprotein homolog	
1l8d{AB}	1c17		1jgcAC	RAD50 Atpase	ATP synthase subunit C
1mso{AC}	1mso *(?)*	P	6rlxAB	Insulin like growth factor A-chain	Insulin like growth factor B-chain
1ixm{AB}	1k75{AB}		1fuuAB	Sporulation response regulatory protein	L-Histidinol dehydrogenase
1iesB	1ecm{AB}		1iesAB	Ferritin	Endooxabicyclic transition state analogue
1ju5C	1uff	B	1azeAB	Abl	Intersectin 2
1osh	1fm6E		1fm6DE	Bile acid receptor	Steroid receptor coactivator

[a] The letters B, D, and P in the verified column correspond to verification in BIND, DIP, and PDB databases. TNF, tumor necrosis factor.

Table 2
Number of Verified Predictions (January 2004)

Interaction database	Unique verifications	Practical maximum verifications
DIP	597	4107
BIND	431	1739
PDB	1094	1497

A reasonable number of predictions were verified in DIP and BIND interaction databases. We do not expect that all predicted interactions can be verified since not all target structures are cross-referenced to DIP or BIND databases. **Table 2** displays the number of verified interactions out of cross-referenced interactions for three interaction databases (as of January 2004). The second column in the table the number of verified (target1, target2) interactions. The third column is the maximum number of predictions that could be verified due to available cross-referenced data in the corresponding database. The results display a good balance of verified and unverified predictions. Verified interactions prove the reliability of our algorithm, whereas unverified ones may correspond to unobserved interactions that actually occur in nature or may synthetically be realized in laboratory conditions. We believe these unverified predictions may have important implications regarding drug design.

5. Conclusions

As large amounts of protein structure data become available, predictive methods to detect and characterize protein–protein interactions are becoming increasingly important venues toward defining new foundations of systems biology. We have developed a novel algorithm for the automated prediction of protein–protein interactions that employs a bottom-up approach combining structure and sequence conservation in protein interfaces, and developed a web server for the analysis of protein–protein interfaces and the resulting predictions.

Starting from a nonredundant dataset that represents structurally available interfaces in protein–protein interactions, some 60,000 predictions were obtained, some of which were verified in interaction databases. The datasets and prediction results can be searched using the Prism web server. Another service provided by Prism is the interactive prediction. This is done by running the algorithm for the user input structures. At present, the online prediction of an interaction for a user input protein and all the structures in our target dataset is possible. Currently, Prism server is being improved both by updating interface and target datasets and by providing more advanced online calculations such as classification of predictions as crystal–crystal interactions or biological interactions.

Acknowledgments

The authors would like to thank A. Selim Aytuna and Utkan Ogmen for the development and implementation of Prism. This project has been funded in whole or in part by a TUBITAK Research Grant (104T504) and by federal funds from the National Cancer Institute, National Institutes of Health, under contract number NO1-CO-12400. This research was supported (in part) by the Intramural Research Program of the NIH, National Cancer Institute, Center for Cancer Research. The content of this publication does not necessarily reflect the views or the policies of the Department of Health and Human Services, nor does mention of trade names, commercial products, or organizations imply endorsement by the U.S. Government. O. Keskin acknowledges the Turkish Academy of Sciences Young Scientist Award (TUBA-GEBIP).

References

1. Bogan, A. A. and Thorn, K. S. (1998) Anatomy of hot spots in protein interfaces. *J. Mol. Biol.* **280,** 1–9.
2. Chakrabarti, P. and Janin, J. (2002) Dissecting protein-protein recognition sites. *Proteins* **47,** 334–343.
3. Jones, S. and Thornton, J. M. (1997) Analysis of protein-protein interaction sites using surface patches. *J. Mol. Biol.* **272,** 121–132.
4. Lo Conte, L., Chothia, C., and Janin, J. (1999) The atomic structure of protein-protein recognition sites. *J. Mol. Biol.* **285,** 2177–2198.
5. Keskin, O., Ma, B., Rogale, K., Gunasekaran, K., and Nussinov, R. (2005) Protein-protein interactions: organization, cooperativity and mapping in a bottom-up systems biology approach. *Phys. Biol. 2,* S24–S35.
6. Glaser, F., Steinberg, D. M., Vakser, I. A., and Ben-Tal, N. (2001) Residue frequencies and pairing preferences at protein-protein interfaces. *Proteins* **43,** 89–102.
7. Ito, T., Tashiro, K., Muta, S., Ozawa, R., Chiba, T., Nishizawa, M., Yamamoto, K., Kuhara, S., and Sakaki, Y. (2000) Toward a protein-protein interaction map of the budding yeast: A comprehensive system to examine two-hybrid interactions in all possible combinations between the yeast proteins. *Proc. Natl. Acad. Sci. USA* **97,** 1143–1147.
8. Jones, S. and Thornton, J. M. (1995) Protein-protein interactions: a review of protein dimer structures. *Prog. Biophys. Mol. Biol.* **63,** 31–65.
9. Neuvirth, H., Raz, R., and Schreiber, G. (2004) ProMate: a structure based prediction program to identify the location of protein-protein binding sites. *J. Mol. Biol.* **338,** 181–199.

10. Zhou, H. X. and Shan, Y. (2001) Prediction of protein interaction sites from sequence profile and residue neighbor list. *Proteins* **44,** 336–343.
11. Clackson, T. and Wells, J. A. (1995) A hot spot of binding energy in a hormone-receptor interface. *Science* **267,** 383–386.
12. DeLano, W. L. (2002) Unraveling hot spots in binding interfaces: progress and challenges. *Curr. Opin. Struct. Biol.* **12,** 14–20.
13. Keskin, O., Ma, B., and Nussinov, R. (2005) Hot regions in protein-protein interactions: the organization and contribution of structurally conserved hot spot residues. *J. Mol. Biol.* **345,** 1281–1294.
14. Ma, B., Wolfson, H. J., and Nussinov, R. (2001) Protein functional epitopes: hot spots, dynamics and combinatorial libraries. *Curr. Opin. Struct. Biol.* **11,** 364–369.
15. Uetz, P., Giot, L., Cagney, G., Mansfield, T. A., Judson, R. S., Knight, J. R., Lockshon, D., Narayan, V., Srinivasan, M., Pochart, P., Qureshi-Emili, A., Li, Y., Godwin, B., Conover, D., Kalbfleisch, T., Vijayadamodar, G., Yang, M., Johnston, M., Fields, S., and Rothberg, J. M. (2000) A comprehensive analysis of protein-protein interactions in Saccharomyces cerevisiae. *Nature* **403,** 623–627.
16. Zhu, H., Bilgin, M., Bangham, R., Hall, D., Casamayor, A., Bertone, P., Lan, N., Jansen, R., Bidlingmaier, S., Houfek, T., Mitchell, T., Miller, P., Dean, R. A., Gerstein, M., and Snyder, M. (2001) Global analysis of protein activities using proteome chips. *Science* **293,** 2101–2105.
17. Kortemme, T. and Baker, D. (2004) Computational design of protein-protein interactions. *Curr. Opin. Chem. Biol.* **8,** 91–97.
18. Keskin, O., Tsai, C. J., Wolfson, H., and Nussinov, R. (2004) A new, structurally nonredundant, diverse data set of protein-protein interfaces and its implications. *Protein Sci.* **13,** 1043–1055.
19. Winter, C., Henschel, A., Kim, W. K., and Schroeder, M. (2006) SCOPPI: a structural classification of protein-protein interfaces. *Nucleic Acids Res.* **34,** D310–314.
20. Aytuna, A. S., Gursoy, A., and Keskin, O. (2005) Prediction of protein-protein interactions by combining structure and sequence conservation in protein interfaces. *Bioinformatics* **21,** 2850–2855.
21. Murakami, Y. and Jones, S. (2006) SHARP2: protein-protein interaction predictions using patch analysis. *Bioinformatics* **22,** 1794–1795.
22. Aloy, P., Bottcher, B., Ceulemans, H., Leutwein, C., Mellwig, C., Fischer, S., Gavin, A. C., Bork, P., Superti-Furga, G., Serrano, L., and Russell, R. B. (2004) Structure-based assembly of protein complexes in yeast. *Science* **303,** 2026–2029.
23. Ogmen, U., Keskin, O., Aytuna, A. S., Nussinov, R., and Gursoy, A. (2005) PRISM: protein interactions by structural matching. *Nucleic Acids Res.* **33,** W331–336.
24. Berman, H. M., Westbrook, J., Feng, Z., Gilliland, G., Bhat, T. N., Weissig, H., Shindyalov, I. N., and Bourne, P. E. (2000) The Protein Data Bank. *Nucleic Acids Res.* **28,** 235–242.
25. Nussinov, R. and Wolfson, H. J. (1991) Efficient detection of three-dimensional structural motifs in biological macromolecules by computer vision techniques. *Proc. Natl. Acad. Sci. USA* **88,** 10495–10499.

26. Ma, B., Elkayam, T., Wolfson, H., and Nussinov, R. (2003) Protein-protein interactions: structurally conserved residues distinguish between binding sites and exposed protein surfaces. *Proc. Natl. Acad. Sci. USA* **100,** 5772–5777.
27. Shatsky, M., Nussinov, R., and Wolfson, H. J. (2004) A method for simultaneous alignment of multiple protein structures. *Proteins* **56,** 143–156.
28. Hubbard, S. J. and Thornton, J. M. (1993) "NACCESS," computer program. Department of Biochemistry and Molecular Biology, University College, London.
29. Altschul, S. F., Madden, T. L., Schaffer, A. A., Zhang, J., Zhang, Z., Miller, W., and Lipman, D. J. (1997) Gapped BLAST and PSI-BLAST: a new generation of protein database search programs. *Nucleic Acids Res.* **25,** 3389–3402.
30. Xenarios, I., Salwinski, L., Duan, X. J., Higney, P., Kim, S. M., and Eisenberg, D. (2002) DIP, the Database of Interacting Proteins: a research tool for studying cellular networks of protein interactions. *Nucleic Acids Res.* **30,** 303–305.
31. Bader, G. D., Betel, D., and Hogue C. W. (2003) BIND: the Biomolecular Interaction Network Database. *Nucleic Acids Res.* **31,** 248–250.

31

Prediction of Protein Interaction Based on Similarity of Phylogenetic Trees

Florencio Pazos, David Juan, Jose M. G. Izarzugaza, Eduardo Leon, and Alfonso Valencia

Summary

Computational methods for predicting protein interaction partners are becoming increasingly popular. Many of them are mature enough to be widely used by molecular biologists who can look for proteins related to the protein of interest in order to infer information about its context in the cell. In this chapter we describe the use of the *mirrortree* set of programs and related software for predicting protein interactions. They are all based on the idea that interacting or functionally related proteins tend to show similar phylogenetic trees due to coevolution. The basic *mirrortree* program can be used to calculate the similarity between the phylogenetic trees implicit in the multiple sequence alignments of two protein families. The ECID database contains protein interactions and relationships from different computational and experimental sources for the model organism *Escherichia coli*, including the ones generated with *mirrortree*. Finally, the TSEMA server uses the concept of tree similarity between interacting families to look for the best mapping between two families of interacting proteins: which member in one family interacts with which member in the other.

Key Words: Protein interaction; protein functional relationship; coevolution; similarity of phylogenetic trees; mirrortree.

1. Introduction

Numerous methods for predicting protein interactions and functional relationships from sequence and genomic information are now available [*see* ***(1–3)*** for reviews]. These methods, apart from being faster and cheaper than their experimental counterparts, have similar levels of accuracy and are not subject to some

From: *Methods in Molecular Biology, vol. 484: Functional Proteomics: Methods and Protocols*
Edited by: J. D. Thompson et al., DOI: 10.1007/978-1-59745-398-1, © Humana Press, Totowa, NJ

of their drawbacks (intrinsic due to their experimental nature) *(4)*. These computational techniques are now fully incorporated in the bioinformatics toolbox of many researchers. Many of them are mature from scientific and technical points of view: they have been exhaustively tested and tuned and they are implemented in friendly programs and web interfaces that enable them to be used by the community.

Predicting which proteins interact with or are functionally related to a given protein provides much information about the protein's functional context. This tactic, known as "context-based prediction," is orthogonal to the classical "sequence-similarity-based" approach for inferring information for a given sequence, and hence these approaches complement each other. The most popular repository of context-based information for proteins is STRING *(5)*.

One of these methods for detecting interaction partners and functional associations from sequence information is based on the idea that interacting families tend to show phylogenetic trees with topologies that are more similar than expected. The hypothesis for explaining such a relationship involves coevolution and coadaptation of these interacting proteins. This relationship was first qualitatively observed for some families [i.e., insulins and insulin receptors *(6)*] and later quantified, with a correlation coefficient between the distance matrices represented by the trees and statistically evaluated *(7,8)*. This simple and intuitive method was followed by many authors who developed variations of it [see references in *(9)*] and applied it to many protein families (i.e. **10**).

In this chapter, we describe in detail the use of a set of available programs and web resources for the prediction of protein interactions using the idea of tree similarity. We start with the basic *mirrortree* program *(8)*, which takes the multiple sequence alignments of two protein families as input and calculates the similarity between the implicit phylogenetic trees as the correlation between the corresponding distance matrices. Then, we describe the ECID system, which contains predicted and experimental context information for the proteins of the model organism *Escherichia coli* and which can be accessed through a web interface. Finally, we describe TSEMA *(11)*, another web interface that implements a system for the interactive prediction of the mapping between the members of two interacting protein families, that is used to predict which protein within one family interacts with which protein in the other (i.e., a family of ligands and their corresponding receptors).

2. Materials

1. *Mirrortree* is distributed as a stand-alone command-line program. Binary versions are available for many different platforms and operative systems. The distribution includes documentation, examples, etc.

http://pdg.cnb.uam.es/pazos/mirrortree provides information on how to obtain this software.

2. ECID is available at http://pdg.cnb.uam.es/ecid.
3. TSEMA is available at http://tsema.bioinfo.cnio.es.

Mirrortree and TSEMA use multiple sequence alignments as input. For general information on how to generate multiple sequence alignments *see* **Note 1**.

3. Methods

3.1. Mirrortree

Mirrortree calculates the similarity between the phylogenetic trees implicit in two multiple sequence alignments as previously described *(8)*.

3.1.1. Preparing the Multiple Alignments for Running the Program

To calculate the similarity between the trees of proteins (families) *A* and *B*, the first things we need are the multiple sequence alignment with the orthologs of *A* in different species (*A1, A2, A3, …*) and the corresponding alignment for *B*. A simple way to detect the ortholog of a protein in another organism (i.e., detect *A2* given *A1*, distinguishing it from other paralogs [*A2'*, *A2"*, …]) is the "BLAST best bidirectional hit" (*see* **Note 2**). There are also repositories of orthologs, such as COG *(12)*. An additional advantage of these repositories is that they also provide the multiple sequence alignments for these sets of orthologs.

Once we have the multiple sequence alignments with the orthologs of proteins *A* and *B*, we have to merge them in a single file "concatenating" the sequences of *A* and *B* in the same species, that is, *A1* with *B1*, *A2* with *B2*, etc. This is the way to inform the program about the species correspondence, which is needed to compare the right distances. If one of the proteins is present in one species but the other is not (i.e., *A1* exists but *B1* does not), this "unpaired" sequence (*A1*) is discarded. Concatenating *A1–B1, A2–B2*, etc., is trivial in alignment formats such as PIR or FASTA (just pasting one sequence after the other). Most multiple sequence alignment programs can generate PIR and FASTA formats (*see* **Note 1**). The *mirrortree* distribution also includes a program for doing this, providing the proteins in the individual alignments are labeled with the species to which they belong. This concatenated alignment represents the multiple sequence alignment of a hypothetical "polyprotein" *AB*. It is important to preserve the original alignment of the individual proteins. That is, do not realign this concatenated alignment. Do not use alignments with less than 10 sequences. The program distribution includes examples of such "concatenated" alignments.

3.1.2. Running the Program

The command line for running the program in a terminal looks like the following:

```
mirrortree alignment(HSSP,PIR/FASTA) matrix naa1 naa2
```

The name of the executable (`mirrortree`) program may be different depending on the operating system (`mirrortree_linux32, MIRROR TREE.EXE, mirrortree_osx, ...`).

The main input for the program is the concatenated alignment of the two protein families as described in **Subheading 3.1.1**. HSSP, PIR, and FASTA formats are accepted. The second argument is an amino acid substitution matrix in Maxhom format. It is also included in the distribution. The last two arguments are the lengths of both proteins in the concatenated alignment, which is used to indicate which portion of the alignment corresponds to the first protein and which one to the second. Gaps are included in this numbering.

3.1.3. Output

The program returns a value between –1.0 and +1.0, which indicates the similarity between the distance matrices of both families, and hence reflects the similarity of the corresponding trees. High values have been shown to be related to interactions and functional relationships. Full details about this calculation are provided by Pazos and Valencia *(8)*. Values lower than –1.0 (i.e., –2.0, –3.0, ...) are used as flags to indicate that the calculation could not be done.

There is also an extension of *mirrortree*, *tol-mirrortree*, which corrects the background similarity between trees due to the underlying speciation events (*see* **Note 3**).

3.2. Escherichia coli Interactions Database (ECID)

This web resource can be used to look for different types of relationships between *E. coli* proteins (*see* **Note 4**). The relational database behind the interface includes predicted interactions coming from four different computational methods: *mirrortree* (described above), *in silico two-hybrid* ***(13)***, *phylogenetic profiling* ***(14)***, and *gene neighborhood* ***(15)***. A short description of these methods is given in **Note 5**. It also includes protein relationships extracted from KEGG pathways ***(16)***, experimental annotated interactions, protein complexes, regulatory pathways, and relationships extracted from the literature with the iHOP system ***(17)*** (**Note 6**). In total, it contains 15 different sources of information on protein relationships.

3.2.1. Looking for a Given Protein

Use the main web page of the system, also accessible in the "Home" tab, to search for a given protein. You can enter either the protein name, gene id, SWISS-PROT id, etc., or the sequence. In the last case, a BLAST search is used to find the protein. In this page, there are also examples with which you can play.

A list of proteins matching your search criteria will appear. For these proteins, the "EciD" link takes you to the database record with the information on that protein, including a link to the corresponding entry in SWISS-PROT. The "Interactions" link allows you to access all the relationships stored for that protein in the database

3.2.2. Browsing the List of Protein Interactions and Relationships

Following the "Interactions" link for the protein in which you are interested, you obtain a summary table with all the stored interactions. The rows are the *E. coli* proteins for which some interaction with yours is stored, and the columns are the methods (*see* above). The table shows which method(s) support a given interaction or relationship (**Fig. 1a**).

You can switch between this global summary table and the ones showing only the interactions for a given method using the upper row. In the last case, additional information on the interactions is included, such as scores of the prediction methods and name of the pathways for the KEGG/EcoCyc relationships. This additional information includes, in many cases, links to the original source of information to obtain more details on this particular interaction/relationship.

In addition, in the summary table, each protein related to yours (rows) has an "i" link that takes you to detailed information on the method(s) supporting that interaction. This includes information such as the scores of the prediction methods and links to the original sources of information for that interaction in a manner similar to that previously described.

3.2.3. Graphic Representation

Below the summary table, an interactive Java applet shows a network representation of all the interactions shown in the table (**Fig. 1b**).

The nodes in this network (proteins) can be dragged in order to look for a clear representation. Clicking one of these nodes will take you to the corresponding summary table with the interactions stored for that protein (**Subheading 3.2.2**). This allows you to navigate all of the interaction network, jumping from the interactions of one protein to the ones of another. The edges of the network

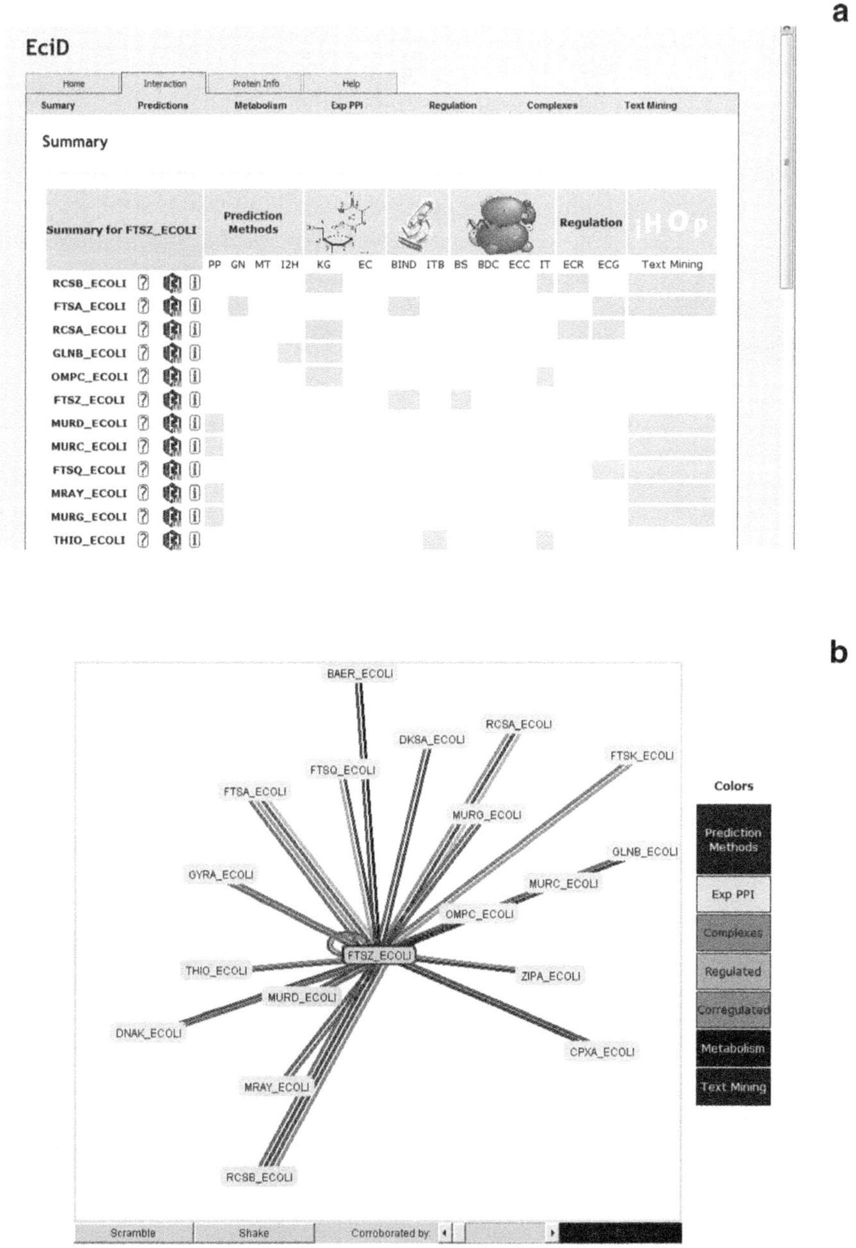

Fig. 1. The ECID web interface. (**a**) Summary table with a list of predicted and annotated interactions and relationships for FTSZ_ECOLI (SWISS-PROT ID). The gray boxes represent the methods that support the interactions/relationships. (**b**) Interactive graphic representation of the network of interactions and relationships.

represent the different methods supporting a given interaction/relationship. Each method is associated with a color according to the legend on the right. Clicking a given edge would take you to a detailed description of that relationship, as described in **Subheading 3.2.2**. The slidebar at the bottom makes it possible to filter the representation in order to show only the relationships supported by a minimum number of methods, which are supposed to be the more reliable ones.

3.3. The Server for Efficient Mapping Assessment (TSEMA)

This server implements a modified version of Ramani and Marcotte's method ***(18)*** for predicting the mapping between the members of two interacting families: which protein within one family interacts with which one in the other. This method looks for the best mapping based on the idea that it will be the one maximizing the similarity of the trees of the two families. A short description of the method is given in **Note 7**. The server makes it possible to interactively modify that initial mapping and assess whether these modifications really improve the mapping ***(11)***.

The web page of the server includes a help file, a detailed tutorial enabling you to become familiar with the system, and some precomputed examples.

The general process for using this system is as follows. First, you submit the two protein families you want to map. The initial mapping is returned by email. In a second step, this initial mapping is submitted back to the server to start the interactive analysis part (modification and improvement of the mapping). These two steps have been separated because the first one can take a long time to run (*see* **Note 7**).

3.3.1. Initial Job Submission

The "New Job" button at the top of the page allows you to submit the two protein families you want to map.

You can either submit the multiple sequence alignments of the families (*see* **Note 1**), in a format compatible with ClustalW ***(19)***, or the phylogenetic trees in newick format. In case multiple sequence alignments are submitted, the corresponding trees are generated using the *neighbor joining* algorithm implemented in ClustalW (*see* **Note 8**). The other required fields are the job name (to help you track different jobs) and the email address to which the results will be returned.

There is a set of advanced options that allows you to control the generation of the initial mapping. These options are intentionally blurred since you normally would not need to change them. You can activate them and change their values. A short description of these options is given in **Note 9**.

Once the initial mapping is calculated you will receive an email with the raw results of this mapping (compressed in a .gz file). You can unpack the file to access these raw results or submit it as it is to the interactive analysis part (next point).

3.3.2. Interactive Analysis and Modification of the Mapping

Since the process for obtaining the mapping is heuristic (*see* **Note 7**), it does not ensure the best solution to be found, but only a "locally" good solution. This is why it is important to inspect this mapping and eventually modify it using any source of information you might have. This manual interactive part could allow you to find better solutions not explored by the heuristic algorithm.

You can start this analysis by pressing the "New Analysis" button and submitting the .gz file with the results of the initial mapping sent to you by email (**Subheading 3.3.1**). The interactive analysis interface (**Fig. 2**) shows a list of predicted pairs of interacting proteins according to the initial mapping. For each pair, four scores are shown: "reliability," representing the percentage of mappings in which that pair appears (*see* **Note 7**), and "segregation," which measures the difference between the reliability of that pair and the second best reliability. The reliability for pair *AB* could be different from that of the pair *BA*, since *A* and *B* might be confronted with different sets of proteins. This is why there are two values of reliability and segregation for each pair. The coincidence matrix (**Fig. 2**) shows the number of repetitions of the heuristic approach (*see* **Note 7**) where these two proteins are linked. There is a color code for the scores from red (bad) to blue (good). The entropies of the trees of the two families are also shown (*see* **Note 10**).

A graphic representation of the two trees showing the predicted interacting pairs of proteins corresponding to the current mapping is also shown in this page (**Fig. 2**). The color of the links corresponds to the *AB* reliability score in the list of pairs. The bootstrap values of the nodes of the trees are shown in this representation, if present in the trees provided by you as input (*see* **Note 11**). If you submit multiple sequence alignments, the system generates bootstrap trees. The initial layouts of the trees are calculated with NJPlot *(20)*.

At the bottom of the interface you can see the distance correlation plots corresponding to the current mapping and other mappings. On the left the correlation plot of the current mapping superposed on that of the immediately previous mapping is shown; the correlation plot of the current mapping compared with that of the original mapping is shown on the right. These plots can be used to assess whether a given change in the mapping affects many distances, or whether a given mapping produces an overall good score but with some outliers. These correlation plots are generated with GNUPlot (www.gnuplot.info).

In this interactive interface, you can change links in the list of predicted pairs and assess how these changes affect the scores. Whenever you change a link, the

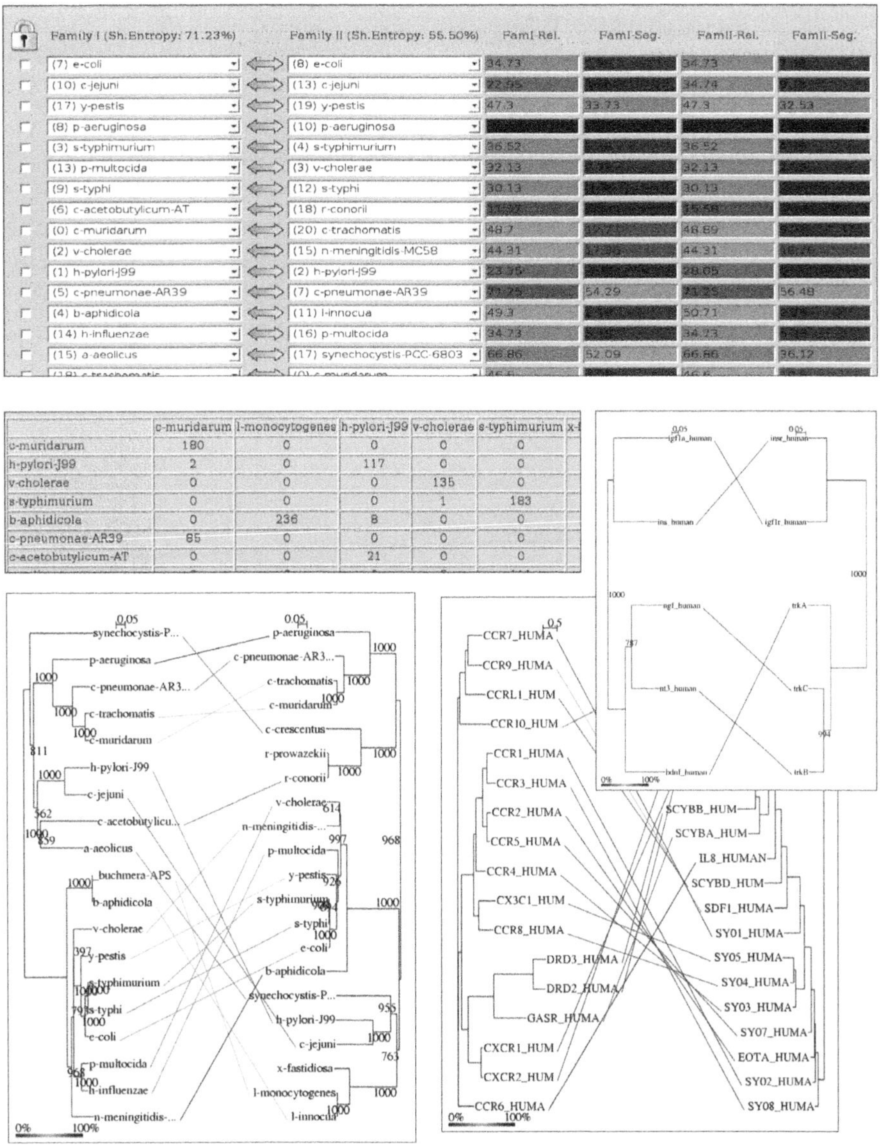

Fig. 2. TSEMA results pages. The top panel shows the list of predicted links between the members of the two families and their associated scores. These links can be interactively changed. These links are also represented in the corresponding trees (below). The table in the middle represents part of the coincidence matrix.

new mapping incorporating that change is represented in the trees and in the correlation plots. You can revert changes to the previous mapping or load the original (first) mapping by pressing the corresponding buttons. The links with which you are more confident can be "locked" to avoid changing them.

The idea of this interface is to interactively explore alternative mappings by applying some changes and to assess their quality graphically and by the scores. A good starting point for guessing possible changes in the mappings is the coincidence matrix (**Fig. 2**). A "stable" pair (found in most of the mappings generated in the different runs) might not be present in the overall highest scoring mapping (the initial one). In this case, it would be worth forcing that pair in the mapping to see whether it makes sense (scores, tree representation, etc.). You can also incorporate expert information in this process, e.g., by forcing some pairs known or suspected to interact.

4. Notes

1. The standard way of generating a multiple sequence alignment for a given protein is to retrieve homologous sequences using, for example, BLAST and to align them with a multiple alignment program such as ClustalW *(19)*. Both programs can be accessed through web interfaces around the world or installed locally. Moreover, systems such as SRS (http://srs.ebi.ac.uk) incorporate the possibility of automatically running ClustalW with the results of a BLAST search. There are also many databases of precalculated multiple sequence alignments with different characteristics. One of the most popular ones is Pfam *(21)*.
2. The "best bidirectional hit" method for finding the ortholog of a given protein *A1* in another organism (*A2*) consists basically in "BLASTing" *A1* against all the proteins in organism 2 and taking the first hit as the ortholog only if "BLASTing" it back against all the proteins in organism 1; the original *A1* is found as the first hit.
3. Any pair of trees has a background similarity due to the underlying speciation events, independent of the interaction or lack of interaction of the corresponding proteins. Correcting that similarity has been shown to improve the performance of the protein interaction prediction based on tree similarity *(9,22)*. In the same *mirrortree* page (*see* **Subheading 2**) there is information on how to obtain *tol-mirrortree*, the extension of *mirrortree* that corrects this speciation signal from the trees.
4. Many methods whose predictions are stored in this database (including *mirrortree*) in fact predict relationships between families (alignments), not individual proteins, and their assumption is that all the proteins within one alignment interact with the corresponding proteins in the other. For this reason, although the database has *E. coli* as the reference organism, it also implicitly contains information on interactions between proteins from other bacteria (through the corresponding *E. coli* orthologs).

5. There are other computational methods for predicting interaction partners apart from *mirrortree*. The *in silico two-hybrid* method looks for an accumulation of correlated mutation signals between the positions of two multiple sequence alignments ***(13)***. Interacting proteins tend to have more correlations between them. The *phylogenetic profiling* method assesses the similarity between the patterns of presence/absence of two proteins in a set of genomes (phylogenetic profiles). Two proteins showing similar phylogenetic profiles are expected to interact or to be functionally related since they tend to appear together in the same set of organisms and to be absent together in the complementary set ***(14)***. The *gene neighborhood* method looks for pairs of genes that are close in the genomes of a set of organisms ***(15)***. The relationship between conservation of gene closeness and functional interaction is related to bacterial operons. The *gene fusion* method looks for pairs of proteins that appear fused in a single polypeptide in one or more organisms ***(23)***. This fusion event is indicative of a functional interaction or functional relationship.
6. iHOP uses genes and proteins as links between PubMed abstracts ***(17)***. In this way, much information contained in the literature can be represented in this network format and navigated (http://www.ihop-net.org).
7. The method of Ramani and Marcotte predicts the mapping between the members of two interacting families based on similarity of phylogenetic trees ***(18)***. It is easy to see that swapping two columns, *A* and *B* (and the corresponding rows), in one of the distance matrices representing the trees is equivalent to interchanging the mappings of these two proteins (link *A* with all the proteins previously linked to *B* and vice versa). The exhaustive approach would hence consist of trying all possible row swappings, and for each one evaluating the similarity between the two resulting matrices. The best mapping would be the one maximizing this similarity. Since this exhaustive exploration is not feasible, the method uses a Monte Carlo algorithm to avoid the complete exploration of the space of solutions. The drawback is that this algorithm does not ensure that the globally best solution is found, only a locally good one. So, different runs of the algorithm usually lead to different solutions (local minima in the space of solutions). Usually, the algorithm is run many times and the consistency of the solutions is evaluated (i.e., in how many of the runs a given link between two proteins appears).
8. *Neighbor joining* is a very fast and convenient way of generating a phylogenetic tree. Nevertheless, there are more reliable techniques for doing that (such as *Parsimony* or *Bayesian trees*), which are normally time consuming and partially manual. These state-of-the-art techniques should be used whenever possible.
9. TSEMA advanced options for the generation of the initial mapping. The number of Monte Carlo runs (*see* **Note 7**) can be specified. Although the detection of the submitted data type (alignment or tree) is done automatically, you can also force the type in case you receive unexpected errors regarding problems with formats. The default scoring function for measuring the similarity between the trees (distance matrices) is Pearson's *T* correlation coefficient. However, you can

also use Pearson's R or RMSD (root mean square deviation) as alternative scoring functions.

10. The entropy of a tree is a measure of its topological complexity. As the tree is more complex, it is easier to "match" it to similar trees since they have distances in the whole range (from low to high) with which to compare. There is more information to compare. If the complexity is low and most of the distances within each tree are very similar, it is more difficult to match these two sets of distances (most of the mappings would produce the same score). This is why the complexity of the trees provides an idea of how good the results you can expect are.
11. The bootstrap value of a node in a tree represents the number of alternative trees (generated "modifying" the input alignment slightly) in which that node appears. Hence, it provides an idea of the "confidence" or stability of that node. Many wrong pairings are associated with internal nodes with low bootstrap support.

Acknowledgments

The authors want to thank the members of the Computational Systems Biology Group (CNB-CSIC) and the Structural Bioinformatics Group (CNIO) for interesting discussions and support. This work was in part funded by the BIO2006-15318, BIO2004-00875, and PIE 200620I240 projects from the Spanish Ministry for Education and Science, and the GeneFun EU project (LSHG-CT-2004-503567). Part of this work was also supported by the Spanish National Bioinformatics Institute (INB, www.inab.org), a platform of "Genoma España."

References

1. Salwinski, L. and Eisenberg, D. (2003) Computational methods of analysis of protein-protein interactions. *Curr. Opin. Struct. Biol.* **13,** 377–382.
2. Valencia, A. and Pazos, F. (2002) Computational methods for the prediction of protein interactions. *Curr. Opin. Struct. Biol.* **12,** 368–373.
3. Huynen, M., Snel, B., Lathe, W., and Bork, P. (2000) Predicting protein function by genomic context: quantitative evaluation and qualitative inferences. *Genome Res.* **10,** 1204–1210.
4. von Mering, C., Krause, R., Snel, B., Cornell, M., Oliver, S. G., Fields, S., and Bork, P. (2002) Comparative assessment of large scale data sets of protein-protein interactions. *Nature* **417,** 399–403.
5. von Mering, C., Huynen, M., Jaeggi, D., Schmidt, S., Bork, P., and Snel, B. (2003) STRING: a database of predicted functional associations between proteins. *Nucleic Acids Res.* **31,** 258–261.
6. Fryxell, K.J. (1996) The coevolution of gene family trees. *Trends Genet.* **12,** 364–369.

7. Goh, C.-S., Bogan, A. A., Joachimiak, M., Walther, D., and Cohen, F.E. (2000) Co-evolution of proteins with their interaction partners. *J. Mol. Biol.* **299,** 283–293.
8. Pazos, F. and Valencia, A. (2001) Similarity of phylogenetic trees as indicator of protein-protein interaction. *Protein Eng.* **14,** 609–614.
9. Pazos, F., Ranea, J. A. G., Juan, D., and Sternberg, M. J. E. (2005) Assessing protein co-evolution in the context of the tree of life assists in the prediction of the interactome. *J. Mol. Biol.* **352,** 1002–1015.
10. Labedan, B., Xu, Y., Naumoff, D. G., and Glansdorff, N. (2004) Using quaternary structures to assess the evolutionary history of proteins: the case of the aspartate carbamoyltransferase. *Mol. Biol. Evol.* **21,** 364–373.
11. Izarzugaza, J. M., Juan, D., Pons, C., Ranea, J. A., Valencia, A., and Pazos, F. (2006) TSEMA: interactive prediction of protein pairings between interacting families. *Nucleic Acids Res.* **34,** W315–319.
12. Tatusov, R. L., Koonin, E. V., and Lipman, D. J. (1997) A genomic perspective of protein families. *Science* **278,** 631–637.
13. Pazos, F. and Valencia, A. (2002) In silico two-hybrid system for the selection of physically interacting protein pairs. *Proteins* **47,** 219–227.
14. Pellegrini, M., Marcotte, E. M., Thompson, M. J., Eisenberg, D., and Yeates, T. O. (1999) Assigning protein functions by comparative genome analysis: protein phylogenetic profiles. *Proc. Natl. Acad. Sci. USA* **96,** 4285–4288.
15. Dandekar, T., Snel, B., Huynen, M., and Bork, P. (1998) Conservation of gene order: a fingerprint of proteins that physically interact. *Trends Biochem. Sci.* **23,** 324–328.
16. Kanehisa, M., Goto, S., Kawashima, S., Okuno, Y., and Hattori, M. (2004) The KEGG resource for deciphering the genome. *Nucleic Acids Res.* **32,** D277–280.
17. Hoffmann, R. and Valencia, A. (2004) A gene network for navigating the literature. *Nat. Genet.* **36,** 664.
18. Ramani, A. K. and Marcotte, E. M. (2003) Exploiting the co-evolution of interacting proteins to discover interaction specificity. *J. Mol. Biol.* **327,** 273–284.
19. Chenna, R., Sugawara, H., Koike, T., Lopez, R., Gibson, T. J., Higgins, D. G., and Thompson, J. D. (2003) Multiple sequence alignment with the Clustal series of programs. *Nucleic Acids Res.* **31,** 3497–3500.
20. Perrière, G. and Gouy, M. (1996) WWW-Query: an on-line retrieval system for biological sequence banks. *Biochimie* **78,** 364–369.
21. Bateman, A., Coin, L., Durbin, R., Finn, R. D., Hollich, V., Griffiths-Jones, S., Khanna, A., Marshall, M., Moxon, S., Sonnhammer, E. L., *et al.* (2004) The Pfam protein families database. *Nucleic Acids Res.* **32,** D138–141.
22. Sato, T., Yamanishi, Y., Kanehisa, M., and Toh, H. (2005) The inference of protein-protein interactions by co-evolutionary analysis is improved by excluding the information about the phylogenetic relationships. *Bioinformatics* **21,** 3482–3489.
23. Marcotte, E. M., Pellegrini, M., Ho-Leung, N., Rice, D. W., Yeates, T. O., and Eisenberg, D. (1999) Detecting protein function and protein-protein interactions from genome sequences. *Science* **285,** 751–753.

32

Large Multiprotein Structures Modeling and Simulation: The Need for Mesoscopic Models

Antoine Coulon, Guillaume Beslon, and Olivier Gandrillon

Summary

Recent observational techniques based upon confocal microscopy make it possible to observe cells at a scale that has never been probed before: the mesoscopic scale. In the eukaryotic cell nucleus, many objects demonstrating phenomena occurring at this scale, such as nuclear bodies, are current subjects of investigations. But from a modeling perspective, this scale has not been widely explored, and hence there is a lack of suitable models for such studies. By reviewing higher and lower scale modeling techniques, we analyze their relevance in the context of mesoscale phenomena. We emphasize important characteristics that should be included in a mesoscopic model: an explicit continuous three-dimensional space with discrete simplified molecules that still have the characteristics of steric volume exclusion and realistic distant interaction forces. Then we present 3DSPI, a model dedicated to studies of nuclear bodies based on a simple formalism inspired from molecular dynamics and coarse-grained models: particles interacting through a potential energy function and driven by an overdamped Langevin equation. Finally, we present the features expected to be included in the model, pointing out the difficulties that might arise.

Key Words: Coarse grained modeling and simulation; protein–protein interactions; mesoscopic scale; nuclear bodies; cell simulation; pair potential energy; overdamped Langevin dynamics; molecular crowding.

1. Introduction

Inside a cell, various biological phenomena occur at very different functionally connected scales: from phosphorylation of an amino acid residue to changes in the cellular architecture along the cell cycle or during a differentiation process. Each of these phenomena relies on processes that take place

From: *Methods in Molecular Biology, vol. 484: Functional Proteomics: Methods and Protocols*
Edited by: J. D. Thompson et al., DOI: 10.1007/978-1-59745-398-1, © Humana Press, Totowa, NJ

at a lower scale (for instance, a signaling pathway relies on molecular recognition properties and biochemical reactions). To understand a biological process, we can abstract the reality of this lower scale and, from its properties, try to determine how the phenomena being studied can occur. This is the principle of modeling.

To do so, the scale at which we are working defines what is admitted and what is not. For example, the *macroscopic* scale considers matter as being continuous and is based on average notions such as concentration, reaction rates, and temperature. Typical biological modeling based upon this scheme is concerned with regulation networks and signaling or metabolic pathways. On the other hand, *atomic* and *nanoscopic* scales consider atoms (or small groups of atoms) separately and focus on the behavior of one or a few molecules. At this scale, the previous average ideas have no meaning. This is the scale commonly used for protein folding and macromolecular complex assembly studies. Between these two, an intermediate scale, the *mesoscopic* scale (from the Greek word *meso*: ''in-between''), refers to the scale at which average concepts such as density and temperature still apply but where we nevertheless need to consider individual macromolecules (or large domains of them) and observe the behavior of large multiprotein structures.

For several years life science studies have been provided with observational tools at the macroscopic scale (i.e., optic microscopy and all derived techniques) and at the atomic scale (e.g., X-ray crystallography, nuclear magnetic resonance [NMR] spectroscopy, and DNA sequencing), providing knowledge concerning both cellular organization and molecular structures or interactions. Therefore modeling studies have generally focused on these scales and developed tools for either cellular or molecular simulation. But because of the difficulty involved in observing hundreds or several thousands of proteins, the mesoscopic scale remains a blind spot in our understanding of the interdependency between scales. Recently, new techniques have allowed us to discover a wide variety of mesoscale objects, the cell nucleus being one of the most striking examples *(1)*. Yet, it is becoming more and more evident that the explanation of many cellular processes depends on an understanding of mesoscale-level phenomena. However, to better understand these processes, new observation tools must be complemented by new modeling approaches that fill the gap in the range of available models. In other words, we need new modeling tools that enable us to study the properties of both the macroscopic world and atomic objects.

2. The Nucleus: A Mesoscopic Goldmine

For many years, while we knew much about the structure and workings of both cytoplasmic organelles and ribosomes or polymerases, we were barely aware of the existence of nucleoli and Cajal bodies. The nucleus was considered to be as it appears using optic microscopy: an unstructured space with uniform protein, DNA, and RNA distribution (with the exception of the nucleolus). Precise positions of molecules inside the nucleus were considered irrelevant; thus macroscopic modeling approaches were considered to be precise enough. However, the recent use of a fluorescent protein tagging technique combined with confocal microscopy ***(2,3)*** provided a precise, dynamic, and three-dimensional (3D) protein species distribution within the nucleus. These experiments revealed the existence of several membrane-free regions of the nuclear space with a particular protein composition ***(1,4)*** but without any important variation of global density (explaining the blindness of optic microscopy). Important examples of these numerous structures, collectively designated as *nuclear bodies* (or *nuclear compartments*), are the nucleolus, Cajal (or coiled) bodies, nuclear speckles, and PML bodies. For most of these, few of their functions have been identified.

Contrary to static immunofluorescence techniques, fluorescent protein tagging makes it possible to measure proteins dynamics through photobleaching experiments such as FLIP and FRAP (*see* **Note 1**). Many recent studies using theses techniques have reported important diffusion coefficients of proteins either inside, outside, or entering and leaving nuclear bodies ***(2–4)***. These results highlight an important property of nuclear bodies: because they are membrane-free regions their shape and size are directly determined by ingoing and outgoing rates of proteins at the body interface, which are themselves influenced by protein functions (such as binding, degradation, and recruitment in a complex); these in turn are influenced by the nuclear body. So, in addition to influencing protein mobility and function (like every organelle), a nuclear body is also greatly affected in its structure. The accumulation of evidence indicating that *dynamics*, *structure*, and *function* are intimately coupled supports the hypothesis that nuclear bodies are formed and maintained by principles of self-organization ***(5,6)***.

Further organization at a higher scale has also been discovered in the nucleus regarding chromosomes. Each of them does not appear to be randomly distributed in the nuclear space, but rather to occupy a very precise region of space referred to as a *chromosome territory* ***(7,8)***. The segregation of chromosomes with very complex interwoven interfaces and the reproducible evolution of the repartition of the chromosome territories in the nucleus during the cell cycle or through differentiation raise additional questions concerning self-organization.

All these studies require the development of a new vision of the nucleus that takes into account the properties of the objects (such as proteins, DNA, and RNA) as well as their 3D dynamic distribution over the nuclear space. This is typically what is used in a mesoscopic approach. However, considering the difficulties of observing nuclear compartments *in vivo* and analyzing their dynamics, the development of dedicated models will be essential to investigations of nuclear dynamics. In this context, 4 years ago, noticing the lack of relevant models, we began a multidisciplinary study to develop such a tool. We present here the current status of our work and reflections on the approaches to mesoscopic nuclear models and simulations.

3. Above and Below the Mesoscale

An ideal model can be thought of as a molecular dynamics (MD) simulation of all the molecules in the nucleus. However, on the one hand, such a simulation is clearly unrealistic. On the other hand, it is necessary to keep in mind that the aim of a model is to enable practitioners to develop new insights on a particular object. In this context, an MD model of the nucleus would probably be too complicated for our understanding. Thus, to build a useful model at the mesoscopic level, we have to consider global, phenomenological, properties of the microscopic objects (accounting for our knowledge of them). Similarly, some macroscopic properties or macroscopic objects will obviously be introduced explicitly (i.e., temperature, nuclear membrane) to keep the model both understandable and computable. Hence, considering a hypothetical hierarchy of models, the definition of mesoscopic models needs to be rooted in both atomic and macroscopic ones: considering a particular scientific question, we have to define which objects/properties will be explicitly described (and the precision of the description) and which will be neglected or, at least, implicitly described. In fact, this distinction between an implicit and explicit description of objects corresponds to macroscopic and atomic descriptions.

This is why we first need to consider other models used for higher and lower scale studies and to understand the different choices made based on the scientific question, as well as the advantages and drawbacks of these choices for our purpose.

3.1. Macroscopic Models

One of the highest level cellular models describes the behavior of a set of biochemical reactions (including catalysis and inhibition) by a set of differential equations to study metabolic and signaling pathways, as well as interaction and regulatory networks. The bioreactions are commonly treated as regular chemical reactions, i.e., concentrations of reactants are assumed to be uniform (at least inside a compartment) and sufficiently high so that the stochasticity of reaction events can be ignored. This approximation seems to be a serious drawback, as it is becoming more and more clear that stochasticity plays a major role in many cellular processes ***(9–13)***. Indeed, it is known to be partly due to the finite number of molecules involved in bioreactions and therefore to the discreteness of matter. Stochasticity-based models are now moving to the front of the stage ***(12,14,15)***.

Another disadvantage of macroscopic of models is the lack of integration of an explicit space. While in many models time is considered to be important, space has been ignored for a long time. However, many recent studies insist on the fact that the spatial localization of molecules is of the utmost importance in both regulation and molecular pathways ***(16–18)***. Indeed, the behavior of a protein (in terms of mobility and biochemical activity) is very dependent on its physical context. In differential equation models, it can be argued that the definition of separated compartments with particular membrane porosity accounts for the effect of space. But this is only an implicit and—more problematically—arbitrary space that does not allow for the flexibility of the physical world and the feedback of bioreactions on structures. In other words, it does not allow for the necessary interdependence of *dynamics*, *structure*, and *function*, known to be very important for nuclear bodies ***(5,6)***. This drawback of compartmental approaches is strikingly illustrated by the recent discovery that chromosomes are dynamically organized in chromosome territories in the nucleus ***(7,8)***, revoking the common hypothesis of regulation studies that considers the nucleus as a simple compartment with a uniform distribution of molecules. This is why many authors argue for cellular models integrating an explicit and realistic (3D and continuous) space ***(18–20)***.

Paradoxically, a 55-year-old differential equation-based model—Alan Turing's model of morphogenesis ***(21)***—can provide a response to these latter drawbacks by integrating space and allowing for feedback structuring. But lacking physical support, this structuring is too temporary for our scale of interest. Recent studies use a similar approach for modeling nuclear body-related phenomena ***(22)***, but they usually need to make assumptions about preexistent structures (i.e., nuclear scaffold). Although valid for several macroscopic phenomena, as it remains on a continuous description of matter, it is not suitable for mesoscale studies.

3.2. Atomic and Nanoscopic Models

Below the mesoscopic scale are *all-atom* models used to predict very precise protein and complex features such as folding, assembly, and dynamics. In these approaches, every atom of each molecule is considered individually in a continuous 3D space. A *potential energy* is defined as a function of the conformation, taking into account both bonded and nonbonded interactions between pairs of atoms. It is used to derive the resulting force applied to each atom, which in turn is used along with Newton's second law to compute their motion. This approach, called *molecular dynamics* (MD), is often used to study the temporal dynamics of already folded proteins (i.e., spontaneous or ligand-induced conformational changes).

There are many MD software packages based on different potential energy functions, usually derived from various pioneering work *(23,24)*. The most widely used are the CHARMM, AMBER, and GROMOS programs. The potential energy functions can be obtained either *ab initio* through quantum mechanics calculations or empirically and they are intended for particular molecular types (i.e., amino acids and nucleic acids) *(25)*. However, the computational load resulting from the consideration of every atom (even with the exception of hydrogen) limits the size of the system and imposes a very short simulation time (up to a few tens of nanoseconds).

To overcome these drawbacks, other models, referred to as *coarse-grained* models, use a slightly less precise description of molecules *(26)*. The principle of these methods is to regroup multiple atoms in single beads (or grains) of roughly identical size. The level of coarsening varies from one to six beads per residue. In this range of models, the coarser the description is, the more the force field between entities tends to be biased toward the native conformation (obtained by X-ray crystallography) to compensate for the loss of precision due to the diminution of the number of parameters. For instance, in the family of Go-like models, mainly used to study protein folding pathways in different contexts *(27–29)*, the energy function is quite similar to all-atoms MD but with a simpler parameterization and with the attractive nonbonded term applying only between residue pairs known to be in contact in the native state. Residues that do not interact in the native conformation have a purely repulsive interaction. Another important example is the elastic network models (ENMs) used to reproduce large period vibration modes of proteins *(30–32)*. This consists of a set of beads (*see* **Note 2**) connected by linear springs of rest length corresponding to the distance between beads in the native state. Any pair of beads (below a certain distance threshold) is connected regardless of whether they are bonded or in contact in the native state. These two families of coarsened models have a purpose very different from ours: they focus on the transition to or the vibration around a known final state, and so they are biased toward it. In contrast, our model has to determine the possible final states of the system without any knowledge of it, so it cannot be biased toward any objective state. Some other models focus on defining coarse-grained potentials with more physical motivations *(33)*, such as potentials of mean force obtained by knowledge-based methods *(34)* and effective potentials derived from MD simulations *(35)*. These potentials are more generic in their definition, but they still present some sort of bias: in the former case, the potential is biased by the selection of existing structure from the Protein Data Bank, and in the latter case, the resulting potential is very dependent on the all-atoms MD simulation used to generate it (composition, temperature, structures, etc.) and cannot be used in other conditions ***(36)***.

Because they reduce the computational load, coarse-grained approaches are commonly used to increase the simulated time with respect to all-atoms MD. However, simultaneously increasing the number of molecules would again imply very short simulation times, preventing any reliable study. Hence, the size of the simulated system remains limited to a few macromolecules, usually of different types (with the exception of water molecules when simulated explicitly), and cannot treat mesoscale protein-based phenomena without achieving a higher level of coarsening.

3.3. Approaching the Protein Mesoscale

Although using coarse grains of a size of the same order as previously used, some other models can be attributed to mesoscale studies. Indeed, in contrast to the previous models that deal with the assembly of a small number of molecules of different types, these models, mainly concerned with phospholipid membranes, involve a significantly greater number of small molecules of the same type (up to a few thousands of phospholipids) and their interaction with one or several membrane proteins ***(36–38)***. But here, it is because of the small size of phospholipids that a large number of molecules can be simulated. The underlying description of the model is similar to some of the coarsest models of the previous section. So it is only in their object of interest that these models can be considered as being at a mesoscale level; there is therefore still no suitable model for protein-based mesoscale phenomena.

We can mention the existence of several models approaching this scale by implicating individual molecules situated in an explicit space (in contrast to macromolecular models; cf. **Subheading 3.1**). For instance, from the artificial life community, there exist many models consisting of stereotyped molecular entities moving and interacting with formal rules in a discrete (square lattice) and usually in 2D space ***(39)***. But the aim of these highly simplified and unrealistic models is not to simulate a biological system but rather to extract the fundamental principle of life, and they are not suitable for our purpose. On the other hand, there is a certain number of agent-based modeling (ABM) (*see* **Note** 3) studies focusing on molecular biology questions ***(40–42)***. In particular, D. Bray's team has developed a very promising model of individual punctual molecules diffusing freely and reacting with simple rules in a 3D continuous space ***(43)***. This model can start to address some of the protein-level mesoscale questions, but it still lacks physical properties: proteins are punctual (there is no steric volume; a radius is defined only for bioreactions) and do not interact with any force. As we will see in the next section, this necessarily prevents the model from being able to reproduce many aspects of mesoscopic phenomena.

3.4. Important Physical Properties at the Mesoscale

Indeed, some physical properties of proteins are known to play an important role in many mesoscopic phenomena. For instance, the fact that water molecules represent only 20% of the mass of the nucleus, a property known as *molecular crowding* ***(44)***, provokes volume exclusion and molecular confinement that have an important influence on many phenomena: folding ***(28)***, aggregation ***(45)***, anomalous diffusion ***(46)***, and bioreaction kinetics enhancement ***(44)***. A model with punctual molecules can definitely not account for all these phenomena. So the model needs to include an explicit non–null steric volume for proteins.

Moreover, it is also known that in addition to contact forces, electrostatic (and electrostatic-induced) forces play an important role in binding and aggregation. Indeed, polar and apolar regions of the protein surface induce forces between proteins: both direct Coulomb forces and indirect hydrophilic and hydrophobic forces resulting from the presence of water molecules. These distant forces are clearly determining for the spatial organization of biomolecules and have to be taken into account.

It is with all these constraints in mind that we can now define a model for studying protein mesoscale phenomena.

4. 3DSPI, a Model for Nuclear Bodies

The model we define here is dedicated to the study of nuclear bodies. As argued previously, it has to rely on physical properties of proteins. So it is inspired from MD models (all-atoms and coarse-grained models) and adapts them to our higher scale of interest.

Obviously, we will not use an atomic description of molecules. Our model will rely on the description of molecular *behaviors* and *interactions*. Moreover, we will not describe all the nuclear molecules: only the molecules of interest will be considered explicitly. The others (in fact, most of them) will be modeled implicitly, considering only their actions on the first ones. This approach, focusing on the description of entities and interactions, is close to the ABM methodology.

4.1. A Probabilistic Version of the Model

A first version of the model has been developed as a proof of concept and demonstrates interesting behaviors that can be compared with biological observations *(47)*.

Proteins are represented by spheres, assigned a given mass, moving according to Newton's second law, which takes into account the effect of implicit molecules (other proteins and water) through a viscosity force and a noise factor accounting for Brownian activity. When two proteins collide, they have a certain probability of binding—defined by the *coefficient of stickiness* (COS)—and at every subsequent time step they have the same probability of remaining bounded. When colliding without binding, proteins behave as hard spheres, mimicking infinitely hard material.

However, although this model correctly reproduces aggregate dynamics *(47)*, it presents a certain number of physical irrelevances. First, the law of motion that is used is not really adapted for this scale (this point is developed in the next section). Second, as pointed out as a drawback in **Subheading 3.4**, there are no distant forces between proteins. Finally, the use of a hard sphere model for contacts and rigid binding for local interactions does not account for the inherent flexibility of proteins and aggregates. Hence, an insufficient degree of realism of this model drove us to define a new model with better physical relevance.

4.2. An Energetic Version of the Model

The real inspiration from MD models starts with this version as we use a part of its framework for defining the physical interaction of proteins. But being at a higher scale, we can make some simplifying approximations, particularly on the law of motion, that make the model much simpler.

4.2.1. The Protein–Protein Pair Potential

Proteins are represented by their volumic barycenter and interact through a potential energy whose shape accounts for both the steric volume (equivalent of contact forces) and distant forces. The potential energy function between proteins i and j is given by

$$V_{ij} = \varepsilon\left[\left(\frac{\sigma}{r_{ij}}\right)^{12} - 2\left(\frac{\sigma}{r_{ij}}\right)^{6}\right] + \frac{Q}{r_{ij}}e^{-r_{ij}/r^*} \qquad (1)$$

in which ε, σ, and Q are parameters defined for every pair of protein species, and r^* depends on the solvent (such as ionic conditions, pH, and temperature).

The expression of V_{ij} is inspired from the noncovalent interaction term of classical all-atom MD models *(23,24)* (**Fig. 1a**). In this potential energy function

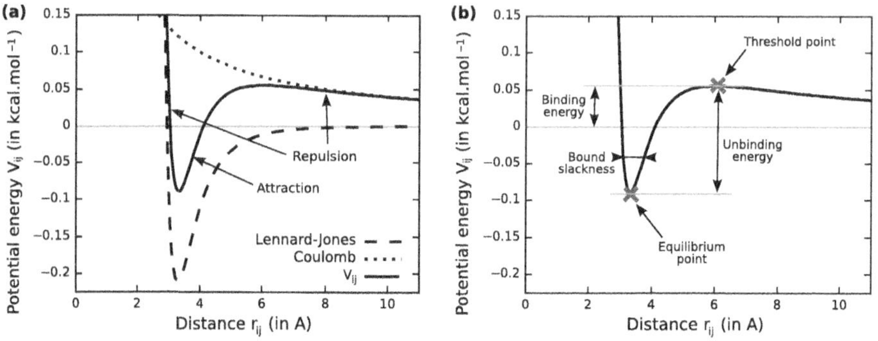

Fig. 1. Typical nonbounded potential energy of MD models between two atoms as a function of the separation distance r_{ij} The corresponding function is equivalent to Eq. (1) without e^{-r_{ij}/r^*}. (**a**) The interatomic force at a given distance is attractive or repulsive depending on the slope of this function. (**b**) The behavior of the system is well characterized by the positions of the *equilibrium point* of binding and the *threshold point* that delimits the two basins of attraction.

between two atoms separated by a distance r_{ij}, the first two terms (respectively, repulsive at a very short distance and attractive at a longer distance) correspond to the 6–12 Lennard–Jones empirical potential accounting for van der Waals interactions, and the third term is the Coulomb interaction (here, the term e^{-r_{ij}/r^*} accounts for implicit solvent screening (*see* **Note** 4), the fact that the Coulomb force tends to vanish with distance because of ion clouds forming around charged domains).

The force $\mathbf{F}_{\mathbf{ij}}$ of particle j on particle i is the opposite of the gradient (i.e., the derivative) of their potential energy as a function of the position $\mathbf{x}_\mathbf{i}$ of i.

$$\mathbf{F}_{\mathbf{ij}} = -\nabla \left[V_{ij}(\mathbf{x}_\mathbf{i}) \right] \tag{2}$$

In other words, a negative (respectively positive) slope of V_{ij} corresponds to a repulsive (respectively attractive) force (**Fig. 1a**). This implies a tendency for particles to minimize their interaction energy. In any case of parameterization, the potential presents a well corresponding to the *equilibrium point* of binding (**Fig. 1b**). When the Coulomb force is repulsive, the interaction presents a second basin of attraction delimited by a *threshold point* (**Fig. 1b**). These two points (four degrees of freedom) nonetheless fully define the parameter set (four parameters), but are also quite representative of the interaction: their position defines the energy necessary for particles to bind and unbind as well as the bound slackness (**Fig. 1b**). When the Coulomb force is attractive, the binding energy is null and the unbinding energy is the depth of the potential well.

This potential energy function is able to reproduce qualitatively the shape of many physically motivated coarse-grained potentials *(33–35)* in terms of attraction basin equilibrium points and energy barriers. But the idea to port this potential function to the protein scale and approximate the whole macromolecular behavior to this single potential raises new questions about how to render protein dynamics, softnes, and flexibility *(48)*.

There are two options to determine protein–protein potentials from the atom–atom potentials (**Fig. 2**). The first one is to consider the core atoms of the proteins as forming a rigid and undeformable body and consider only the surface atoms. This strategy is used, for instance, to study the influence of a binding partner (modeled rigidly) on the folding process of a protein (modeled classically). In our case, r_{ij} would represent the distance between the two protein surfaces, resulting in a shift of the energy function (**Fig. 2a**). But this choice does not reproduce the characteristic softness of proteins *(48)*.

A simple way to render protein softness is to have r_{ij} represent the distance between protein centers and to scale this function in distance (**Fig. 2b**) by simply placing the equilibrium and threshold point consequently. Indeed, viewing the superimposition of atoms as a superimposition of nonlinear springs with the energy function of Eq.(1), the resulting protein–protein interaction is reduced to such a scaled energy function assuming the condition of having a single basin of attraction. When this is not the case (i.e., the Coulomb force is repulsive) some hysteresis behavior could appear (**Fig. 3**), but it may be counteracted (*see* **Note 5**) by the presence of noise in the system (cf. **Subheading 4.2.2**). As a result, the representation of a soft protein–protein interaction by a single potential is a correct approximation.

Therefore, ignoring the anisotropy of molecular recognition, which depends on the atom–atom matching of the two protein surfaces, we represent every protein–protein interaction as being isotropic but species pair dependent. In an all-atom MD, parameters are defined for every species and pair potential parameters are derived from them. In contrast, and like many physically motivated coarse-grained approaches *(34–36)*, pair potential parameters directly constitute the parameter set in order to overcome the loss of information due to coarsening. Thus, in our model, a pair of species can have a specific interaction through topological matching or mismatching (deep or shallow Lennard–Jones potential well ε, respectively) and/or electrostatic surface charge matching or mismatching (negative or positive important value of Q, respectively). Nonspecific interactions would correspond to a relatively shallow depth of the Lennard–Jones potential well ε and a neutral ($Q \approx 0$) Coulomb interaction (*see* **Note 6**). The variety of interactions that can be obtained with different parameters results in a good diversity of situations that can be expressed with this model.

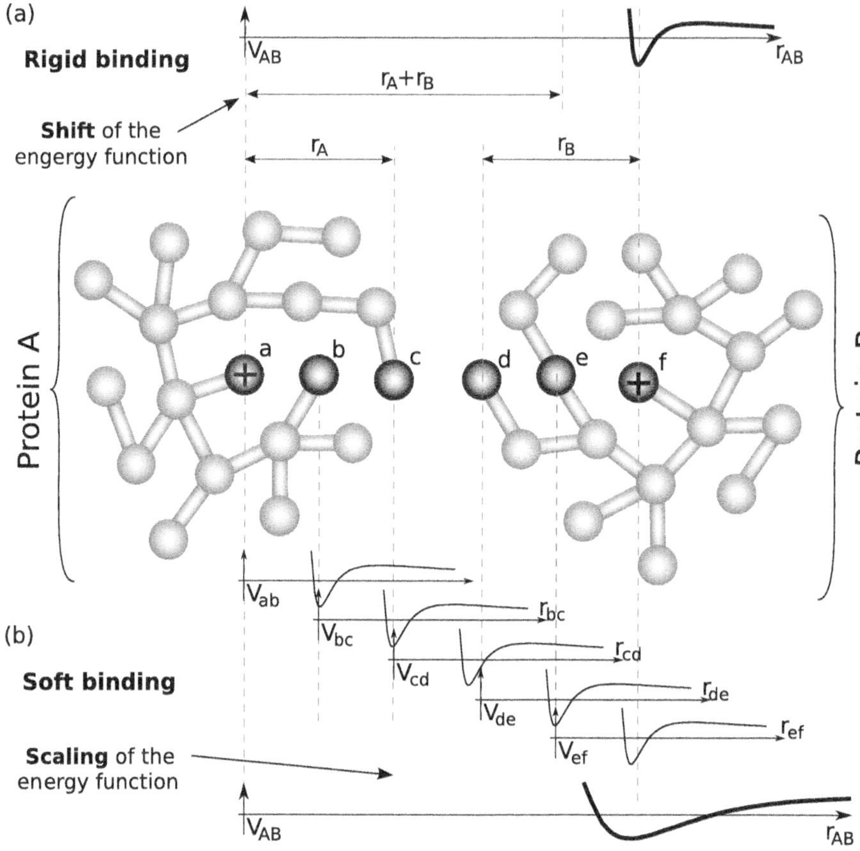

Fig. 2. Protein–protein potential energy can be derived from atom–atom potentials in two ways: (**a**) by shifting the energy function of surface atoms with the two protein radii if buried atoms are considered to form a rigid body, or (**b**) by scaling the energy function to account for the softness of the proteins.

Provided with a potential accounting for many types of interactions, finding realistic parameters remains a difficult task, as for many coarse-grained models. For physically motivated potentials (in opposition to final state-motivated potentials such as Gō models and ENMs), a classical solution is to reproduce some known physical characteristics of the system (such as density, surface tension, and radial distribution function), obtained either experimentally or from all-atom MD simulations, with an appropriate method for inferring parameters ***(36,49)***. In our case, the known physical measures from which protein-scale parameters (i.e., binding and unbinding energy; **Fig. 1b**) can be derived can be, for instance, biacore measures ***(50)*** of association and dissociation constants between any pair

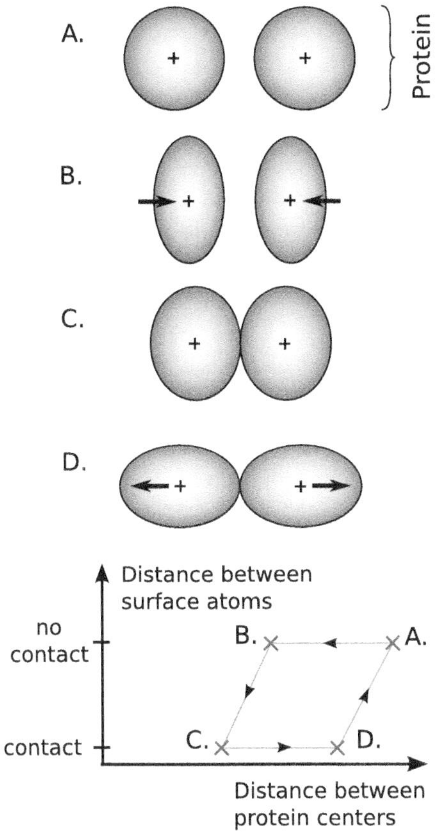

Fig. 3. Hysteresis in soft protein binding appears when surface atoms have to pass a potential barrier. Because of distortion, when two such proteins get closer, they make contact at a distance shorter than the distance at which they break this contact when they are taken away.

of protein species in different conditions (such as concentration, ionic conditions, and temperature). Another approach to the problem is not to set parameters at a precise value, but rather to explore intervals of realistic values and characterize the behavior of the system for the different points of this phase space ***(51)***.

4.2.2. The Law of Motion

Biological observations of protein mobility in the nucleus have revealed energy-independent (in opposition to active transport observed in the cytoplasm; i.e., tubulin filaments) normal diffusion ***(5,18)***. In accordance with these observations, we use a law of motion known as the *overdamped Langevin equation*

$$\lambda_i \dot{\mathbf{x}}_\mathbf{i} = \sum_{j \neq i} \mathbf{F}_{\mathbf{ij}} + \boldsymbol{\xi}_\mathbf{i} \tag{3}$$

It describes the motion of protein i at position $\mathbf{x_i}$ with an implicit solvent accounting for both a Brownian noise $\boldsymbol{\xi}_\mathbf{i}$ (corresponding to the random collisions of solvent molecules on the protein) and a dissipation force of a viscosity coefficient λ_i (being the resistance of solvent molecules against the protein motions). Since proteins are considered spheres, the viscosity coefficient is

$$\lambda_i = 6\pi\, r_i \eta \tag{4}$$

where r_i is the protein radius and η is the dynamic viscosity of the solvent (which depends on temperature). The Brownian force $\boldsymbol{\xi}_\mathbf{i}$ is a 3D protein- and time-decorrelated (i.e., white) Gaussian noise such that

$$\langle \boldsymbol{\xi}_\mathbf{i}(t) \rangle = \mathbf{0} \tag{5}$$

$$\left\langle \boldsymbol{\xi}_\mathbf{i}(t) \boldsymbol{\xi}_\mathbf{i}(t') \right\rangle = 6\lambda_i k_B T \delta_{ij} \delta(t - t') \tag{6}$$

where k_B is the Boltzmann constant, T is the temperature, δ_{ij} is the Kronecker delta, and $\delta(t)$ is the Dirac function. This microscopic description of the Brownian motion as a random walk can be very simply related to the macroscopic theories of diffusion (i.e., Fick's law) *(52)*.

Equation (3) is derived from the classical Langevin equation—Newton's second law on a particle of mass m_i —usually used in MD:

$$m_i \ddot{\mathbf{x}}_\mathbf{i} = \sum_{j \neq i} \mathbf{F}_{\mathbf{ij}} - \lambda_i \dot{\mathbf{x}}_\mathbf{i} + \boldsymbol{\xi}_\mathbf{i} \tag{7}$$

Such a particle is known to have a diffusion-driven behavior on long time scales (then $m_i \ddot{\mathbf{x}}_\mathbf{i}$ is negligible) and to demonstrate inertial motions only for short periods of time (then $\lambda_i \dot{\mathbf{x}}_\mathbf{i}$ is negligible). But noting that $m_i \propto r_i^3$ and $\lambda_i \propto r_i$ at our scale (almost the smallest that have a fully implicit solvent) we have $m_i << \lambda_i$ (more than 10 orders of magnitude). So, the inertia term $m_i \ddot{\mathbf{x}}_\mathbf{i}$ remains negligible compared to the viscosity term $\lambda_i \dot{\mathbf{x}}_\mathbf{i}$ down to very short time scales (**Fig. 4**).

This yields the much simpler equation of motion (3), which does not depend on acceleration $\ddot{\mathbf{x}}_\mathbf{i}$ but only on velocity $\dot{\mathbf{x}}_\mathbf{i}$. This very simple difference has important consequences. For instance, it is simpler and faster to integrate. Indeed, models based on Eq. (7) need to use expensive integrators and/or renormalization algorithms to prevent energy divergence through the inertia term due

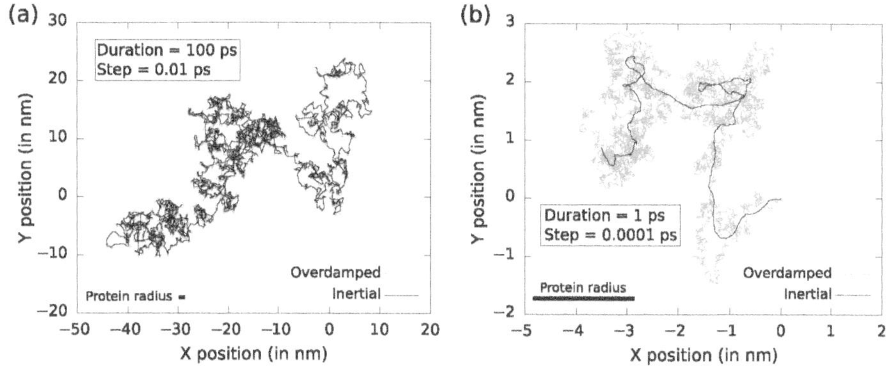

Fig. 4. Comparison between trajectories of a protein using the overdamped version [Eq. (3); gray lines] and the inertial version [Eq. (7); black lines] of the Langevin equation. Parameters correspond roughly to the hemoglobin protein. (**a**) For time steps down to the hundredth of picoseconds (10^{-14}s), the two equations have very similar behavior. (b) For shorter time steps the two trajectories start to differ significantly.

to integration approximations. In our case, in addition to having only one level of integration (positions from velocities and not velocities from accelerations), the integration is easier because we can use a simple Euler integrator without any energy divergence: the expectation of $\sum_{j \neq i} \mathbf{F}_{ij}$ being null, the temperature of the systems (defined by the velocities distribution) is directly determined by ξ_i, which is defined for a fixed temperature T in Eq.(6).

Moreover, systems driven by Eq.(3) are easier to study analytically because they are Markov systems and there are many more mathematical tools for their analysis.

As a result, this energetic version provides a simple way to represent proteins with physical characteristics that are important for mesoscale studies: steric volume exclusion with soft contact, distant forces of realistic shape (multiple basins of attraction), and flexible binding. This simplified formalism of interactions, coupled with the efficient law of motion, reduces the computational load of simulations and allows protein systems to be studied at a mesoscopic size.

4.2.3. Applications

The second version of the model is implemented and has been used to carry out some preliminary experiments (*see* **Note 7**). It has already demonstrated physically realistic behavior (i.e., liquid–gas phase transition) as well as a spontaneous organization in helix and sheet structures *(53)*.

We present here a simple experiment showing how changing the binding and unbinding energies (**Fig. 1b**) can influence the dynamics of disruption of a small

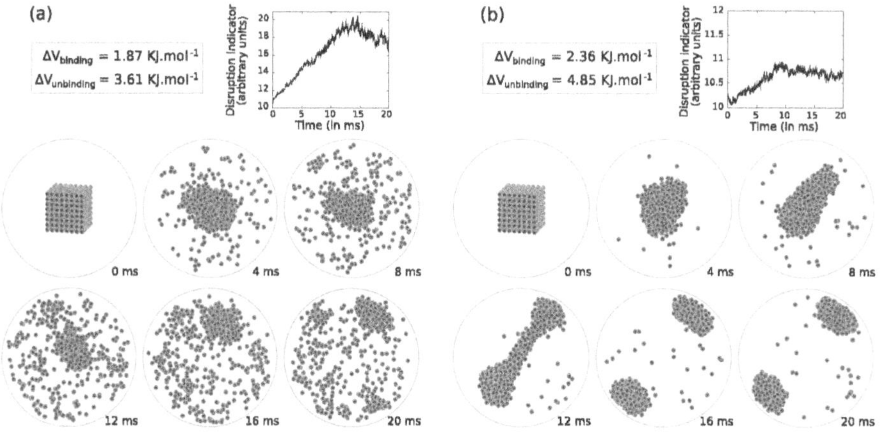

Fig. 5. Influence of binding and unbinding energies (cf. **Fig. 1b**) on the disruption process of a small multiprotein structure. (**a, b**) The stabilization of the disruption indicator in both cases shows that the proportion of aggregated and disaggregated proteins reaches equilibrium within the simulation time. Note that the split of the aggregate in (**b**) is not a deterministic behavior.

simple multiprotein structure (**Fig. 5**). The steady-state concentration outside the aggregates is very dependent on these energies and is maintained by the presence of the aggregate. So this experiment reproduces a behavior that has been suggested to be a role for nuclear bodies *(6)*: the presence of an aggregate of certain protein species ensures the conservation of a precise concentration of these species all over the nuclear space. This is typically a mesoscopic behavior that could not be reproduced with a smaller system (such as those of MD and coarse-grained studies) or without incorporating physical properties of proteins (such as distant forces and Brownian motion).

Moreover, in these experiences, tracing the trajectory of a given protein (data not shown) (*see* **Note 7**) reveals that aggregates are very dynamic (proteins enter and leave the aggregate constantly) and fluid (proteins have high mobility within the aggregates), still in accordance with observations and hypotheses *(2–5)*.

These first results constitute an interesting starting point and call for further investigations to propose or test hypothesis regarding nuclear dynamics.

4.3. Forthcoming Features and Perspectives

With this model, we reached a probably sufficient degree of physical realism for our mesoscale purpose. However, if we need to attain greater physical realism, there would be some additional improvement regarding forces that could be of importance. For instance, an improvement that would pose much difficulty is to include more realistic solvent-induced forces. In the majority of coarsened models, hydrophobic forces are not treated as such; they are treated by simply adjusting the potential energy parameters so that grains known to be hydrophobic are attracted. More realistic methods for implicit solvents have been known for a long time (Poisson–Boltzmann equation, generalized Born approximation, surface accessible solvent area) but they are quite costly and are not simple to apply to a force field (nonpairwise additivity). A number of teams are currently working on the question of coarse-grained models and are proposing some simplified solutions *(33)*.

Related to this problem is the question of partially implicit molecular crowding. If we do not want to simulate explicitly every protein in the studied region we have to consider their effect in a fashion that is the same as that for an implicit solvent. But the difficulty comes from the fact that some of the proteins are implicit and some are explicit. Indeed, the rules applied to a protein should be different based on whether the protein is or is not surrounded by other explicit proteins.

Although we can already address several mesoscale biological questions, we can envisage some improvements toward attaining a better biological realism. The most important of these is to introduce anisotropy in protein interactions in order to account for the structural nature of specific binding. There could be several ways to achieve this. One possibility would be to introduce a direction-dependent potential as inspired from studies *(54,55)* on interresidue anisotropic potentials. Another more computationally expensive solution would be to represent proteins by a set of potential sources instead of a single one. But this raises the question of how to maintain them together. Possible solutions could be either by rigid bonds (forces and momentum would still apply to the protein barycenter) or, once again inspired from smaller scale works, by MD- or Go-like bonds, or by a set of springs as in ENMs. The computational overhead will be significant in any case, so choices have to be made carefully in order to minimize it.

An additional feature that would enable another category of experiments would be to include biochemical reactions and transformations in the model, such as in *(43)*. But in the case of steric proteins in a crowded context, the question of where new proteins appear after the reaction is a crucial issue. For example, a reaction in which products have a greater volume than reactants (i.e., proteosynthesis, because in our model amino acids are not present explicitly) will necessarily make products appear in the excluded volume of neighboring proteins, inducing very important and unrealistic forces. The use of a smooth fading from reactants to products could be a solution to this problem. We can also note that there is a need to distinguish a product that is created and a product that is transformed from a reactant either spontaneously (i.e., denaturation, degradation) or because of another molecule (i.e., phosphorylation, enzyme catalysis, change in conformation upon binding). Indeed, the transformed molecule has to remain at the same place before and after the transformation in order not to perturb artifactually the structure in which the protein is involved. The challenge is to add this feature to the model in a realistic way while retaining a simple formalism.

All these future improvements raise a certain number of questions and would necessarily complicate the model. In our approach to modeling, it is important that the model remains sufficiently simple so that we can understand the causes of the observed phenomena and explain them through mathematical analysis. Indeed, a phenomena demonstrated by a very complicated model cannot really be explained in the sense that it is difficult to assess its precise cause. So improvements have to be considered with much attention to this point.

5. Conclusions

To fill in the gap corresponding to the mesoscopic scale in the spectrum of models from atomic to macroscopic scales, we first quickly reviewed higher and lower scale models and highlighted their advantages and disadvantages in addressing questions concerning mesoscale phenomena. Basically, higher scale models lack some important features of proteins that are needed to understand some of the known mesoscale phenomena: discreteness of matter, spatial situation of molecules, steric volume, and distant forces. On the other hand, lower scale models are either too precise (all-[heavy]-atoms MD) and do not allow the simulation of sufficiently many proteins for our interest or, in the case of less precise models (coarse-grained models such as Go and ENMs), many of them make assumptions on interactions that are relevant for their questions (such as folding and vibrational dynamics) but not for ours. More physically motivated coarse-grained models can be of greater interest.

We then developed our own model inspired from MD and coarse-grained model formalisms and achieved a further level of abstraction. Interactions are based on a pair potential energy function equivalent to noncovalent interactions in MD models, ported to the protein scale (accounting for protein softness in both contact and binding) and protein species pair-specific. Proteins are moved according to a simplified law of motion that takes into account the effect of the solvent. Perspectives for improvement are the inclusion of biochemical reactions and transformations, anisotropy in the interactions, and more realistic implicit molecule-induced forces (either hydrophobic or crowding forces).

This provides us with protein models realistic enough to address mesoscale questions, but simple enough for simulation within a reasonable time and realistic enough to explain phenomena.

6. Notes

1. Fluorescence loss in photobleaching and fluorescence recovery after photobleaching *(2,3)*.
2. ENMs classically assign one residue per bead, but some studies have used larger beads.
3. Agent-based modeling (ABM) or multiagent systems (MAS) constitute a framework from theoretical computer science and complexity theory. It consists of the simulation of a large number of simple entities (with basic behavior and interaction rules) demonstrating a particular collective behavior not deducible from the local rules.
4. There are various expressions for solvent screening. For instance, in a previous publication *(53)* we used the term Q/(rij+r0) for Coulomb energy.
5. Noise in a hysteretic system is known to flatten the hysteresis.
6. The Go strategy adopts a purely repulsive interaction between nonnatively contacting residues pairs. However, and more physically relevant, our nonspecific interactions still have a (although weak) basin of attraction.
7. Refer to http://bsmc.insa-lyon.fr/moceme/.

Acknowledgments

The authors wish to thank all the members of the BSMC (http://bsmc.insa-lyon.fr/) and SMABio (http://www.phys-mito.u-bordeaux2.fr/mitowiki/index.php?title=GT_SMA_%26_Simulation_Cellulaire) groups for stimulating discussions, as well as the IN2P3 computing center for providing access to their resources. This work is funded by the ACI IMPBio program and by the Rhône-Alpes region through a Ph.D. grant from the cluster ''Informatique, signal, logiciels embarqués" to A.C.

References

1. Matera, A. G. (1999) Nuclear bodies: multifaceted subdomains of the interchromatin space. *Trends Cell Biol.* **9**, 302–309.
2. McNally, J. G., Müller, W. G., Walker, D., Wolford, R., and Hager, G. L. (2000) The glucocorticoid receptor: rapid exchange with regulatory sites in living cells. *Science* **287**, 1262–1265.
3. Phair, R. D., Scaffidi, P., Elbi, C., Vecerova, J., Dey, A., Ozato, K., Brown, D. T., Hager, G., Bustin, M., and Misteli, T. (2004) Global nature of dynamic protein-chromatin interactions in vivo: three-dimensional genome scanning and dynamic interaction networks of chromatin proteins. *Mol. Cell. Biol.* **24**(14), 6393–6402.
4. Handwerger, K. E. and Gall, J. G. (2006) Subnuclear organelles: new insights into form and function. *Trends Cell Biol.* **16**, 19–26.
5. Misteli, T. (2001) Protein dynamics: implications for nuclear architecture and gene expression. *Science* **291**, 843–847.

6. Misteli, T. (2005) Concepts in nuclear architecture. *BioEssays* **27**, 477–487.
7. Cremer, T., Cremer, M., Dietzel, S., Müller, S., Solovei, I., and Fakan, S. (2006) Chromosome territories—a functional nuclear landscape. *Curr. Opinion Cell Biol.* **18**, 307–316.
8. Branco, M. R. and Pombo, A. (2006) Intermingling of chromosome territories in interphase suggests role in translocations and transcription-dependent associations. *PLoS Biol.* **4(5),** 0780–0788.
9. Kupiec, J.-J. (1997) A Darwinian theory for the origin of cellular differentiation. *Mol. Gen. Genet.* **255**, 201–208.
10. Blake, W. J., Kærn, M., Cantor, C. R., and Collins, J. J. (2003) Noise in eukaryotic gene expression. *Nature* **422**, 633–637.
11. Levsky, J. M. and Singer, R. H. (2003) Gene expression and the myth of the average cell. *Trends Cell Biol.* **13**, 4–6.
12. Kærn, M., Elston, T. C., Blake, W. J., and Collins, J. J. (2005) Stochasticity in gene expression: from theories to phenotypes. *Nat. Rev. Genet.* **6**, 451–464.
13. Sigal, A., Milo, R., Cohen, A., Geva-Zatorsky, N., Klein, Y., Liron, Y., Rosenfeld, N., Danon, T., Perzov, N., and Alon, U. (2006) Variability and memory of protein levels in human cells. *Nature* **444**, 643–646.
14. Halford, S. E. and Marko, J. F. (2004) How do site-specific DNA-binding proteins find their targets? *Nucleic Acids Res.* **32(10),** 3040–3052.
15. van Zon, J. S., Morelli, M. J., Tanase-Nicola, S., and ten Wolde, P. R. (2006) Diffusion of transcription factors can drastically enhance the noise in gene expression. *Biophys. J.* **91**, 4350–4367.
16. Amar, P., Ballet, P., Barlovatz-Meimon, G., Benecke, A., Bernot, G., Bouligand, Y., Bourguine, P., Delaplace, F., Delosme, J.-M., Demarty, M., Fishov, I., Fourmentin-Guilbert, J., Fralick, J., Giavitto, J.-L., Gleyse, B., Godin, C., Incitti, R., Képès, F., Lange, C., Sceller, L. L., Loutellier, C., Michel, O., Molina, F., Monnier, C., Natowicz, R., Norris, V., Orange, N., Pollard, H., Raine, D., Ripoll, C., Rouviere-Yaniv, J., Jr., M. S., Soler, P., Tambourin, P., Thellier, M., Tracqui, P., Ussery, D., Vincent, J.-C., Vannier, J.-P., Wiggins, P., and Zemirline, A. (2002) Hyperstructures, genome analysis and Icell. *Acta Biotheor.* **50**(4), 357–373.
17. Chambeyron, S. and Bickmore, W. A. (2004) Chromatin decondensation and nuclear reorganization of the HoxB locus upon induction of transcription. *Genes Dev.* **18**, 1119–1130.
18. Bork, P. and Serrano, L. (2005) Towards cellular systems in 4D. *Cell* **121**, 507–509.
19. Takahashi, K., Arjunan, S. N. V., and Tomita, M. (2005) Space in systems biology of signaling pathways—towards intracellular molecular crowding in silico. *FEBS Lett.* **579**, 1783–1788.
20. Lemerle, C., Ventura, B. D., and Serrano, L. (2005) Space as the final frontier in stochastic simulations of biological systems. *FEBS Lett.* **579**, 1789–1794.
21. Turing, A. M. (1952) The chemical basis of morphogenesis. *Phil. Trans. Royal Soc. Lond. B* **327**, 37–72.
22. Carrero, G., Hendzel, M. J., and de Vries, G. (2005) Modelling the compartmentalization of splicing factors. *J. Theor. Biol.* **239**(3), 298–312.

23. Cornell, W. D., Cieplak, P., Bayly, C. I., Gould, I. R., Merz Jr., K. M., Ferguson, D. M., Spellmeyer, D. C., Fox, T., Caldwell, J. W., and Kollman, P. A. (1995) A second generation force field for the simulation of proteins, nucleic acids, and organic molecules. *J. Am. Chem. Soc.* **117**, 5179–5197.
24. MacKerell, A. D., Jr., Wiórkiewicz-Kuczera, J., and Karplus, M. (1995) An all-atom empirical energy function for the simulation of nucleic acids. *J. Am. Chem. Soc.* **117**, 11946–11975.
25. Hobza, P., Kabeláč, M., Šponer, J., Mejzlík, P., and Vondrašêk, J. (1997) Performance of empirical potentials (AMBER, CFF95, CVFF, CHARMM, OPLS, POLTEV), semiempirical quantum chemical methods (AM1, MNDO/M, PM3), and ab initio Hartree-Fock method for interaction of DNA bases: comparison with nonempirical beyond Hartree-Fock results. *J. Comp. Chem.* **18**(9), 1136–1150.
26. Tozzini, V. (2005) Coarse-grained models for proteins. *Curr. Opinion Struct. Biol.* **15**, 144–150.
27. Koga, N. and Takada, S. (2001) Roles of native topology and chain-length scaling in protein folding: a simulation study with a Go-like model. *J. Mol. Biol.* **313**, 171–180.
28. Takagi, F., Koga, N., and Takada, S. (2003) How protein thermodynamics and folding mechanisms are altered by the chaperoning cage: molecular simulations. *Proc. Natl. Acad. Sci. USA* **100**, 11367–11372.
29. Levy, Y., Caflisch, A., Onuchic, J., and Wolynes, P. (2004) The folding and dimerization of HIV-1 protease: evidence for a stable monomer from simulations. *J. Mol. Biol.* **340**, 67–79.
30. Chacon, P., Tama, F., and Wriggers, W. (2003) Mega-dalton biomolecular motion captured from electron microscopy reconstructions. *J. Mol. Biol.* **326**, 485–492.
31. Delarue, M. and Dumas, P. (2004) On the use of low-frequency normal modes to enforce collective movements in refining macromolecular structural models. *Proc. Natl. Acad. Sci. USA* **101**, 6957–6962.
32. Tama, F., Miyashita, O., and Brooks, C. I. (2004) Normal mode based flexible fitting of high-resolution structure into low-resolution experimental data from cryo-EM. *J. Struct. Biol.* **147**, 315–326.
33. Head-Gordon, T. and Brown, S. (2003) Minimalist models for protein folding and design. *Curr. Opinion Struct. Biol.* **13**(2), 160–167.
34. Jiang, L., Gao, Y., Mao, F., Liu, Z., and Lai, L. (2001) Potential of mean force for protein-protein interaction studies. *Proteins: Struct. Funct. Genet.* **46**(2), 190–196.
35. Lyubartsev, A. P. (2005) Multiscale modeling of lipids and lipid bilayers. *Eur. Biophys. J.* **35**, 53–61.
36. Nielsen, S. O., Lopez, C. F., Srinivas, G., and Klein, M. L. (2004) Coarse grain models and the computer simulation of soft materials. *J. Phys. Condens. Matter* **16**, R481–R512.
37. Shelley, J. C., Shelley, M. Y., Reeder, R. C., Bandyopadhyay, S., and Klein, M. L. (2001) A coarse grain model for phospholipid simulations. *J. Phys. Chem. B* **105(16),** 4464–4470.

38. Shelley, J. C., Shelley, M. Y., Reeder, R. C., Bandyopadhyay, S., Moore, P. B., and Klein, M. L. (2001) Simulations of phospholipids using a coarse grain model. *J. Phys. Chem. B* **105**(40), 9785–9792.
39. Dittrich, P., Ziegler, J., and Banzhaf, W. (2001) Artifcial chemistries—a review. *Artificial Life* **7**, 225–275.
40. Ballet, P., Zemirline, A., and Marce L. (2004) The BioDyn language and simulator. Application to an immune response and E. coli and phage interaction. *J. Biol. Phys. Chem.* **4**(2), 93–101.
41. Amar, P., Bernot, G., and Norris, V. (2004) HSIM: a simulation programme to study large assemblies of proteins. *J. Biol. Phys. Chem.* **4(2)**, 79–84.
42. Lales, C., Parisey, N., Mazat, J.-P., and Beurton-Aimar, M. (2005) Simulation of mitochondrial metabolism using multi-agents system. *Proc. MAS*BIOMED'05*, 137.
43. Andrews, S. S. and Bray, D. (2004) Stochastic simulation of chemical reactions with spatial resolution and single molecule detail. *Phys. Biol.* **1**, 137–151.
44. Ellis, R. J. (2001) Macromolecular crowding: obvious but underappreciated. *Trends Biochem. Sci.* **26(10),** 597–603.
45. Hancock, R. (2004) A role for macromolecular crowding effects in the assembly and function of compartments in the nucleus. *J. Struct. Biol.* **146**, 281–290.
46. Banks, D. S. and Fradin, C. (2005) Anomalous diffusion of proteins due to molecular crowding. *Biophys. J.* **89**, 2960–2971.
47. Soula, H., Robardet, C., Perrin, F., Gripon, S., Beslon, G., and Gandrillon, O. (2005) Modeling the emergence of multi-protein dynamic structures by principles of self-organization through the use of 3DSpi, a multi-agent-based software. *BMC Bioinform.* **6**, 228.
48. Zaccai, G. (2000) How soft is a protein? A protein dynamics force constant measured by neutron scattering. *Science* **288(5471),** 1604–1607.
49. Nielsen, S. O., Lopez, C. F., Srinivas, G., and Klein, M. L. (2003) A coarse grain model for n-alkanes parameterized from surface tension data. *J. Chem. Phys.* **119(14),** 7043–7049.
50. Fivash, M., Towler, E. M., and Fisher, R. J. (1998) BIAcore for macromolecular interaction. *Curr. Opin. Biotechnol.* **9(1),** 97–101.
51. Hoang, T. X., Trovato, A., Seno, F., Banavar, J. R., and Maritan, A. (2004) Geometry and symmetry presculpt the free-energy landscape of proteins. *Proc. Natl. Acad. Sci. USA* **101(21),** 7960–7964.
52. Berg, H. C. (1993) *Random Walks in Biology*, 2nd ed. Princeton University Press, Princeton, NJ.
53. Coulon, A., Soula, H., Mazet, O., Gandrillon, O., and Beslon, G. (2007) Modélisation cellulaire pour l'émergence de structures multiprotéiques auto-organisées. *Tech. Sci. Inform.* **26**, 123–148.
54. Buchete, N.-V., Straub, J. E., and Thirumalai, D. (2004) Orientation-dependent coarse-grained potentials derived by statistical analysis of molecular structural databases. *Polymer* **45(2),** 597–608.
55. Mukherjee, A., Bhimalapuram, P., and Bagchi, B. (2005) Orientation-dependent potential of mean force for protein folding. *J. Chem. Phys.* **123**, 014901-1-11.

33

Dynamic Pathway Modeling of Signal Transduction Networks: A Domain-Oriented Approach

Holger Conzelmann and Ernst-Dieter Gilles

Summary

Mathematical models of biological processes become more and more important in biology. The aim is a holistic understanding of how processes such as cellular communication, cell division, regulation, homeostasis, or adaptation work, how they are regulated, and how they react to perturbations. The great complexity of most of these processes necessitates the generation of mathematical models in order to address these questions. In this chapter we provide an introduction to basic principles of dynamic modeling and highlight both problems and chances of dynamic modeling in biology. The main focus will be on modeling of s transduction pathways, which requires the application of a special modeling approach.

A common pattern, especially in eukaryotic signaling systems, is the formation of multi protein signaling complexes. Even for a small number of interacting proteins the number of distinguishable molecular species can be extremely high. This combinatorial complexity is due to the great number of distinct binding domains of many receptors and scaffold proteins involved in signal transduction. However, these problems can be overcome using a new domain-oriented modeling approach, which makes it possible to handle complex and branched signaling pathways.

Key Words: Mathematical model; signal transduction pathways; multi protein complex formation; binding domains; combinatorial complexity; model reduction; detailed balance.

1. Introduction

The complexity of cellular reaction networks most often does not facilitate an intuitive understanding of how genes, proteins, metabolites, and other cellular substances work together. In the field of systems biology, mathematical

From: *Methods in Molecular Biology, vol. 484: Functional Proteomics: Methods and Protocols*
Edited by: J. D. Thompson et al., DOI: 10.1007/978-1-59745-398-1, © Humana Press, Totowa, NJ

models are used to access the complexity of these networks quantitatively. In this chapter we focus on *dynamic* modeling, which makes it possible to describe the transient behavior of biological networks. The possibilities that are opened through dynamic modeling of biological networks are enormous. All kinds of *in silico* experiments are feasible, which in reality would be time consuming, expensive, or even impossible to accomplish. This includes deletion or addition of components and interactions, or the change of kinetic properties. Additionally, systems theory provides a broad spectrum of mathematical analysis tools, which may provide numerous suggestions for experimental design or drug target identification.

However, a requirement for high-quality contributions from theory is the existence of a well-founded mathematical model. Such models are mostly formulated using ordinary differential equations (ODEs). In **Section 2** we will provide a brief introduction to the underlying mathematics, which, however, does not claim to be a complete depiction. The interested reader may find much additional information about mathematical tools in systems biology in Klipp *et al.* ***(1)***.

The focus of this chapter is on modeling signal transduction pathways, since they are, as well as regulation networks, highly dynamic processes. At the same time, there are special problems in modeling these systems. Receptors and scaffold proteins, which usually possess a large number of distinct binding domains, induce the formation of large multiprotein signaling complexes. Because of combinatorial reasons the number of distinguishable species grows exponentially with the number of binding domains and can easily reach several million ***(2)***. Most models published in the literature do not account for this combinatorial variety. Their focus is on small subsets of reactions and complexes. The main difficulty with such reduced model structures is in having to decide which reactions and complexes can be neglected and which are essential ***(3)***. In particular, heuristic reductions may lead to incorrect models, as we will show below.

A more systematic approach has been suggested by Blinov *et al.*, who introduced the software tool BioNetGen ***(4)***. BioNetGen allows a rule-based model formulation to be translated into a complete ODE model accounting for all feasible reactions and species. However, even with a strongly limited number of components and binding domains, the resulting models are already very large and barely manageable. For these reasons a novel approach should be preferred to deal with this combinatorial complexity ***(5,6)***. This approach is based on the belief that protein domains and not individual multiprotein species are the fundamental elements of signal transduction. According to this, the conventional mechanistic description of all feasible multiprotein complexes is replaced by a more macroscopic one. Occupancy levels and other characteristics of individual domains, e.g., the phosphorylation states of these sites, are chosen as new variables. A model using these macroscopic quantities also accounts for limitations in current experimental techniques to measure concentrations of individual multiprotein species. The results of common biological measurements (e.g., immunoprecipitation followed by Western blotting) correspond to cumulative quantities such as levels of occupancy or degrees of phosphorylation. Thus, the introduction of these and similar quantities into modeling simplifies the comparison of model variables with experimental readouts. Besides these considerations, this macroscopic description also provides a number of mathematical benefits. The method permits an accurate description of macroscopic quantities in strongly reduced models. As described below, the only required precondition for this reduction is the presence of domains that do not interact allosterically with each other.

In **Heading 4** some general problems of modeling reaction networks as well as problems using the domain-oriented modeling approach shall be discussed. Additionally, we provide an introduction to parameter identification. Parameter identification is a highly mathematical subject dealing with questions such as which and how many measurements are required to identify, i.e., to assign values to, kinetics. Here a more descriptive overview of the possibilities and limitations in parameter identification is given.

2. Materials

2.1. Differential Equations and Kinetic Modeling

Dynamic models describe the transient behavior of a system using ordinary differential equations (ODEs). One of the most simple ODEs probably is

$$\dot{x} = \frac{dx}{dt} = k \cdot x$$

With initial condition $x(0) = x_0$ its general solution is $x(t) = x_0 e^{kt}$. This could be a model describing both exponential growth (e.g., growth of a bacterial culture) or exponential decay (e.g., radioactive decay) depending on the sign of k. Dynamic models describing signaling networks naturally are much more complicated, including a large number of coupled ODEs, whose solution usually cannot be analytically determined. However, the ODE models can be numerically solved and the transient behavior therefore can be simulated. The setting up of ODEs for reaction networks is based on the mathematical formulation of reaction rates. This shall be demonstrated considering a simple reversible reaction

$$\mathrm{A} + \mathrm{B} \leftrightharpoons \mathrm{C}$$

The concentrations of the participating species are denoted as [A], [B], and [C] and are usually given in moles per liter. Mostly the volume of the modeled system is assumed to be constant. In this case the ODE of a certain concentration can be determined by summing up the rates of all reactions in which this component is involved, multiplied by its stoichiometric coefficient. By definition the stoichiometric coefficients of products are positive, while those of reactants are negative. This is due to the fact that in each molecular reaction step the reactants are consumed, while the products are produced. The considered example includes one reaction and therefore can be described using one reaction rate r. The stoichiometric coefficients of the reactants A and B are –1, while the stoichiometric coefficient of the product C is +1. With these definitions it is possible to formulate a dynamic model describing the transient behavior of [A], [B], and [C] in the following example:

$$\frac{d[A]}{dt} = \frac{d[B]}{dt} = -r, \frac{d[C]}{dt} = r$$

To obtain a set of ODEs that can be simulated, it is necessary to find a mathematical expression for the reaction rates depending on concentrations and kinetic parameters. Principally, a reaction can have various kinetic properties and there exist mathematical expressions for a large class of kinetics, e.g., Michaelis–Menten or Hill kinetics. However, in the following we will assume that all reactions can be formulated using the law of mass action, which usually will be justified when considering elementary, i.e., non decomposable, reactions. Therewith, the reaction rate for the example discussed above is

$$r = k_1 \cdot [A] \cdot [B] - k_{-1} \cdot [C]$$

where k_1 describes the velocity of the forward reaction and k_{-1} the reverse reaction. A more detailed introduction to the basic principles of kinetic modeling can be found in Atkins and Paula *(7)*.

A frequently used abbreviated mathematical formulation of the model equations is a vector representation

$$\dot{\vec{x}} = \frac{d}{dt}\begin{pmatrix}[A]\\ [B]\\ [C]\end{pmatrix} = \vec{f}(\vec{x}) = \begin{pmatrix}-r\\ -r\\ r\end{pmatrix}$$

in which the vector $\vec{x}$ describes all dynamic variables. The vector function $\vec{f}(\vec{x})$ represents the sums of reaction rates related to the variables of $\vec{x}$. Usually the rate vector also depends on vectors $\vec{u}$ and $\vec{p}$, which represent external stimulations (inputs) of the system and kinetic parameters, respectively. To simulate the derived model equations we finally need initial conditions for all dynamic variables, i.e., the concentrations of each component at the beginning of the process.

2.2. A Dynamic Model

The reason why dynamic modeling is important to really understand a cellular reaction pathway is illustrated in the following. We consider a relatively small reaction network, which, however, shows an unexpected dynamic behavior. The system includes the following reactions, which are assumed to be irreversible:

$$A \rightarrow X \qquad r_1 = k_1 \cdot [A] \qquad B + X \rightarrow Y + C \qquad r_3 = k_3 \cdot [B]\,[X]$$
$$2X + Y \rightarrow 3X \qquad r_2 = k_2 \cdot [X]^2\,[Y] \qquad X \rightarrow D \qquad r_4 = k_4 \cdot [X]$$

Additionally, we assume that the concentrations [A] and [B] are kept constant and that we are interested only in the concentrations [X] and [Y]. The simulation of the resulting ODEs

$$\frac{d\,[X]}{dt} = r_1 + r_2 - r_3 - r_4$$
$$\frac{d\,[Y]}{dt} = -r_2 + r_3$$

shows an oscillating behavior of the reaction system, which probably would not have been predicted intuitively (*see* **Fig. 1**). Thus, we see that even very small reaction networks may provide unexpected phenomena. Such oscillations can also be observed in biological systems. The most prominent examples probably are the cell cycle and circadian rhythms.

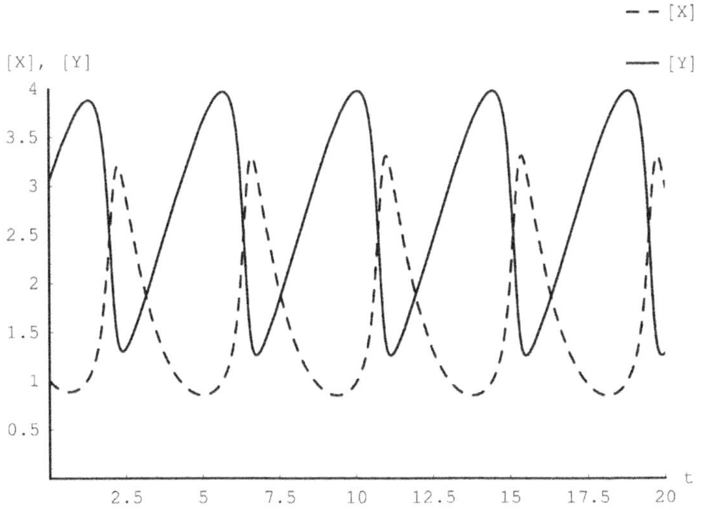

Fig. 1. Simulation results of the famous Brusselator model. Simulations are performed using the kinetic parameters $k_1 = 1$, $k_2 = 1$, $k_3 = 0.5$, and $k_4 = 1$, the fixed concentrations [A] = 1.6, [B] = 8, and the initial conditions $[X](0) = 1$, $[Y](0) = 3$. All values are given in arbitrary units.

2.3. Choice of Coordinates

Before a mathematical model can be generated it is necessary to think about the dynamic variables in which the considered process shall be described. These dynamic variables are referred to as coordinates of the system. Hence, the predefinition of coordinates, i.e., the choice of dynamic variables of the model, is an inevitable prerequisite. Most often they are determined customarily. In the case of a reaction network (also see the examples above) frequently the concentrations of all participating components are chosen. However, this choice is not obligatory and in many cases even not convenient. This especially holds true for modeling signal transduction pathways. Alternatively, any other set of coordinates might be introduced, which makes it possible to recalculate the natural ones, i.e., the individual component concentrations. In the case of a reaction network, alternative coordinates could, e.g., be the chemical potential of each component. Two such mathematical representations of a process can be mutually converted into each other using a transformation.

A transformation shall be exemplified reconsidering the simple example of **Subheading 2.1,** in which the original coordinates are given by the concentrations [A], [B], and [C]. Now we want to introduce other, alternative, variables to describe the process.

$$\begin{aligned} z_{AC} &= [A] + [C] & [A] &= z_{AC} - z_C \\ z_{BC} &= [B] + [C] & [B] &= z_{BC} - z_C \\ z_C &= [C] & [C] &= z_C \end{aligned}$$

Importantly, it is necessary that the transformation is invertible, i.e., it must also be possible to compute the old variables from the new ones as shown above. Observe that the same equations are also used to transform the initial conditions of the model. The procedure to derive the transformed ODEs is quite simple:

1. Differentiate the transformation equations and replace the resulting derivatives of [A], [B], and [C] using the original ODEs.

$$\begin{aligned} \frac{dz_{AC}}{dt} &= \frac{d[A]}{dt} + \frac{d[C]}{dt} = -r + r = 0 \\ \frac{dz_{BC}}{dt} &= \frac{d[B]}{dt} + \frac{d[C]}{dt} = -r + r = 0 \\ \frac{dz_C}{dt} &= \frac{d[C]}{dt} = r = k_1 \cdot [A][B] - k_{-1} \cdot [C] \end{aligned}$$

2. Replace the old variables by the new ones using the inverse transformation shown above.

$$\frac{dz_C}{dt} = k_1 (z_{AC} - z_C)(z_{BC} - z_C) - k_{-1} \cdot z_C$$

Interestingly, two of the three variables do not change over time, since their derivative equals zero. This is due to existing conservation relations within the system. Hence, in this mathematical representation one ODE is sufficient to completely characterize the dynamics of this process. Such transformations will also play an important role in modeling signal transduction pathways.

3. Methods

3.1. Problems of Common Modeling Approaches

Many existing models evade the problem of combinatorial variety by replacing the complete mechanistic network structure by a reduced and heuristic one focusing on a restricted number of molecular species and reactions. To illustrate the problems associated with this heuristic approach, we will show that even in a simple example seemingly reasonable simplifications may lead to a wrong model. The example we will discuss is a receptor, denoted as R, with three binding domains. These are an extracellular domain 1 and two intracellular domains 2 and 3. We assume that extracellular ligand binding induces conformational changes, which greatly increase the affinity of the intracellular domains toward their binding partners (for assumed kinetic parameters *see* **Table 1**). A complete mechanistic model comprises 11 different molecular species (extracellular ligand L, intracellular effectors E and F, and receptor species R000, RL00, R0E0, R00F, RLE0, RL0F, R0EF, and RLEF) and 12 binding reactions (four reactions describing L binding to R000, R0E0, R00F, and R0EF, four describing E binding to R000, RL00, R00F, and RL0F, and four describing F binding to R000, RL00, R0E0, and RLE0). For a reduced model we make some heuristic but reasonable assumptions. Since the affinity of the intracellular domains are extremely low for an unoccupied extracellular domain, it seems reasonable to

Table 1
Kinetic Parameters of the Example

Affinity of domain	k_{on} $[M^{-1}\ \text{min}^{-1}]$	k_{off} $[\text{min}^{-1}]$	$K_{\text{equilibrium}}$ $[M^{-1}]$
1 (always)	3×10^5	6	5×10^4
2 (domain 1 unoccupied)	1	18	5.6×10^{-2}
2 (domain 1 occupied)	5×10^7	24	2.1×10^6
3 (domain 1 unoccupied)	1	12	8.3×10^{-2}
3 (domain 1 occupied)	1×10^5	60	1.7×10^3

neglect the related reactions. After extracellular ligand binding the resulting affinity as well as the resulting association constant of domain 2 are several hundred-fold higher than the affinity and the association constant of the third domain. Hence, we additionally assume that the effector E in the majority of cases will bind before F and the reduced model has to include only the following three reactions

$$R000 + L \leftrightharpoons RL00 \quad RL00 + E \leftrightharpoons RLE0 \quad RLE0 + F \leftrightharpoons RLEF$$

and the seven states L, E, F, R000, RL00, RLE0, and RLEF. The model is parameterized with the related kinetic constants shown in **Table 1**. This represents a commonly performed simplification. In the model of EGF signaling presented in Schoeberl *et al.* *(8)*, the two effectors GAP and Shc bind consecutively to the receptor after stimulation, although the EGF receptor provides two distinct binding domains for these proteins similar to our example. To compare the predictions of the reduced model with a complete one accounting for all feasible molecular species and reactions, we consider the total concentration of receptors with occupied domains 1, 2, and 3 (the levels of occupancy of each domain). A comparison of the simulation results shows that the predictions of the reduced model are incorrect, which reveals how problematic such heuristic approaches are (**Fig. 2**.) The results also emphasize the necessity of an alternative systematic approach, which will be described in the following.

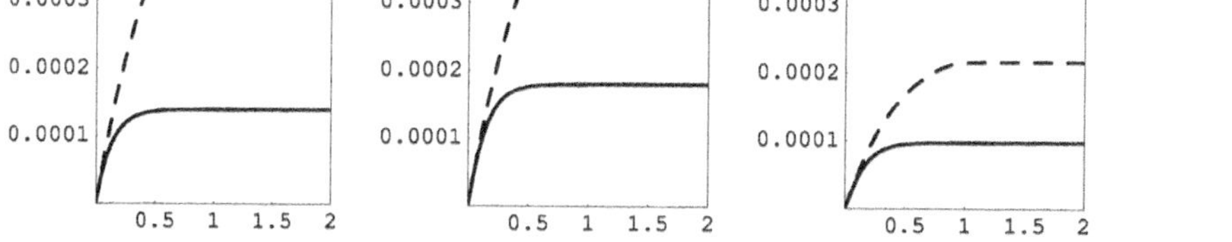

Fig. 2. Simulation results of the two described models. The three graphs show the three levels of occupancy of the considered receptor. The left graph shows the level of occupancy of domain 1, the graph in the middle the one of domain 2, and the graph on the right the one of domain 3. Both models are simulated using the same kinetic parameters and initial conditions. The concentrations are plotted in *M*.

3.2. Domain-Oriented Modeling Approach

3.2.1. Macroscopic Point of View

One way to solve the problem of combinatorial variety in modeling signal transduction pathways is to create complete models including all molecular species and reactions. The modeling tool BioNetGen *(4)* was developed to automatically generate such a complete ODE model. However, most real examples cannot be handled using this approach. As an example, a still simplified model of insulin signaling accounting for only a restricted number of proteins and binding domains would grow to over 145 million equations *(9)*. This huge number of equations not only exceeds the current computational possibilities, since the number of insulin receptors per cell is much lower than 145 millions, it also shows that the majority of individual species will not occur in the cell, and the concentrations of others are too low to have any significance. An alternative approach is to focus on more macroscopic and also measurable quantities such as levels of occupancy or the phosphorylation state of domains. These cumulative quantities correspond to experimental readouts and therefore have a much greater significance and are of greater interest than concentrations of individual multiprotein species. Hence, the aim of a modeling technique should be the creation of a manageable model describing the transient behavior of these macroscopic quantities.

3.2.2. Process Interactions

Besides the structure of a model, its kinetic parameters play a crucial role in determining the system dynamics and the functionality of a pathway. Considering scaffold proteins, kinetic parameters define which binding domains or related processes, e.g., binding of a ligand or phosphorylation of a domain, interact with each other. In the following we will refer to such interactions as process interactions.

A variety of cases exist, such as noninteracting processes or the existence of one controlling process influencing the others (compare the example above). Three different possibilities of process interactions are distinguished, which shall be exemplified considering a protein with two domains that bind the ligands L and E. In this case the reaction system consists of four reactions (two describing L binding to R00 and R0E and two describing E binding to R00 and RL0), for which the following reaction rates can be formulated:

$$\begin{aligned} r_1 &= k_1\,[L]\cdot[R00] - k_{-1}\,[RL0] \\ r_2 &= k_2\,[L]\cdot[R0E] - k_{-2}\,[RLE] \\ r_3 &= k_3\,[E]\cdot[R00] - k_{-3}\,[R0E] \\ r_4 &= k_4\,[E]\cdot[RL0] - k_{-4}\,[RLE] \end{aligned}$$

The theoretically possible process interactions include the following:

1. Noninteracting processes. Complete independence implies that the kinetic association and dissociation constants of one domain do not change upon ligand binding on the other domain. Hence, it follows for the parameters $k_2 = k_1$, $k_{-2} = k_{-1}$, $k_4 = k_3$, and $k_{-4} = k_{-3}$.
2. Unidirectionally interacting processes. The binding of one ligand, e.g., ligand L, is not influenced by binding of the other one. However, L binding does change the kinetic properties of the other domain. In this case only the conditions $k_2 = k_1$ and $k_{-2} = k_{-1}$ have to be fulfilled.
3. Mutually interacting processes. This is the most general case. Binding of a ligand influences binding of the other ligand and vice versa. In this case all parameters can have different values. However, note that there exist some thermodynamic constraints for the kinetic parameters that should be considered (*see* **Section 4**).

Interestingly, this qualitative information about process interactions not only helps to reduce the number of distinct parameters but it also helps to reduce the number of equations, as we will show in the following. Quantitative knowledge about the model parameters (measured equilibrium constants or kinetic association and dissociation constants) reduces the complexity of parameter identification (*see* **Section 4**).

3.2.3 Mathematical Point of View

Although we argued before that a complete mechanistic model describing the formation of signaling complexes most often is unmanageable or at least cumbersome, it theoretically has one very remarkable advantage: it properly describes the system structure. The transient behavior of the macroscopic quantities of interest is computable by summing up the concentrations of individual species provided by a complete model. The main problem is the immense number of equations. An alternative idea is a domain-oriented modeling approach, which allows an adequate and preferably exact description of the macroscopic quantities with a smaller number of equations. This aim can be achieved by transforming the model equations to new more convenient variables including the macroscopic ones. Obviously, a model reduction is possible if some variables of the system have no influence or only a small influence on these macroscopic variables. In systems theory variables that do not influence the measured quantities are called unobservable. Mathematically, these unobservable variables can be separated from the other ones by a transformation as introduced above. However, note that the described procedure alone is not an adequate solution for the problem of combinatorial complexity, especially in very large systems such as the insulin signaling pathway. The problem is that it is first necessary to have a complete mechanistic model, which subsequently is transformed and reduced. This problem can be overcome by separating independent interaction paths. In the following we will discuss this modeling process step by step. Each step will be illustrated considering the example presented in **Fig. 3**.

1. Definition of all proteins, binding domains, and processes that shall be included in the model. **Example:** Considers molecules A, B, C, and D with their binding domains as shown in **Fig. 3A**. The processes that are occurring, which are numbered in **Fig. 3A**, are (1) binding of A, (2, 3, and 7) phosphorylation of B at different domains, (4) binding of C, (5 and 6) phosphorylation of C at different domains, and (8) binding of D.

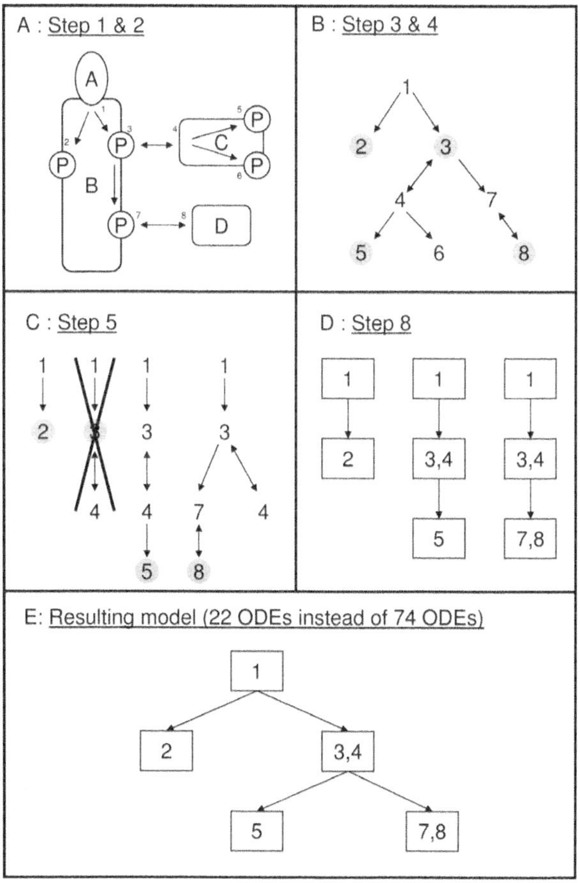

Fig. 3. (**A–E**) From interactions to the mathematical model. A detailed explanation can be found in the text.

2. It is necessary to define all possible process interactions (noninteracting, unidirectional, or mutually interacting) on the basis of knowledge about the kinetic properties of the proteins involved. Since a mathematical model requires a complete definition for all interactions, mostly fragmentary knowledge has to be completed by assumptions. **Example:** In **Fig. 3A** the processes that are occurring are indicated by arrows. Processes (1 and 2), (1 and 3), (3 and 7), (4 and 5), and (4 and 6) are assumed to interact unidirectionally and processes (3 and 4) and (7 and 8) mutually; all other processes do not interact.
3. The interaction pattern of the system has to be translated into an interaction graph, in which the processes are nodes and the interactions are represented by directed edges (arrows) pointing to the process that is influenced. **Example:** *See* **Fig. 3B**.
4. Output processes are defined. An output process is a process that can be measured or that is of special interest. The aim is a model describing these output processes as accurately as possible. Other processes will be of interest only if they influence one of the output processes. **Example:** In **Fig. 3B** we choose, for example, processes 2, 3, 5, and 8 as output processes. They are marked by gray circles.
5. The interaction graph can be divided into output subgraphs. An output subgraph contains all nodes from which the considered output can be reached following the directed edges. Hence, an output subgraph comprises all processes that influence the considered output process directly or indirectly. If a node does not occur in any output subgraph the corresponding process cannot influence any of the output processes and can be completely omitted in the following. Finally, it is necessary to eliminate redundant information, i.e., subgraphs completely found in other larger subgraphs. **Example:** The graph shown in **Fig. 3B** can be divided into four output subgraphs (as shown in **Fig. 3C**). In this example, process 6 does not influence any of the considered output processes and can be omitted in the following considerations. The subgraph for output process 3 is completely found in two other subgraphs and therefore can be eliminated.
6. Each of the subgraphs describes an autonomous signaling path that can be modeled separately. Hence, the next step is to create complete mechanistic models for each subgraph. Processes not part of a subgraph are not included in the corresponding model. **Example:** The modeling shall be illustrated by considering the smallest subgraph comprising processes 1 and 2. The mathematical model is given by

$$\begin{aligned} r_1 &= k_1\,[A]\cdot[B00] - k_{-1}\,[BA0] \\ r_2 &= k_1\,[A]\cdot[B0P] - k_{-1}\,[BAP] \\ r_3 &= k_2\,[B00] - k_{-2}\,[B0P] \\ r_4 &= k_3\,[BA0] - k_{-3}\,[BAP] \end{aligned} \qquad \frac{d}{dt}\begin{pmatrix}[A]\\ [B00]\\ [BA0]\\ [B0P]\\ [BAP]\end{pmatrix} = \begin{pmatrix}-r_1 - r_2\\ -r_1 - r_3\\ r_1 - r_4\\ -r_2 + r_3\\ r_2 + r_4\end{pmatrix}$$

in which the rates r_1 and r_2 describe the binding of A to the scaffold protein B (process 1), and the rates r_3 and r_4 describe the phosphorylation of B (process 2).
7. The model equations have to be transformed into new more convenient variables; this makes it possible to eliminate redundant information still included in the subgraphs. This redundancy is due to the fact that some processes are involved

in several subgraphs. Mathematical analyses showed that the most convenient choice of new variables follows a hierarchical pattern. The new variables of the model can be divided into tiers describing different levels of detail. We start with the overall concentrations of all participating proteins. The overall concentration of a certain protein corresponds to the sum of all protein complex concentrations of which the considered protein is part. If the mathematical model does not consider production and/or degradation of these proteins, the overall concentrations will be constant. The next tier describes the transient behavior of the levels of occupancy or phosphorylation degrees. The level of occupancy of a certain receptor domain is defined as the sum of all receptor species to which the related ligand has bound. The following level of detail comprises what we call second-order levels of occupancy, which correspond to the sums of all species that have concurrently bound two specific ligands. This is followed by higher order levels of occupancy until the last tier, in which all ligands have bound and all binding domains are phosphorylated, is reached. It can be proved that such a transformation is always invertible. **Example:** Here we again consider only the smallest subgraph comprising processes 1 and 2. The first tier in this example includes the overall concentrations of molecules A and B. The new variables are given by

$$\begin{aligned} z_{A,tot} &= [A] + [BA0] + [BAP] \\ z_{B,tot} &= [B00] + [BA0] + [B0P] + [BAP]\,. \end{aligned}$$

The next tier contains all levels of occupancy and phosphorylation degrees, which are given by

$$\begin{aligned} z_{AB} &= [BA0] + [BAP] \\ z_{BP} &= [B0P] + [BAP]\,. \end{aligned}$$

In this example there is one further tier describing the second-order levels of occupancy given by

$$z_{ABP} = [BAP]$$

This transformation can be inverted by solving the given five linear equations for the concentrations [A], [B00], [BA0], [B0P], and [BAP]. The transformed model equations can be obtained following the steps described in **Subheading 2.3**. First, it is necessary to differentiate the transformation equations and replace the derivatives of the concentrations by the original ODEs.

$$\begin{aligned} \dot{z}_{A,tot} &= -r_1 - r_2 + r_1 - r_4 + r_2 + r && = 0 \\ \dot{z}_{B,tot} &= -r_1 - r_3 + r_1 - r_4 - r_2 + r_3 + r_2 + r_4 && = 0 \\ \dot{z}_{AB} &= r_1 - r_4 + r_2 + r_4 && = r_1 + r_2 \\ \dot{z}_{BP} &= -r_2 + r_3 + r_2 + r_4 && = r_3 + r_4 \\ \dot{z}_{ABP} &= r_2 + r_4 && = r_2 + r_4 \end{aligned}$$

In the second step it is necessary to replace the old variables in the reaction rates by the new ones using the inverse transformation.

$$
\begin{aligned}
\dot{z}_{A,tot} &= 0\\
\dot{z}_{B,tot} &= 0\\
\dot{z}_{AB} &= k_1\left(z_{A,tot} - z_{AB}\right)\left(z_{B,tot} - z_{AB}\right) - k_{-1}z_{AB}\\
\dot{z}_{BP} &= k_2\left(z_{B,tot} - z_{AB} - z_{BP} + z_{ABP}\right) - k_{-2}\left(z_{BP} - z_{ABP}\right) + k_3\left(z_{AB} - z_{ABP}\right) - k_{-3}z_{ABP}\\
\dot{z}_{ABP} &= k_1\left(z_{A,tot} - z_{AB}\right)\left(z_{BP} - z_{ABP}\right) - k_{-1}z_{ABP} + k_3\left(z_{AB} - z_{ABP}\right) - k_{-3}z_{ABP}
\end{aligned}
$$

8. The transformed model equations can be divided into modules, which are characterized by unidirectional communication with other modules. Processes that directly or indirectly interact mutually form one module. If some processes are included in more than one subgraph, the models of these subgraphs will contain identical modules. Multiple copies of modules can be eliminated and the remaining modules can be merged into a complete model. **Example:** The transformed ODEs for the discussed smallest subgraph (see the above equations) have a special structure. The variables $z_{A,tot}$ and $z_{B,tot}$ are constant and equal their initial concentration. The corresponding ODEs are not required. Additionally, the ODE for z_{AB} does not depend on z_{BP} and z_{ABP}, which is due to the unidirectional process interaction between A binding to B and phosphorylation of B. Hence, the remaining three ODEs can be divided into two modules. One module comprises only the ODE for z_{AB}, which describes the dynamics of process 1. The second module comprises the other two ODEs, which describe the dynamics of process 2. The ODEs deduced from the two remaining output subgraphs shown in **Fig. 3C** can be divided into six more modules as indicated in **Fig. 3D**. Each box represents a set of ODEs. The modules are labeled with the process numbers that are described by the appropriate ODEs. Two copies of module (1) and one of modules (3 and 4) can be eliminated here. The resulting model, which consists of only 22 ODEs, is schematically shown in **Fig. 3E**. A complete mechanistic model of the exemplified network would comprise 74 ODEs.
9. In a last step the unknown model parameters have to be identified. Some basic ideas and problems of parameter identification are discussed in **Heading 4**.

4. Notes

1. Reaction networks are thermodynamic systems, and a dynamic model of biological pathways should be consistent with the laws of thermodynamics. The first point to be addressed is the number of thermodynamic state variables. A thermodynamic system can be completely described by temperature, pressure, volume, and the concentrations of all participating components. Using the modeling formalism described above, temperature, pressure, and volume are assumed to be constant. Another implicit assumption of modeling is that the cytoplasm is an approximately ideal mixture. In biological systems these assumptions are mostly justified. However, if they are not fulfilled it is necessary to adjust the modeling approach:

a. The case in which volume does not stay constant is probably the most important one. It plays a crucial role, e.g., modeling cell cycle processes in which cellular growth diminishes the protein concentrations within the cytoplasm. Some simple considerations make it possible to incorporate this decrease in a model. Importantly, the mole balances are given by

$$\frac{dn_i}{dt} = \sum_j r_{iJ} V$$

where r_{ij} describe all j reactions in which the component i is involved. This mole balance is inserted into the differentiated relation $n_i = c_i V$. From this it directly follows
that

$$\frac{dc_i}{dt} \sum_j r_{ij} - \frac{c_i}{V} \cdot \frac{dV}{dt}$$

b. The kinetic parameters usually depend on temperature, pressure, and due to the thermodynamic nonideality of cytoplasm also on the concentrations of the components. Pressure plays a smaller role in biological systems, since it usually neither changes over a large magnitude nor is of great influence on reactions in a liquid phase. If temperature changes, the kinetic rate constants can be adjusted using the Arrhenius equation. The formerly constant parameter is replaced by

$$k(T) = k_0 e^{\frac{E}{RT}}$$

in which E represents the activation energy of the reaction, R is the gas constant, and T is the temperature. Effects due to nonideal solutions generally could be included by considering the activity of the participating components *(7)*. However, these effects are mostly neglected, since reliable data, especially for proteins, are not available and the effects are mostly inferior. The most important nonideal effects in biological systems are given by changes in the pH and the ionic strength.

2. Commonly used modeling techniques do not automatically guarantee that the resulting mathematical models fulfill the laws of thermodynamics. To avoid the violation of thermodynamic laws the model has to fulfill the Wegscheider condition (also called the detailed balance constraint) ***(10)***. The Wegscheider condition requires that the product of all equilibrium constants along a reaction cycle must be equal to one. This follows from the requirement that a closed system finally has to reach thermodynamic equilibrium, which is characterized by vanishing reaction rates. Naturally, living biological systems usually operate far from equilibrium. However, if an organism is completely isolated from the environment, it will die and end up in thermodynamic equilibrium. A model violating the Wegscheider condition will not show this characteristic and will

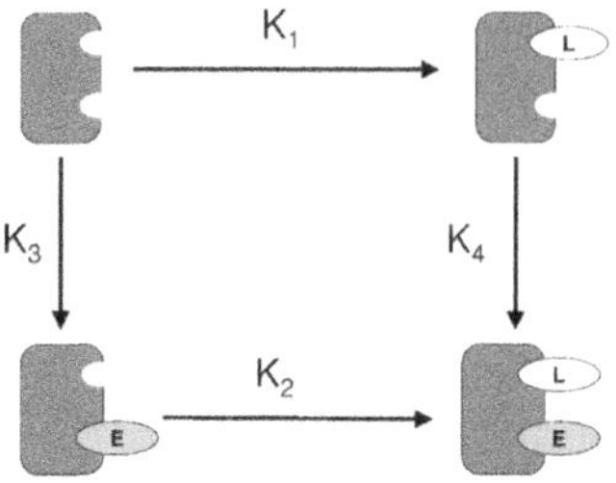

Fig. 4. Model of a protein with two binding domains. All reactions are reversible. K_i represent the equilibrium constants of the four reactions.

never reach thermodynamic equilibrium even if the modeled system is isolated from the environment. Using the Wegscheider condition it is possible to deduce some interesting conclusions about process interactions in signaling networks. This can be shown by considering the simple reaction network depicted in **Fig. 4**. Note, that all reactions are reversible. The arrows define only which species are considered as reactants and which as products. The resulting Wegscheider condition is given by $K_1 K_4 K_2^{-1} K_3^{-1} = 1$. From this it follows that only two different kinds of interactions are possible:

a. Noninteracting processes. Kinetic parameters realizing noninteracting processes such as those discussed above always fulfill the Wegscheider condition.
b. Allosterically interacting processes. Not all kinetic parameters realizing mutually interacting processes fulfill the Wegscheider condition. If ligand binding to a receptor increases its affinity to an effector by a certain factor ($K_2 = aK_1$), the Wegscheider condition requires that effector binding in turn increases the affinity to the ligand by exactly the same factor ($K_4 = aK_3$).

However, it is possible to show that other kinds of process interactions, in particular, unidirectional interactions, may be used to replace more complex molecular mechanisms, which approximately realize unidirectionality. Promising candidates for approximate unidirectional interactions are given by energy-consuming processes. The energy consumption can be used to decouple two processes and to realize a unidirectional interaction between them. **Figure 5** shows one way in which phosphorylation of a binding domain can be unidirectionally influenced. We consider a protein with two domains. One domain can bind a ligand L, while the other one can be phosphorylated. The complete mechanistic model would involve four phosphorylation reactions, two reactions in which ATP is converted to ADP and two in which free phosphates bind to the domain. We assume that ligand binding is not influenced by phosphorylation. Hence, from the Wegscheider condition it follows that both ATP reactions and both P reactions must have the same equilibrium constants, respectively. It could be assumed that

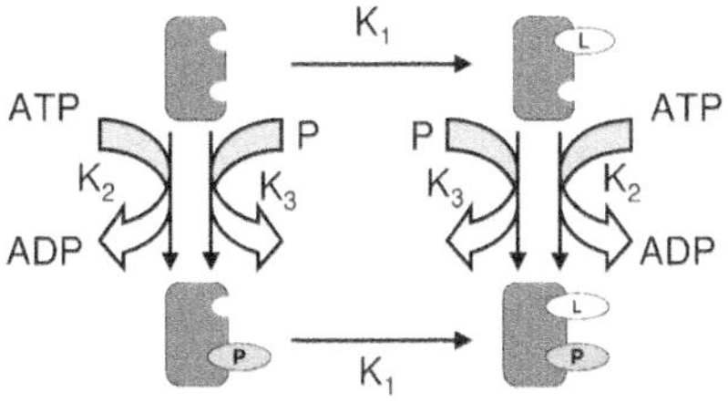

Fig. 5. All reactions shown are reversible reactions. The arrows indicate which reactions are considered to be the forward reactions and which are the backward ones. In this example it is necessary to fulfill two Wegscheider conditions, namely, $K_1K_2K_1^{-1}K_2^{-1} = 1$ and $K_1K_3K_1^{-1}K_3^{-1} = 1$, which obviously are fulfilled. The negative exponent in the formulas is due to the fact that in going around one cycle it is necessary to go in a direction that is opposite to two of the arrows.

the ligand functions as some kind of enzyme, not changing the equilibrium of the phosphorylation reactions, but changing their velocities by certain factors x and y.

$$r_1 = k_2\,[ATP]\cdot[R00] - k_{-2}\,[ADP]\cdot[R0P]$$

$$r_2 = x\cdot(k_2\,[ATP]\cdot[RL0] - k_{-2}\,[ADP]\cdot[RLP])$$

$$r_3 = k_3\,[P]\cdot[R00] - k_{-3}\,[R0P]$$

$$r_4 = y\cdot(k_3\,[P]\cdot[RL0] - k_{-3}\,[RLP])$$

Assuming that the concentrations of ATP, ADP, and P remain constant, which is generally true in living and healthy cells, the four reactions can be merged into two virtual reaction rates realizing a unidirectional interaction as discussed in **Subheading 3.2.2.**

$$r_{1,3} = r_1 + r_3 = \underbrace{(k_2\,[ATP] + k_3\,[P])}_{k_{1,3}}[R00] - \underbrace{(k_{-2}\,[ADP] + k_{-3})}_{k_{-1,3}}[R0P]$$

$$r_{2,4} = r_2 + r_4 = \underbrace{(x\cdot k_2\,[ATP] + y\cdot k_3\,[P])}_{k_{2,4}}[RL0] - \underbrace{(x\cdot k_{-2}\,[ADP] + y\cdot k_{-3})}_{k_{-2,4}}[RLP]$$

Using this knowledge to model receptor activation, it becomes clear that ligand binding and processes such as receptor dimerization have to interact to fulfill the Wegscheider condition, while unidirectional interactions might approximate the subsequent phosphorylation of intracellular domains very well.

3. The domain-oriented modeling approach helps to create reduced mathematical models that accurately describe the dynamics of the considered levels of occupancy. The basic principle is to eliminate all variables that do not have any influence on these levels of occupancy. Naturally, such an elimination leads to a certain loss of information. In most cases the lost variables will provide information only of low significance and interest. The reduced model does not allow the

concentrations of all individual species to be retrieved. Such information can be approximated only by using the assumption of rapid equilibrium *(6)*. Additionally, it is necessary to state that the method does not always provide a reduced model. The quality of reduction, i.e., how many ODEs can be neglected, greatly depends on the process interactions. If a network in which all processes interact with each other is considered, the model cannot be reduced. In these cases it is necessary to either consider complete models or use different reduction methods *(9)*.

4. Parameter identification, i.e., the determination of kinetic model parameter values using measurement data, most often is neglected in discussions about modeling. However, it is a very important step, since a mathematical model without determined parameter values is not very useful. For lack of space we will not discuss identification algorithms and techniques, but will provide an overview and introduction to the potentials and limitations of parameter identification by considering a simple example. This note will discuss the types of measurements and how many measurements are required to identify kinetic parameters of dynamic models. More detailed information is available *(11)*.

As an example, a protein X_1 is steadily produced with a constant rate k_0. Once it is produced the protein changes its conformation. This change from the first conformation X_1 to the second X_2 can be modeled as a reversible reaction parameterized by k_1 and k_{-1}. Additionally, the degradation rates of both protein species are proportional to their concentration and are modeled using the rate constant k_2. The resulting ODEs are

$$\frac{d[X_1]}{dt} = k_0 - k_1[X_1] + k_{-1}[X_2] - k_2[X_1]$$

$$\frac{d[X_2]}{dt} = k_1[X_1] - k_{-1}[X_2] - k_2[X_2]$$

To identify kinetic parameters it is necessary to measure at least one dynamic state of the system. This can be either one of the modeled concentrations or any other state that uniquely depends on one or both of them (e.g., $y = 3[X_1] + 5[X_2]$). It is important to understand that even an infinite number of measured values of just one dynamic quantity normally will not make it possible to identify all kinetic parameters. Each dynamic variable can be assigned to a set of parameters that theoretically can be identified by measuring it. Whether the resulting measurements are sufficient to identify all parameters of this set depends on their number and their error bound.

In our example, we assume that the applied measurement technique cannot distinguish between the two different protein conformations. Hence, the measured quantity is $y = X_1 + X_2$. The first question is which parameters can be identified theoretically by measuring this quantity. In general, answering this question can be rather complicated. However, in this example it is quite simple to deduce an ODE for the measured variable. All parameters occurring in this ODE influence this quantity and therefore are identifiable. Differentiation of the expression $y =$

$X_1 + X_2$ and subsequent replacement of the derivatives of $[X_1]$ and $[X_2]$ using the ODEs mentioned above yield

$$\dot{y} = k_0 - k_2([X_1][X_2]) = k_0 - k_2 y$$

The kinetic parameters k_1 and k_{-1} do not occur in this equation. Hence, only the parameters k_0 and k_2 can be identified here.

The next question is what type of measurements and how many measuring points are required to allow an identification and to guarantee small identification errors. It is important that the measuring data resolve the dynamic behavior of the measured variable. If, for example, only stationary measurements (i.e., $\dot{y} = 0$ or $y = \text{constant}$) are provided, it will follow from

$$\dot{y} = 0 = k_0 - k_2 y \quad \Leftrightarrow \quad y = \frac{k_0}{k_2}$$

that the quotient of the two parameters can be identified, but never the exact value of both individual parameters. The number of required measuring values on the other side strongly depends on the error bounds of the measurements. The theoretically minimal number of measurement values required to determine the identifiable parameters uniquely corresponds to the number of identifiable parameters. However, often a much larger amount of measurement data is required to provide good identification results. This can be exemplified assuming a relative error bound of approximately 20% in the measurements of our example. An extensive simulation study, which was analyzed using statistical methods, shows that about 20 equidistant measurement values are required to provide a 95% guarantee that the two parameters are identified with an error smaller than 10%. For comparison, using 10 measuring points only an error smaller than approximately 20% can be guaranteed.

In a second scenario we assume that it is possible to measure $y = X_2$. Twofold differentiation of this equation, subsequent replacement of the $[X_1]$ and $[X_2]$ derivatives using the ODEs mentioned above, and elimination of $[X_1]$ and $[X_2]$ yield the following ODE describing the dynamics of this quantity:

$$\ddot{y} + (k_1 + k_{-1} + 2k_2)\dot{y} + k_2(k_1 + k_{-1} + k_2)y = k_0 k_1$$

This time all parameters occur in the equation. However, it is possible to identify only the following three parameter combinations:

$$k_1 + k_{-1} + 2k_2 \quad k_2(k_1 + k_{-1} + k_2) \quad k_0 k_1$$

These make it possible to determine a value for only k_2; the other three parameters cannot be identified without further measurements. If we assume that the data about the other measurements $y = X_1 + X_2$ are also available, it will be possible to identify all kinetic parameters.

Thus, it becomes apparent that the answer to the question of how many and which kinetic parameters can be identified by certain measurements strongly depends on the structure of the considered pathway. The amount of required measurement data depends on the system's structure and on which quantities are measured. Instead of increasing the number of measured quantities or measuring frequency, it is often more convenient to consider the system behavior for different stimulations or to investigate mutations if possible. Close cooperation and mutual exchange between modelers and biologists are needed to minimize the number of required measurements and to identify the maximum number of parameters. These are needed to obtain good results from parameter identification and therefore to obtain reliable models.

References

1. Klipp, E., Herwig, R., Kowald, A., Wierling, C., and Lehrach, H. (eds.) (2005) *Systems Biology in Practice*. Wiley-VCH, New York.
2. Hlavacek, W. S., Faeder, J. R., Blinov, M. L., Perelson, A. S., and Goldstein, B. (2004) The complexity of complexes in signal transduction. *Biotechnol. Bioeng.* **84**, 783–794.
3. Faeder, J. R., Blinov, M. L., Goldstein, B., and Hlavacek, W.S. (2005) Combinatorial complexity and dynamical restriction of network flows in signal transduction. *IEE Syst. Biol.* **2**, 5–15.
4. Blinov, M. L., Faeder, J. R., Goldstein, B., and Hlavacek, W. S. (2004) BioNetGen: software for rule-based modeling of signal transduction based on the interactions of molecular domains. *Bioinformatics* **20**, 3289–3291.
5. Conzelmann, H., Saez-Rodriguez, J., Sauter, T., Kholodenko, B. N., and Gilles, E. D. (2006) A domain-oriented approach to the reduction of combinatorial complexity in signal transduction networks. *BMC Bioinform.* **7,** 34.
6. Borisov, N. M., Markevich, N. I., Hoek, J. B., and Kholodenko, B. N. (2005) Signaling through receptors and scaffolds: independent interactions reduce combinatorial complexity. *Biophys. J.* **89**, 951–966.
7. Atkins, P. and Paula, J. (eds.) (2006) *Physical Chemistry*. Oxford University Press, New York.
8. Schoeberl, B., Eichler-Jonsson, C., Gilles, E. D., and Müller, G. (2002) Computational modeling of the dynamics of the MAP kinase cascade activated by surface and internalized EGF receptors. *Nat. Biotechnol.* **20(4),** 370–375.
9. Koschorreck, M., Conzelmann, H., Ebert, S., Ederer, M., and Gilles, E. D. (2007) Reduced modeling of signal transduction. *BMC Bioinform.* **8,** 336.
10. Heinrich, R. and Schuster, S. (eds.) (1996) *The Regulation of Cellular Systems*. Chapman & Hall, New York, pp. 102–112.
11. Nelles, O. (ed.) (2001) *Nonlinear System Identification*. Springer, New York.

Index

D

GPSR Compliance

The European Union's (EU) General Product Safety Regulation (GPSR) is a set of rules that requires consumer products to be safe and our obligations to ensure this.

If you have any concerns about our products, you can contact us on ProductSafety@springernature.com

In case Publisher is established outside the EU, the EU authorized representative is:

Springer Nature Customer Service Center GmbH
Europaplatz 3
69115 Heidelberg, Germany

Batch number: 10755461

Printed by Printforce, the Netherlands